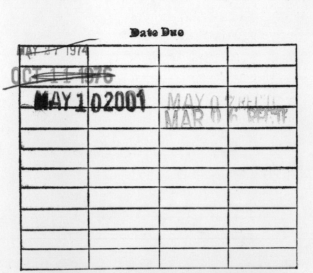

BIOPHYSICS AND PHYSIOLOGY

OF EXCITABLE MEMBRANES

Biophysics and Physiology of Excitable Membranes

Edited by

WILLIAM J. ADELMAN, JR.

 VAN NOSTRAND REINHOLD COMPANY

New York   Cincinnati   Toronto   London   Melbourne

Van Nostrand Reinhold Company Regional Offices:
New York Cincinnati Chicago Millbrae Dallas

Van Nostrand Reinhold Company International Offices:
London Toronto Melbourne

Copyright © 1971 by Litton Educational Publishing, Inc.

Library of Congress Catalog Card Number: 78-151256

MANUFACTURED IN THE UNITED STATES OF AMERICA

Published by Van Nostrand Reinhold Company
450 West 33rd Street, New York, N.Y. 10001

Published simultaneously in Canada by Van Nostrand Reinhold Ltd.

15 14 13 12 11 10 9 8 7 6 5 4 3 2 1

CONTRIBUTORS

WILLIAM J. ADELMAN, Jr., Ph.D.
Professor of Physiology, University of Maryland School of Medicine
Baltimore, Maryland

LAURENCE D. CARNAY, M.D.
Research Associate, National Institute of Mental Health
Bethesda, Maryland

LAWRENCE B. COHEN, Ph.D
Associate Professor of Physiology, Yale University School of Medicine
New Haven, Connecticut

KENNETH S. COLE, Ph.D., Sc.D.
Staff Member, Laboratory of Biophysics
National Institute of Neurological Diseases and Stroke
Bethesda, Maryland

GERALD EHRENSTEIN, Ph.D.
Staff Member, Laboratory of Biophysics
National Institute of Neurological Diseases and Stroke
Bethesda, Maryland

DANIEL L. GILBERT, Ph.D.
Head, Section of Cellular Biophysics, Laboratory of Biophysics
National Institute of Neurological Diseases and Stroke
Bethesda, Maryland

DAVID E. GOLDMAN, Ph.D.
Professor of Physiology, Biophysics and Anatomy,
Woman's Medical College of Pennsylvania
Philadelphia, Pennsylvania

HARRY GRUNDFEST, Ph.D.
Professor of Neurology, College of Physicians and Surgeons
Columbia University
New York, New York

RITA GUTTMAN, Ph.D.
Professor of Biology, Brooklyn College
Brooklyn, New York

BERTIL E. HILLE, Ph.D.
Associate Professor of Physiology and Biophysics, University of Washington
Seattle, Washington

DAVID LANDOWNE, Ph.D.
Post-Doctoral Research Fellow, Department of Physiology
Yale University School of Medicine
New Haven, Connecticut

JOHN W. MOORE, Ph.D.
Professor of Physiology and Pharmacology, Duke University Medical Center
Durham, North Carolina

TOSHIO NARAHASHI, Ph.D.
Professor and Head of Department of Physiology
Duke University Medical Center
Durham, North Carolina

YORAM PALTI, M.D., Ph.D.
Professor of Physiology, The Technion School of Medicine
Haifa, Israel

TOBIAS SCHWARTZ, Ph.D.
Associate Professor of Regulatory Biology, University of Connecticut
Storrs, Connecticut

RAYMOND A. SJODIN, Ph.D.
Professor of Biophysics, University of Maryland School of Medicine
Baltimore, Maryland

ICHIJI TASAKI, M.D., Ph.D.
Chief, Laboratory of Neurobiology, National Institute of Mental Health
Bethesda, Maryland

C. ROY WORTHINGTON, Ph.D.
Professor of Chemistry and Physics
Carnegie-Mellon University
Pittsburgh, Pennsylvania

PREFACE

This book is based on a series of lectures given at the Marine Biological Laboratory during the summer of 1969 to pre-doctoral and post-doctoral trainees in a program concerned with the biophysics and physiology of excitable membranes. The primary emphasis of the program was to review those approaches that make use of the conceptual ideas and methodology of biophysics so as to provide an understanding of the function of excitable membranes. In addition, the program attempted to provide basic insights into the foundations of modern neurophysiology and neuropharmacology. The program in part was supported by a grant from the National Science Foundation.

The book which has resulted from these lectures is thus an introduction to the study of excitable membranes. As in the lectures, the book is not a compilation of scientific data or a study of all types of excitable membranes but rather a guide to certain scientific means whereby excitable membranes can be examined.

The reader will soon notice that some of the approaches bear the mark of individual investigators. Thus, the book does not present a uniform approach to understanding excitable membranes. The area of common interest among the contributors to this book has been the study of nerve. It is hoped that the reader will gain an insight into not only how the research efforts of this group of investigators have progressed but how techniques, theories and personal research styles help to develop a field of scientific endeavor.

The editor wishes to express his appreciation to each of the contributors. The assistance of Mrs. Nancy Laue is gratefully acknowledged. In addition, the role of the Marine Biological Laboratory in supporting the program on which this book is based is hereby acknowledged. Special thanks is due to

Dr. H. Burr Steinbach for his help in initiating the program, and Drs. Kenneth S. Cole and Lorin J. Mullins for their advice and guidance during its development.

WILLIAM J. ADELMAN, JR.

Baltimore, Md.
May, 1971

CONTENTS

BIOPHYSICS AND PHYSIOLOGY

OF EXCITABLE MEMBRANES

X-RAY ANALYSIS OF NERVE MYELIN

1

C. R. WORTHINGTON

X-ray diffraction studies on biological tissues have a long history in that some of the early x-ray patterns were recorded soon after the discovery of x-ray diffraction. In the case of nerve, wide-angle x-ray diffraction patterns were obtained before 1930, while the first low-angle x-ray diffraction pattern of nerve was obtained in 1935. The low-angle x-ray pattern contained important new information relating to the dimensions and the contents of a macro-repeating unit in live myelinated nerve. The main purpose of an x-ray analysis of the low-angle pattern of nerve is to account for the observed x-ray intensities in terms of the electron density distribution within the macro-repeating unit. Until recently no rigorous x-ray analysis of the low-angle x-ray data from nerve had been attempted. There were two main reasons for this failure: only a few x-ray reflections were recorded and valid procedures for structure analysis did not exist. This situation is no longer true in the case of nerve. Many additional x-ray reflections have been recorded and improved methods of structure analysis have been derived. Hence one can expect to obtain a fairly detailed electron density description of nerve structure. Before describing the details of the x-ray analysis, it is convenient to give an introduction to the structure and function of myelinated nerve.

Nerve myelin refers to the sheath of lipoprotein which surrounds the axon of myelinated nerve at periodic intervals along its length. Regularly spaced at intervals between the myelin are the nodes of Ranvier. The myelin sheath measures about 1 mm along the axis of the nerve fiber, while the node of Ranvier is much shorter in length and measures about 1 μ. According current views on nerve conduction, the nerve pulse which travels down the axon does so by jumping from node to node with the myelin sheath providing a low capacity insulation path between nodes. The myelin sheath then provides a great improvement in the speed and the economy of signal transmission. However, this lack of activity on the part of the myelin sheath during nerve conduction has not led to any loss of intrinsic interest in its molecular structure. This is because nerve fibers are a convenient source and are of optimum size for structural studies and, furthermore, they have a well-defined multi-layered arrangement which is eminently suited for analysis by x-ray diffraction. Because of these advantages, nerve myelin is often regarded as a model membrane for structural studies.

In the early work on nerve myelin the relationship between cell membranes and nerve myelin was not clear, although one gains the impression that a one-to-one relationship was often inferred. The direct relationship was established in 1954. In an electron microscopy study Geren (1954) first described the mode of formation of nerve myelin. The electron micrographs showed that the myelin sheath was derived from a spiral wrapping of the Schwann cell membranes around the axon. Therefore, the lipo-protein layers of nerve myelin were, in fact, biological membranes. Thus a knowledge of any one kind of cell membrane and, in particular, nerve myelin should lead to a better understanding of the biological function of membranes in general.

The central theme underlying cell membrane structure since 1925 has been, and still is, the idea that lipid bilayers occur in cell membranes. This idea is supported by electron microscopy. The results of many other experimental studies on membranes are also in support of the occurrence of lipid bilayers. Nevertheless direct evidence of the lipid bilayer configuration in membranes has been difficult to obtain. It is of interest to review the early history of lipid bilayers in cell membranes.

EARLY HISTORY OF LIPID BILAYERS IN CELL MEMBRANES

Gorter and Grendel (1925) first proposed that the red blood cell membrane contained a lipid bilayer. The kind of lipid bilayer Gorter and Grendel had in mind was the classical extended lipid bilayer configuration. It consisted of two mono-molecular lipid films with the hydrocarbon chains inside the bilayer and the polar ends of the lipid molecules on the outside. Langmuir (1917) had already shown that lipid molecules form a mono-molecular film on the surface of the Langmuir trough: the hydrocarbon chains were packed side-by-side with the polar ends attached to the water surface of the trough. Gorter and Grendel proposed the lipid bilayer on finding that the surface area of the extracted lipid molecules was twice the surface area of the red blood cell, which indicated that the cell surface was covered by a layer of lipid material which was two molecules thick.

The lipid composition of cell membranes and the theoretical dimensions of the component lipid molecules were known. In order to gain an insight into the possible lipid arrangement in the bilayer, an estimate of the thickness of a cell membrane is needed. However, it was not easy to make such a measurement. Grendel (1929), from a study of the surface areas of the lipid molecules measured using a Langmuir trough and from a knowledge of the lipid composition in the red blood cell membrane, computed a thickness of 31 Å for the lipid bilayer. In an earlier study, Fricke (1925) was able to obtain a thickness of a cell membrane from a measurement of its electrical capacity: his value of the thickness of the red blood cell membrane was 33 Å. The thickness of a lipid bilayer as deduced from x-ray studies was much larger than 31–33 Å. Bear, Palmer and Schmitt (1941) measured the long x-ray spacings of a lipid extract from nerve as well as those from individual lipid molecules and mixtures of nerve lipids. The lipids were thought to be in an extended bilayer configuration. The air-dried spacing of the lipid extract was 64 Å, but larger spacings of up to 127 Å were obtained on adding water. Note that the thickness of the lipid bilayer in the wet state is not directly measured because the long x-ray spacing contains both the bilayer and a water layer between the adjacent bilayers. A thickness for the lipid bilayer can be estimated by assuming that the bilayer thickness does not change with hydration. Hence Schmitt, Bear and Palmer (1941) in a following paper adopted a thickness of 67 Å for the lipid bilayer in nerve.

Danielli and Davson (1935) proposed a model for cell membranes which could explain the permeability and electrical properties of cell membranes. The membrane model had a lipid bilayer of undefined thickness and was covered on each side by a mono-molecular layer of globular proteins. The Danielli-Davson model currently enjoys a measure of popularity because it is, more or less, in agreement with the familiar triple-layered membrane unit seen in electron micrographs.

EARLY HISTORY OF STRUCTURAL STUDIES ON NERVE

Early in the 20th century it was known that nerve myelin was birefringent and, hence, that there was a certain measure of molecular order within the myelin sheath. Schmidt (1936) proposed a model for nerve myelin which could explain the birefringence of nerve. The model had successively repeating lipid and protein layers which were roughly parallel to the surface of the myelin sheath. Thus, the model had a radial repeating unit consisting of one lipid bilayer and one layer of protein material. The hydrocarbon chains within the lipid bilayer were required to be orientated along the radial direction; that is, the chains were at right angles to the surface of the myelin sheath.

Supporting evidence that the hydrocarbon chains of the lipid molecules in nerve are aligned radially comes from x-ray diffraction. Early x-ray studies on nerve, for example Boehm (1931), had shown a meridionally accentuated ring at 4.7 Å. Schmitt, Bear and Clark (1935) recorded wide-angle x-ray patterns of nerve, measured the intensity of the 4.7 Å reflection and, on the basis of a model, calculated the intensity distribution of the 4.7 Å reflection. The model chosen had fully extended hydrocarbon chains orientated radially and the chains had a mean side-by-side separation which gave rise to the 4.7 Å reflection. The intensity distribution from the model gave only fair agreement with the observed intensity, but better agreement was expected by assuming that the hydrocarbon chains were not in a strict regular array. So far no quantitative study relating the degree of orientation of the hydrocarbon chains to the observed intensity has been made. Nevertheless, the qualitative explanation of the 4.7 Å reflection has not been questioned. In summary, the conclusion drawn from the results of the polarized light and wide-angle x-ray diffraction experiments, namely, that the hydrocarbon chains lie predominantly in a radial direction, remains unchallenged.

Schmitt, Bear and Clark (1935) showed that nerve myelin had a well-defined radial repeating unit. They succeeded in obtaining a low-angle x-ray diffraction pattern from frog sciatic nerve: diffraction orders $h = 2, 3, 4$ and 5 of a radial repeat distance $d = 171$ Å were recorded. This was an important observation and it was somewhat surprising considering the chemical composition. Nerve myelin contained protein and lipid molecules with a high water content. There were a variety of different lipid molecules and, together with the high percentage of water, it could be argued that a regular arrangement of lipo-protein layers in nerve was unlikely. However, the birefringence studies of nerve had already suggested a certain measure of molecular order. The Schmidt (1936) model had a regular repeating unit, but this regular arrangement was assumed as it could not have been deduced from the birefringence studies.

In the first low-angle x-ray study Schmitt et al. (1935) suggested that the radial repeat distance of nerve contained an end-to-end association of eight lipid molecules which were orientated radially. Schmitt et al. (1941) in a subsequent study of nerve myelin obtained additional diffraction data. They recorded the first five diffraction orders $h = 1–5$ of $d = 171$ Å for frog sciatic nerve and $d = 184$ Å for dog and cat spinal roots; the corresponding air-dried periods were $d = 144$ Å and $d = 158$ Å respectively. These authors adopted a thickness of 67 Å for a lipid bilayer in nerve from previous studies on lipid extracts and hence suggested that the radial repeat distance of nerve contained a total of four lipid mono-layers. No x-ray analysis was attempted, but an estimate was made of the thickness of the protein and water components by writing:

normal period — air-dried period = water thickness
air-dried period — 134 Å   = protein thickness.

It is clear that the above method gave only a rough estimate and, in fact, could not be taken seriously. However, the method gave reasonable answers for mammalian nerve, but ran into trouble with frog sciatic nerve. The authors proposed four possible arrangements of lipid and protein layers, but unless either an x-ray analysis was attempted, or else additional information was available, there was no way they could have deduced the correct arrangement of the lipo-protein layers in nerve.

In the early 1950's new information was obtained from both electron microscopy and x-ray diffraction. The x-ray studies were carried out by Finean and co-workers, who recorded x-ray patterns of nerve when the structure was modified by various physical and chemical treatments. The electron microscope observations on nerve are now described.

Sjostrand (1953) first demonstrated the myelin period in electron micrographs using ultra-thin sections and osmium fixed material. According to Sjostrand (1960) a mean value for the period of osmium fixed material was about 130 Å, a value which was even smaller than the air-dried period recorded by x-ray diffraction. The electron micrographs showed limited detail: there was a strongly stained dense line and a moderately stained intraperiod line per myelin period.

The mode of formation of peripheral nerve myelin has been described by Geren (1954). Nerve myelin is formed from a spiral wrapping of Schwann cell membranes around the axon. The cytoplasmic surfaces of the Schwann cell fuse together and form the dense lines seen in the electron micrographs, while the extracellular surfaces of the Schwann cell form the intraperiod lines. The Schwann cell membranes in common with other cell membranes show the familiar triple-layered membrane units, but, after myelination, the cytoplasmic and extracellular surfaces stain differently. An important consequence of Geren's observation is that the radial repeating unit of nerve myelin contains two triple-layered membrane units and that there is a plane of symmetry at the center of both the dense and the intraperiod lines. This observation of a center of symmetry within the radial repeating unit is an important consideration for x-ray analysis.

According to electron microscopy it can be argued that the myelin period contains two lipid bilayers and two protein layers which differ in some way. If the two protein layers are denoted D and I respectively, then the myelin period contains in sequence the following layers: D, lipid bilayer, I and another lipid bilayer. In fact, the above arrangement describes a composite model thought to explain the appearance of nerve myelin in electron micrographs. This kind of model for nerve has been proposed by Sjostrand (1960), Fernandez-Moran and Finean (1957) and Robertson (1959). It is of interest that Finean (1954) had already proposed exactly this kind of model for nerve myelin from his x-ray studies of nerve modified in various ways.

Finean (1954) deduced that the thickness of the lipid bilayer in frog sciatic nerve was about 55 Å and the radial repeating unit of his model contained two lipid bilayers and two protein layers. Each lipid bilayer was interspaced with a layer of protein of about 30 Å in thickness. However, this model had only one-half the observed myelin period and, in order to overcome this difficulty, Finean assumed that one-half of the basic unit was slightly different from the other half in some undefined way. Finean (1958) later identified his so-called "difference factor" with the protein layers, in agreement with Geren's electron micrographs.

As described in the above section, the early models for nerve myelin generally contained lipid bilayers and protein layers in some sequence. It was not possible to be more specific, although various thicknesses were assigned on the basis of some theory. However, direct evidence of structural details was lacking. Now, in principle, x-ray analysis of the low-angle x-ray data should lead to a knowledge of the structural parameters of nerve. Some progress has now been made in obtaining an electron density description of nerve myelin structure (Worthington and Blaurock, 1969b; Worthington, 1969b). Before giving an account of the structural information which has been obtained, an introduction to x-ray diffraction theory as it relates to nerve myelin is presented.

X-RAY DIFFRACTION THEORY

The basic principles underlying x-ray diffraction studies of membrane-type structures can be described either in general terms or formally. The description given here is in general terms. Some readers, however, might wish to see a formal treatment of x-ray diffraction theory as it relates to nerve; accordingly an appropriate treatment is given in the Appendices.

In order to illustrate the principles involved, consider a familiar triple-layered membrane unit. A drawing of this three-dimensional structure is shown in Fig. 1-1. Let t(r) represent the electron density of the triple-layered unit and let $T(\mathbf{R})$ represent its Fourier transform where $\mathbf{r}$ and $\mathbf{R}$ are real and reciprocal space vectors respectively. Note, t(r) and $T(\mathbf{R})$ are a pair of Fourier transforms, each is obtained from the other by Fourier transformation which

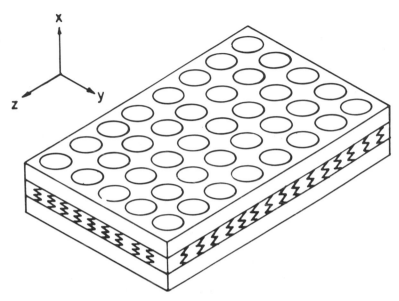

Fig. 1-1 *A three-dimensional drawing of a triple-layered membrane unit. The one-dimensional lamellar structure refers to the electron density distribution along x. Possible sub-unit structure is shown in the top layer of the membrane unit.*

is denoted by $t(\mathbf{r}) \rightleftharpoons T(\mathbf{R})$ (see Appendix I). The immediate objective of an x-ray diffraction experiment is to record $|T(\mathbf{R})|$, that is, the magnitude of the Fourier transform of the membrane structure. The Fourier transform $T(\mathbf{R})$ can be written $T(\mathbf{R}) = |T(\mathbf{R})| \exp(i\beta)$, where $\{\beta\}$ are the phases associated with the diffraction amplitudes $T(\mathbf{R})$. If the phases are known, then a Fourier transformation allows the electron density, $t(\mathbf{r})$, to be obtained. The determination of the electron density, $t(\mathbf{r})$, is the desired end result.

In practice, it is convenient to study the lamellar and sub-unit structure separately instead of studying the three-dimensional structure. The lamellar structure of the triple-layered unit in Fig. 1-1 refers to the one-dimensional electron density distribution in a direction at right angles to the membrane surface, that is, along x. The sub-unit structure refers to the structure within the plane of the membrane, for example, if the sub-units were in the top layer of the triple-layered unit shown in Fig. 1-1, then the sub-unit structure refers to the two-dimensional electron density distribution within this layer. For argument's sake, we may write $t(\mathbf{r}) = t(x)m(y,z)$, where $t(x)$ refers to the lamellar structure and $m(y,z)$ the sub-unit structure. The corresponding Fourier transforms are $T(X)$ and $M(Y,Z)$ and $T(\mathbf{R}) = T(X)M(Y, Z)$. Therefore for a study of lamellar structure, $|T(X)|$ is needed, whereas for a study of sub-unit structure, $|M(Y,Z)|$ is needed.

In the case of myelinated nerve the low-angle x-ray diffraction data relates only to the lamellar structure. So far, no x-ray data on sub-unit structure within the nerve myelin membranes have been recorded. However, x-ray evidence of sub-unit structure has been obtained in the case of other membrane-type structures. For example, x-ray data on sub-units within retinal photoreceptor disk membranes have been recorded (Blasie, Dewey, Blaurock and Worthington, 1965). These sub-units have now been identified with the photopigment molecules (Blasie, Worthington and Dewey, 1969) and the arrangement of the sub-units within the disk membranes has been determined (Blasie and Worthington, 1969). Thus the existence of sub-unit structure in one kind of biological membrane is established. But, in the case of nerve, at the present time, the presence of sub-unit structure has not been detected by x-ray diffraction. There is, however, extensive low-angle x-ray diffraction data on the lamellar structure of nerve and this aspect is treated in some detail.

Therefore, concentrating our attention on the lamellar structure of nerve, first the Fourier transform $|T(X)|$ is needed and then the phases of $T(X)$ are required in order to obtain the lamellar structure of nerve. Fortunately, it has been shown that the myelin period has a center of symmetry and therefore the phases are $\{\pm\}$. Even so, the determination of phases presents a formidable problem. However, before the phase problem can be treated, the first step is to obtain $|T(X)|$. An x-ray experiment provides an intensity pattern $I(X)$ on the x-ray film. The Fourier transform $|T(X)|$ is then obtained from the observed intensities on the basis of diffraction theory.

A formal treatment of x-ray diffraction theory as it relates to nerve is given in the Appendices. In order to treat the case of nerve the single triple-layered unit is replaced by a myelinated nerve specimen, but this is done in stages as follows: Appendix I treats the single triple-layered unit, Appendix II treats a planar multi-layered array of membranes, while Appendices III and IV treat the case of nerve myelin. The main features of this treatment are summarized. Nerve myelin gives rise to discrete x-ray reflections according to Bragg's Law:

$$2d\sin\theta_h = h\lambda, \tag{1-1}$$

where d is the radial period, $2\theta_h$ is the diffraction angle, h is the diffraction order and λ is the wavelength of the X-radiation. The relation between the observed x-ray intensities $I(h)$ and the Fourier transform $|T(h)|$ is

$$[hI(h)] = \varepsilon |T(h)|^2, \tag{1-2}$$

where ε is a proportionality constant. In most cases ε is an unknown quantity. Experimentally, a set of corrected intensities, $J_{obs}(h)$ is obtained and

$$J_{obs}(h) = [h\,I(h)]. \tag{1-3}$$

This set of intensities $J_{obs}(h)$ is said to be on a relative scale. Ideally, if ε is known, then an absolute scale for $|T(h)|$ is established. The Fourier series representation for nerve t''(x) which can be computed from the observed low-angle x-ray data is given by

$$t''(x) = (2/d) \sum_1^h \{\pm\}[J_{obs}(h)]^{1/2} \cos 2\pi hx/d. \qquad (1\text{-}4)$$

The phase problem then concerns the choice of $\{\pm\}$ phases. However, the treatment of the phase problem is delayed until after an account of the various x-ray diffraction patterns of myelinated nerve is given.

X-RAY DIFFRACTION PATTERNS OF NERVE

X-ray experiments have been performed on fresh nerve specimens maintained in a living condition within a glass capillary to record whether any structural changes occur as a function of time, but no changes are apparent even after exposure times of up to 24 hours (Blaurock and Worthington, 1969). Fresh

Fig. 1-2 A wide-angle x-ray diffraction pattern of frog sciatic nerve in Ringer's solution. The pattern was taken using 30 KV copper Kα nickel-filtered radiation, a specimen-to-film distance of 6.4 cm and an exposure time of 6½ hours. The prominent meridionally accentuated ring at 4.7 Å is visible. The equatorial low-angle x-ray reflections are visible on either side of the beam stop together with the 15–16 Å equatorial reflection.

nerve specimens maintained in a living condition have been referred to in various ways (live, native, intact, normal, wet, fresh) but the term "live" nerve is used here when necessary. The convention is adopted that x-ray diffraction patterns of nerve refer specifically to live nerve unless otherwise stated.

Schmitt et al. (1935) have adequately described the wide-angle x-ray diffraction pattern of myelinated nerve recorded with the x-ray beam at right angles to the fiber axis of the nerve specimen. A typical wide-angle x-ray diffraction pattern from frog sciatic nerve is shown in Fig. 1-2. The main reflection is the 4.7 Å meridionally accentuated ring which, as noted earlier, arises from the side-to-side packing of the hydrocarbon chains of the lipid molecules. There is a broad diffuse ring at about 3.3 Å which arises from the surplus water (Ringer's solution). This reflection is outside of the 4.7 Å reflection and is only partially shown in Fig. 1-2. Unresolved low-angle x-ray reflections can be seen on either side of the beam stop. The equatorial reflection at 15–16 Å has been identified as belonging to the low-angle x-ray diffraction pattern (Blaurock and Worthington, 1969).

Schmitt et al. (1941) succeeded in recording the first five diffraction orders of the low-angle x-ray diffraction pattern of peripheral nerve myelin. Surprisingly it was not until 1969 that more than five diffraction orders were reported for peripheral nerve myelin (Blaurock and Worthington, 1969). Some improvements in experimental technique have now been made (Worthington, 1969c). These improvements have led to the faster recording of low-angle x-ray patterns which have good resolution between diffraction orders. The improved low-angle x-ray patterns of nerve (which are described here) were obtained using an optically focusing x-ray camera (Elliott and Worthington, 1963). Line-focus patterns (slit collimation) were obtained by using only one mirror. A typical line-focus low-angle x-ray diffraction pattern of frog sciatic nerve is shown in Fig. 1-3. The first five diffraction orders ($h = 1$–5) of the radial repeat $d = 171$ Å are shown. Note the sharpness of the diffraction lines. According to diffraction theory, this sharpness indicates that the radial ordering of the lipo-protein layers is remarkably precise.

There is a wide variety of myelinated nerves in nature: the two main types of nerves are the peripheral and the central nervous system nerves. Both types of nerves show essentially the same wide-angle x-ray diffraction pattern but they show different low-angle x-ray patterns. The low angle patterns differ in the size of the radial period and the patterns also have a different intensity variation between the first five diffraction orders (Blaurock and Worthington, 1969). Schmitt et al. (1941) first recorded the low-angle pattern from peripheral nerve myelin while Finean (1958) first recorded the low-angle pattern from central nervous system nerve myelin. The peripheral nerve pattern consisted of the first five orders ($h = 1$–5) of a radial period $d \approx 170$–

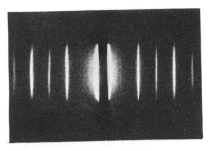

Fig. 1-3 *A line-focus low-angle x-ray diffraction pattern of frog sciatic nerve in Ringer's solution. The pattern was taken using 30 KV copper Kα nickel-filtered radiation, a specimen-to-film distance of 5.5 cm, an exposure time of 1 hour and the radial period was d = 171 Å. The first five orders of diffraction, h = 1–5 are visible. The first order h = 1 is better resolved on the original x-ray film as the reproduction tends to magnify the central x-ray beam scatter.*

185 Å and the central nervous system pattern consisted of only two orders ($h = 2$ and 4) of a radial period $d \approx 150$–160 Å. Additional diffraction orders from both types of nerves have now been recorded. Blaurock and Worthington (1969) have reported $h \approx 1$–12 for three live peripheral nerve myelins and $h \approx 1$–11 for four live central nervous system myelins. In recent work further additional diffraction orders $h \approx 1$–18 have been recorded in the low-angle x-ray pattern of peripheral nerve myelin (Worthington, 1969c).

The myelinated nerves of fish have also been studied by x-ray diffraction. The wide-angle x-ray diffraction is similar to that shown by other myelinated nerves but the low-angle x-ray diffraction pattern from fish myelin differs from that shown by both peripheral and central nervous system nerves. Finean (1960) first recorded the low-angle x-ray pattern from fish myelin. The pattern consisted of three orders ($h = 2$, 3 and 4) of the myelin period $d = 150$–160 Å. Hoglund and Ringertz (1961) have recorded the same three diffraction orders from five varieties of fish. Additional orders ($h \approx 1$–11) have now been reported (Blaurock and Worthington, 1969).

The normal pattern of nerve refers to the low-angle x-ray pattern which is recorded from live nerve. However, other kinds of x-ray patterns can be obtained by exposing the nerve specimen to various physical and chemical treatments such as air-drying, freezing, freezing and thawing, and chemical fixation. These modified nerve patterns differ from the normal pattern in that there is either a change in radial period or a change in the x-ray diffraction intensities of the first few orders, or both. The reason for studying modified nerve is to try to deduce structural information from the differences between the normal and modified patterns. An account of the various modified nerve patterns will not be given because these modified patterns have yet explained and, so far, they have failed to provide any definite information on nerve structure. This situation is not true, however, in the case of swollen nerve. The swollen peripheral nerve pattern is obtained by immersing the nerve specimen in diluted Ringer's solution or in various other hypotonic solutions.

Finean and Millington (1957), in a low-angle x-ray experiment, first recorded a much larger than normal radial repeat period for frog sciatic nerve in diluted Ringer's solution. Robertson (1958), in an electron microscope study, has shown that the major part of this swelling takes place at the outer surfaces of the membrane pair (the two adhering plasma membranes). That is, the swelling takes place at the intraperiod line. This property of swelling is important, for it suggests an obvious x-ray method of structure analysis which was first used with hemoglobin crystals by Boyes-Watson and Perutz (1943). This method requires that the structure of the membrane pair remains unchanged and that the swollen nerve myelin differs from the normal nerve myelin only in the width of the fluid layer between the adjacent membrane pairs. This x-ray method was first tried by Moody (1963) and Finean and Burge (1963), who recorded low-angle x-ray patterns of swollen nerve.

A fairly complete x-ray study of the swelling behavior of peripheral nerve myelin has been made. A detailed account of this work and a full description of the low-angle x-ray data from swollen nerve has been given (Worthington and Blaurock, 1969a and 1969b). For the present purposes only a concise description of these x-ray patterns is given. Various sets of low-angle x-ray data were obtained from frog sciatic nerve swollen in three different immersion media: distilled water, 0.24 M sucrose and 0.82 M sucrose solution. A number of experiments were performed using the same immersion medium; the resulting patterns had well-defined but different radial periods. The low-angle x-ray data $J_{obs}(h)$ were placed on the same relative scale using the procedures described by Worthington and Blaurock (1969b). The moduli of the amplitudes $[J_{obs}(h)]^{1/2}$ were plotted against reciprocal space coordinate X. Three sets of low-angle x-ray data which were obtained by swelling frog sciatic nerve in 0.24 M sucrose solution are shown in Fig. 1-4. A continuous curve connects the experimental points and defines regions I, II, and III. The peak height in region II exceeds that of region III while region I is not complete in the region of X where X approaches 0 because the $h = 0$ reflection was not recorded. The continuous curve has three well-defined minima. Three sets of low-angle x-ray data which were obtained by swelling frog sciatic nerve in distilled water are shown in Fig. 1-5. The moduli $[J_{obs}(h)]^{1/2}$ in Figs. 1-4 and 1-5 are all on the same relative scale. The continuous curves are similar in shape and each curve shows three well-defined minima but the curves differ in detail.

Because of this difference between the two curves it can be immediately deduced that the molecular structure of the myelin membranes is different in

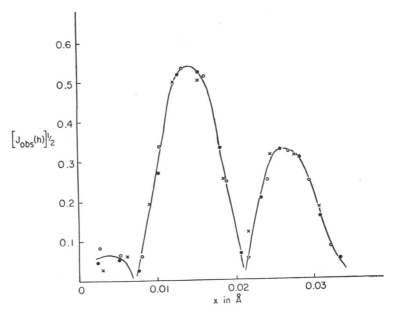

Fig. 1-4 *The moduli of the amplitudes* $[J_{obs}(h)]^{1/2}$ *for three sets of low-angle x-ray data are plotted against X, the reciprocal space coordinate. The three sets of data are on the same relative scale. The continuous curve is fitted by eye to the experimental points (solid circle) the 0.24 M sucrose data,* $d = 388$ Å *(open circle) the 0.24 M sucrose with 4 mM NaCl data,* $d = 373$ Å *(cross) the 0.24 M sucrose with 4 mM KCl data,* $d = 326$ Å. *(From Worthington and Blaurock, 1969b.)*

each of the two different immersion media. Hence the swelling method which was used in the study of hemoglobin crystals cannot be used with nerve myelin. However, the observation of three minima is important because it immediately simplifies the phase problem of swollen nerve. Also, assuming that the $J_{obs}(h)$ values of live nerve lie close to the continuous curves of swollen nerve, then the phase problems of live and swollen nerve run closely parallel. Further, by assuming that region II has a + phase then there are only four choices remaining: these are (±, +, ±). Moody (1963) and Finean and Burge (1963) reached this same conclusion from a study of their composite x-ray data. However, at that time there was no obvious way of deciding which one of these four choices was in fact the correct choice.

The fact that the different sets of low-angle x-ray data which were obtained with the same swelling medium all fitted on a continuous curve was reassuring and it led to confidence in the reliability of the x-ray data. Accordingly, it was decided to try to rigorously derive the electron density distribution of swollen nerve. Three sets of low-angle x-ray data were chosen for analysis: the distilled water data, $d = 252$ Å, $h = 1$–8; the 0.24 M sucrose data, $d = 388$ Å,

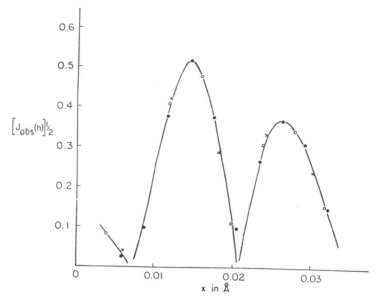

Fig. 1-5 *The moduli of the amplitudes* $[J_{obs}(h)]^{1/2}$ *for three sets of low-angle x-ray data are plotted against X, the reciprocal space coordinate. The three sets of data are on the same relative scale. The continuous curve is fitted by eye to the experimental points; (solid circle) the distilled water data, $d = 342$ Å (open circle) the distilled water data, $d = 252$ Å; (cross) the subnormal nerve data, $d = 166$ Å. (From Worthington and Blaurock, 1969b.)*

$h = 1$–13; and the 0.82 M sucrose data, $d = 359$ Å, $h = 1$–11. The $J_{obs}(h)$ values have been reported (Worthington and Blaurock, 1969*b*). Before describing the details of the x-ray analysis it is necessary to give an account of the theory of x-ray analysis as it relates to nerve.

X-RAY ANALYSIS: THEORY

An x-ray analysis of the low-angle x-ray pattern shown in Fig. 1-3 is dependent on solving the phase problem. Each diffraction order can have ($\pm$) phases, hence there are 32 possible solutions for $t''(x)$ as defined by Eq. (1–4). The problem is to distinguish which one of the 32 possible solutions is the correct solution. It is easy to see why a structure analysis of this pattern was not attempted until recently.

In the section on x-ray diffraction theory, Eq. (1-4) defines a Fourier series representation $t''(x)$ for nerve which can be computed provided that the phases are known. This is the Fourier method which has been very successful with crystals. However, in the case of nerve, the Fourier method has some definite drawbacks. Even if the phases can be found, the resulting Fourier synthesis has low resolution and is very difficult to interpret in terms of molecular structure (Worthington, 1969a). In order to circumvent these difficulties the model approach was adopted in the present x-ray analysis. Both methods have much in common because in the model approach all possible models must be examined and this runs parallel to examining the various phase solutions. The model approach has distinct advantages in the study of membrane-type structures and consequently this method was used in the study of nerve structure. The first step in this method is to find a suitable electron density strip model. The use of electron density strip models in interpreting the low-angle x-ray data from membrane-type structures has been described elsewhere (Worthington, 1969a) and a much shorter account is given here.

Before a start can be made, a model of some kind is needed. Of course, there is no good basis for choosing a model for nerve myelin, although, as will be explained, once a model is proposed it can be soon verified or rejected. In retrospect, the first model chosen for swollen nerve was influenced by electron microscopy, and by an earlier membrane-type model for mitochondria (Worthington, 1960). The model for nerve contained two symmetric triple-layered membrane units end-to-end with a fluid layer between the adjacent membrane pairs (Worthington and Blaurock, 1968). The model is shown in Fig. 1-6. The familiar triple-layered membrane unit is thought to consist of a

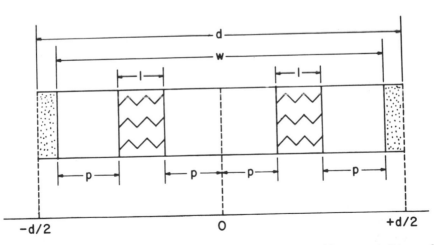

Fig. 1-6   *The model for swollen nerve is centrosymmetrical and has repeat distance **d** and membrane pair thickness **w**. The membrane pair contains two symmetric membrane units (**p, l, p**) end-to-end. The clear regions refer to electron density P, the zigzag lines refer to electron density L and the dotted regions refer to the fluid electron density F.*

layer of lipid with the surfaces covered by a layer of nonlipid or protein. However, the model refers specifically to electron densities rather than to chemical composition. Let the central layer have thickness l and electron density L, and the outer surfaces have thicknesses p and electron densities P. The radial repeat distance is d and there are two triple-layered units per radial repeat. The width of a membrane pair is $w = 2(2p + l)$ and this leaves an interspace of width $d - w$ which contains fluid of electron density F. The model is centrosymmetric. There are five model parameters w, l, P, L and F.

In order to test whether the model is a suitable choice, a comparison is made between the theoretical diffraction from the model and the observed diffraction. The comparison is made by computing the R-value. The R-value is an index of how closely the theoretical diffraction matches the observed diffraction. The theoretical diffraction is obtained by writing down the Fourier transform of the model. The observed diffraction refers to the $J_{obs}(h)$ values and does not include the zero-order reflection which is not observed. Hence, there is no reason to compute the Fourier transform at $X = 0$. It is convenient to write $T(X) = T(0) + T'(X)$ and the Fourier transform of the model $T'(X)$ then gives the appropriate theoretical diffraction. The derivation of $T'(X)$ and the definition of the R-value are given in Appendix V. The R-value calculations are made using the moduli of the amplitudes $[J_{obs}(h)]^{1/2}$ but only after the moduli values have been normalized (see Eq. (1-30) in Appendix V). Before normalization the five model parameters can be rearranged as w, l, α, $(P - L)$ and F where α is the ratio $\alpha = (P - F)/(P - L)$. After normalization the parameter $(P - L)$ is lost (see Appendix V). Further, if the value of F is assumed to be the same as for the immersion medium, then the model has three operational parameters w, l, α.

The average electron density of the membrane pair in swollen nerve can be derived from a study of two sets of low-angle x-ray data which were obtained using two different values of F. The average electron density is denoted M; an estimate of M can be made on the basis of the following theory (see Appendix V for a formal treatment). Consider a simplified model which contains a membrane pair with uniform electron density M and a fluid layer F between the adjacent membrane pairs. The theoretical diffraction from the model and the corresponding theoretical diffraction from the simplified model are amost identical in the region of small X values where X approaches zero. That is, the two models have approximately the same Fourier transform at very small angles of diffraction (Worthington, 1969a). Thus in the case of swollen nerve, the x-ray diffraction at small values of X will depend on the value of $(M - F)$. If it is assumed that M does not vary (appreciably) with different values of F then, from a study of two sets of low-angle x-ray data at small values of X, a value for M can be obtained.

An absolute electron density scale for the electron density model implies knowledge of the electron densities P, L and F in electrons per cubic angstrom (electrons/$Å^3$). The value of F is known and a value for α is obtained from R-value considerations. However, the term $(P - L)$ is not known. The determination of an absolute electron density scale therefore depends on obtaining a value for $(P - L)$. This value is obtained by using the result that the Fourier transform at $X = 0$ for the model is equal to the Fourier transform at $X = 0$ for the simplified model with uniform electron densities M and F. This equality provides a relation between model parameters w, l, α, M and F (see Appendix V) and hence a value for $(P - L)$ is obtained.

The Patterson function requires no phase information and can be directly computed either using the observed x-ray data or directly from the model. A formal treatment of the Patterson function is given in Appendix VI. Until recently, there had been the very real possibility that an interpretation of the Patterson function could have led to a new model for nerve. But no such interpretation was forthcoming. An interpretation can now be given because the relationship between the electron density strip models and their corresponding Patterson functions has been established (Worthington, 1969a). Hence it is now possible to account for the observed Patterson function of some membrane-type structure in terms of a definite model. However, in the case of swollen nerve, the Patterson function had only a secondary role because the model for nerve was first chosen on the basis of other considerations. Nevertheless, the study of the Patterson function of swollen nerve has proven to be extremely useful in verifying the correctness of the proposed model. In later work a comparison of the Patterson function computed using the observed intensities $J_{obs}(h)$ and the Patterson function computed directly from the model has led to the development of a more complicated model for nerve myelin (Worthington, 1969b).

STRUCTURE ANALYSIS OF SWOLLEN NERVE

The Fourier transform $T'(X)$ for the model shown in Fig. 1-6 is given by Eq. (1-27) in Appendix V. The following properties are noted. $T'(X)$ has zeroes at $X = n/w$, where n is an odd integer and the first three zeroes occur precisely at $X = 1/w$, $3/w$, $5/w$. This feature is shown to a good approximation by the experimental $[J_{obs}(h)]^{1/2}$ curves in Fig. 1-4 and 1-5 and hence a value of w can be assigned to each curve. The peak height in Region II occurs close to $X \approx 2/w$ and that of Region III occurs close to $X \approx 4/w$. Also there are two values of the model parameter l which will give the same peak heights in Regions II and III. Hence the parameters w, l can be directly assigned to the curves although l has two possible values. However, the model parameters were not assigned on this basis: they were assigned by matching all experimental values of a particular data set.

There are eight possible models to consider, but if it is assumed that $P > L$, that is, the low electron density occurs in the central region of the triple-layered unit, then only four possible models remain. The four kinds of models are as follows:

$$
\begin{array}{ll}
\text{I} & M - F > 0, l \text{ small} \\
\text{II} & M - F < 0, l \text{ small} \\
\text{III} & M - F > 0, l \text{ large} \\
\text{IV} & M - F < 0, l \text{ large.}
\end{array}
$$

The model parameters w, l, α were systematically varied and a search was made for the minimum R-value for each of the four kinds of model. The R-values for the four kinds of models are shown in Table 1-1 for the three sets of x-ray data from swollen nerve.

Table 1-1 Model parameters and the corresponding R-values for the swollen frog sciatic nerve

Solution	Model	w	l	α	R
		Å	Å		%
Distilled water	I	145	17.5	0.34	10
	II	145	18.0	0.22	7
	III	142	51.0	0.84	32
	IV	142	52.0	0.70	33
0.24 M sucrose	I and II	142	21.0	0.28	6
	III	141	49.0	0.74	29
	IV	140	48.0	0.68	29
0.82 M sucrose	I	142	27.0	0.58	34
	II	140	27.5	0.16	9
	III	143	43.0	0.80	36
	IV	140	41.5	0.43	21

From Table 1-1 models I and II have approximately the same w, l with l small and models III and IV have approximately the same w, l with l large. Only models I and/or II give good agreement with the low-angle x-ray data. On the other hand, models III and IV give poor agreement with the low-angle x-ray data, and hence the kind of model which has a large central region of comparatively low uniform electron density is not in agreement with the intensity data from swollen peripheral nerve myelin. Therefore, the Schwann cell membrane in swollen nerve does not contain a classical extended lipid bilayer, but it contains only a narrow region of low uniform electron density in the central region of the symmetric membrane unit. Any choice between models I and II is delayed until an estimation of the actual value of the average electron density of membrane pair (M) has been obtained.

In Fig. 1-4 and 1-5 the continuous curves are not very different in Regions II and III but differences are apparent in Region I. The differences in Region I mainly reflect the value of ($M - F$) which is the difference between the average electron density of the membrane pair M and the fluid layer F. The electron densities for distilled water, 0.24 and 0.82 M sucrose solutions are 0.334, 0.343 and 0.366 electrons/Å^3, respectively. In Fig. 1-5 the $[J_{obs}(h)]^{1/2}$ values are small in the region of X where X approaches zero, hence $M \approx F$, that is, $M \approx 0.343$ electrons/Å^3. Then, in the case of distilled water $M > F$ while for 0.82 M sucrose solution $M < F$. A study of the distilled water and 0.82 M sucrose data in the region of X where X approaches zero provides a value for M. A value of $M = 0.343$ electrons/Å^3 was obtained (Worthington and Blaurock, 1969b). This value is consistent with the value deduced from the 0.24 M sucrose data. Thus the average electron density of the membrane pair in swollen nerve is 0.343 electrons/Å^3.

Model parameters for nerve myelin swollen in three different media can now be assigned. Model I is the correct choice for the distilled water data: $w = 145$ Å, $l = 17.5$ Å, $\alpha = 0.34$ with $R = 10\%$. The corresponding phases are $(+, +, -)$. Model II is the correct choice for the 0.82 M sucrose data: $w = 140$ Å, $l = 27.5$ Å, $\alpha = 0.16$ with $R = 9\%$. The corresponding phases are $(-, +, -)$. In the case of the 0.24 M sucrose data, a choice between models I and II cannot be made: $w = 142$ Å, $l = 21.0$ Å, $\alpha = 0.28$ with $R = 6\%$. The corresponding phases are either $+$ or $-$ in Region I while the phases in Regions II and III are $(+, -)$.

An absolute electron density scale for swollen nerve is derived by substituting the appropriate model parameters in Eq. (1-35) in Appendix V. The following three relations are obtained:

distilled water data: $(0.343 - 0.334)145 = (P - L)(+14)$
0.24 M sucrose data: $(0.343 - 0.343)142 = (P - L)(-1)$
0.82 M sucrose data: $(0.343 - 0.366)140 = (P - L)(-33)$.

The relation for the 0.24 M sucrose data cannot be relied upon as the value of ($M - F$) approaches zero. The distilled water data gives a value of ($P - L$) $= 0.093$ electrons/Å^3, the 0.82 M sucrose data gives a value of ($P - L$) $= 0.098$ electron/Å^3 and a value of ($P - L$) $= 0.095$ electrons/Å^3 is interpolated for the 0.24 M sucrose data. A knowledge of ($P - L$) and α provides the following values of P and L: nerve swollen in distilled water, $P = 0.368$ and $L = 0.275$ electrons/Å^3, nerve swollen in 0.24 M sucrose, $P = 0.370$ and

$L = 0.274$ electrons/$Å^3$, nerve swollen in 0.82 M sucrose, $P = 0.382$ and $L = 0.284$ electrons/$Å^3$.

STRUCTURE ANALYSIS OF LIVE NERVE

It is reasonable to ask if this same kind of model might also prove to be an appropriate model for live nerve. Nerve myelin shows only five orders of diffraction within the same reciprocal space interval as swollen nerve; the reciprocal space interval of swollen nerve includes Regions I, II and III as shown in Figs. 1-4 and 1-5. The peripheral nerve data when placed on the same relative scale (Worthington and Blaurock, 1969b) lie only moderately close to the continuous curves of swollen nerve and this reflects some structural differences between live and swollen nerve. Model parameters were obtained by computing the R-values using the first six orders ($h = 1$–6) of diffraction for the peripheral nerves and the first five diffraction orders ($h = 1$–5) for the central nervous system nerves. The corresponding $J_{obs}(h)$ values have been reported (Worthington and Blaurock, 1969b). The following model parameters were obtained and are listed in Table 1-2.

Table 1-2 Model parameters and the corresponding R-values for three live peripheral nerves and four live central nervous system nerves

	d	w	l	R
	Å	Å	Å	%
Peripheral nerve myelin				
Frog sciatic nerve	171	155	19.5	11
Rat sciatic nerve	176	157.5	16.0	14
Chicken sciatic nerve	182	166	19.0	12
Central nervous system myelin				
Frog optic nerve	154	146	21.0	7
Rat optic nerve	159	150.5	21.5	9
Chicken optic nerve	155	148.5	24.5	7
Frog spinal cord	153	144	20.0	6

From Table 1-2 the agreement for the peripheral nerves $R = 11$–14% is not as good as the swollen nerves where $R = 7$–9%. The agreement for the central nervous system nerves $R = 6$–9% is quite good but only ($h = 5$) orders were used in the comparison, whereas ($h = 6$) orders were used in the peripheral nerve comparison. The present model which gives good agreement for swollen nerve gives only fair agreement for the live nerves. Some modifications to this model are needed in order to obtain better agreement with the low-angle x-ray data from live nerves.

The Patterson function for live peripheral nerve (Worthington and Blaurock, 1969b; Worthington, 1969b) shows two definite peaks: a major peak occurs at $d/2$ and a minor peak occurs at $d/4 \pm 1$ Å. A Patterson function for frog sciatic nerve computed using Eq. (1-36) in Appendix VI and the first five values of $J_{obs}(h)$ is shown in Fig. 1-7. The Patterson functions computed using

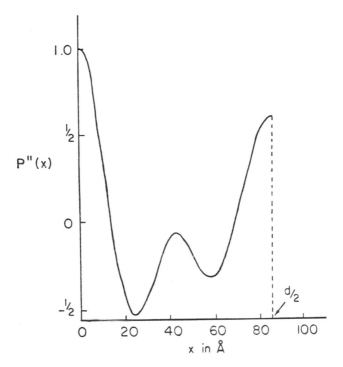

Fig. 1-7 *The Patterson function for frog sciatic nerve, $d = 171$ Å. The Patterson function was computed using Eq. (1–36) in Appendix VI and the first five intensity values. (From Worthington and Blaurock, 1969b.)*

the low-angle x-ray data from the three peripheral nerves and the four central system nerves all showed a major peak at $d/2$ and a minor peak at $d/4$. A Patterson function was computed directly from the model using Eq. (1-37) in Appendix IV. In this Patterson a major peak also occurred at $d/2$ and hence this peak was nicely accounted for by the model. A minor peak occurred at

$p + l$ (Worthington, 1969a) but the position of this peak in the model Patterson differed from $d/4$ by about 5 Å. Later it was noted that the minor peak position in the model Patterson function could be restored to $d/4$ by assuming that a cytoplasmic fluid layer was present within the membrane pairs. Accordingly, a more complicated model for nerve was examined.

Two additional parameters were added to the model shown in Fig. 1–6. A cytoplasmic fluid layer of width c was included between the two triple-layered units of the membrane pair. Also the triple-layered unit was allowed to be asymmetric, that is, the inner (cytoplasmic) layer of high electron density P had thickness p and the outer (extracellular) layer of high electron density P had thickness q. The asymmetry of the membrane unit is denoted σ where $\sigma = q - p$. The model shown in Fig. 1-6 has five parameters; the new

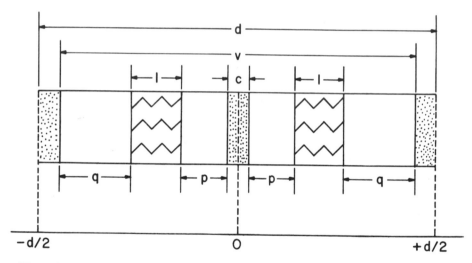

Fig. 1-8 *The model for live nerve is centrosymmetrical and has repeat distance **d** and membrane pair thickness **v**. The membrane pair contains two asymmetric membranes units (**p, l, q**) with a fluid layer of width **c** between the two units. The clear regions refer to electron density P, the zigzag lines refer to electron density L, and the dotted regions refer to the fluid electron density F.*

model, shown in Fig. 1-8, now has seven parameters. The low electron density region has electron density L and thickness l, the higher electron density layers have the same electron density P and the fluid layers have electron density F. The electron density of the fluid is assumed to be the same in both the cytoplasmic layer of thickness c and the extracellular layer of thickness $d - v$. The membrane pair has width $v = 2p + 2q + 2l + c$. The seven parameters are v, c, p, l, P, L and F. Note that if $c = 0$ and $\sigma = 0$, then this non-symmetric membrane model transforms to the earlier symmetric membrane model.

The same ratio α as defined earlier is adopted and the fluid layer F is assumed to be the same as the immersion medium. Again, the term $(P - L)$ is lost in the normalization procedure. Hence, the model has only five operational parameters v, c, p, l and α. The Fourier transform $T'(X)$ for this non-symmetric membrane model is given by Eq. (1-28) and (1-32) in Appendix V.

Model parameters were assigned to the three live peripheral nerves using the first seven orders of diffraction ($h = 1-7$). Significant improvement in R-values was obtained and the assigned model parameters are shown in Table 1-3. The low-density regions within the plasma membrane components

Table 1-3   Model parameters and the corresponding R-values for three live peripheral nerves and two live central nervous system nerves

	d	v	$d-v$	c	p	l	q	α	R
	Å	Å	Å	Å	Å	Å	Å		%
Frog sciatic nerve	171	156	15	6	24 3/4	22 1/2	28	0.30	8.2
Rat sciatic nerve	176	159	17	10	24	22	28 1/2	0.34	9.2
Chicken sciatic nerve	182	165	17	8	26 1/4	23	29	0.30	5.2
Frog optic nerve	154	148	6	6	21 1/4	24	25 3/4	0.36	2.0
Frog spinal cord	153	147	6	8	20 1/2	23	26	0.25	1.4

are 22–23 Å wide for the three varieties of sciatic nerve; these values are somewhat larger than those derived previously (see Table 1-2) using the symmetric membrane model. There is, however, a cytoplasmic fluid layer of 6–10 Å and an extracellular fluid layer of 15–17 Å. The high-density layers of the plasma membrane components are not symmetric: the width of the outer extracellular layer is greater and the asymmetry $\sigma = 3-5$ Å for the three sciatic nerves.

Model parameters were assigned to two nerves of the central nervous system: frog optic nerve and frog spinal cord. The first six orders of diffraction ($h = 1-6$) were used in the R-value calculations. Very low R-values were obtained, which to some extent reflected the fact that the number of diffraction orders was only slightly greater than the number of operational parameters, that is, six compared to five. The low-density regions within the plasma membrane components are 23–24 Å wide for the two nerves of the central nervous system; again, these values are somewhat larger than those derived using the symmetric membrane model. There is a cytoplasmic fluid layer of

6–8 Å and this layer has about the same width as the extracellular fluid layer. The outer high-density layers of the plasma membrane components are not symmetric and, like the case of peripheral nerve, the outer layer is wider than the inner layer. In the two nerves of the frog's central nervous system, the asymmetry $\sigma = 5$–6 Å and is slightly greater than that shown by the peripheral nerves.

In later work (Worthington and King, 1970) the restriction that the inner and outer high density layers had the same density value P was removed. The new model then contained electron densities P and Q for the inner and outer layers, respectively. The model had eight parameters v, c, p, l, P, Q, L and F, but only six operational parameters could be assigned as the result of R-value comparisons. Only the low-angle x-ray data from frog sciatic nerve has been analyzed. Model parameters were assigned using the previously mentioned additional orders of diffraction ($h \approx 1$–18). The model parameters v, c, p, l and α were similar to those derived using the seven-parameter model; some changes were noticed as additional orders ($h > 7$) were used in the comparison. The value of Q was found to be slightly less than the value of P. That is, the electron density of the outer extracellular layer is significantly less than the electron density of the inner cytoplasmic layer.

SUMMARY OF RESULTS

The low-angle x-ray patterns from swollen nerve have been interpreted in terms of a model. The model contains two triple-layered units end-to-end with only a small, if any, cytoplasmic space between the units. There is a wide extracellular fluid layer between the adjacent membrane pairs. The triple-layered units are symmetric when sucrose solutions are used as the immersion media but, in the case of distilled water, there is a slight asymmetry $\sigma = 3$ Å. The thicknesses of the single membrane units in swollen nerve are 70–72.5 Å and are less than the 75 Å thickness for live nerve (frog sciatic nerve). The widths of the low-density regions within the triple-layered membrane units are unexpectedly low. The actual values of the widths depend on the swelling media; they range from 17.5–27.0 Å. This variation in the widths of the low-density regions demonstrates that different swelling media lead to molecular changes within the nerve membranes.

A value of the average electron density of the membrane pair of swollen nerve has been obtained. This value is 0.343 electrons/Å^3. An absolute electron density scale for the high and low densities within the triple-layered unit of swollen nerve has been established. The values for the high density regions are 0.37–0.38 electrons/Å^3 while the values for the low density regions in the central parts of the membrane units are 0.27–0.28 electrons/Å^3.

The low-angle x-ray patterns from live nerves have been interpreted in terms of a seven-parameter model. The model contains two triple-layered units end-to-end but with a cytoplasmic fluid layer between the units. There is an extracellular fluid layer between adjacent membrane pairs. The triple-layered units are asymmetric. The asymmetry of the high density layers is $\sigma = 3$–5 Å for the three peripheral nerves and $\sigma = 5$–6 Å for the two central nervous system nerves. The electron density of the inner cytoplasmic layer exceeds that of the outer extracellular layer in the one case studied so far, namely, frog sciatic nerve. The thicknesses of the single membrane units are 74.5–78.5 Å for the three peripheral nerves and 69.5–71.0 Å for the two central nervous system nerves. The peripheral nerves have cytoplasmic fluid layers of 6–10 Å within the membrane pairs and extracellular fluid layers of 15–17 Å between the adjacent membrane pairs. The two central nervous system nerves also have cytoplasmic and extracellular fluid layers and these layers have approximately the same width, namely, 6–8 Å.

The widths of the central low-density regions within the membrane pairs are approximately the same for both types of nerve. The actual values of the widths are unexpectedly low; the values are 22–23 Å for the three peripheral nerves and 23–24 Å for the two central nervous system nerves.

The three varieties of peripheral nerve have radial repeat distances varying from 171–182 Å. The differences in the radial repeat distances are accounted for by the differences in the thicknesses of the membrane units and by the differences in the widths of the fluid layers.

DISCUSSION

A structure analysis of nerve myelin has been carried out using the model approach. In this method a search is made for a simple model which will give good agreement with the observed x-ray pattern at the low resolution. If sufficient orders of diffraction are available, then more detailed models are necessary. This is the present situation with live nerve because additional orders of diffraction have been recorded and this means that more complicated models for nerve can now be studied. So far an eight-parameter model has been proposed for frog sciatic nerve and seven-parameter models have been proposed for a variety of peripheral and central nervous system nerves.

A five-parameter model was first proposed to account for the swollen peripheral nerve patterns. This model gave excellent agreement with the observed low-angle x-ray data. As a result of finding the correct model from the examination of the possible models, the phase problem of swollen nerve was rigorously solved. However, the knowledge of the phases of swollen nerve within regions I, II and III is only academic for it leads to no new structural information because the favored model was already in excellent agreement with the observed x-ray data.

Although the five-parameter model was an appropriate choice for swollen nerve, it did not prove to be a good choice for live nerve. A more complicated seven-parameter model was needed to give good agreement with the observed low-angle x-ray data. In retrospect, it was reasonable to expect that the molecular structure of swollen and normal nerve should have been different. However, what was not expected was the excellent agreement which the first model gave with the swollen nerve x-ray patterns. This led to confidence in this kind of model and, furthermore, values for M, P and L were obtained in electrons/$Å^3$. Such values are entirely reasonable and might even have been anticipated. It is of course realized that the description of the swollen nerve membrane having uniform densities P and L is unlikely to be strictly correct. However, at low resolution, it is a valid description of the membrane structure within swollen nerve although the possibility of using other kinds of models has not been studied. A property of either the five- or the seven-parameter model is that both have theoretical diffraction which extends out to moderately large values of X. Higher orders of diffraction have now been recorded from nerve and as this property is matched by the models, it suggests that the electron density strip model might also be valid at moderate resolution.

As noted in the survey of earlier work, Finean (1954) had proposed a model which expressed the possible arrangement of protein and lipid molecules in the radial repeat. The model contained two triple-layered units end-to-end but there was no fluid layer either cytoplasmic or extracellular. In a later study Finean and Burge (1963) computed a Fourier series representation of peripheral nerve using the phases $(-, +, +, -, -)$ for the first five orders of diffraction. The last four phases $(+, +, -, -)$ have been shown to be correct, but some doubt concerning the first order remains (Worthington and Blaurock, 1969b). The Fourier synthesis for nerve computed by Finean and Burge (1963) was incorrect in that they used $[I(h)]^{1/2}$ instead of $[J_{obs}(h)]^{1/2}$ as the Fourier coefficient (see Appendix IV). Nevertheless, the resulting Fourier showed approximately the correct distribution because the phase information was correct. There were two peaks on either side of a deep trough of electron density. The deep trough was identified with the central region of a lipid bilayer and the two peaks were identified with phosphorus atoms on either side of the bilayer. This Fourier synthesis of nerve has been reproduced in many articles and its correlation with Finean's (1954) model seems to have been established. However, on the basis of diffraction theory, this correlation is certainly not justified. The difficulty of interpreting a low resolution Fourier synthesis has been discussed previously (Worthington, 1969a). The correctness of the identification of the phosphorus-to-phosphorus distance across the bilayer with the separation of the corresponding peaks of electron density is now examined. A phosphorus-to-phosphorus distance of about 50 Å is quoted for frog sciatic nerve by Finean and Burge (1963). The Patterson function of frog sciatic nerve shown in Fig. 1-7 has a minor peak at $d \approx 43$ Å but there is no peak in the vicinity of 50 Å. This suggests that either this distance does not exist, or else it cannot be detected by x-ray diffraction. For argument's sake, consider a one-to-one phospholipid: cholesterol complex with a surface area of about 100 Å^2 and let the phosphorus atoms all line up as in the Finean (1954) model but allow a variation of ± 2 Å. The phosphorus atoms then contribute an electron density of only 0.04 electrons/Å^3, which is small and it is doubtful if this electron density fluctuation could be detected by x-ray analysis at this time. Also, there is no basis for identifying certain features in the Fourier synthesis of nerve with the protein layers in the 1954 model. Therefore, the Fourier computed by Finean and Burge (1963) does not support a phosphorus-to-phosphorus distance on either side of the bilayer and it does not support the presence of protein layers. Hence the above Fourier does not lend support to the 1954 model apart from the deep troughs which are thought to be some kind of lipid bilayer.

The five-parameter model for swollen nerve and the seven-parameter model for live nerve are the first nerve models to have been described in detail and they are the first models which are known to be in agreement with the observed x-ray patterns. It is, therefore, only fair to draw attention to the fact that there are two reasons why the 1954 model is obsolete. First, the 1954 model does not show a narrow width of uniform low electron density and second, it does not show any fluid layers. Finean (1969) has noted that his 1954 model does not have this narrow region of low density and has stated that alternative molecular arrangements need to be considered.

No evidence of protein layers occurring in nerve myelin has been obtained. If there were protein layers on the outside of a lipid bilayer then, because of the known high density of protein molecules, the presence of discrete protein layers should have been detected, either in model comparisons, or from a study of difference Fourier syntheses. A non-symmetric triple-layered unit with uniformly high density values on the outside of the central low density region was deduced from the x-ray analysis. The thicknesses of the membrane unit in the case of the three peripheral nerves were 74.5–78.5 Å (75 Å for frog sciatic nerve) and the widths of the high density regions range from 24–29 Å. Thus, either the protein molecules have approximately the same electron density as the polar parts of the lipid molecules, or else the protein molecules do not occur in a discrete layer but are dispersed in some way within the high density regions. The value of $P = 0.37$–0.38 electrons/Å^3 is consistent with that expected from the protein molecules, the polar parts of the lipid molecules and an unknown amount of water.

The seven-parameter model for nerve has discrete fluid layers between the membrane units. The width of the extracellular fluid layer is about twice that of the cytoplasmic layer in the peripheral nerves whereas the widths of the fluid layers in the central nervous system nerves are approximately equal. The idea of discrete water layers in nerve had been suggested earlier by Schmitt (1958) who described possible arrangements for peripheral nerve which had two equal independent water layers per radial repeat. However, there were other considerations which did not support the existence of discrete water layers (Finean, 1969). As noted earlier in the survey of structural studies, Schmitt's assignment was based on the difference between the normal period of peripheral nerve and the corresponding air-dried period. This assignment was not rigorous because structural changes do occur when nerve myelin is air-dried. The present assignment of fluid layers in nerve myelin has been deduced from a careful study of model parameters using improved low-angle x-ray patterns of nerve.

In peripheral nerve the main contribution to the so-called "difference factor" arises because the extracellular fluid layer is appreciably wider than the cytoplasmic fluid layer. The asymmetry of each membrane unit provides only a small "difference factor". In the central nervous system nerves, this asymmetry is slightly more pronounced and in this case the so-called "difference factor" arises solely from the asymmetry of the membrane unit.

The actual values of the widths of the uniformly low density regions in the membrane units of nerve range from 22–24 Å for both types of nerve. As noted earlier these values are smaller than might have been anticipated. Previous expected values were closer to 35–40 Å in keeping with earlier ideas on extended lipid bilayers [for a discussion of this point, see Finean (1969)]. Nerve myelin contains about 80% lipid molecules (dry weight) and the lipid composition has been grouped together as phospholipid, cholesterol and cerebroside in the molecular ratios of 2:2:1 (Finean, 1962). Nerve myelin therefore has a high proportion of cholesterol, a molecule which is considerably different from the phospholipid molecules. Cholesterol is about 20 Å long (Vanderheuvel, 1963) and, on the basis of size alone, could easily fit into each of the three layers (p, l, q) in the membrane unit. Cholesterol has an electron density of about 0.36 electrons/Å^3 (calculated) and therefore the actual location of this molecule in the nerve membrane will strongly influence the observed electron density distribution.

Some preliminary calculations can be made on the possible location of cholesterol molecules in the non-symmetric membrane unit. The value of $L \approx 0.28$ electrons/Å^3 for the electron density of the low-density region is consistent with that expected from the hydrocarbon parts of the lipid molecules (Worthington and Blaurock, 1969b). If it is assumed that the hydrocarbon chains are predominately aligned radially, in agreement with the results of wide-angle x-ray diffraction and birefringence, then from knowledge of L, the average surface area per single hydrocarbon chain is about 23 Å^2. The phospholipid and cholesterol molecules are expected to form a one-to-one molecular complex (Finean, 1953; Vanderheuvel, 1963). The surface area of a phospholipid:cholesterol complex measured using the Langmuir trough is about 100 Å^2 and it is assumed that this one-to-one complex will have the same surface area in the nerve membrane. Therefore, if the cholesterol molecules are end-to-end in the central part of the low-density region, then two considerations are apparent. The width of the uniform central region will be twice the length of the cholesterol molecule, that is, about 40 Å. The electron density of the central region will be somewhat higher than the derived value of 0.28 electrons/Å^3. Hence, this possibility seems unlikely. However, if the cholesterol molecules are in the high density layers, then, on the basis of size alone, the cholesterol molecules could easily fit into these layers as the widths range from 24–29 Å. Then, allowing two hydrocarbon chains per phospholipid molecule, the surface area per hydrocarbon chain is about 50 Å^2 in the low density region. This value for the surface area per hydrocarbon chain is high by a factor of two. Therefore, if the cholesterol molecules are in the high density layers, the simplest solution is to assume that the hydrocarbon chains from either side of the bilayer overlap within the low-density region. That is, the hydrocarbon chains from the extracellular side interdigitate with the hydrocarbon chains from the cytoplasmic side. This arrangement then has an electron density very close to the derived low electron density value of 0.28 electrons/Å^3. Hence this possibility is a reasonable one and merits further consideration.

A calculation can also be made on the average side-by-side distance between hydrocarbon chains and the number of nearest neighbors. The number of nearest neighbors refers to the number of hydrocarbon chains which surround any particular hydrocarbon chain at the average distance of separation. First, it is assumed that there is a Gaussian distribution of nearest neighbors and hence the relation $rR = 1.12$ is obtained where $R = 1/d$ and r = average separation distance. The value of R is known from $d = 4.7$ Å and hence $r = 5.3$ Å. The average separation distance between hydrocarbon chains is 5.3 Å. The number of nearest neighbors is then obtained from knowledge of L and this number is about 5.6. Each hydrocarbon chain therefore has an average of about 5.6 neighbors at the mean separation distance of 5.3 Å. Thus the hydrocarbon chains in the central region of the nerve membrane show a tendency to pack hexagonally; a hexagonal array has 6 nearest neighbors.

It is interesting to calculate the electrical capacity of the seven-parameter model for nerve. The capacity of each layer C is given by

$$C = \varepsilon_o A(\mathcal{E}/d) \qquad (1\text{-}5)$$

in MKS units, where ε here is the dielectric constant, d is the width of the layer and A is the surface area of the layer. The capacity C_r of the radial repeat is given by

$$C_r = \varepsilon_o A[(d - v + c)/X + 2(p + q)/Y + 2l/Z]^{-1}, \qquad (1\text{-}6)$$

where X, Y and Z are the dielectric constants corresponding to F, P and L. The values $X \approx 80$, $Y \approx 50$ and $Z \approx 2.2$ are arbitrarily assigned. Using the parameters for frog sciatic nerve in Table 1-3, the expression in brackets in Eq. (1-6) reduces to $[\frac{1}{4} + 2 + 20]^{-1}$ and hence $C_r \approx 4 \times 10^{-2}$ farads per square meter. The capacity per nerve membrane is twice this value and expressed in the more usual units of microfarads per square centimeter, a capacity of $0.8 \mu f/cm^2$ is obtained. This is close to the observed value. The consequence of this kind of calculation is that, if membrane capacities of the order of $1 \mu f/cm^2$ are measured, then only a narrow region of hydrocarbon chains is needed to account for this measurement. Fricke's (1925) determination of thickness of the red blood cell membrane can now be interpreted as being simply a slightly high estimate of the width of the hydrocarbon chain region. Most cell membranes apparently have about the same electrical capacity and this suggests that all cell membranes have a narrow width of low density components in the central part of the membrane. For the same reasons, the lower electrical capacities of artificial membrane preparations tend to suggest that the widths of the hydrocarbon regions are greater than the corresponding widths in biological membranes.

The lamellar structure of retinal photoreceptors has also been studied by x-ray diffraction. A seven-parameter model similar to that shown in Fig. 1-8 has been proposed as a model of photoreceptors (Gras and Worthington, 1969). The thickness of the photoreceptor membrane unit is about 75 Å and is similar to nerve. The widths of the high density layers are considerably different but there is a narrow width of low density material about 16 Å in between the high density layers. Hence the two kinds of biological membranes which have been studied by x-ray diffraction, namely, nerve myelin and retinal photoreceptors both show a narrow region of uniformly low density in the central region of the membrane unit.

Retinal photoreceptors do not contain the same high proportion of cholesterol molecules as nerve. The proportion of cholesterol molecules per lipid composition is about one-fifth that of nerve and hence the location of the cholesterol molecule will not strongly influence the electron density distribution in photoreceptors. It is noted that both nerve and photoreceptors have narrow regions of low density material inside a triple-layered membrane unit even though the proportion of cholesterol molecules is considerably different in nerve and photoreceptors. This implies that cholesterol molecules do not contribute to the low density regions and hence are likely to be found in the high density regions. Admittedly, this is a comparatively weak implication, nevertheless, comparison between different membranes should provide useful information, if correctly interpreted.

In summary, improved low-angle x-ray patterns of nerve have been recorded and a structure analysis has been carried out on the low-angle x-ray data. A seven-parameter model has been proposed for nerve myelin which is in good agreement with the observed x-ray data. The model describes nerve myelin in terms of uniform electron densities within layers. The dimensions of the layers and the corresponding electron densities are derived directly from the x-ray analysis. In general terms, the model can be described as some kind of lipid bilayer model and hence no new conceptual ideas on membrane structure have been presented. The model can also be arbitrarily identified with the familiar triple-layered unit often seen in electron micrographs. However, the details of the present model had not been previously anticipated and furthermore the model contains some new and surprising features which are likely to be relevant to biological function. The seven-parameter model will predictably have a short lifetime because additional orders of diffraction have now been recorded and hence, in future work, more detailed models will soon be proposed. Nevertheless the prominent and distinctive electron density variations are likely to still be present in later models.

ADDENDUM

Electron Density Profiles and the Refinement of Model Parameters of Nerve Myelin

This section was written eight months after the completion of the preceding chapter. During this period of time a refinement of nerve myelin structure using the additional orders of diffraction (h ≈ 12 and h ≈ 18) from frog sciatic nerve has been in progress. The phases used in this refinement are the same as the phases derived from the eight-parameter model. Therefore a Fourier series representation of nerve myelin t″(x) could have been presented in the chapter because the phases from the eight-parameter model were known. However, at this earlier time the only supporting evidence in favor of this choice of phases came from model-building considerations. Various other sets of phases for live nerve myelin have now been studied using methods of analyses which are independent of model building considerations (King and Worthington, to be published). As a result of this study there are good reasons for believing that the phases given by the eight-parameter model are correct. Hence one-dimensional electron density profiles of frog sciatic nerve can now be presented with a certain measure of confidence that the electron density profiles are the correct ones.

The Fourier series representation $t''(x)$ for nerve myelin is given by Eq. (1-25) in Appendix IV. Fourier syntheses were calculated using the corrected observed intensities, that is, using the amplitudes $[J_{obs} (h)]^{1/2}$ as the Fourier coefficients with $h = 5$, $h = 12$ and $h = 18$ reflections, respectively, and the three Fouriers are shown in Fig. 1-9, 1-10 and 1-11. The corresponding

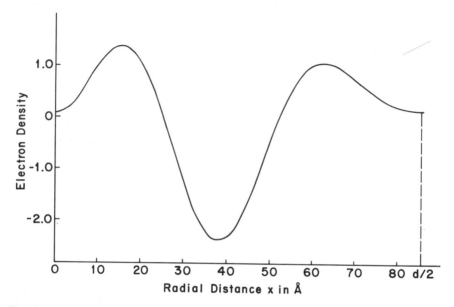

Fig. 1-9 *Fourier series representation for the myelin layers of frog sciatic nerve computed using the first five reflections. The resolution of the synthesis is* 17 Å.

effective resolutions Δx of these Fouriers are 17 Å, 7 Å and 4.8 Å, respectively; Δx is given by $\Delta x = d(2h)^{-1}$ where h is the number of reflections (either 5, 12 or 18).

The electron density profile for frog sciatic nerve computed using $h = 5$ reflections and the phases $(-, +, +, -, -)$ is shown in Fig. 1-9. A similar Fourier synthesis for nerve myelin ($h = 5$) has been previously described and compared to the seven-parameter model (Worthington, 1969c). The low-resolution electron density profile in Fig. 1-9 shows two definite peaks with a pronounced minimum. The two major peaks refer to the two high density regions; the peak at $x \approx 16$ Å refers to the cytoplasmic high density region and the peak at $x \approx 62$ Å refers to the extracellular high density region. The pronounced minimum at $x \approx 38$ Å refers to the narrow low-density region and the value of the parameter $l \approx 20$ Å is noted. This feature, $l \approx 20$ Å, was first described by Worthington and Blaurock (1968) in reference to the five-

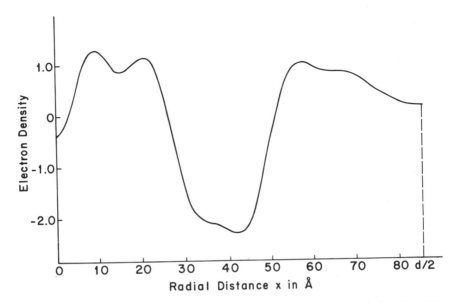

Fig. 1-10 *Fourier series representation for the myelin layers of frog sciatic nerve computed using the first twelve reflections. The resolution of the synthesis is 7 Å.*

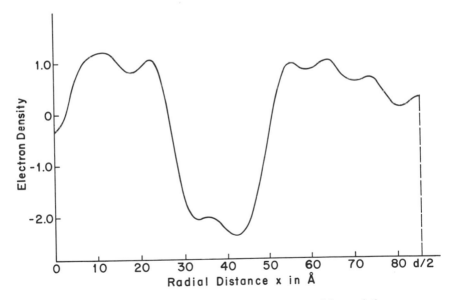

Fig. 1-11 *Fourier series representation for the myelin layers of frog sciatic nerve computed using the first eighteen reflections. The resolution of the synthesis is 4.8 Å.*

parameter model. Note, the electron density profile in Fig. 1-9 is not the same as the earlier curve obtained by Finean and Burge (1963). This earlier curve was in error as it was computed using uncorrected amplitudes $[I(h)]^{1/2}$ instead of corrected amplitudes, $[J_{obs}(h)]^{1/2}$. However, Finean and Burge (1963) used the same set of phases for the $h = 5$ reflections and, although their curve had approximately the correct electron density distribution, nevertheless, there were important differences. For example, the earlier curve did not suggest a narrow low-density region and, consequently, the 1963 electron density profile was thought to be in agreement with earlier ideas on extended lipid bilayers.

Electron density profiles for frog sciatic nerve were computed using $h = 12$ and $h = 18$ reflections together with the phases from the eight-parameter model and are shown in Fig. 1-10 and 1-11, respectively. These profiles show the two major peaks and the pronounced minimum in more detail. The high density regions have more shape and the narrow low density region $l \approx 20\,\text{Å}$ now begins to trace out a narrow uniform step of low electron density. Note, in Fig. 1-10 and 1-11 the identification of discrete fluid layers on either side of the nerve membrane cannot be directly made from a study of the electron density profiles alone. The cytoplasmic high density region shows a ripple contour consisting of two minor peaks in Fig. 1-10 and 1-11. The extracellular high density region shows a ripple contour consisting of two minor peaks in Fig. 1-10 whereas in Fig. 1-11 this same region shows three minor peaks.

Attention has been already drawn to the widely accepted view that the separation distance of the two peaks of electron density shown in Fig. 1-9 corresponds to a phosphorus-to-phosphorus distance across a lipid bilayer. In the discussion section it has been argued on the basis of diffraction theory that this interpretation is unsound. The correctness of this conclusion can be readily seen if one tries to locate the position of the phosphorus atoms in Fig. 1-9, 1-10 and 1-11. It is seen that the major peaks are single in Fig. 1-9; they are double in Fig. 1-10 and are both double and triple in Fig. 1-11. This ripple contour is in no way related to the presence of phosphorus atoms; it arises from the fact that each of the Fourier syntheses has limited resolution.

Although the attaining of the three electron density profiles of frog sciatic nerve is an achievement in itself, nevertheless it does not represent the end result. This is because the profiles have only moderate resolution and they require interpretation in terms of a model. Note, the main thesis of the chapter has been not to rely on the Fourier approach, but to directly obtain an estimate of the molecular parameters from model-building considerations. A comparative study of the electron density profiles obtained using the observed amplitudes and using the amplitudes calculated from the eight-parameter model can suggest changes in the original model. Accordingly, a more complicated model with twelve-parameters has been formulated (King and Worthington, to be published). The additional parameters come from dividing the cytoplasmic high density layer into two strips and the extracellular high density layer into three strips. This twelve-parameter model which has ten operational parameters is in good agreement with the x-ray data, $h = 12$ and $h = 18$, obtained from a variety of nerves. Although the determination of the optimum values for the various model parameters is still in progress, the probable values of some of the previously defined model parameters can be given. These parameters are as follows: v is the thickness of the membrane pair and $v = 2m + c$, where m is the thickness of the single nerve membrane unit; l is the thickness of the low-density region; c and $d - v$ are the widths of the cytoplasmic and extracellular fluid layers, respectively; $\sigma = q - p$ is the asymmetry of the single nerve membrane unit. Note, because regions p and q contain two and three strips, respectively, σ now becomes approximately equal to the thickness of the uniform strip of moderate electron density which is on the outside of the nerve membrane unit facing the extracellular fluid layer. A list of these model parameters is given in Table 1-4 for frog and rat sciatic nerves and for frog optic nerve.

Table 1-4 Model parameters and the corresponding R values for two live peripheral nerves and one live optic nerve

	d	m	c	$d - v$	l	σ	R
	Å	Å	Å	Å	Å		%
Frog sciatic nerve	171	75	10	11	21	10	6
Rat sciatic nerve	176	75	13	13	21	10	6
Frog optic nerve	154	70	10	4	21	6	8

A comparison of the model parameters in Tables 1-3 and 1-4 shows that some changes in the model parameters have occurred. This is to be expected as Table 1-3 was obtained from a study of $h = 7$ reflections using five operational parameters whereas Table 1-4 was obtained from a study of $h = 12$ reflections using ten operational parameters. However, the thickness of the single membrane unit $m = 75$ Å for the sciatic nerves and $m = 70$ Å for frog optic nerve remains relatively unaltered. Furthermore, the value of l is also very close to earlier estimates using less complicated models; the value of l now becomes 21 Å for the three nerves. Changes in the parameters c, $d - v$ and σ are noticeable. Parameter c has increased by 3–4 Å over the values given by the seven-parameter model, σ is larger by 2–7 Å whereas $d - v$ becomes smaller by 2–4 Å.

In the discussion section the main contributions to the so-called "difference factor" of nerve myelin have been given for peripheral and central nervous system nerves. This description remains valid for the seven-parameter model. However, the parameters c, $d - v$ and σ for the twelve-parameter model show some systematic changes and, although these changes are small, nevertheless, they are sufficient to modify the contributions to the so-called "difference-factor". The "difference factor" essentially accounts for the odd orders of diffraction. In the case of peripheral nerves the widths of the cytoplasmic and extracellular fluid layers are about the same and the "difference factor" arises from the asymmetry of the nerve membrane unit. In the case of frog optic nerve the asymmetry of the nerve membrane is smaller but the nerve membrane is unevenly placed within the fluid layers such that $c \approx \sigma + (d - v)$, that is, the asymmetry σ compensates for the smaller extracellular space to give a small "difference factor".

In conclusion, a brief description of nerve myelin structure can be given. The nerve membrane unit consists of some kind of lipid bilayer which has a narrow central region of low electron density. The width of this low density region is about 21 Å for both peripheral and central nervous system nerves. The thickness of the nerve membrane unit is about 75 Å for peripheral nerves and about 70 Å for the central nervous system nerves. The nerve membrane unit is non-symmetric. The asymmetry arises because there is a layer of moderate electron density on only one side of the nerve membrane unit, the extracellular side. This layer is about 10 Å wide in peripheral nerves and about 6 Å wide in central nervous system nerves. In peripheral nerves the nerve membrane units are evenly spaced from each other with discrete fluid layers of about 10–13 Å in between the nerve membrane units. On the other hand in central nervous system nerves the nerve membrane units are unevenly spaced and have unequal fluid channels between the nerve membrane units; the cytoplasmic fluid layer is wider than the extracellular fluid layer.

September, 1970

Note added in proof: **Molecular Models of Nerve Myelin.** The relationship between possible molecular models of nerve myelin and the derived electron density strip models is now considered. The earlier electron density models had only a few parameters and knowledge of these parameters did not lead to any unique way of assembling the molecular components. This situation has now improved because a twelve-parameter model for nerve myelin has been derived and knowledge of these additional parameters now limit the number of possible molecular arrangements which can be proposed. The procedure is to assemble molecular models of the various membrane components in such a way that the electron density contour of the resulting molecular arrangement is in agreement with the profile of the twelve-parameter electron density model. One possible molecular arrangement for frog sciatic nerve will be described.

In the case of frog sciatic nerve certain parameters of the twelve-parameter model are given in Table 1-4 while the parameters for the seven-parameter model are given in Table 1-3. In the seven-parameter model, the thickness of a single membrane unit is $m = p + l + q$ and $m \approx 75A$. In the twelve-parameter model the nerve membrane unit is divided into a total of six strips: l remains as before, p is divided into two strips and q is divided into three strips. The six strips are numbered from p to q. The six strips have the following electron density variation: (1) high, (2) moderately high, (3) low, (4) moderately high, (5) high, and (6) moderate. Strips (2) and (4) have the same electron density while strips (1) and (5) have equal electron densities. Strip (1) faces the cytoplasmic fluid layer and strip (6) faces the extracellular fluid layer.

The additional parameters from the twelve-parameter model led to a surprising result, namely, the widths of strips (2) and (4) are approximately equal and the widths of strips (1) and (5) are also approximately equal. It follows that the first five strips approximate a centrosymmetrical unit. Because of this result, it is tempting to assign a symmetric lipid bilayer construction to the first five strips and to assign the protein components to the remaining strip. A molecular model was assembled with this idea in mind.

A total of 10 lipid molecules consisting of 2 lecithins, 2 sphingomyelins, 4 cholesterols and 2 cerebrosides were constructed using CPK atomic models. The molecular model for the lipid bilayer arrangement within frog sciatic nerve is shown in Fig. 1-12 together with a scale showing the widths of the six strips given

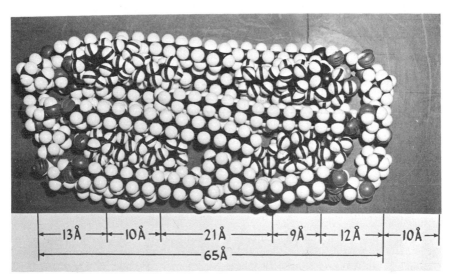

Fig. 1-12 A molecular model for the lipid bilayer construction within frog sciatic nerve. The width of this model is 65 Å. The protein component (not shown) is assigned to the 10 Å layer. The widths of the six strips given by the twelve-parameter electron density model are shown. The hydrogen atoms of the cholesterol molecules have a black line on a white background.

by the twelve-parameter model. The molecular model was constructed with the hydrocarbon chains interdigitated within the central low density region. The phosphorus-to-phosphorus distance shown by the model is 58 Å. The cholesterol molecules are easily identified because a black line has been added to the normal white hydrogen atoms. The cholesterol molecules are evenly distributed on each side of the bilayer and they are located within strips (2) and (4) of width 10 Å and 9 Å on either side of the central low density region. The methyl groups of the cholesterol molecules are within the low density region whereas the head groups of the cholesterols are in the high density regions which also contain the head groups of the phospholipid and cerebroside molecules. This molecular model has an electron density contour which resembles that of the twelve-parameter electron density model.

In the model the hydrocarbon chains are shown to be more or less straight. It is clear that the hydrocarbon chains within strips (2) and (4) are required to be more or less straight because of the fairly close packing with the cholesterol molecules. On the other hand the hydrocarbon chains which are interdigitated within the central strip are not packed closely together. It is likely that the hydrocarbon chains within this region may, in fact, be liquid-like.

In building the model it has been assumed that the protein component is present in the 10 Å strip facing the extracellular space. Whether all the protein component is actually within this layer is not known. Nevertheless this possibility merits further consideration. The possibility that there are protein molecules within the high density layers in addition to the 10 Å layer facing the extracellular layer has not been fully examined. At this time it would be helpful to know exactly the proportion of protein present in nerve myelin.

In conclusion a molecular model for the lipid bilayer construction of frog sciatic nerve has been described. A very similar molecular model also applies to rat sciatic nerve and to frog optic nerve. It is of interest that the molecular model tends to support the arguments put forward in the discussion section, namely, the hydrocarbon chains are interdigitated and the cholesterol molecules are located within the high density regions. The molecular model has the cholesterol molecules evenly distributed on each side of the low density region.

March, 1971

APPENDICES

Appendix I: X-ray Diffraction of a Single Membrane

Consider a single membrane which has thickness x_o. The lamellar structure of the single membrane is denoted by t(x) and its Fourier transform is denoted by $T(X)$ where x, X are real and reciprocal space co-ordinates. t(x) and $T(X)$ are a pair of Fourier transforms, denote t(x) $\rightleftharpoons T(X)$ where

$$T(X) = \int_{-\infty}^{+\infty} t(x) \exp(i2\pi xX)dx, \tag{1-7}$$

$$t(x) = \int_{-\infty}^{+\infty} T(X) \exp(-i2\pi xX)dX. \tag{1-8}$$

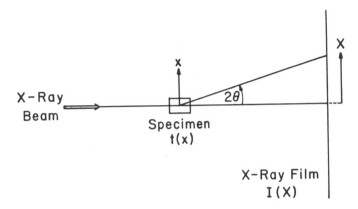

Fig. 1-13 *Diagram of a typical x-ray diffraction experiment. The x-ray specimen has electron density t(x) where x is the real space co-ordinate. The intensity recorded on the x-ray film refers to I(X) where X is the reciprocal space coordinate. The diffraction angle is 2θ and refers to the angle between the incident x-ray beam and the diffracted x-ray beam.*

Note, Eq. (1-7) and (1-8) are general formulae and for the single membrane the integration limits in Eq. (1-7) are from 0 to x_0.

A typical x-ray experiment is shown in Fig. 1-13. The collimated x-ray beam is incident parallel to the surface of the membrane. The diffraction pattern is recorded on x-ray film and shows continuous variation over all X. The observed x-ray intensity is denoted $I(X)$. The relationship between $I(X)$ and the Fourier transform is given by

$$I(X) = \varepsilon \Delta(X) |T(X)|^2, \tag{1-9}$$

where ε is a proportionality constant and $\Delta(X)$ is a function which depends on the experimental set-up and the specimen. If ε and $\Delta(X)$ are known, then $|T(X)|$ is obtained. The Fourier transform $T(X)$ can be expressed as $T(X) = |T(X)| \exp(i\beta)$ where β are the phases associated with $T(X)$. If the phases are known, then t(x) can be reconstructed from Eq. (1-8).

Appendix II: X-ray Diffraction from a Planar Multilayered Array of Membranes

The x-ray experiment using the single membrane is likely to fail because of the long exposure time required to record the diffraction. The exposure time is considerably reduced if a number of these single membranes can be obtained in a multilayered form. Let the single membranes be regularly stacked on top of each other and let the linear repeating distance be *d*.

It is convenient to adopt the following notation:

$$a(x) * b(x) = \int_{-\infty}^{+\infty} a(x')b(x - x') \, dx', \tag{1-10}$$

where $*$ is the convolution symbol. The convolution theorem states that

$$a(x) * b(x) \rightleftharpoons A(X)B(X), \qquad (1\text{-}11)$$

where $a(x) \rightleftharpoons A(X)$ and $b(x) \rightleftharpoons B(X)$ and the Fourier transformation ($\rightleftharpoons$) is defined by Eq. (1-7) and (1-8). The multilayered membrane structure can now be regarded as being formed as the result of a convolution between a linear array of points and a single membrane. Denote the multilayered structure, single membrane and the linear array of points by $m(x)$, $t(x)$ and $\phi(x)$, respectively. Hence $m(x)$ is given by

$$m(x) = t(x) * \phi(x), \qquad (1\text{-}12)$$

where $\phi(x) = \sum_{H} \delta(x - Hd)$ and δ is a delta function. The Fourier transform of multilayered structure is given by

$$M(X) = T(X)\Phi(X), \qquad (1\text{-}13)$$

where $m(x) \rightleftharpoons M(X)$, $t(x) \rightleftharpoons T(X)$ and $\phi(x) \rightleftharpoons \Phi(X)$. Now $\Phi(X) = \sum_{h} \delta(X - h/d)$ and therefore $M(X)$ only has values at $X = h/d$, h an integer. Experimentally a set of intensities $I(h)$ is recorded and from Eq. (1-9)

$$I(h) = \varepsilon\Delta(h)|M(h)|^2. \qquad (1\text{-}14)$$

Because the delta function has a constant value for each reflection, this constant value can be included into the proportionality constant and hence

$$I(h) = \varepsilon\Delta(h)|T(h)|^2. \qquad (1\text{-}15)$$

If ε and $\Delta(h)$ are known, then $|T(h)|$ is obtained. Now again if the phases associated with $T(h)$ are known then from Eq. (1-8), a Fourier series representation for the electron density variation $t(x)$ is given by

$$t(x) = (1/d)\sum_{-\infty}^{+\infty} T(h)\exp\,(-i2\pi x h/d). \qquad (1\text{-}16)$$

Note if $t(x)$ has a center of symmetry, that is, $t(x) = t(-x)$ then Eq. (1-16) reduces to

$$t(x) = (1/d)\sum_{-\infty}^{+\infty} \{\pm\}\,|T(h)|\cos\,2\pi x h/d. \qquad (1\text{-}17)$$

Appendix III: X-Ray Diffraction from Nerve Myelin

Nerve myelin consists of an approximately concentric mutilayered array of membranes with a well defined radial repeat distance d. Let $t(x)$ be the electron density along a radial direction within one radial repeating unit and let $T(X)$ be the corresponding Fourier transform. According to theory, $t(x)$ can be reconstructed from a knowledge of $T(X)$. An x-ray experiment with the x-ray beam incident at right angles to the nerve axis provides an x-ray intensity $I(X)$. A relation between $I(X)$ and $|T(X)|$ is needed. This relation was first given by Blaurock and Worthington (1966). The original analysis tends to be long and complicated and for this reason it is not reproduced here. However, a similar but shorter analysis is presented.

A single concentric layer of myelin can be regarded as being formed as the result of a convolution between a single membrane structure of thickness d and a delta function ring. Nerve myelin is then formed by adding additional layers. Let the nerve have an axon of radius a and let there be N layers so that the outer radius is a + Nd. If g(x) describes the N layers of nerve structure then

$$g(x) = \sum_{n}^{N} \delta [x - (a + nd)] * t(x), \qquad (1\text{-}18)$$

and the Fourier transform of the N layers of nerve structure is given by

$$G(X) = \sum_{n}^{N} 2\pi(a + nd) J_o[2\pi X(a + nd)] T(X). \qquad (1\text{-}19)$$

Now the following approximation can be made for the zero-order Bessel function

$$J_o(2\pi xX) \approx \pi^{-1}(Xx)^{-1/2} \exp(i2\pi xX - i\pi/4). \qquad (1\text{-}20)$$

Substitution of Eq. (1-20) into Eq. (1-19) gives

$$G(X) = 2X^{-1/2} \gamma(X)\Omega(X)T(X), \qquad (1\text{-}21)$$

where $\gamma(X)$ is a phase factor, that is $|\gamma(X)|^2 = 1$ and $\Omega(X)$ is an interference function which only has appreciable values at $X = h/d$, $h =$ integer. Therefore

$$|G(h)|^2 = (\text{constant}) \, h^{-1} |T(h)|^2. \qquad (1\text{-}22)$$

It can then be argued that the relation between $I(h)$ the observed intensity and $|T(h)|$ the required Fourier transform is given by

$$[h \, I(h)] = \varepsilon |T(h)|^2. \qquad (1\text{-}23)$$

Appendix IV: Fourier Series Representation for Nerve Myelin

If ε and the phases associated with the Fourier transform $T(h)$ are known, then a Fourier series representation t(x) of nerve is given by Eq. (1-17). In practice, this Fourier series representation cannot be computed. There are generally three reasons which are as follows:

(1) Only finite h is recorded.
(2) The zero order reflection ($h = 0$) is not recorded.
(3) The proportionality constant is not known.

Because of the above it is convenient to write the Fourier transform $T(X)$ as $T(X) = T'(X) + T(0)$ and $T(h) = T'(h) + T(0)$. That is, $T'(h)$ refers to the Fourier transform of t(x) at values of $X = h/d$, $h =$ integer but the $h = 0$ value is not included. Experimentally a set of intensities $I(h)$ is measured from the x-ray film and from Eq. (1-23) a set of corrected intensities $J_{obs}(h)$ is obtained where

$$J_{obs}(h) = [hI(h)]. \qquad (1\text{-}24)$$

$J_{obs}(h)$ is said to be on a relative scale. However, if a normalization constant K can be found such that $K \, J_{obs}(h) = |T'(h)|^2$, then the $J_{obs}(h)$ data are said to be on an absolute scale. The Fourier series representation for nerve myelin that can be directly computed from the observed low-angle x-ray data, provided the phases {±} are known, is denoted t″(x) where

$$t''(x) = (2/d)\sum_{1}^{h}\{\pm\}[J_{obs}(h)]^{1/2}\cos 2\pi xh/d. \tag{1-25}$$

Appendix V: Model Calculations

The Fourier transform $T(X)$ of a centro-symmetric electron density strip model with a repeating period d is given by

$$T(X) = 2\int_{0}^{d/2} t(x)\cos 2\pi xX \, dx. \tag{1-26}$$

By integrating strip-by-strip, convenient formulae are obtained (Worthington, 1969a). However, as explained in Appendix IV, only $T'(X)$ is required. $T'(X)$ for the five-parameter model (shown in Fig. 1-6) can be expressed as

$$T'(X) = (P - L)[\alpha w \text{ sinc } \pi wX - 2l \text{ sinc } \pi lX \cos \pi wX/2], \tag{1-27}$$

where $X = h/d$, h an integer and sinc $\theta = (\sin \theta)/\theta$. It is convenient to use the Fourier transform $S(X)$ where

$$S(X) = T'(X)(P - L)^{-1}, \tag{1-28}$$

and $S(X)$ becomes

$$S(X) = \alpha w \text{ sinc } \pi wX - 2l \text{ sinc } \pi lX \cos \pi wX/2, \tag{1-29}$$

where $S(X)$ only has values $X = h/d$. $S(h)$ then contains three operational parameters w, l, α.

The R-value is an index of how closely the theoretical diffraction $|S(h)|$ compares with the observed diffraction $[J_{obs}(h)]^{1/2}$. The normalization procedure is to find a constant k such that

$$\sum_{h}[k \, J_{obs}(h)]^{1/2} = \sum_{h}|S(h)|. \tag{1-30}$$

The R-value is computed from the formula

$$R = \sum_{h}\left||S(h)| - [kJ_{obs}(h)]^{1/2}\right| \bigg/ \sum_{h}[kJ_{obs}(h)]^{1/2}. \tag{1-31}$$

Model parameters w, l, α are obtained by computing the R-values for all possible w, l, α values.

The Fourier transform of the seven-parameter model $S(X)$ is given by

$$S(X) = \alpha(v - c)\text{sinc } \pi(v - c)X/2 \cos \pi(v + c)X/2 - 2l\text{sinc}\pi lX \cos \pi(2p + l + c). \tag{1-32}$$

Model parameters (v, c, p, l, α) are obtained by computing the R-values for all possible (v, c, p, l, α) values.

A simplified model for either the five- or the seven-parameter model can be used at very small angles of diffraction. Consider the five-parameter model (a similar analysis also applies to the seven-parameter model) and let the membrane unit (p, l, p) have uniform electron density (M). The Fourier transform $T'(X)$ of this model is given by

$$T'(X) = (M - F) w \text{ sinc } \pi wX. \tag{1-33}$$

Note $T'(X)$ from Eq. (1-33) and $T'(X)$ from Eq. (1-27) are approximately the same at small values of X where X approaches zero. Consider two sets of data obtained from swollen nerve using two different swelling media F_1 and F_2. It is assumed that M is the same for both media and if the two x-ray data sets are on the same relative scale, then for a very small value of X

$$[J_{obs}(X)]_1^{1/2} : [J_{obs}(X)]_2^{1/2} = (M - F_1) : (M - F_2). \tag{1-34}$$

Hence a value for M is obtained from Eq. (1-34).

An absolute electron density scale for the five-parameter model (a similar analysis also applies to the seven-parameter model) depends on the determination of $(P - L)$. Model parameters (w, l, α) are obtained from R-value calculations and knowledge of F and M are assumed. From Eqs. (1-33) and (1-27) the values of $T'(X)$ at $X = 0$ provides the identity

$$(M - F) w = (P - L)(\alpha w - 2l). \tag{1-35}$$

Hence a value for $(P - L)$ is obtained from Eq. (1-35). This means that from the knowledge of M and F and the parameters (w, l, α) electron densities P and L can be specified in terms of electrons/Å³.

Appendix VI: Patterson Functions of Nerve Myelin

The Patterson function $P''(x)$ of nerve myelin can be directly evaluated from the observed data without any knowledge of the phases. $P''(x)$ is given by

$$P''(x) = (2/d) \sum_1^h J_{obs}(h) \cos 2\pi x \, h/d. \tag{1-36}$$

In diffraction theory the Patterson function $P(x)$ is related to the Fourier transform $|T(X)|$ according to $P(x) \rightleftharpoons |T(X)|^2$ and, in terms of the electron density $t(x)$, $P(x)$ is given by

$$P(x) = t(x)*t(-x). \tag{1-37}$$

Therefore, given a particular model, the Patterson function $P(x)$ can be computed using Eq. (1-37). This model Patterson function can then be compared with the Patterson function $P''(x)$ calculated using the observed low-angle x-ray data. Now a study of $P''(x)$ can lead to the derivation of a model for membrane-type structures (Worthington, 1969a). On the other hand, if a model is already available, then a comparison between the two Patterson functions can lead to a new model or a refinement of the existing model.

REFERENCES

BEAR, R. S., K. J. PALMER and F. O. SCHMITT. 1941. X-Ray Diffraction Studies of Nerve Lipids. *J. Cell. Comp. Physiol.* 17:355.

BLASIE, J. K., M. M. DEWEY, R. E. BLAUROCK and C. R. WORTHINGTON. 1965. Electron Microscope and Low-angle X-ray Diffraction Studies on Outer Segment Membranes from the Retina of the Frog. *J. Mol. Biol.* 14:143.

BLASIE, J. K., C. R. WORTHINGTON and M. M. DEWEY. 1969. Molecular Localization of Frog Retinal Receptor Photopigment by Electron Microscopy and Low Angle X-Ray Diffraction. *J. Mol. Biol.* 39:407.

BLASIE, J. K., and C. R. WORTHINGTON. 1969. Planar Liquid-Like Arrangement of Photopigment Molecules in Frog Retinal Receptor Disk Membranes. *J. Mol. Biol.* 39:417.

BLAUROCK, A. E., and C. R. WORTHINGTON. 1966. Treatment of Low-Angle X-Ray Data from Planar and Concentric Multilayered Structures. *Biophys. J.* 6:305.

_____. 1969. Low-Angle X-Ray Diffraction Patterns from a Variety of Myelinated Nerves. *Biochim. Biophys. Acta.* 173:419.

BOEHM, G. 1931. Kurzzeitige Rontgeninterferenazufnahmen als Neue Physiologische Untersuchungs Methode. *Z. Biol.* 91:203.

BOYES-WATSON, J., and M. F. PERUTZ. 1943. X-Ray Analysis of Haemoglobin. *Nature.* 151:714.

DANIELLI, J. F., and J. DAVSON. 1935. A Contribution to the Theory of Permeability of Thin Films. *J. Cell. Comp. Physiol.* 5:495.

ELLIOTT, G. F., and C. R. WORTHINGTON. 1963. A Small Angle Optically Focusing X-Ray Diffraction Camera in Biological Research. Part I. *J. Ultrastruct. Res.* 9:166.

FERNANDEZ-MORAN, J., and J. B. FINEAN. 1957. Electron Microscope and Low-Angle X-Ray Diffraction Studies of the Nerve Myelin Sheath. *J. Biophys. Biochem. Cytol.* 3:725.

FINEAN, J. B. 1953. Phospholipid-Cholesterol Complex in the Structure of Myelin. *Experimentia.* 9:17.

_____. 1954. X-Ray Analysis of the Structure of Peripheral Nerve Myelin. *Nature.* 173:549.

_____. 1958. X-Ray Diffraction Studies of the Myelin Sheath in Peripheral and Central Nerve Fibres. *Exp. Cell Res. Suppl.* 5:18.

_____. 1960. X-Ray Diffraction Analysis of Nerve Myelin. *In* J. N. Cumings [ed.] *Modern Scientific Aspects of Neurology.* Edward Arnold, Ltd., London.

_____. 1962. The Nature and Stability of the Plasma Membrane. *Circulation.* 26:1151.

_____. 1969. Biophysical Contributions to Membrane Structure. *Quart. Rev. Biophys.* 2:1.

FINEAN, J. B., and R. E. BURGE. 1963. The Determination of Fourier Transform of the Myelin Layer from a Study of Swelling Phenomena. *J. Mol. Biol.* 7:672.

FINEAN, J. B., and P. M. MILLINGTON. 1957. Effects of Ionic Strength of Immersion Medium on the Structure of Peripheral Nerve Myelin. *J. Biophys. Biochem. Cytol.* 3:89.

FRICKE, H. 1925. The Electric Capacity of Suspensions with Special Reference to Blood. *J. Gen. Physiol.* 9:137.

GEREN, B. B. 1954. The Formation From the Schwann Cell Surface of Myelin in the Peripheral Nerves of Chick Embryos. *Exp. Cell. Res.* 7:558.

GORTER, E., and F. GRENDEL. 1925. On Bimolecular Layers of Lipoids of the Chromocytes of the Blood. *J. Exp. Med.* 41:439.

GRAS, W. J., and C. R. WORTHINGTON. 1969. X-Ray Analysis of Retinal Photo-receptors. *Proc. Nat. Acad. Sci.* 63:233.

GRENDEL, F. 1929. Uber die Lipoidschrift der Chromocyten bein Schaf. *Biochem. Z.* 214:231.

HOGLUND, G., and H. RINGERTZ. 1961. X-Ray Diffraction Studies on Peripheral Nerve Myelin. *Acta Physiol. Scand.* 51:290.

LANGMUIR, I. 1917. The Constitution and Fundamental Properties of Solids and Liquids. II. Liquids. *J. Am. Chem. Soc.* 39:1848.

MOODY, M. F. 1963. X-Ray Diffraction Pattern of Nerve Myelin: A Method for Determining the Phases. *Science.* 142:1173.

ROBERTSON, J. D. 1958. New Observations of the Ultrastructure of the Membranes of Frog Peripheral Nerve Fibres. *J. Biophys. Biochem. Cytol.* 3:1043.

_____. 1959. The Ultrastructure of Cell Membranes and their Derivatives. *Biochem. Soc. Symp.* (Cambridge, England) 16:3.

SCHMIDT, W. J. 1936. Doppelbrechung und Feinbau der Markscheide der Nerven-fasern. *Z. Zellforsch. Mikroskop. Anat. Abt. Histochem.* 23:657.

SCHMITT, F. O. 1958. Axon-Satellite Cell Relationships in Peripheral Nerve Fibers. *Exp. Cell Res. Suppl.* 5:33.

SCHMITT, F. O., R. S. BEAR and G. L. CLARK. 1935. X-Ray Diffraction Studies on Nerve. *Radiology.* 25:131.

SCHMITT, F. O., R. S. BEAR and K. S. PALMER. 1941. X-Ray Diffraction Studies on the Structure of the Nerve Myelin Sheath. *J. Cell Comp. Physiol.* 18:31.

SJOSTRAND, F. S. 1953. The Lamellated Structure of the Nerve Myelin Sheath as Revealed by High Resolution Electron Microscopy. *Experimentia.* 9:68.

_____. 1960. Electron Microscopy of Myelin and of Nerve Cells and Tissue. *In* J. N. Cumings [ed.] *Modern Scientific Aspects of Neurology.* Edward Arnold, Ltd., London.

VANDERHEUVEL, F. A. 1963. The Lipids in the Myelin Sheath of Nerve. *J. Am. Oil Chemists' Soc.* 40:455.

WORTHINGTON, C. R. 1960. Discrete Low Angle X-Ray Diffraction from Air-Dried Mitochondria. *J. Mol. Biol.* 2:327.

_____. 1969a. The Interpretation of Low-Angle X-Ray Data from Planar and Concentric Multilayered Structures: The Use of One-Dimensional Electron Density Strip Models. *Biophys. J.* 9:222.

————. 1969*b*. Structural Parameters of Nerve Myelin. *Proc. Nat. Acad. Sci.* 63:604.

————. 1969*c*. Low-angle X-Ray Diffraction of Biological Membranes. Johnson Foundation Conference, April, 1969.

WORTHINGTON, C. R., and A. E. BLAUROCK. 1968. Electron Density Model for Nerve Myelin. *Nature.* 218:87.

————. 1969*a*. A Low-Angle X-Ray Diffraction Study of the Swelling Behavior of Peripheral Nerve Myelin. *Biochim. Biophys. Acta.* 173:427.

————. 1969*b*. A Structural Analysis of Nerve Myelin. *Biophys. J.* 9:970.

WORTHINGTON, C. R., and G. KING. 1970. Refinement of Nerve Myelin Structure. *Abstr. Biophys. Soc. 14th Annu. Meet.* 50*a*.

THE THERMODYNAMIC FOUNDATIONS OF MEMBRANE PHYSIOLOGY

2

T. L. SCHWARTZ

One fruitful approach to the problem of membrane permeability has considered the membrane interior as a liquid-like diffusion regime. Descriptive relationships can then be derived using thermodynamics. Some valuable results have been obtained from this model even though it is, at best, only a first approximation to the real membranes. In this chapter we will briefly reexamine, from this viewpoint, the most common classical relationships dealing with excitable membranes.*

The laws governing the transport of substances across living membranes can be investigated only if this process is first separated into two complementary sets of diffusion phenomena: the passive and the active. Passive phenomena are those that are produced by the classical diffusion forces: gradients of concentration, hydrostatic pressure, or electrical potential. Facilitated and exchange diffusion, coupling between fluxes, or solvent drag

---

* It is not my intent to review the large body of pertinent literature. My purpose is to summarize and, where possible, reexamine this material at an appropriate level. References are given to the extent that they serve this end. A literature review may be found in Lakshminarayanaiah (1969). Cole (1968) deals with a broad spectrum of additional material of both scientific and historical interest.

effects are variations on the same theme.* They are also passive processes.

But transport processes have been discovered in living membranes which cannot be explained by such diffusion forces. Biologists have been forced to the conclusion that such membranes can somehow use metabolically derived energy to generate additional transporting forces. The creation of these forces of course involves chemical reactions. The resultant diffusion phenomena are called active. Note that it is not the use of metabolic energy to maintain structure, and thus mobilities, that is important here. This occurs in all living systems even during passive diffusion. The specific requirement for an active regime is that metabolic energy must be utilized to generate a phenomenological diffusion force.†

This chapter will be mainly concerned with non-carrier mediated passive phenomena. In a few cases, the effects of active fluxes will also be considered.

THE FLUX EQUATION

The theory of nonequilibrium thermodynamics tells us directly that the phenomenological force for the movement of electrically charged matter in constant temperature, continuous systems is the negative gradient of the electrochemical potential.‡ However, it is best to get some "physical feel" for this statement.

The electrochemical potential of a system can be defined from the thermodynamic relationship.

$$dG = -SdT + Vdp + \sum_i \tilde{\mu}_i dn_i. \tag{2-1}$$

---

* A number of diffusion phenomena important to biologists are briefly described by Park (1961). He indicates that in facilitated diffusion "the transported molecule combines reversibly with a carrier in the membrane and the complex oscillates between the surfaces to release or pick up molecules on either side." The carrier can cross the membrane in either the complexed or the uncomplexed form. But, in exchange diffusion it can cross only in the complexed form (see also Stein, 1967). In neither case is metabolic energy required. Solvent drag occurs when a bulk flow of solvent—as in a porous membrane—carries solutes along. Coupling between diffusive fluxes is related to momentum transfer between diffusing components. The distinction between active and passive transport is important to the working out of mechanisms because different driving forces are involved. But a clear distinction is difficult to make. This problem has been reviewed by, for instance, Heinz (1963), Katchalsky and Curran (1965) and Curran and Schultz (1968).

† When the diffusion is observed on a macroscopic level, it appears to be produced by a force. This can result from the action of a real force: for instance, an electric field acting on an ion. Or it can represent a "statistical force" such as a concentration gradient. Both are included in this usage of the term "phenomenological".

‡ Katchalsky and Curran (1965) and Snell et al. (1965) discuss nonequilibrium thermodynamics in a manner designed to be both rigorous and helpful to physically oriented biologists. A good physical chemistry reference, such as Guggenheim (1959), should be consulted. Helfferich (1962) contains additional useful material.

Here G is the Gibbs free energy, T is the absolute temperature, S is the entropy, V is the system volume, p is its hydrostatic pressure, and n_i are the number of moles of the ith substance present. The electrochemical potential is denoted by $\tilde{\mu}_i$, where

$$\tilde{\mu}_i = \frac{\partial G}{\partial n_i}\bigg]_{T,p,n_j}. \qquad (2\text{-}2)$$

It describes how the energy of the system changes with alterations in the ith constituent, while T, p and the amounts of all other constituents are held constant. Suppose we bring two such systems into contact so that they can interchange matter (Fig. 2-1). Under what thermodynamic conditions would dn_j moles of the jth constituent diffuse from system A to system B?

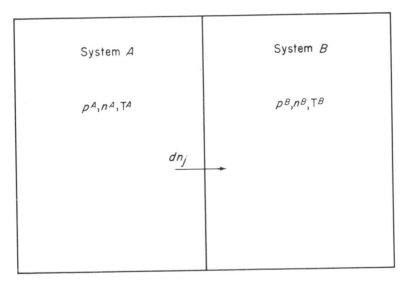

System A

p^A, n^A, T^A

System B

p^B, n^B, T^B

dn_j

Fig. 2-1 *Natural transfer of matter in a discontinuous regime.*

The total Gibbs free energy is

$$dG = dG^A + dG^B, \qquad (2\text{-}3)$$

or

$$dG = -S^A dT^A - S^B dT^B + V^A dp^A + V^B dp^B + \sum_i \tilde{\mu}_i^A dn_i^A + \sum_i \tilde{\mu}_i^B dn_i^B, \qquad (2\text{-}4)$$

where the superscript denotes the subsystem or phase.

If no changes in temperature or pressure in either of the two phases accompany the transfer,

$$dG = \tilde{\mu}_j{}^A dn_j{}^A + \tilde{\mu}_j{}^B dn_j{}^B. \tag{2-5}$$

But

$$-dn_j{}^A = dn_j{}^B \equiv dn_j, \tag{2-6}$$

so that

$$dG = (\tilde{\mu}_j{}^B - \tilde{\mu}_j{}^A)dn_j = \Delta\tilde{\mu}_j dn_j. \tag{2-7}$$

The second law of thermodynamics states that

$$dG < 0 \tag{2-8}$$

if such a process occurs naturally. Therefore

$$(\tilde{\mu}_j{}^B - \tilde{\mu}_j{}^A)dn_j < 0. \tag{2-9}$$

Since

$$dn_j > 0, \tag{2-10}$$

$$(\tilde{\mu}_j{}^B - \tilde{\mu}_j{}^A) < 0. \tag{2-11}$$

Thus, for matter to flow from phase A to phase B, it is necessary that

$$\tilde{\mu}_j{}^A > \tilde{\mu}_j{}^B. \tag{2-12}$$

Matter will diffuse from a region of higher to a region of lower potential. In a discontinuous system the difference in potentials thus appears as a driving force. But in a continuous system the situation is somewhat different.

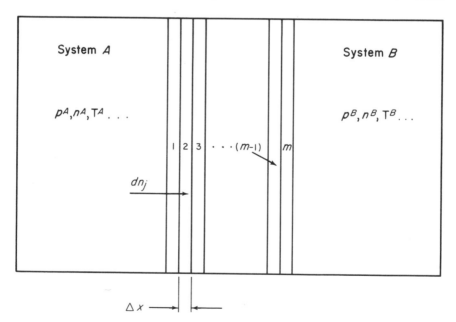

Fig. 2-2 *Diffusion between two compartments connected by a continuum.*

The transition to a *continuum* may be made if A and B are separated and the region between them is populated with a set of m further subsystems, each of thickness $(\Delta x)_l$ through which dn_j must pass.* Eq. (2-7) then takes the form

$$dG = \sum_{l=0}^{m} \Delta \tilde{\mu}_j{}^l dn_j$$

(2-13)

or

$$dG = dn_j \sum_{l=0}^{m} \left(\frac{\Delta \tilde{\mu}_j{}^l}{\Delta x_l} \right) (\Delta x)_l.$$

In the limit of $(\Delta x)_l \to 0$

$$dG = dn_j \int_A^B \left(\frac{\partial \tilde{\mu}_j}{\partial x} \right) dx$$

(2-14)

is the change in the free energy of the system resulting from the diffusion. This expression has the differential form

$$\frac{d}{dx} \left(\frac{dG}{dn_j} \right) = \frac{\partial \tilde{\mu}_j}{\partial x},$$

(2-15)

where the left-hand term represents the free energy decrease per pole per unit path length. If x_j denotes the internal phenomenological, molar, diffusion force,

$$x_j = -\frac{\partial \tilde{\mu}_j}{\partial x}$$

(2-16)

since Eq. (2-14) and (2-15) are analogous with the usual $\int x dx$ for an energy change accomplished through the action of some force, x.

If a diffusion regime is not too far from equilibrium, convection of the entire fluid is subtracted, and coupling between fluxes can be neglected, then the average diffusion velocity of a constituent can be assumed proportional to its driving force. Thus

$$v_i = -u_i \frac{\partial \tilde{\mu}_i}{\partial x},$$

(2-17)

* A unidimensional notation is used wherever this does not result in the loss of information.

where u_i the mobility, is a proportionality factor that is not a function of the force.* If c_i is the concentration of the diffusing species in moles per unit volume, then its flux density is

$$j_i = c_i v_i, \qquad (2\text{-}18)$$

so that

$$j_i = -u_i c_i \frac{\partial \tilde{\mu}_i}{\partial x}. \qquad (2\text{-}19)$$

This expression includes only that part of the flux of the ith constituent caused by its own potential gradient. But the flows of all the constituents present will actually be coupled to each other. Eq. (2-19) thus contains only the first of a number of terms occurring in the complete system of flux equations. In general, for a system containing m diffusing constituents, we should write the set of equations

$$
\begin{aligned}
j_1 &= L_{11}X_1 + L_{12}X_2 + \cdots + L_{1m}X_m \\
j_2 &= L_{21}X_1 + L_{22}X_2 + \cdots + L_{2m}X_m \\
&\quad \cdot \qquad\qquad\qquad\qquad \cdot \\
&\quad \cdot \qquad\qquad\qquad\qquad \cdot \\
&\quad \cdot \qquad\qquad\qquad\qquad \cdot \\
j_m &= L_{m1}X_1 + L_{m2}X_2 + \cdots + L_{mm}X_m.
\end{aligned} \qquad (2\text{-}20)
$$

The diffusion force, x_i, is conjugate to the flux, j_i. This means that

$$X_i = -\frac{\partial \tilde{\mu}_i}{\partial x}$$

and

$$L_{ii} = -u_i c_i.$$

$$(2\text{-}21)$$

The L_{ii} are ordinary coupling coefficients comparable to the ordinary diffusion constants. But the $L_{ij, i \neq j}$ are cross-coupling coefficients that take into account the effects on j_i of the non-conjugate forces such as $-(\partial \tilde{\mu}_j / \partial x)_{i \neq j}$.

By assuming negligible cross-coupling in the derivation of Eq. (2-19), all

---

* Throughout this chapter flux densities and velocities are referred to the membrane component of the diffusion regime, following Kirkwood's (1954) notation. The mobility is defined accordingly. For many biological systems this will be equivalent to a volume fixed frame of reference. A discussion of other possible reference velocities is beyond the scope of this chapter. There is another related mobility that is frequently used. It is defined from the molar diffusion force produced by an electrical potential gradient. The relationship between the two mobilities is

$$\Omega_i = |z_i| F u_i.$$

Table 2-1 Coupling coefficients for NaCl solutions. The coefficient units are 10^{-21} moles^2/dyne sec cm^2. Note that L_{11} and L_{22} are approximately proportional to concentration, as is expected from Eq. (2-21)

[Table reproduced from Katchalsky and Curran (1965). The data used in calculating L_{ij} were obtained from R. A. Robinson and R. H. Stokes, 1959, *Electrolyte Solutions*, Academic Press, New York, and from H. S. Harned and B. B. Owen, 1958, *The Physical Chemistry of Electrolyte Solutions*, Reinhold, New York.]

Coupling coefficient	Molar concentration						
	0.01	0.02	0.05	0.10	0.20	0.50	1.00
L_{11}	5.2	10.4	25.4	49.8	97.3	232.9	440.2
L_{22}	8.0	15.8	38.9	76.2	148.9	356.0	669.3
L_{12}	0.2	0.6	2.3	5.5	13.6	42.4	93.1

of the $L_{ij, i \neq j}$ were made zero. This is an approximation. But it is useful for dilute solutions (Table 2-1). In such solutions

$$\tilde{\mu}_i = \mu_i^\circ(p,T) + RT \ln c_i + z_i F\phi, \tag{2-22}$$

where z_i is the valence, R is the gas constant, F is Faraday's number, ϕ is the electrical potential, and $\mu_i^\circ(p,T)$ is the chemical potential in a standard state.* If the diffusion regime is both isothermal and isobaric

$$\frac{\partial \tilde{\mu}_i}{\partial x} = RT \frac{\partial}{\partial x} \ln c_i + z_i F \frac{\partial \phi}{\partial x}. \tag{2-23}$$

Combining with Eq. (2-19)

$$j_i = -u_i c_i \left[RT \frac{\partial}{\partial x} \ln c_i + z_i F \frac{\partial \phi}{\partial x} \right] \tag{2-24}$$

gives the relationship between the flux density and the gradients of concentration and electrical potential. This is the Nernst-Planck equation. It is the starting point for much of the following discussion.

* More generally

$$\tilde{\mu}_i = \mu_i^\circ(p, T) + RT \ln(\gamma_i c_i) + z_i F\phi,$$

where γ_i is the activity coefficient. It is defined so that

$$\lim_{c_i \to 0} \gamma_i = 1.$$

Both γ_i and μ_i° change if other units of concentration are used, but the expression for the electrochemical potential retains the same form.

SOME SIMPLE CONSEQUENCES

If the diffusing species is non-ionic $z_i = 0$ and Eq. (2-24) reduces to

$$j_i = -u_i RT \frac{\partial c_i}{\partial x},\qquad (2\text{-}25)$$

so that

$$j_i = -D_i \frac{\partial c_i}{\partial x} \qquad (2\text{-}26)$$

when the diffusion coefficient

$$D_i = u_i RT. \qquad (2\text{-}27)$$

Eq. (2-26) is a statement of Fick's Law for an undissociated substance. It is one obvious consequence of Eq. (2-24).

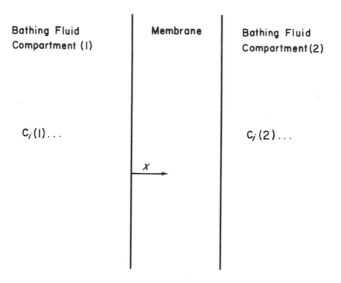

Fig. 2-3 *Unidimensional diffusion regime. Solutions are well stirred and baths are infinite.*

If the diffusing species is ionic and is in equilibrium across a membrane (Fig. 2-3),

$$j_i = 0 = -u_i c_i \left[RT \frac{\partial}{\partial x} \ln c_i + z_i F \frac{\partial \phi}{\partial x} \right]. \qquad (2\text{-}28)$$

Integration from one membrane boundary to the other yields

$$\phi(2) - \phi(1) = -\frac{RT}{z_i F} \ln \frac{c_i(2)}{c_i(1)}. \tag{2-29}$$

The subscripts 1 and 2 denote the boundaries of the membrane phase. This is a statement of the Nernst equilibrium potential. In its derivation we have sidestepped the question of the relationship between concentrations in the bathing media and those just inside the membrane by restricting our discussion to the membrane phase itself. This phase is assumed to be in equilibrium with the neighboring solutions. We shall continue to make this assumption in most of this chapter.

The electrical potential that appears across a membrane due to the free diffusion of a salt may also be deduced from Eq. (2-24). Using the subscripts a for anion and k for cation,

$$j_k = -u_k c_k \left[RT \frac{\partial}{\partial x} \ln c_k + z_k F \frac{\partial \phi}{\partial x} \right]$$

and

$$j_a = -u_a c_a \left[RT \frac{\partial}{\partial x} \ln c_a - |z_a| F \frac{\partial \phi}{\partial x} \right], \tag{2-30}$$

where $|z_a|$ has been written as the magnitude of the anionic valence.

Electroneutrality in the membrane requires that*

$$c_k z_k = c_a |z_a|. \tag{2-31}$$

The salt dissociates into ν_a anions and ν_k cations. If c_s is the salt concentration,

$$c_a = \nu_a c_s$$

and

$$c_k = \nu_k c_s. \tag{2-32}$$

Substituting into Eq. (2-30),

$$z_k j_k = -u_k (c_k z_k) \left[RT \frac{\partial}{\partial x} \ln c_s + z_k F \frac{\partial \phi}{\partial x} \right]$$

and

$$|z_a| j_a = -u_a (c_k z_k) \left[RT \frac{\partial}{\partial x} \ln c_s - |z_a| F \frac{\partial \phi}{\partial x} \right]. \tag{2-33}$$

Since there can be no net charge transfer in free diffusion,

---

* There is actually some charge separation because of differences between anionic and cationic mobilities. But the effect is small enough to be negligible everywhere except within a few angstroms of the membrane-solution interface.

$$z_k j_k = |z_a| j_a, \tag{2-34}$$

and

$$\frac{\partial \phi}{\partial x} = -\frac{u_k - u_a}{u_k z_k + u_a |z_a|} \frac{RT}{F} \frac{\partial}{\partial x} \ln c_s. \tag{2-35}$$

If we assume that the mobility does not vary with x,* we may integrate to obtain

$$\phi(2) - \phi(1) = -\frac{u_k - u_a}{u_k z_k + u_a |z_a|} \frac{RT}{F} \ln \frac{c_s(2)}{c_s(1)} \tag{2-36}$$

as the resultant electrical potential. Note that this potential and its polarity depend on the difference in mobility between the anion and the cation.

THE USSING-TEORELL UNIDIRECTIONAL FLUX RATIO

The experimental identification of the various types of diffusion processes common to living membranes is often difficult. For example, the transport of a substance from a region of lower to a region of higher electrochemical potential suggests the presence of an active process. But without further evidence such a conclusion is unwarranted, because this flux could be the result of diffusive coupling to the movement of some other species in the sense of Eq. (2-20). This kind of interaction does not involve metabolic chemical reactions in the membrane and is therefore a passive phenomenon. Even a flux which is reduced by a metabolic inhibitor may be passive and not active since the inhibitor may affect the membrane's permeability as well as its metabolism.

Ussing (1949) and Teorell (1949) have made an important contribution towards the partial resolution of this problem. Working independently, they each derived a relationship between the tracer fluxes across a membrane which Ussing showed could be used to distinguish experimentally between simple, non-coupled diffusion and all other transport phenomena. Active diffusion processes fall into the latter category. Their presence in certain diffusion regimes can at least be ruled out with this test.

---

* This assumption is not trivial and may be invalid for most biological membranes. It certainly fails for epithelial membranes.

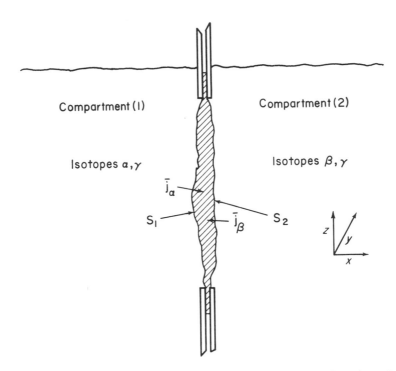

Fig. 2-4 *Inhomogeneous, irregularly shaped membrane mounted to determine the unidirectional flux ratio. Compartments are infinite and well stirred.*

Suppose the isolated, mounted membrane and its infinite, well-stirred, bathing media are in a steady state (Fig. 2-4). Suppose further, in this "Gedanken experiment", that the species of interest have three isotopes that are identical for all purposes except tracing. Let compartment (1) contain only isotope α of this ion while compartment (2) contains only isotope β. Assume that the system is isothermal and isobaric, and that solvent fluxes are negligible. Ussing's derivation, Teorell's derivation, and subsequent derivations by other authors have required the assumption that the membrane is homogeneous in all directions except x. This constraint on the result is annoying in general, and catastrophic for epithelial membranes. Fortunately it is unnecessary (Schwartz, 1971a) and will not be used in the following derivation.

The three dimensional equivalent of Eq. (2-24) gives

$$\bar{\jmath}_\alpha = -u_\alpha c_\alpha [RT\nabla \ln c_\alpha + zF\nabla\phi]$$
$$\bar{\jmath}_\beta = -u_\beta c_\beta [RT\nabla \ln c_\beta + zF\nabla\phi],$$

(2-37)

where $\bar{j}_\alpha$ and $\bar{j}_\beta$ are vector flux densities.* Making use of the identity

$$\nabla c_i + c_i \frac{zF}{RT} \nabla \phi = e^{-zF\phi/RT} \nabla [c_i e^{zF\phi/RT}]; \; i = \alpha, \beta \tag{2-38}$$

and the fact that

$$u_\alpha = u_\beta, \tag{2-39}$$

because of the identical diffusion properties of the tracers, Eq. (2-37) can be rearranged to yield

$$\bar{j}_\alpha \cdot \nabla [c_\beta e^{zF\phi/RT}] = \bar{j}_\beta \cdot \nabla [c_\alpha e^{zF\phi/RT}]. \tag{2-40}$$

A steady state in the membrane means that

$$\frac{\partial c_i}{\partial t} = -\nabla \cdot \bar{j}_i = 0; \; i = \alpha, \beta, \tag{2-41}$$

so that mass is conserved. This, the vector identity

$$\nabla \cdot [c_\beta e^{zF\phi/RT} \bar{j}_\alpha] = c_\beta e^{zF\phi/RT} \nabla \cdot \bar{j}_\alpha + \bar{j}_\alpha \cdot \nabla [c_\beta e^{zF\phi/RT}], \tag{2-42}$$

---

* The opening chapters of Lass (1950) are one good reference for vector calculus. The analogous derivation of the expression for the flux ratio that requires the homogeneity constraint is unidimensional and simpler. Eq. (2-24) is written for the fluxes of both isotopes. Using the identity

$$\frac{\partial c_i}{\partial x} + c_i \frac{zF}{RT} \frac{\partial \phi}{\partial x} = e^{-(zF\phi/RT)} \frac{\partial}{\partial x} (c_i e^{zF\phi/RT}); \; i = \alpha, \beta$$

and the equality of the two mobilities [Eq. (2-39)],

$$\frac{j_\alpha}{j_\beta} = -\frac{\dfrac{\partial}{\partial x}(c_\alpha e^{zF\phi/RT})}{\dfrac{\partial}{\partial x}(c_\beta e^{zF\phi/RT})}$$

results. If the membrane is homogeneous in the y-z plane, the steady state condition, Eq. (2-41),

$$\frac{\partial c_i}{\partial t} = -\nabla \cdot \bar{j}_i, \; i = \alpha, \beta$$

becomes

$$\frac{\partial c_i}{\partial t} = -\frac{\partial j_i}{\partial x}, \; i = \alpha, \beta,$$

where j_i is the x component of j_i. This enables us to integrate and obtain

$$\frac{j_\alpha}{j_\beta} = \frac{c_\alpha(1)}{c_\beta(2)} \exp\left\{\frac{zF}{RT}[\phi(1) - \phi(2)]\right\},$$

when Eq. (2-44) is used. This is the analog of Eq. (2-45). These fluxes can be used to define new J_α and J_β as in Eq. (2-46), provided the membrane surface lies in the y-z plane.

and the matching identity for $[c_\alpha e^{zF\phi/RT} \bar{J}_\beta]$ may be used to obtain

$$\nabla \cdot [c_\beta e^{zF\phi/RT} \bar{J}_\alpha] = \nabla \cdot [c_\alpha e^{zF\phi/RT} \bar{J}_\beta] \qquad (2\text{-}43)$$

from Eq. (2-40). Integration over the volume of the membrane under the experimentally imposed boundary conditions of

$$c_\beta(1) = 0$$

and $\qquad (2\text{-}44)$

$$c_\alpha(2) = 0,$$

and Gauss' divergence theorem yield

$$c_\alpha(1)e^{zF\phi(1)/RT} \int_{s_1} \bar{J}_\beta \cdot \overline{ds} = c_\beta(2)e^{zF\phi(2)/RT} \int_{s_2} \bar{J}_\alpha \cdot \overline{ds}. \qquad (2\text{-}45)$$

Here s_1 and s_2 are the irregular bounding surfaces of the membrane at which uniform concentrations and electrical potentials exist. The integral on the left is the net flux of isotope β into compartment (1), while that on the right is the net flux of isotope α into compartment (2). If

$$J_\alpha \equiv \int_{s_2} \bar{J}_\alpha \cdot \overline{ds},$$

and $\qquad (2\text{-}46)$

$$J_\beta \equiv \int_{s_1} \bar{J}_\beta \cdot \overline{ds},$$

it follows that

$$\frac{J_\alpha}{J_\beta} = \frac{c_\alpha(1)}{c_\beta(2)} \exp\left\{\frac{zF}{RT}[\phi(1) - \phi(2)]\right\}. \qquad (2\text{-}47)$$

This statement holds for steady state non-coupled diffusion processes. This implies, for example, the absence of carriers or the "single-file" diffusion defined by Hodgkin and Keynes (1955). It is independent of the three-dimensional structure of the membrane interior.

During a real double-label experiment, the tracer isotopes are present only in small amounts and are used to trace the movement of an isotope present in bulk. The specific activities of the tracer isotopes, α and β, can then be defined as

$$a_\alpha(1) \equiv \frac{c_\alpha(1)}{c_\alpha(1) + c_\beta(1) + c_\gamma(1)}$$

$$(2\text{-}48)$$

$$a_\beta(2) \equiv \frac{c_\beta(2)}{c_\alpha(2) + c_\beta(2) + c_\gamma(2)},$$

where c_γ refers to the concentration of the bulk isotope. Since γ is present in great excess, and because of Eq. (2-44),

$$a_\alpha(1) \cong \frac{c_\alpha(1)}{c_\gamma(1)}$$

$$\tag{2-49}$$

$$a_\beta(2) \cong \frac{c_\beta(2)}{c_\gamma(2)}.$$

Thus

$$\frac{[J_\alpha/a_\alpha(1)]}{[J_\beta/a_\beta(2)]} = \frac{c_\gamma(1)}{c_\gamma(2)} \exp\left\{\frac{zF}{RT}[\phi(1) - \phi(2)]\right\}. \tag{2-50}$$

This is the usual expression for the unidirectional flux ratio. All of its terms can be determined experimentally.

The terms $[J_\alpha/a_\alpha(1)]$ and $[J_\beta/a_\beta(2)]$ are often called unidirectional fluxes. When applied to the tracee during a nonsteady state, this usage can cause confusion because local concentrations in the membrane are still changing and there is no unique unidirectional flux. But in the steady state, J_{12}, the net flux of tracee from compartment (1) to compartment (2) is

$$J_{12} = [J_\alpha/a_\alpha(1)] - [J_\beta/a_\beta(2)], \tag{2-51}$$

without regard to the structure of the membrane compartment (Kedem and Essig, 1965; Schwartz, 1966; Schwartz and Snell, 1968). If J_{12} is zero, an equilibrium exists and Eq. (2-50) yields

$$\phi(1) - \phi(2) = \frac{RT}{zF} \ln \frac{c_\gamma(2)}{c_\gamma(1)}, \tag{2-52}$$

the Nernst potential, Eq. (2-29), which must occur at equilibrium.

Deviations from the predicted ratio are expected in the presence of either active or passive carrier-mediated diffusion, "single file" diffusion, or coupling between fluxes. In some cases experiments in which the direction of deviation is measured can yield information on the nature of the fluxes. Eq. (2-50) may be rewritten as

$$RT \ln \frac{[J_\alpha/a_\alpha(1)]}{[J_\beta/a_\beta(2)]} = [RT \ln c_\gamma(1) + zF\phi(1)] - [RT \ln c_\gamma(2) + zF\phi(2)] \tag{2-53}$$

or

$$RT \ln \frac{[J_\alpha/a_\alpha(1)]}{[J_\beta/a_\beta(2)]} = \tilde{\mu}_\gamma(1) - \tilde{\mu}_\gamma(2). \tag{2-54}$$

Exchange diffusion (Park, 1961) will tend to force the flux ratio to unity without regard to the magnitude of the thermodynamic driving force, $[\tilde{\mu}_\gamma(1) - \tilde{\mu}_\gamma(2)]$. In this case

$$\left| RT \ln \frac{[J_\alpha/a_\alpha(1)]}{[J_\beta/a_\beta(2)]} \right| < \left| [\tilde{\mu}_\gamma(1) - \tilde{\mu}_\gamma(2)] \right| \tag{2-55}$$

so that the magnitude of the logarithm of the measured ratio will be smaller than that predicted by Eq. (2-50).

In the presence of active transport, one of the fluxes will be larger than its passive value. Thus when it is found that

$$\left| RT \ln \frac{[J_\alpha/a_\alpha(1)]}{[J_\beta/a_\beta(2)]} \right| > \left| [\tilde{\mu}_\gamma(1) - \tilde{\mu}_\gamma(2)] \right|, \tag{2-56}$$

that is, the magnitude of the logarithm of the measured ratio is larger than that predicted by Eq. (2-50), the presence of active transport is suspected.

It follows from Eq. (2-28) and (2-29) that if a flux is passive and the membrane potential is readjusted by the passage of an electrical current from an external source so that the relationship

$$\phi(2) - \phi(1) = -\frac{RT}{z_i F} \ln \frac{c_i(2)}{c_i(1)} \tag{2-29}$$

is satisfied, the net flux of that species should vanish. Hodgkin and Keynes (1955) found that the measured flux ratio of potassium in the metabolically poisoned membrane of the giant axon of the cuttlefish, *Sepia officinalis*, was larger than predicted by Eq. (2-50) even though the fluxes obeyed the above passive relationship. They reasoned that these ions may be constrained to cross the membrane in some sort of single file in which they would necessarily interact. The effective valence of such a "complex" would have to be greater than 1. Under such circumstances, Eq. (2-50) yields

$$\left| \ln \left[\frac{\text{Actual Ratio}}{\text{Theoretical Ratio}} \right] \right| = \left| [\phi(1) - \phi(2)](z' - 1)\frac{F}{RT} \right|, \tag{2-57}$$

where z' is the effective valence. Hodgkin and Keynes found its magnitude to be between 2 and 3. This phenomenon has been called "single file diffusion".

OSMOTIC PRESSURE

Intracellular volume changes depend strongly on solvent diffusion and, therefore, on osmotic phenomena. Suppose that two well-stirred aqueous media are separated by a membrane capable of withstanding a hydrostatic pressure difference, but permeable only to the solvent (Fig. 2-5). If both compartments initially contain nothing but water, and some solute is subsequently added to compartment (2), what are the features of the new equilibrium which will emerge?

The solute is excluded from the equilibrium because it cannot penetrate the membrane. But, for the solvent

$$\tilde{\mu}_w(1) = \tilde{\mu}_w(2) \tag{2-58}$$

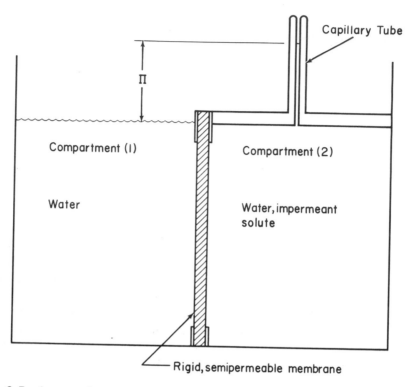

Fig. 2-5 *An osmotic equilibrium. (Both compartments are well stirred in the non-equilibrium, leaky membrane case when the solute is not impermeant.)*

at equilibrium. The subscript, w, refers to water. Eq. (2-22) then yields (Helfferich, 1962, p. 110)

$$\tilde{\mu}_w = \mu^\circ(p^\circ, T) + RT \ln X_w + \bar{v}_w(p - p^\circ), \qquad (2\text{-}59)$$

where $\bar{v}_w$ is the partial molal volume of the solvent, p is the pressure, and p° is the standard reference pressure. The pressure dependence has been isolated in the last term. It is convenient to use X_w, the mole fraction, as the unit of concentration.

Since only compartment (2) contains solute

$$\tilde{\mu}_w(1) = \mu^\circ(p^\circ, T) + \bar{v}_w[p(1) - p^\circ]$$

and

$$\tilde{\mu}_w(2) = \mu^\circ(p^\circ, T) + RT \ln X_w(2) + \bar{v}_w[p(2) - p^\circ].$$

(2-60)

We have made the reasonable assumption that $\bar{v}_w$ has no strong functional dependence on pressure. Equilibrium will occur when [Eq. (2-58)]

$$\bar{v}_w[p(1) - p^\circ] = RT \ln X_w(2) + \bar{v}_w[p(2) - p^\circ] \qquad (2\text{-}61)$$

or

$$\Pi \equiv [p(2) - p(1)] = -\frac{RT}{\bar{v}_w} \ln X_w(2). \qquad (2\text{-}62)$$

Since $X_w(2) < 1$, compartment (2) will develop a higher pressure than compartment (1), and fluid will rise in the capillary tube. This pressure difference is defined as the osmotic pressure of an ideal solution. If there is an impermeant solute also present in compartment (1), the analogous expression will be

$$\Delta\Pi = -\frac{RT}{\bar{v}_w} \ln \frac{X_w(2)}{X_w(1)}. \qquad (2\text{-}63)$$

Eq. (2-62) can be rewritten in terms of the solute concentration. The solute mole fraction is

$$X_s = 1 - X_w. \qquad (2\text{-}64)$$

Therefore

$$\ln X_w = \ln(1 - X_s) = -\left[X_s + \frac{X_s^2}{2} + \frac{X_s^3}{3} + \cdots\right]. \qquad (2\text{-}65)$$

For a dilute solution

$$X_s(2) \ll 1, \qquad (2\text{-}66)$$

$$\ln X_w(2) \cong -X_s(2) = -\frac{n_s(2)}{n_s(2) + n_w(2)} \cong -\frac{n_s(2)}{n_w(2)}, \qquad (2\text{-}67)$$

and

$$\Pi \cong \frac{RT}{\bar{v}_w} \frac{n_s(2)}{n_w(2)}. \qquad (2\text{-}68)$$

The compartmental volume is

$$V(2) = n_s(2)\bar{v}_s + n_w(2)\bar{v}_w \cong n_w(2)\bar{v}_w. \qquad (2\text{-}69)$$

It follows that

$$\Pi \cong RT \frac{n_s(2)}{V(2)}. \qquad (2\text{-}70)$$

or

$$\Pi \cong RT \, c_s(2). \qquad (2\text{-}71)$$

This is van't Hoff's law for the osmotic pressure of dilute solutions. The equivalent expression derived from Eq. (2-63) is

$$\Delta\Pi \cong RT[c_s(2) - c_s(1)]. \tag{2-72}$$

Note the formal similarity between Eq. (2-70) and the ideal gas law,

$$pV = nRT. \tag{2-73}$$

OSMOTIC PHENOMENA IN LEAKY MEMBRANES

The description of osmotic events becomes a more complex task if the membrane is not perfectly selective for the solvent, but also admits solute. Many cellular membranes are leaky to probing molecules. To understand the osmometric behavior of cells we must examine the attendant phenomena.

When solute is added to compartment (2) (Fig. 2-5) under these new conditions, the height of fluid in the capillary tube will no longer simply rise until it attains a final level. It will rise at first in response to rapid solvent diffusion. But as solute movement into compartment (1) slowly eliminates concentration differences, solvent diffusion will reverse and the height of fluid in the capillary will fall. The system will finally reach an equilibrium in which the two compartments are identical in pressure and composition.

Suppose that the capillary tube (Fig. 2-5) is sealed at the surface of compartment (2) by the insertion of a pressure gage. This must be done carefully to exclude air, and quickly to avoid solute diffusion. The flow of volume from one compartment to another is then zero, and it is possible to determine the associated pressure difference. But this experimentally determined osmotic pressure cannot be equated with the theoretical osmotic pressure discussed in the previous section, even prior to solute diffusion. Indeed, Staverman (1951) demonstrated that the two pressures are related by the expression

$$\sigma = \frac{\Pi_{\text{experimental}}}{\Pi_{\text{theoretical}}}. \tag{2-74}$$

The constant, σ, is a property of the membrane-solute system and is called the reflection coefficient. If

$$\sigma = 1$$

$$\Pi_{\text{experimental}} = \Pi_{\text{theoretical}} \tag{2-75}$$

and the membrane is perfectly semipermeable. If

$$\sigma < 1 \tag{2-76}$$

the membrane is leaky to solute.

Some insight into the physical significance of this coefficient may be obtained by examining its relationship to the parameters of the flux equation. A transfer of volume across the membrane would involve the fluxes of both the solvent and the solute. The volume of one compartment is, in general

$$V = \sum_i n_i \bar{v}_i, \qquad (2\text{-}77)$$

so that

$$\frac{dV}{dt} = \sum_i \bar{v}_i \frac{dn_i}{dt} \qquad (2\text{-}78)$$

and

$$j_V = \sum_i \bar{v}_i j_i, \qquad (2\text{-}79)$$

where j_V is the flux density of volume flow and the j_i are the usual molar flux densities. We should then write

$$j_s = -L_{ss}\frac{\partial \tilde{\mu}_s}{\partial x} - L_{sw}\frac{\partial \tilde{\mu}_w}{\partial x}$$

$$\qquad (2\text{-}80)$$

$$j_w = -L_{ws}\frac{\partial \tilde{\mu}_s}{\partial x} - L_{ww}\frac{\partial \tilde{\mu}_w}{\partial x},$$

for the system that we have been discussing. For simplicity we will assume a regime where the coupling is weak enough for us to write*

$$L_{sw} \cong 0$$

$$\qquad (2\text{-}81)$$

$$L_{ws} \cong 0.$$

It follows that

$$j_V = -(u_w\bar{v}_w)c_w\frac{\partial \tilde{\mu}_w}{\partial x} - (u_s\bar{v}_s)c_s\frac{\partial \tilde{\mu}_s}{\partial x} \qquad (2\text{-}82)$$

or

$$j_V = -(u_w\bar{v}_w)c_w\left[\frac{\partial \tilde{\mu}_w}{\partial x} + \frac{u_s\bar{v}_s}{u_w\bar{v}_w}\frac{c_s}{c_w}\frac{\partial \tilde{\mu}_s}{\partial x}\right]. \qquad (2\text{-}83)$$

The differential form of Eq. (2-63),

$$\frac{\partial \Pi}{\partial x} = -\frac{RT}{\bar{v}_w}\frac{\partial}{\partial x}\ln X_w \qquad (2\text{-}84)$$

*A discussion of more complex regimes, including electrolytes, can be found in Katchalsky and Curran (1965) and Helfferich (1962).

and the derivative of the solvent electrochemical potential,

$$\frac{\partial \tilde{\mu}_w}{\partial x} = \bar{v}_w \left[\frac{RT}{\bar{v}_w} \frac{\partial}{\partial x} \ln X_w + \frac{\partial p}{\partial x} \right],$$ (2-85)

can be combined to yield

$$\frac{\partial \tilde{\mu}_w}{\partial x} = \bar{v}_w \left[\frac{\partial p}{\partial x} - \frac{\partial \Pi}{\partial x} \right].$$ (2-86)

If the solute is uncharged

$$\frac{X_s}{X_w} \frac{\partial \tilde{\mu}_s}{\partial x} = \bar{v}_s \left[\frac{RT}{\bar{v}_s} \frac{X_s}{X_w} \frac{\partial}{\partial x} \ln X_s + \frac{X_s}{X_w} \frac{\partial p}{\partial x} \right].$$ (2-87)

But

$$X_s + X_w = 1.$$ (2-88)

Differentiating

$$\frac{X_s}{X_w} \frac{\partial}{\partial x} \ln X_s = -\frac{\partial}{\partial x} \ln X_w,$$ (2-89)

and

$$\frac{X_s}{X_w} \frac{\partial \tilde{\mu}_s}{\partial x} = \bar{v}_s \left[\frac{X_s}{X_w} \frac{\partial p}{\partial x} + \frac{\bar{v}_w}{\bar{v}_s} \frac{\partial \Pi}{\partial x} \right].$$ (2-90)

Since

$$\frac{c_s}{c_w} = \frac{X_s}{X_w},$$ (2-91)

Eq. (2-86) and (2-90) can be substituted into Eq. (2-83) to give

$$j_V = (u_w \bar{v}_w) c_w \left[\frac{\partial p}{\partial x} \left(\bar{v}_w + \frac{u_s \bar{v}_s}{u_w \bar{v}_w} \frac{c_s}{c_w} \bar{v}_s \right) - \frac{\partial \Pi}{\partial x} \left(1 - \frac{u_s \bar{v}_s}{u_w \bar{v}_w} \right) \bar{v}_w \right].$$ (2-92)

Since the volume flow vanishes,

$$\frac{\dfrac{\partial p}{\partial x}}{\dfrac{\partial \Pi}{\partial x}} = \frac{\left(1 - \dfrac{u_s \bar{v}_s}{u_w \bar{v}_w} \right)}{\left(1 + \dfrac{u_s \bar{v}_s}{u_w \bar{v}_w} \dfrac{c_s \bar{v}_s}{c_w \bar{v}_w} \right)}.$$ (2-93)

Eq. (2-93) can be integrated to yield

$$\sigma = \frac{\Pi_{\text{experimental}}}{\Pi_{\text{theoretical}}} = \frac{\left(1 - \dfrac{u_s \bar{v}_s}{u_w \bar{v}_w}\right)}{\left(1 + \dfrac{u_s \bar{v}_s}{u_w \bar{v}_w}\left[\overline{\dfrac{c_s \bar{v}_s}{c_w \bar{v}_w}}\right]\right)} \qquad (2\text{-}94)$$

for a homogeneous membrane. This utilizes the definition of Π in Eq. (2-62), and Eq. (2-94) thus applies at $t = 0$ before solute has diffused into compartment (1). The bar over $\overline{[c_s \bar{v}_s / c_w \bar{v}_w]}$ indicates an average value.* Thus

$$\sigma = 0 \text{ if } u_s \bar{v}_s = u_w \bar{v}_w, \qquad (2\text{-}95)$$

$$\sigma = 1 \text{ if } u_s \bar{v}_s = 0.$$

This simple membrane has lost all selectivity when the "mobilities" for volume flow [see Eq. (2-82)] are equal, and is perfectly selective when the solute mobility vanishes. A negative σ results if

$$u_s \bar{v}_s > u_w \bar{v}_w. \qquad (2\text{-}96)$$

The reflection coefficients of several biological membranes to some small non-ionic solutes have been determined (Table 2-2). The magnitude of σ has even been used to investigate possible membrane imposed steric limitations to the diffusion of nonlipid-soluble, nonionic molecules (Goldstein and Solomon, 1960) and has thus contributed to the idea that membranes may contain "pores". A simplified rationale for such an experiment would be that as the radius of the test molecule increases to that of the equivalent pore, membrane impermeability should increase and σ should approach 1. The equivalent pore size in the human erythrocyte determined in this manner was 4.2 Å.

Strong independent evidence for the existence of membrane "pores" had emerged as a result of earlier osmotic experiments. It had been found that the water permeability of a number of membranes as measured by the diffusion of labelled water was lower than the permeability measured in the presence of an osmotic gradient (see, for instance, Hevesy, Hofer and Krogh, 1935; Koefoed-Johnsen and Ussing, 1953; Mauro, 1957; Paganelli and Solomon, 1957). This contradiction cannot be resolved if water diffuses through these membranes by dissolving in them. But the problem vanishes if

---

* From the mean value theorem of integral calculus

$$\int_1^2 \left[1 + \left(\frac{u_s \bar{v}_s}{u_w \bar{v}_w}\right) \frac{c_s \bar{v}_s}{c_w \bar{v}_w}\right] \frac{\partial p}{\partial x}\, dx = [p(2) - p(1)]\left[1 + \left(\frac{u_s \bar{v}_s}{u_w \bar{v}_w}\right) \frac{c_s(\xi) \bar{v}_s}{c_w(\xi) \bar{v}_w}\right],$$

where $x(1) \leq \xi \leq x(2)$. All functions must be continuous and $\partial p / \partial x$ must be nonnegative.

Table 2-2 Staverman reflection coefficients in biological membranes.

Table modified from Katchalsky and Curran (1965)

Membrane	Solute	Reflection coefficient
Toad skin*	Acetamide	0.89
	Thiourea	0.98
Nitella translucens†	Methanol	0.50
	Ethanol	0.44
	Isopropanol	0.40
	Urea	1
Human red blood cell‡	Urea	0.62
	Ethylene glycol	0.63
	Malonamide	0.83

* Andersen, B., and H. H. Ussing. 1957. *Acta Physiol. Scand.* 39:228.
† Dainty, J., and B. Z. Ginzburg. 1964. *Biochim. Biophys. Acta.* 79:102, 112, 122, 129.
‡ Goldstein, D. G., and A. K. Solomon (1960).

there exist regions in the membrane through which water can move in response to a pressure gradient in an associated, hydrodynamic manner.

If these regions are idealized as right circular cylinders of radius r penetrating the membrane, and there are n such cylinders, the area available for the diffusion of labelled water is proportional to nr^2. But the laminar flow of water through such pores under a pressure gradient is described by Poiseuille's law (Paganelli and Solomon, 1957).* The flow is then proportional to nr^4. If an osmotic difference between the bathing media produces a pressure difference across the membrane it follows that the water permeability in an osmotic experiment should be larger than that determined in a diffusion experiment, as had been observed.

An equivalent pore size can be obtained from such a set of experiments. This was done by Paganelli and Solomon (1957) for the human red cell. The close correspondence between the 3.5 Å that they found and the 4.2 Å obtained by Goldstein and Solomon (1960) by measuring σ lent additional credence to the pore concept. It also suggested that the small, nonionic molecules used in the latter experiments move through the same "pores" as does water. But Diamond and Wright (1969) have pointed out that the permselective properties of membranes for small ions are unrelated to sieving of this kind.

* This is true when the diffusing molecule is much smaller than the pore. Otherwise corrections must be made (Paganelli and Solomon, 1957).

Mauro (1957, 1965) has suggested and experimentally demonstrated a mechanism for the transformation of an osmotic difference between the boundary solutions into a pressure difference in a porous membrane. The electrochemical potential of the permeant solvent must be continuous at the mouth of a pore, for infinite forces would otherwise appear.* Thus

$$\tilde{\mu}_w(2) = \tilde{\mu}_w(m2) \tag{2-97}$$

at the membrane-compartment (2) interface. The bracketed ($m2$) denotes the interior of a pore in this region. But the larger solute molecules are excluded from the pore interior. It follows that

$$\tilde{\mu}_w(2) = \mu°(p°,T) + RT \ln X_w(2) + \bar{v}_w[p(2) - p°],$$
$$\tilde{\mu}_w(m2) = \mu°(p°,T) + \bar{v}_w[p(m2) - p°], \tag{2-98}$$

and

$$[p(2) - p(m2)] = -\frac{RT}{\bar{v}_w} \ln X_w(2), \tag{2-99}$$

by reasoning similar to that used for Eq. (2-60), (2-61) and (2-62). The discontinuity in concentrations thus generates a pressure discontinuity at the mouth of the pore. There is no such problem on the pure solvent side so that

$$p(1) = p(m1) = p(2) \tag{2-100}$$

if both baths are at the same pressure. Thus

$$[p(m1) - p(m2)] = -\frac{RT}{\bar{v}_w} \ln X_w(2), \tag{2-101}$$

and a pressure difference drives solvent from side 1 to side 2 in the pore. Note the formal similarity between this pressure and the osmotic pressure, Eq. (2-62). The osmotic pressure, which is defined at equilibrium, thus effectively appears as a hydrostatic pressure in the porous membrane phase in this non-equilibrium regime.

THE DONNAN EQUILIBRIUM

A different type of osmotic equilibrium appears across a membrane when the impermeant species is ionic. It is characterized by an electrical potential difference in addition to the usual osmotic pressure, and is called the Donnan equilibrium. Any study of living membranes must consider these effects

---

* The assumption has been made here that a thermodynamic electrochemical potential can be defined at the microscopic level of a pore.

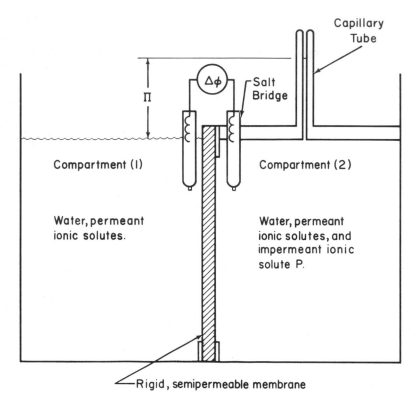

Fig. 2-6 *The Donnan equilibrium. The Donnan potential is measured with reversible electrodes connected to the baths by salt bridges, as shown. The combination is irreversible. The impermeant solute has valence z_p and mole fraction X_p.*

because large, impermeant, organic ions are common intracellular constituents. For example, such ions exist in concentrations of 155 mM in the intracellular fluid of mammalian muscle cells (Ruch, et al., 1965, p. 4), and 265 and 304 mM in the giant axons of the squid *Loligo* and *Docidicus*, respectively (Deffner, 1961). However, living cells are never at equilibrium with their surroundings. They are in a steady state in which active transport processes play an important role. The formalism of the Donnan equilibrium must therefore be used cautiously for such systems.

A typical experimental Donnan regime is depicted in Fig. 2-6. At equilibrium

$$\tilde{\mu}_w(1) = \tilde{\mu}_w(2)$$

and

$$\tilde{\mu}_i(1) = \tilde{\mu}_i(2)$$

for the solvent and each of the permeant solutes. Then the behavior of solvent may be described by

$$\Delta\Pi = [p(2) - p(1)] = -\frac{RT}{\bar{v}_w}\ln\frac{X_w(2)}{X_w(1)} \tag{2-63}$$

as in the derivation of Eq. (2-62) and (2-63), and the behavior of each solute may be described by

$$RT\ln X_i(1) + z_iF\phi(1) + \bar{v}_i[p(1) - p^\circ] = RT\ln X_i(2) + z_iF\phi(2) + \bar{v}_i[p(2) - p^\circ]. \tag{2-103}$$

We have once again assumed $\bar{v}_i$ to be independent of pressure. The equilibrium membrane potential then is

$$\Delta\phi = [\phi(2) - \phi(1)] = \frac{1}{Fz_i}\left[RT\ln\frac{X_i(1)}{X_i(2)} - \bar{v}_i(\Delta\Pi)\right]. \tag{2-104}$$

When the concentration of the impermeant ion is low, $\bar{v}_i(\Delta\Pi)$ is so small that Eq. (2-104) becomes

$$\Delta\phi \cong \frac{RT}{Fz_i}\ln\frac{X_i(1)}{X_i(2)}. \tag{2-105}$$

Thus for any pair of permeant ions

$$\left[\frac{X_j(1)}{X_j(2)}\right]^{1/z_j} = \left[\frac{X_k(1)}{X_k(2)}\right]^{1/z_k}. \tag{2-106}$$

The approximation symbol has been abandoned for convenience. In particular, if one ion is an anion and the other a cation

$$[X_a(1)]^{1/|z_a|}[X_k(1)]^{1/z_k} = [X_a(2)]^{1/|z_a|}[X_k(2)]^{1/z_k}, \tag{2-107}$$

where $|z_a|$ denotes the magnitude of the anionic valence. For uni-univalent salts Eq. (2-107) yields the constant product relationship.

$$X_a(1)X_k(1) = X_a(2)X_k(2). \tag{2-108}$$

Thus if the salt concentration in compartment (1) is altered in such a manner as to maintain a constant product of permeant ions, the only ionic fluxes will be those resulting from small osmotic changes. Investigators will sometimes follow this strategy in an attempt to minimize the disturbance of intracellular conditions.

Some of the aspects of this equilibrium can be examined more simply if the discussion is confined to a uni-univalent salt. The relationship

$$r \equiv \frac{X_k(1)}{X_k(2)} = \frac{X_a(2)}{X_a(1)} \tag{2-109}$$

then defines the Donnan ratio and

$$\Delta\phi = \frac{RT}{F}\ln r \qquad (2\text{-}110)$$

follows from Eq. (2-105). From the conditions for electrical neutrality,

$$X_k(1) - X_a(1) = 0,$$
$$z_p X_p + X_k(2) - X_a(2) = 0, \qquad (2\text{-}111)$$

where the subscript p refers to the impermeant ion. It follows that

$$X_k(1) = X_a(1) = X(1) \qquad (2\text{-}112)$$

is the salt concentration in compartment (1), and that

$$X_k(2) = X_a(2) - z_p X_p. \qquad (2\text{-}113)$$

Dividing Eq. (2-113) by Eq. (2-112),

$$\frac{X_k(2)}{X_k(1)} = \frac{X_a(2)}{X_a(1)} - \frac{z_p X_p}{X(1)} \qquad (2\text{-}114)$$

so that

$$\frac{1}{r} = r - \frac{z_p X_p}{X(1)}. \qquad (2\text{-}115)$$

The resulting quadratic equation in r has the solution*

$$r = \left(\frac{z_p X_p}{2X(1)}\right) + \left[\left(\frac{z_p X_p}{2X(1)}\right)^2 + 1\right]^{1/2}. \qquad (2\text{-}116)$$

Only if $(z_p X_p)$ vanishes can r become unity. An electrical potential difference will exist at equilibrium whenever a charged impermeant ion is present. This potential difference will be sensitive to pH and will pass through zero when the impermeant ion is at its isoelectric point. For positive z_p, $r > 1$. Compartment (2) is then positive with respect to compartment (1) [Eq. (2-110)]. When z_p is negative, $r < 1$ and compartment (1) will then be positive with respect to compartment (2). The sign of the electrical potential difference thus follows the sign of the impermeant ion.

---

* The second root has been discarded because it leads to a negative r devoid of physical reality. It also should be noted that this section, with the exception of the osmotic expressions, could have been developed in much the same fashion in terms of molar concentrations. Then, for instance,

$$r = \left(\frac{z_p c_p}{2c(1)}\right) + \left[\left(\frac{z_p c_p}{2c(1)}\right)^2 + 1\right]^{1/2}.$$

The pressure difference given by Eq. (2-63) and (2-72) becomes

$$\Delta\Pi = RT[c_a(2) + c_k(2) + c_p - 2c(1)], \qquad (2-117)$$

if we make use of the molar form of Eq. (2-112) and the fact that in this case

$$c_s(1) = c_a(1) + c_k(1)$$

and

$$\qquad (2-118)$$

$$c_s(2) = c_a(2) + c_k(2) + c_p.$$

We shall examine Eq. (2-117) to determine the direction of the pressure difference. From the molar form of Eq. (2-108), (2-111), (2-112) and (2-113),

$$c_a(2)c_k(2) = c^2(1)$$

and

$$\qquad (2-119)$$

$$c_k(2) = c_a(2) - z_p c_p.$$

These yield the quadratic

$$c_k^2(2) + (z_p c_p)c_k(2) - c^2(1) = 0 \qquad (2-120)$$

and the one physically significant root

$$c_k(2) = c(1)\left\{ \left[\left(\frac{z_p c_p}{2c(1)}\right)^2 + 1 \right]^{1/2} - \left(\frac{z_p c_p}{2c(1)}\right) \right\}. \qquad (2-121)$$

Substitution into Eq. (2-117) yields

$$\Delta\Pi = RT\left[c(1)\left(g + \frac{1}{g} - 2\right) + c_p \right], \qquad (2-122)$$

where

$$g \equiv \left[\left(\frac{z_p c_p}{2c(1)}\right)^2 + 1 \right]^{1/2} - \left(\frac{z_p c_p}{2c(1)}\right). \qquad (2-123)$$

We can, without difficulty, define a real number

$$\xi \equiv \ln g \qquad (2-124)$$

because $g > 0$. It follows that

$$\Delta\Pi = RT[c(1)(e^\xi + e^{-\xi} - 2) + c_p] \qquad (2-125)$$

or

$$\Delta\Pi = RT\left[4c(1) \sinh^2\left(\frac{\xi}{2}\right) + c_p \right], \qquad (2-126)$$

a quantity that is always positive. The pressure is thus higher on the side containing the impermeant ion, as one might have expected intuitively. It is interesting that a pressure of approximately 2 mm of water, inside higher than outside, has been reported for both sheep and human erythrocytes (Rand and Burton, 1964a, 1964b). These authors examined the relationship of their results to both active transport and osmotic mechanisms. Baker et al. (1962) have reported the revitalization of a squid axon from which the axoplasm has been extruded to be pressure sensitive in the range of only a few centimeters of water. It was necessary to maintain the internal pressure at 2–4 cm of water above the external. But 3-6 cm of water caused damage if maintained. Teorell (1959a, 1959b) has suggested that pressure differences may be involved in the mechanisms of excitation.

A Donnan equilibrium, resulting from the presence of large amounts of intracellular, impermeant anions, is helpful in explaining the maintenance of high levels of K^+ and low levels of Cl^- in the cell interior. But it does not give insight into the low level of intracellular Na^+. Sodium and potassium should distribute identically in such an equilibrium. Active transport mechanisms must be invoked to explain this phenomenon.*

THE GOLDMAN EQUATION

Electrophysiological studies of membrane phenomena attempt to correlate transmembrane electrical potentials and currents, bathing fluid compositions, and ionic fluxes in order to gain insight into the associated permeability mechanisms. Attempts have been made to deduce their theoretical interconnection from applicable membrane models as an aid in the interpretation of experimental data. The expression

$$\phi(2) - \phi(1) = \frac{RT}{F} \ln \frac{\sum_a u_k c_k(1) + \sum_a u_a c_a(2)}{\sum_k u_k c_k(2) + \sum_k u_a c_a(1)} \tag{2-127}$$

due to Goldman (1943) and later rederived by Hodgkin and Katz (1949) has been quite useful.† Fig. 2-7 applies and the subscripts a and k denote anion and cation, respectively.

---

* The discussion of the Donnan equilibrium has here been confined to dilute, ideal solutions. The presence of a large, multivalent ion as an impermeant species will often introduce large deviations from ideality. See, for instance, Overbeek (1961) for examination of some of the related problems. The qualitative conclusions drawn from the ideal case remain, in essence, correct.

† Hodgkin and Katz (1949) introduced permeability coefficients to account both for the partition of each solute between the solution and membrane phases, and for the solute mobility. Thus, P's appear in their equations in place of the u's in Eq. (2-127).

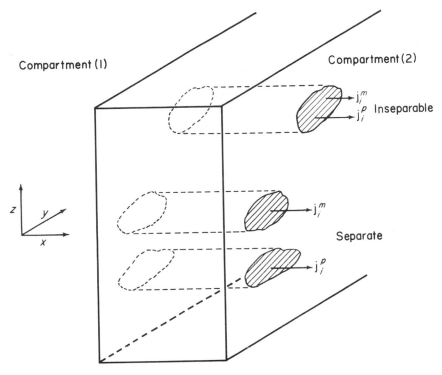

Fig. 2-7 *Three dimensional membrane. Compartments are infinite and well stirred.*

The simple form of the Goldman equation has lead to its extensive use in spite of the following sharp constraints imposed in its derivation.

1) The only permeant charged species are univalent.
2) The membrane is in a free diffusion steady state; that is, there is no current source. The current density, i, is given by

$$i = F\left[\sum_k j_k - \sum_a j_a\right] = 0, \tag{2-128}$$

but the individual ionic fluxes are not necessarily zero.

3) Fluxes resulting from active transport are ignored.
4) Membrane regions through which diffusion occurs are macroscopically homogeneous so that uniform mobilities may be assumed.
5) The electrical field in the membrane is constant.

The first two constraints merely limit the regimes to which the equation may be applied. But the third, fourth, and fifth raise questions about its applicability to living membranes. Its apparently successful use in spite of such limitations has suggested a need for reexamination (Patlak, 1960; Mullins and Noda, 1963; Barr, 1965; Geduldig, 1968; MacGillivray and Hare, 1969; Schwartz, 1969, 1971b).

For univalent species in the presence of active transport, Eq. (2-24) yields

$$j_k = j_k^m - u_k c_k^p \left[RT \frac{\partial}{\partial x} \ln c_k^p + F \frac{\partial \phi}{\partial x} \right],$$

$$j_a = j_a^m - u_a c_a^p \left[RT \frac{\partial}{\partial x} \ln c_a^p - F \frac{\partial \phi}{\partial x} \right]$$

$$(2\text{-}129)$$

for each anion and cation. The superscript p denotes a local concentration of ions free to diffuse passively. The active flux densities, j_k^m and j_a^m, are introduced kinematically; that is, no attempt to specify force-flux relationships is made. Total flux densities are denoted here by j_k and j_a. This differs from the usage in Eq. (2-24) where they refer to the passive portion only.

Since

$$\frac{\partial}{\partial x} \left(c_i e^{(z_i F/RT)\phi} \right) = e^{(z_i F/RT)\phi} \left(\frac{\partial c_i}{\partial x} + c_i \frac{z_i F}{RT} \frac{\partial \phi}{\partial x} \right);$$

$$i = a, k; \, z_k = +1, \, z_a = -1,$$

$$(2\text{-}130)$$

we may write

$$\frac{j_i}{u_i} e^{(z_i F/RT)\phi} = \frac{j_i^m}{u_i} e^{(z_i F/RT)\phi} - RT \frac{\partial}{\partial x} \left(c_i^p e^{(z_i F/RT)\phi} \right).$$

$$(2\text{-}131)$$

Boundary conditions may be introduced by an integration across the membrane. But we can obtain a useful expression only if we assume a steady state in the membrane. This, as usual, requires that membrane concentrations and fluxes no longer vary in time. But this can occur in two distinct ways depending on the nature of the membrane (Schwartz, 1969, 1971b). If the active and passive paths are spatially separate, they may be viewed as parallel. A steady state will then occur when

$$\frac{\partial c_i^m}{\partial t} = -\frac{\partial j_i^m}{\partial x} = 0,$$

$$i = a, k \qquad (2\text{-}132)$$

$$\frac{\partial c_i^p}{\partial t} = -\frac{\partial j_i^p}{\partial x} = 0$$

for each ion in the active and passive regions separately. This is the unidimensional form of Eq. (2-41). The superscript p now denotes all passive quantities. However, if the active and passive paths are spatially inseparable, the two processes may interact locally inside the membrane. Then

$$\frac{\partial c_i^m}{\partial t} = -\frac{\partial j_i^m}{\partial x} + e_i = 0,$$

$$i = a, k \qquad (2\text{-}133)$$

$$\frac{\partial c_i^p}{\partial t} = -\frac{\partial j_i^p}{\partial x} - e_i = 0$$

when c_i^m and c_i^p are the local concentrations of the species of interest present in the active and passive systems, respectively. The term, e_i, gives the local transfer of material from the passive to the active regime. Adding Eqs. (2-133) yields the steady state relationship

$$\frac{\partial c_i}{\partial t} = -\frac{\partial j_i}{\partial x} = 0; \; i = a, k \qquad (2\text{-}134)$$

in which

$$c_i = c_i^m + c_i^p;$$

$$j_i = j_i^m + j_i^p.$$

Thus, in this second type of steady state, it is only the total flux density that does not vary across the membrane.

Considering first this second less constrained type of steady state, Eq. (2-131) can be integrated to yield

$$j_k \int_1^2 \frac{1}{u_k} e^{(F/RT)\phi} dx = \int_1^2 \frac{1}{u_k} j_k^m e^{(F/RT)\phi} dx - RT[c_k(2)e^{(F/RT)\phi(2)} - c_k(1)e^{(F/RT)\phi(1)}];$$

$$j_a \int_1^2 \frac{1}{u_a} e^{(-F/RT)\phi} dx = \int_1^2 \frac{1}{u_a} j_a^m e^{(-F/RT)\phi} dx$$
$$- RT[c_a(2)e^{-(F/RT)\phi(2)} - c_a(1)e^{-(F/RT)\phi(1)}];$$
$$(2\text{-}135)$$

since, at the boundaries,

$$c_i^p = c_i.$$

The transmembrane electrical potential

$$\Delta\phi \equiv \phi(2) - \phi(1) \qquad (2\text{-}136)$$

is introduced by multiplying the cationic equation by $e^{(-F/RT)\phi(1)}$ and the anionic equation by $e^{(F/RT)\phi(2)}$. Then

$$j_k = \frac{1}{Q_k}\left\{M_k - RT\left[c_k(2)e^{(F/RT)\Delta\phi} - c_k(1)\right]\right\}$$

and

$$(2\text{-}137)$$

$$j_a = \frac{1}{Q_a}\left\{M_a - RT\left[c_a(2) - c_a(1)e^{(F/RT)\Delta\phi}\right]\right\}$$

describe the steady state fluxes if

$$Q_k \equiv \int_1^2 \frac{1}{u_k} e^{(F/RT)[\phi - \phi(1)]} dx,$$

$$Q_a \equiv \int_1^2 \frac{1}{u_a} e^{(F/RT)[\phi(2) - \phi]} dx,$$

$$(2\text{-}138)$$

$$M_k \equiv \int_1^2 \frac{j_k^m}{u_k} e^{(F/RT)[\phi - \phi(1)]} dx,$$

$$M_a \equiv \int_1^2 \frac{j_a^m}{u_a} e^{(F/RT)[\phi(2) - \phi]} dx.$$

These flux densities are uniform over the entire functional membrane if the transporting regions are homogeneous in the y–z plane (Fig. 2-7). Assuming no current, Eq. (2-128) is applicable even in the presence of active fluxes.* A summation over the fluxes and a regrouping of terms then yields

$$\Delta\phi = \frac{RT}{F} \ln \frac{\sum_k \frac{c_k(1)}{Q_k} + \sum_a \frac{c_a(2)}{Q_a} + \frac{1}{RT}\left[\sum_k \frac{M_k}{Q_k} - \sum_a \frac{M_a}{Q_a}\right]}{\sum_k \frac{c_k(2)}{Q_k} + \sum_a \frac{c_a(1)}{Q_a}}. \qquad (2\text{-}139)$$

We have, as yet, made no assumptions about either the nature of the electric field or membrane homogeneity in the x direction. But it can already be noted that the active fluxes will alter the membrane potential directly unless

$$\sum_k \frac{M_k}{Q_k} = \sum_a \frac{M_a}{Q_a}, \qquad (2\text{-}140)$$

a condition which will not generally be met.† If the "pumps" transfer no net

---

* When used in this manner, Eq. (2-128) has a different physical content than in the derivations due to Goldman (1943) and Hodgkin and Katz (1949). They constrained the sum of only the passive fluxes to zero whereas we are constraining the sum of the total fluxes. If their procedure is followed in the presence of active fluxes, one can consider only collectively nonelectrogenic "pumping" systems in a steady state membrane.

† The term "directly" implies that the effect is not the result of changes in the bounding concentrations. These would constitute an indirect effect.

charge and can collectively be considered nonelectrogenic

$$\sum_k j_k^m = \sum_a j_a^m \equiv J_m.$$ (2-141)

However, Eq. (2-140) will not even then be satisfied in general, and the "pump" fluxes may continue to alter the membrane potential. Only if the active and passive paths are spatially separate and the corresponding more restrictive Eq. (2-132)—and not Eq. (2-134)—applies, does this situation change. Since the j_k^m and j_a^m are then no longer functions of x, they can be removed from under the integral signs in Eq. (2-138). Eq. (2-140) then simplifies to

$$\sum_k j_k^m = \sum_a j_a^m$$ (2-142)

which is precisely the nonelectrogenic condition. Therefore, if the membrane is one in which the active and passive paths are spatially inseparable, even a nonelectrogenic pump can directly alter the membrane potential. It achieves this by displacing the internal concentration profiles from their passive configurations. It is often incorrectly assumed that only electrogenic pumps can exert a direct influence on the potential. In general, a nonelectrogenic pump will fail to affect the membrane potential directly only if the active and passive paths are separate and, in effect, parallel (Schwartz, 1969, 1971b). We shall subsequently see that further required assumptions regarding the electric field and the mobility will not alter this conclusion.

Eq. (2-139) formally resembles the Goldman equation (2-127) if the reciprocals of the Q's are regarded as permeabilities and the active transport terms are discarded. But these Q's are dependent on the structure of the electric field in the membrane. They therefore are not membrane parameters to be determined by perturbing either the bathing media or the membrane potential as are the u's of Eq. (2-127).* The derivation must be carried further to preserve the simplicity of the Goldman equation.

It is convenient to define the average of the potentials at the membrane boundaries as a reference level. Thus

$$\phi' \equiv \phi - \left[\frac{\phi(1) + \phi(2)}{2}\right],$$ (2-143)

and

$$\phi - \phi(1) = \frac{\Delta\phi}{2} + \phi',$$ (2-144)

$$\phi(2) - \phi = \frac{\Delta\phi}{2} - \phi'.$$

* The permeabilities defined by Patlak (1960) also appear to suffer from this difficulty.

If, in addition, the mobilities are assumed to be independent of x, Eq. (2-138) may be rewritten as

$$Q_k = \frac{1}{u_k} e^{(F/RT)(\Delta\phi/2)} \int_1^2 e^{(F/RT)\phi'} dx \equiv \frac{1}{u_k} e^{(F/RT)(\Delta\phi/2)} N_+,$$

$$Q_a = \frac{1}{u_a} e^{(F/RT)(\Delta\phi/2)} \int_1^2 e^{-(F/RT)\phi'} dx \equiv \frac{1}{u_a} e^{(F/RT)(\Delta\phi/2)} N_-,$$

$$\text{(2-145)}$$

$$M_k = \frac{1}{u_k} e^{(F/RT)(\Delta\phi/2)} \int_1^2 j_k^m e^{(F/RT)\phi'} dx \equiv \frac{1}{u_k} e^{(F/RT)(\Delta\phi/2)} L_k,$$

$$M_a = \frac{1}{u_a} e^{(F/RT)(\Delta\phi/2)} \int_1^2 j_a^m e^{-(F/RT)\phi'} dx \equiv \frac{1}{u_a} e^{(F/RT)(\Delta\phi/2)} L_a.$$

Eq. (2-139) then becomes

$$\Delta\phi = \frac{RT}{F} \ln$$

$$\frac{\dfrac{1}{N_+}\sum_k u_k c_k(1) + \dfrac{1}{N_-}\sum_a u_a c_a(2) + \dfrac{1}{RT} e^{(F/RT)(\Delta\phi/2)}\left[\dfrac{1}{N_+}\sum_k L_k - \dfrac{1}{N_-}\sum_a L_a\right]}{\dfrac{1}{N_+}\sum_k u_k c_k(2) + \dfrac{1}{N_-}\sum_a u_a c_a(1)}.$$

$$\text{(2-146)}$$

If only ions of one sign are permeant and the contribution of the active terms vanishes, an equation of the form

$$\Delta\phi = \frac{RT}{F} \ln\frac{\sum_i u_i c_i(\alpha)}{\sum_i u_i c_i(\beta)};$$

$$\text{(2-147)}$$

$$i = k, \ \alpha = 1, \ \beta = 2; \ i = a, \ \alpha = 2, \ \beta = 1$$

appears. The Goldman equation is then valid without any assumption regarding the nature of the electric field (Mullins and Noda, 1963).

But the more general condition that Eq. (2-146) become the Goldman equation (2-127) with an added active term is that

$$N_+ = N_-, \tag{2-148}$$

that is

$$\int_1^2 [e^{(F/RT)\phi'} - e^{-(F/RT)\phi'}] dx = 0. \tag{2-149}$$

Thus

$$\int_1^2 \sinh\left(\frac{F}{RT}\phi'\right) dx = 0 \qquad (2\text{-}150)$$

is the only necessary constraint on the electric field (Barr, 1965).* Eq. (2-146) can then be written

$$\Delta\phi = \frac{RT}{F}\ln\frac{\sum_k u_k c_k(1) + \sum_a u_a c_a(2) + \frac{1}{RT}e^{(F/RT)(\Delta\phi/2)}\left[\sum_k L_k - \sum_a L_a\right]}{\sum_k u_k c_k(2) + \sum_a u_a c_a(1)}, \qquad (2\text{-}151)$$

an expression identical to the Goldman equation without terms added for the active transport fluxes.

Eq. (2-150) implies the existence of a large group of functions, $\phi'(x)$, which will yield the Goldman equation. This group includes the set of all functions having odd symmetry about a midplane through the membrane.† The ϕ' that specifies a constant field is only one such odd function. But the complete group is by no means confined to odd functions. The constant field constraint is thus much more severe than required. With regard to this constraint, at least, the Goldman equation is therefore more generally applicable than was previously apparent. But the necessity to assume spatially constant u's in the permeable regions of the membrane seemingly invalidates this equation for use with epithelia.

The direct contribution of the active fluxes to the membrane potential depends on the term

$$\sum_k L_k - \sum_a L_a = \int_1^2 [e^{(F/RT)\phi'}\sum_k j_k^m - e^{-(F/RT)\phi'}\sum_a j_a^m]dx. \qquad (2\text{-}152)$$

Consider the nonelectrogenic case. Eq. (2-141) describes this condition. Eq. (2-152) then yields

$$\sum_k L_k - \sum_a L_a = 2\int_1^2 J_m \sinh\left(\frac{F}{RT}\phi'\right)dx. \qquad (2\text{-}153)$$

In this form, Eq. (2-153) describes a membrane with spatially inseparable active and passive regions. It will not, in general, be equal to zero. But if the

---

* Eq. (2-150) differs slightly from Eq. 14 in Barr (1965), which contains an error.

† If ξ measures the distance from a midplane through the membrane and towards its boundary, a function that has the property

$$\phi'(\xi) = -\phi'(-\xi)$$

is said to have odd symmetry, or to be an odd function.

active and passive regions are spatially separate, J_m is not a function of x [Eq. (2-132) and (2-141)]. Eq. (2-153) then becomes

$$\sum_k L_k - \sum_a L_a = 2J_m \int_1^2 \sinh\left(\frac{F}{RT}\phi'\right)dx = 0 \qquad (2\text{-}154)$$

with the aid of the constraint on the potential specified in Eq. (2-150). The active contribution vanishes only under these conditions. The validity of this conclusion, which was drawn earlier in the discussion, has thus not been altered by the subsequent assumptions about ϕ' and u.

It is unfortunately impossible to say too much more about the active contribution to the potential in either type of membrane regime without detailed knowledge of ϕ'. This is evident from Eq. (2-152) and (2-153). The problem is that Eq. (2-148) specifies only that $\int_1^2 e^{(F/RT)\phi'}dx$ and $\int_1^2 e^{-(F/RT)\phi'}dx$ are equal. While this is sufficient to eliminate functions of ϕ' from the passive terms in Eq. (2-151), it is insufficient to do so for the active terms. Permeabilities calculated from the Goldman equation must therefore contain errors unless the active terms vanish. However if

$$\frac{1}{RT}e^{(F/RT)(\Delta\phi/2)}[\sum_k L_k - \sum_a L_a] \ll [\sum_k u_k c_k(1) + \sum_a u_a c_a(2)], \qquad (2\text{-}155)$$

the errors should be small.

A similar problem arises with the integrated flux equations (2-137). Applying the appropriate assumptions about u, these yield

$$j_k = \frac{1}{N_+}\{L_k - u_k RT e^{-(F/RT)(\Delta\phi/2)}[c_k(2)e^{(F/RT)\Delta\phi} - c_k(1)]\};$$

$$\qquad (2\text{-}156)$$

$$j_a = \frac{1}{N_-}\{L_a - u_a RT e^{-(F/RT)\Delta\phi/2}[c_a(2) - c_a(1)e^{(F/RT)\Delta\phi}]\}.$$

If the active and passive regimes are separable, the active terms can be simplified to

$$\frac{L_k}{N_+} = j_k^m;$$

$$\qquad (2\text{-}157)$$

$$\frac{L_a}{N_-} = j_a^m.$$

But N_+ and N_- still remain in the passive terms. If the field constraint of Eq. (2-148) is invoked, we need examine only either N_+ or N_-. For a constant field

$$\phi' = m + \frac{\phi(2) - \phi(1)}{\delta}x, \qquad (2\text{-}158)$$

where m is a constant and δ is the membrane thickness. It follows that

$$d\phi' = \frac{\Delta\phi}{\delta} dx, \tag{2-159}$$

and

$$N_+ = \frac{\delta}{\Delta\phi} \int_{\phi'(1)}^{\phi'(2)} e^{(F/RT)\phi'} d\phi'$$

$$= \frac{RT}{F}\left(\frac{\delta}{\Delta\phi}\right)[e^{(F/RT)\phi(2)} - e^{(F/RT)\phi'(1)}] \tag{2-160}$$

$$= \frac{2RT}{F}\left(\frac{\delta}{\Delta\phi}\right)\sinh\left(\frac{F}{RT}\frac{\Delta\phi}{2}\right).$$

Eqs. (2-156) then reduce to

$$j_k = \frac{F}{\delta RT}\left(\frac{\Delta\phi}{2}\right)\frac{L_k}{\sinh\left[(F/RT)(\Delta\phi/2)\right]} - \frac{(\Delta\phi)Fu_k}{\delta}\frac{[c_k(2)e^{(F/RT)\Delta\phi} - c_k(1)]}{[e^{(F/RT)\Delta\phi} - 1]},$$

and $$\tag{2-161}$$

$$j_a = \frac{F}{\delta RT}\left(\frac{\Delta\phi}{2}\right)\frac{L_a}{\sinh\left[(F/RT)(\Delta\phi/2)\right]} - \frac{(\Delta\phi)Fu_a}{\delta}\frac{[c_a(2) - c_a(1)e^{(F/RT)\Delta\phi}]}{[e^{(F/RT)\Delta\phi} - 1]}$$

when the transport regimes are inseparable. In the separable case, Eq. (2-157) still describes the active terms. With the omission of the active terms, these are the usual Goldman-Hodgkin-Katz constant field, flux equations. Note that their passive portions can be expected to be less generally applicable than the passive part of the Goldman equation (2-151). This is because the flux equations (2-161) require a constant field assumption while the potential equation (2-151) does not.

A steady state, passive, diffusion regime with constant total ionic concentration will automatically develop a constant field in a homogeneous, uncharged membrane.* Using Eq. (2-24) we may write

$$\sum_k \frac{j_k}{u_k} = -\left[RT\frac{\partial}{\partial x}\sum_k c_k + F\frac{\partial\phi}{\partial x}\sum_k c_k\right];$$

$$\sum_a \frac{j_a}{u_a} = -\left[RT\frac{\partial}{\partial x}\sum_a c_a - F\frac{\partial\phi}{\partial x}\sum_a c_a\right] \tag{2-162}$$

for univalent species. Electrical neutrality requires that

$$\sum_k c_k = \sum_a c_a \equiv c; \quad \frac{\partial c}{\partial x} = 0 \tag{2-163}$$

* Finkelstein and Mauro (1963) have demonstrated that this is true even if the membrane contains fixed charges.

if the membrane is uncharged and a constant total concentration prevails. Substitution into Eq. (2-162) and subtraction yields

$$\frac{\partial \phi}{\partial x} = -\frac{1}{2Fc}\left[\sum_k \frac{j_k}{u_k} - \sum_a \frac{j_a}{u_a}\right]. \qquad (2\text{-}164)$$

The terms on the right side of this expression are all both spatially and temporally constant during a steady state. The electric field is therefore constant. The Goldman equation (2-151) without the active terms, is immediately applicable to this regime.

Any terms in Eq. (2-139), (2-146), or (2-151) that are due to a permeant ion which is at equilibrium may be omitted. The flux of such an ion is zero and does not participate in the summation that generates Eq. (2-139).

SIMPLE DIFFUSION REGIMES AND ELECTRICAL EQUIVALENT CIRCUITS

The study of the ionic permeability of living membranes has been greatly advanced through the measurement of their electrical properties. This has been adequately demonstrated by Cole (1968). External currents and voltages can be easily controlled, and their response to a variety of perturbations is readily measured. But the problem that confronts membrane physiologists is that of correlating these external observations with transport mechanisms inside the membrane. The Goldman equation and other, more complex, solutions of the flux equations are typical of one approach to this problem. It attempts a direct connection between the electrical measurements and the membrane diffusion regimes. Another natural, but more indirect, approach has been to correlate the observed voltages and currents with a network of resistors, capacitors, voltage sources, and other such components. The object is to deduce the connection between these components and aspects of the diffusion regime.

This second approach permits a phenomenological description from which clues with regard to mechanism may then be deduced. In this sense, the use of equivalent circuits should not be counterposed to a more direct examination of transport mechanisms. It should rather be regarded as an alternate method for extracting the same information. Hodgkin and Huxley's (1952c) application of an equivalent circuit to their analysis of excitability in squid axon is, perhaps, the most familiar example of this technique (Fig. 2-8). Equivalent circuits of this type have since been utilized in the discussion of

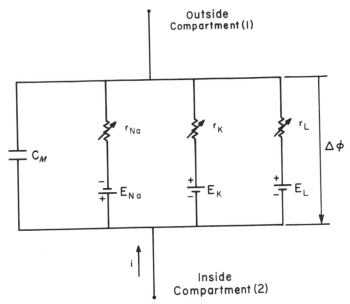

Fig. 2-8 *Equivalent circuit for squid axon [after Hodgkin and Huxley (1952c)]. The sign convention for current and voltage have been changed to the recent general usage: an outward current is positive, and the membrane potential is referenced to the outside. The compartmental notation refers to subsequent discussion. C_M is the membrane capacitance.*

many other excitable membranes. But the validity of this formalism depends on whether the pertinent diffusion equations can be cast into a form amenable to representation by an equivalent circuit.*

Finkelstein and Mauro (1963) discussed this problem in a simple regime without capacitance. They demonstrated that the diffusion of univalent species through a homogeneous membrane can be described by two types of equivalent circuit, each with its own area of applicability.

Both membrane faces are exposed to infinite, well-stirred bathing media denoted as compartments (1) and (2). For univalent species† Eq. (2-24) may

---

* Hodgkin and Huxley (1952b, pp. 480–482) were aware of this problem even though their equivalent circuit was not derived directly from the flux equations. They reasoned from knowledge of the driving forces for passive diffusion (Hodgkin and Huxley, 1952a, p. 461), and tested their conclusions by experiment. The parameters of their equivalent circuit were, thus, by no means based upon arbitrary definitions unrelated to membrane structure, Kornacker (1969) notwithstanding.

† The discussion can be extended beyond univalent species. However, they are sufficient for the point to be made and were used by Finkelstein and Mauro (1963).

be rearranged to yield

$$-\frac{j_k}{Fu_kc_k} = \frac{RT}{F}\frac{\partial}{\partial x}\ln c_k + \frac{\partial\phi}{\partial x};$$

$$-\frac{j_a}{Fu_ac_a} = \frac{RT}{F}\frac{\partial}{\partial x}\ln c_a - \frac{\partial\phi}{\partial x}.$$

(2-165)

It follows from the unidimensional form of Eq. (2-41) that in a steady state

$$\frac{\partial j_i}{\partial x} = 0; \quad i = a, k$$

(2-166)

so that Eq. (2-165) may be integrated to yield

$$-(j_kF)\int_1^2 \frac{dx}{F^2u_kc_k} = \frac{RT}{F}\ln\frac{c_k(2)}{c_k(1)} + [\phi(2) - \phi(1)];$$

$$-(j_aF)\int_1^2 \frac{dx}{F^2u_ac_a} = \frac{RT}{F}\ln\frac{c_a(2)}{c_a(1)} - [\phi(2) - \phi(1)].$$

(2-167)

Compartment (1) will, as usual, be taken as the potential reference. To be consistent with recent electrophysiological convention, it must then correspond to the outside of the membrane. This usage also assumes outward current to be positive.* The quantities, (j_kF) and (j_aF) have the units of current density. The current density carried by each ion may therefore be formally defined as

$$i_k = -j_kF;$$

$$i_a = j_aF.$$

(2-168)

If the reciprocal of the integral conductance and the resistance to each ion respectively are denoted as

$$\frac{1}{g_i} = r_i = \int_1^2 \frac{dx}{F^2u_ic_i}; \quad i = a, k,$$

(2-169)

and its Nernst potential as

$$E_k = -\frac{RT}{F}\ln\frac{c_k(2)}{c_k(1)},$$

or

$$E_a = \frac{RT}{F}\ln\frac{c_a(2)}{c_a(1)},$$

(2-170)

---

* Finkelstein and Mauro (1963) have taken inward currents to be positive and have effectively referenced the membrane potential to the inside solution.

Eq. (2-167) yields

$$i_k = g_k[\Delta\phi - E_k];$$

(2-171)

$$i_a = g_a[\Delta\phi - E_a].$$

These equations and the fact that, according to Kirchoff's rule, the total, external current density must be

$$i = \sum_k i_k + \sum_a i_a$$

(2-172)

suggest the parallel branched equivalent circuit of Fig. 2-9. Its similarity

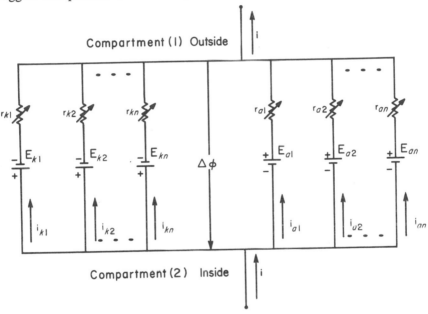

Fig. 2-9 *Equivalent circuit suggested by Eq. (2–171) and (2–172).*

to the Hodgkin and Huxley equivalent circuit without its capacitance is evident.

It is to be noted that the integral conductances [Eq. (2-169)] are concentration dependent and thus may also be current dependent. This results from the fact that intramembrane concentration profiles in a homogeneous regime will, in general, shift with changes in transmembrane current. Even though this equivalent circuit is restricted to a steady state, the possibility of time dependent conductances is clearly implied.

What further insight into the internal state of the diffusion regime can we gain from this circuit? To answer this question, we shall examine a specific case. Suppose we have a freely diffusing biionic salt in which the anion and cation have identical mobilities. Let the salt be at different concentrations in compartments (1) and (2). Since there are no current sources, the net external current must be zero. Electroneutrality requires that

$$c_k = c_a = c_s, \tag{2-173}$$

where c_s is the salt concentration. Eq. (2-169) then yields

$$r = r_k = r_a = \int_1^2 \frac{dx}{F^2 u c_s} \tag{2-174}$$

because

$$u_k = u_a \equiv u.$$

From Eq. (2-170)

$$E_k = -E_a \equiv E. \tag{2-175}$$

Eq. (2-171) and (2-172) then tell us that

$$i_k = -i_a = -\frac{E}{r}$$

and

$$\Delta\phi = 0. \tag{2-176}$$

Compartment (I)

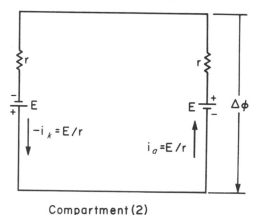

Compartment (2)

Fig. 2-10 *Parallel branched equivalent circuit for a single salt for which* $u_a = u_k$.

We know from Eq. (2-35) and (2-36) that the latter result is correct. But these equations also tell us that there are no potential differences or currents anywhere in this membrane. The equivalent circuit (Fig. 2-10) thus correctly predicts the externally measureable voltage and current. However, the internal picture of separate, opposing Nernst potentials and a circulating current is certainly misleading in comparison with the real membrane.

There is a somewhat different membrane to which this internal picture is quite applicable. If the membrane is composed of patches each of which is internally homogeneous but permeable only to a single ion, circulating currents of this kind are to be expected. Since other current carriers are lacking in each permeable patch, the quantities i_k and i_a then constitute actual currents in the parallel branches of a circuit.* These currents will, in the absence of capacitative effects, be governed by the relationship

$$F\frac{\partial c_i}{\partial t} = \frac{\partial i_i}{\partial x} = 0; \ i = a, k \qquad (2\text{-}177)$$

in a much shorter time than that required to achieve a diffusion steady state. Eq. (2-177) reflects the fact that charge cannot accumulate. Thus, in a patch, or mosaic, membrane the equivalent circuit of Fig. 2-9 is applicable without the constraint of a steady state. While certain experimental difficulties may arise, there is no theoretical reason why the i_a and i_k cannot be measured by suitable manipulation of the ionic environment.

If the membrane is homogeneous and not composed of a mosaic, Eq. (2-177) is valid only during a steady state. This was noted when the conditions for Eq. (2-166) were stated. It follows that in such a membrane the quantities i_k and i_a are not really currents in the branches of a circuit. They are, more properly, fluxes and can be measured by the proper tracer experiments.

The flux equations can be manipulated to yield a second type of equivalent circuit more amenable to use with a homogeneous regime. Eq. (2-24) may be written

$$j_k = -\left[RT\frac{\partial}{\partial x}(u_k c_k) + F(u_k c_k)\frac{\partial\phi}{\partial x}\right]$$

and $$(2\text{-}178)$$

$$j_a = -\left[RT\frac{\partial}{\partial x}(u_a c_a) - F(u_a c_a)\frac{\partial\phi}{\partial x}\right],$$

---

* The patches in such a membrane must, in general, be permselective but uncharged. See Diamond and Wright (1969) for a discussion of possible selectivity mechanisms. In a charged membrane the sites would have to be saturated with the permeating ion for other current carriers to be lacking.

for univalent species. Making use of Kirchoff's rule [Eq. (2-172)], and Eq. (2-168) thus yields,

$$\frac{i}{F^2(\sum_k u_k c_k + \sum_a u_a c_a)} = \frac{RT}{F} \frac{1}{(\sum_k u_k c_k + \sum_a u_a c_a)} \frac{\partial}{\partial x}(\sum_k u_k c_k - \sum_a u_a c_a) + \frac{\partial \phi}{\partial x}.$$

(2-179)

Integration across the membrane results in

$$i\int_1^2 \frac{dx}{F^2(\sum_k u_k c_k + \sum_a u_a c_a)}$$

$$= \frac{RT}{F} \int_1^2 \frac{1}{(\sum_k u_k c_k + \sum_a u_a c_a)} \frac{\partial}{\partial x}(\sum_k u_k c_k - \sum_a u_a c_a) dx + [\phi(2) - \phi(1)].$$

(2-180)

The steady state constraint need not be invoked because, in the absence of capacitative effects,

$$\frac{\partial i}{\partial x} = 0.$$

(2-181)

This is the equivalent of Eq. (2-177). The total integral resistance of the membrane is

$$R = \int_1^2 \frac{dx}{F^2(\sum_k u_k c_k + \sum_a u_a c_a)}.$$

(2-182)

The membrane potential in the absence of current will be

$$E = -\frac{RT}{F} \int_1^2 \frac{1}{(\sum_k u_k c_k + \sum_a u_a c_a)} \frac{\partial}{\partial x}(\sum_k u_k c_k - \sum_a u_a c_a) dx,$$

(2-183)

so that Eq. (2-180) may be rewritten as

$$\Delta\phi = iR + E.$$

(2-184)

This suggests the equivalent circuit in Fig. 2-11.

Both E and R depend on concentration and are thus, in general, current and time dependent. In the case of a freely diffusing biionic salt, we find that

$$\Delta\phi = -\frac{RT}{F} \int_1^2 \frac{1}{u_k c_k + u_a c_a} \frac{\partial}{\partial x}(u_k c_k - u_a c_a) dx$$

(2-185)

But electroneutrality requires that

$$c_k = c_a = c_s,$$

Compartment (1)
Outside

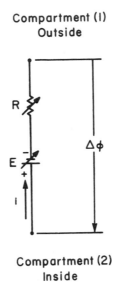

Compartment (2)
Inside

Fig. 2-11 Single branched equivalent circuit representative of a homogeneous membrane.

so that Eq. (2-185) reduces to

$$\Delta\phi = -\frac{RT}{F}\frac{u_k - u_a}{u_k + u_a}\ln\frac{c_s(2)}{c_s(1)}. \qquad (2\text{-}186)$$

This is the diffusion potential previously encountered in Eq. (2-36). This equivalent circuit thus seems nicely able to describe the internal state of a homogeneous regime. But it will obviously be unable to yield the circulating currents in a mosaic membrane. The two equivalent circuits can thus be most useful when each is applied to its appropriate type of diffusion regime.

The total integral resistance of the homogeneous membrane, R, contains the mobility and concentration of every permeant ion in the system [Eq. (2-182)]. It is therefore impossible either to define or to measure a conductance corresponding to a single ion. This is closely related to a statement made earlier in the discussion where it was pointed out that the quantities i_k and i_a, which occur when a parallel branched equivalent circuit is applied to a homogeneous membrane, are not really branch currents. They do have the properties of currents when that equivalent circuit is used to describe a mosaic membrane. Thus, the very fact that Hodgkin and Huxley (1952a, 1952b, 1952c) were able to separate out sodium and potassium currents and conductances in the squid axon by manipulations of the ionic milieu constitutes an independent argument for the mosaic nature of that membrane. This should be added to the accumulating pharmacological evidence summarized by Hille in Chapter 12 which points to the same conclusion.

It is also of interest that in such a mosaic membrane, voltage, and hence time, dependent conductances cannot arise from current dependent ionic profiles (Finkelstein and Mauro, 1963). If there is only a single current carrier in each permeable patch, these profiles will not change in response to currents. The reasoning is similar to that in Eq. (2-177); that is, charge may not accumulate. But time and voltage dependent conductances could occur, for example, as a result of changing mobilities or altered numbers of permeable patches. This would imply that such membrane processes, which are beyond those of simple electro-diffusion, must be involved in the phenomena of electrical excitability.

CONCLUSION

The subject matter of this chapter has ranged from what was undoubtedly familiar to many readers to some previously unpublished material. The latter included a new derivation of the Ussing-Teorell unidirectional flux ratio in which it was demonstrated to be valid in spite of inhomogeneities in the plane of the membrane. A reexamination of the Goldman equation demonstrated that a constant field assumption is unnecessary. Furthermore, a careful definition of the steady state leads to the conclusion that a nonelectrogenic pump can, indeed, alter the membrane potential directly.

The relationship between electrical equivalent circuits and the diffusion regime was reviewed based largely on the work of Finkelstein and Mauro (1963). It is hoped that this discussion will help to resolve a tendency to counterpose what is called the "engineering" approach with the diffusion approach. When properly used, the two should be but different aspects of the same investigation.

ACKNOWLEDGMENT

This work was supported in part by P.H.S. grants number NB08444-01A1, NS08444-02 and 03 from the National Institute of Neurological Diseases and Stroke.

REFERENCES

BAKER, P. F., A. L. HODGKIN and T. I. SHAW. 1962. Replacement of the Axoplasm of Giant Nerve Fibres with Artificial Solutions. *J. Physiol. (London)*. 164:330–354.

BARR, L. 1965. Membrane Potential Profiles and the Goldman Equation. *J. Theoret. Biol.* 9:351–356.

COLE, K. S. 1968. *Membranes, Ions and Impulses.* University of California Press, Berkeley, California.

CURRAN, P. F., and S. G. SCHULTZ. 1968. Transport Across Membranes: General Principles. *In Handbook of Physiology.* Vol. 3. Section 6: Alimentary Canal. Amer. Physiol. Soc., Washington, D.C.

DEFFNER, G. G. 1961. The Dialyzable Free Organic Constituents of Squid Blood: A Comparison with Nerve Axoplasm. *Biochem. Biophys. Acta.* 47:378–388.

DEGROOT, S. R., and P. MAZUR. 1962. *Non-Equilibrium Thermodynamics.* North Holland Publishing Company, Amsterdam.

DIAMOND, J. M., and E. M. WRIGHT. 1969. Biological Membranes: The Physical Basis of Ion and Nonelectrolyte Selectivity. *Ann. Rev. Physiol.* 31:581–646.

FINKELSTEIN, A., and A. MAURO. 1963. Equivalent Circuits as Related to Ionic Systems. *Biophys. J.* 3:215–237.

GEDULDIG, D. 1968. Analysis of Membrane Permeability Coefficient Ratios and Internal Ion Concentrations from a Constant Field Equation. *J. Theoret. Biol.* 19:67–78.

GOLDMAN, D. E. 1943. Potential, Impedance, and Rectification in Membranes. *J. Gen. Physiol.* 27:37–60.

GOLDSTEIN, D. A., and A. K. SOLOMON. 1960. Determination of Equivalent Pore Radius for Human Red Cells by Osmotic Pressure Measurement. *J. Gen. Physiol.* 44:1–17.

GUGGENHEIM, E. A. 1959. *Thermodynamics.* North Holland Publishing Company, Amsterdam.

HEINZ, E. 1963. Physiologie, Biochemie und Energetik des Aktiven Transports. *Naunyn-Schmeideberg's Arch. Exp. Pathol. Pharmakol.* 245:10 28. (English translation available from N.I.H. Library Branch, Translating Unit.)

HELFFERICH, F. 1962. *Ion Exchange.* McGraw-Hill, New York.

HEVESY, G. V., E. HOFER and A. KROGH. 1935. The Permeability of the Skin of Frog to Water as Determined by D_2O and H_2O. *Scand. Arch. Physiol.* 72:199–214.

HODGKIN, A. L., and A. F. HUXLEY. 1952a. Currents Carried by Sodium and Potassium Ions through the Membrane of the Giant Axon of *Loligo.* *J. Physiol. (London).* 116:449–472.

_____. 1952b. The Components of Membrane Conductance in the Giant Axon of *Loligo.* *J. Physiol. (London).* 116:473–496.

_____. 1952c. A Quantitative Description of Membrane Current and its Application to Conduction and Excitation in Nerve. *J. Physiol. (London).* 117:500–544.

HODGKIN, A. L., and B. KATZ. 1949. The Effect of Sodium Ions on the Electrical Activity of the Giant Axon of the Squid. *J. Physiol. (London).* 108:37–77.

HODGKIN, A. L., and R. D. KEYNES. 1955. The Potassium Permeability of a Giant Nerve Fibre. *J. Physiol. (London).* 128:61–88.

KATCHALSKY, A., and P. F. CURRAN. 1965. *Nonequilibrium Thermodynamics in Biophysics.* Harvard University Press, Cambridge.

KEDEM, O., and A. ESSIG. 1965. Isotope Flows and Flux Ratios in Biological Membranes. *J. Gen. Physiol.* 48:1047–1070.

KIRKWOOD, J. G. 1954. Transport through Biological Membranes. *In* H. T. Clarke [ed.] *Ion Transport Across Membranes.* Academic Press, Inc., New York.

KOEFOED-JOHNSEN, V., and H. H. USSING. 1953. The Contributions of Diffusion and Flow to the Passage of D_2O through Living Membranes. *Acta Physiol. Scand.* 28:60–76.

KORNACKER, K. 1969. Physical Principles of Active Transport and Electrical Excitability. *In* R. M. Dowben [ed.] *Biological Membranes.* Little, Brown and Company, Boston.

LAKSHMINARAYANAIAH, N. 1969. *Transport Phenomena in Membranes.* Academic Press, New York.

LASS, H. 1950. *Vector and Tensor Analysis.* McGraw-Hill Book Company, Inc., New York.

MACGILLIVRAY, A. D., and D. HARE. 1969. Applicability of Goldman's Constant Field Assumption to Biological Systems. *J. Theoret. Biol.* 25:113–126.

MAURO, A. 1957. Nature of Solvent Transfer in Osmosis. *Science.* 126:252–253.

———. 1965. Osmotic Flow in a Rigid Porous Membrane. *Science.* 149:867–869.

MULLINS, L. J., and K. NODA. 1963. The Influence of Sodium-free Solutions on the Membrane Potential of Frog Muscle Fibers. *J. Gen. Physiol.* 47:117–132.

OVERBEEK, J. TH. G. 1961. The Donnan Equilibrium. *Progr. Biophys.* 6:57–84.

PAGANELLI, C. V., and A. K. SOLOMON. 1957. The Rate of Exchange of Tritiated Water across the Human Red Cell Membrane. *J. Gen. Physiol.* 41:259–277.

PARK, C. R. 1961. Introduction, pp. 19–21. *In* A. Kleinzeller and A. Kotyk [ed.] *Membrane Transport and Metabolism.* Academic Press, New York City.

PATLAK, C. S. 1960. Derivation of an Equation for the Diffusion Potential. *Nature.* 188:944–945.

RAND, R. P., and A. C. BURTON. 1964a. Mechanical Properties of the Red Cell Membrane. I. Membrane Stiffness and Intracellular Pressure. *Biophys. J.* 4:115–135.

———. 1964b. The Pressure Inside Red Cells and the "Metabolic Pump". Letter to the editor. *Biophys. J.* 4:491–495.

RUCH, T. C., H. D. PATTON, J. W. WOODBURY and A. L. TOWE. 1965. *Neurophysiology.* W. B. Saunders Company, Philadelphia.

SCHWARTZ, T. L. 1966. A Tracer Study of Transients in the Sodium Fluxes in the Toad Urinary Bladder in Response to Sudden Short-circuit. Dissertation. SUNYAB. Dissertation Abstracts: 66–7985.

———. 1969. The Goldman Equation: Its Constraints and its Applicability in the Face of Electrogenic and Non-electrogenic Pumps. *Biol. Bull.* 137:390.

_____. 1971a. The Validity of the Ussing Flux-Ratio Equation in a Three-Dimensionally Inhomogeneous Membrane. *Biophys. J.* In press.

_____. 1971b. Direct Effects on the Membrane Potential Due to Non-Electrogenic "Pumps". Submitted for publication.

SCHWARTZ, T. L., and F. M. SNELL. 1968. Nonsteady-state Three Compartment Tracer Kinetics. I. Theory. *Biophys. J.* 8:805–817.

SNELL, F. M., S. SHULMAN, R. P. SPENCER and C. MOOS. 1965. *Biophysical Principles of Structure and Function.* Addison-Wesley Publishing Company, Inc., Reading, Massachusetts.

STAVERMAN, A. J. 1951. The Theory of Measurement of Osmotic Pressure. *Rec. Trav. Chim.* 70:344–352.

STEIN, W. D. 1967. *The Movement of Molecules Across Cell Membranes.* Academic Press, New York.

TEORELL, T. 1949. Membrane Electrophoresis in Relation to Bioelectrical Polarization Effects. *Arch. Sci. Physiol.* 3:205–219.

_____. 1959a. Elektrokinetic Membrane Processes in Relation to Properties of Excitable Tissues. I. Experiments on Oscillatory Transport phenomena in Artificial Membranes. *J. Gen. Physiol.* 42:831–845.

_____. 1959b. Elektrokinetic Membrane Processes in Relation to Properties of Excitable Tissues. II. Some Theoretical Considerations. *J. Gen. Physiol.* 42:847–862.

USSING, H. H. 1949. The Distinction by Means of Tracers Between Active Transport and Diffusion. *Acta Physiol. Scand.* 19:43–56.

ION TRANSPORT ACROSS EXCITABLE CELL MEMBRANES

3

R. A. SJODIN

The transport of ions across cell membranes is a field currently under intense investigation and re-investigation. It would be impossible, in a chapter of this length, to exhaustively review even the most current research. Instead, a survey of the current status of the field will be made.

The classical approach to an analysis of transport phenomena across biological membranes has been to regard total transport rate as the algebraic sum of three rather distinct processes: passive diffusion, exchange diffusion and metabolically mediated active transport. In general, these processes have been defined as though there is no mutual interdependence among them. It is remarkable how much experimental data can be sorted out fairly satisfactorily according to the notion that the three basic processes are indeed distinct and non-interdependent. Recent findings to be discussed later, however, cast considerable doubt on the validity of this approach in any precise or theoretically correct sense. The safest approach at present seems to be a careful description of fractions of the total measurable transport rate in terms of the experimental operations required to measure them.

MODES OF ION MOVEMENT ACROSS CELL MEMBRANES

Passive Diffusion and Electro-Diffusion

By virtue of its internal structure every biological cell membrane will offer some resistance to the passage of a molecule or ion through it. The average velocity of particles crossing the membrane is directly proportional to the force driving the particles across and inversely proportional to the resistance offered by the membrane. The trans-membrane flux by this mechanism is equal to the membrane concentration of particles at each point times the velocity at each point. The membrane mobility is the velocity per unit force so that Flux = Concentration × Mobility × Force.

It is usual to define the forces producing these fluxes as the negative gradients of free energy. The most common free energy gradients encountered are the gradients in chemical potential in the case of diffusion of a neutral particle and the gradients in electrochemical potential in the case of electro-diffusion or the diffusion of charged particles in the presence of an electrical field. Fluxes that are proportional to such gradients across cell membranes are termed *passive fluxes* and these fluxes constitute passive transport. The fundamental equation for this type of transport is

$$j = -c_M u \nabla \psi, \tag{3-1}$$

where j denotes flux in units of moles/cm^{2}sec, c_M refers to membrane concentration, u refers to absolute mobility in units of cm/sec per dyne and $\nabla \psi$ refers to the gradient in potential energy.

Eq. (3-1) looks deceptively simple. Except for some very special cases, however, the solution of Eq. (3-1) is complicated. A particularly simple case exists for the free diffusion of a neutral molecule across a cell membrane. In this instance $\nabla \psi$ is equal to the gradient in chemical potential and Eq. (3-1) becomes Fick's law. Hence, passive transport due to neutral molecules freely diffusing across a cell membrane is said to be "Fickian". Conversely, any departure from Fick's law for neutral molecules is often taken to indicate nonpassive diffusion processes. The case of Fickian diffusion across a membrane can be exploited further to define some additional useful terms and an important theoretical problem. Eq. (3-1) becomes:

$$j = -RTu\frac{dc_M}{dx} = -D\frac{dc_M}{dx}, \tag{3-2}$$

where R is the gas constant per mole, T is the absolute temperature, x is distance perpendicular to the membrane, D is the diffusion coefficient and other symbols are as previously defined. For a constant concentration gradient across the membrane, dc_M/dx becomes $(c_{M_i} - c_{M_o})/\delta$ where c_{M_i} refers to membrane concentration at the inner membrane surface, c_{M_o} to membrane concentration at the outer membrane surface and δ is the membrane thickness. Substituting these quantities into Eq. (3-2) yields

$$j = \frac{D}{\delta}(c_{M_o} - c_{M_i}). \tag{3-3}$$

The right hand side of Eq. (3-3) does not contain directly measurable quantities. The membrane thickness, δ, can be estimated by methods discussed elsewhere. The membrane concentrations, however, cannot even be estimated by any direct experimental method. The problem is surmounted somewhat by defining a partition coefficient, β (Hodgkin and Katz, 1949) such that the membrane concentrations are given by $c_{M_i} = \beta c_i$ and $c_{M_o} = \beta c_o$ where c_i and c_o refer to concentrations inside and outside the cell respectively. Accordingly, Eq. (3-3) can be transformed into a more useful form:

$$j = P(c_o - c_i). \tag{3-4}$$

From this, it is evident that P, the permeability coefficient is equal to $\beta D/\delta$.

The question arises now, as to how the permeability coefficient, P, can be measured. From Eq. (3-4) it is evident that P can be determined if the flux, j, is known at some concentration difference $(c_o - c_i)$. The flux is usually measured by using a radioactive label on the species involved and measuring the variation of concentration of the labeled species inside the cell as a function of time. P can then be deduced from such measurements.

Since $j = (V/A)dc_i/dt$ where V is the volume and A is the surface area of the cell, Eq. (3-4) is equivalent to

$$\frac{dc_i}{dt} = \frac{A}{V} P(c_o - c_i). \tag{3-5}$$

The solution to Eq. (3-5) for the boundary conditions $c_i = 0$ $\overset{t=0}{}$ and $c_i = C_o$ $\overset{t=\infty}{}$ is

$$c_i = c_o(1 - e^{-kt}), \tag{3-6}$$

where the rate constant, k, is equal to (AP/V). Then, to measure P, one measures c_i as a function of time and determines the rate constant, k, from which P is computed. Though P can be readily determined by such methods, an important fact becomes apparent. Since $P = \beta D/\delta$, it is clear that only the product, βD, is measurable and not β and D directly. Hence, a measurement of the permeability coefficient does not provide direct information about the frictional resistance or mobility within the membrane.

In the case of the movement of charged particles or ions, Eq. (3-1) becomes more complicated. The quantity $\nabla \psi$ is now equal to $\nabla \tilde{\mu}$ where $\tilde{\mu}$ is the electrochemical potential. For ion movement, Eq. (3-1) becomes

$$j = - c_M u \nabla \mu_i - c_M uzF\nabla \phi \tag{3-7}$$

where z refers to ionic valance, F is the Faraday of charge, μ_i is the chemical potential, ϕ is the electrical potential and other symbols have their previous significance. The quantity u is often replaced by the electrical mobility or velocity per unit field, u, where $u = u/F$. In general, Eq. (3-7) must be solved simultaneously for all ions present, subject to the condition of electrical neutrality and zero net electrical current. A complete solution taking into account fixed ionic charges has been made by Teorell (1951). A simpler solution is obtained if the electrical field, $\nabla\phi$, is constant. A solution on this basis was obtained by Goldman (1943) and by Hodgkin and Katz (1949). The equation obtained by the latter authors is:

$$j = P\frac{zEF}{RT}\left(\frac{c_o - c_i e^{zFE/RT}}{e^{zEF/RT} - 1}\right) \tag{3-8}$$

where P has the same significance as in the previous development and where E is the membrane potential. When the flux obeys Eq. (3-8), the permeability coefficient can be obtained by measuring c_i as a function of time and determining P from the relation:

$$c_i = c_o e^{-zEF/RT}(1 - e^{-kT}), \tag{3-9}$$

where

$$k = P\left(\frac{A}{V}\right)\left[\frac{zEF}{RT(1 - e^{-zEF/RT})}\right]$$

(Sjodin, 1959).

Though the determination of permeability coefficients is a useful tool in studying properties of passive ion transport, the most powerful application of electrochemical flux equations to the study of ion transport in cells is the method of flux ratio analysis. For example, Eq. (3-8) can be written in the form

$$j = P f(E) (c_o - c_i e^{zEF/RT}). \tag{3-10}$$

In terms of unidirectional ion fluxes, $j = j_i - j_o$ where j_i is influx and j_o is efflux. From Eq. (3-10), if one assumes that the unidirectional fluxes are independent of one another, then:

$$\frac{j_o}{j_i} = \frac{c_i}{c_o} e^{zEF/RT}. \tag{3-11}$$

If one can measure the unidirectional fluxes, the membrane potential and ion concentrations, it is possible to test for obedience of the data to Eq. (3-11).

It is difficult indeed to make certain that one is measuring a purely passive flux component. The probability that one is dealing with a passive component of ionic flux is high if Eq. (3-8) and (3-11) are obeyed and if the flux component is relatively insensitive to the application of metabolic inhibitors.

Exchange Diffusion

This type of ionic movement involves an electrically neutral, one-for-one exchange of ions of the same species. Its characteristics are that no net transport of ions or electrical current can occur due solely to exchange diffusion and that no energy is required to account for the movement. As no energy is required, the process as initially envisioned (Ussing, 1949) would be expected to be rather independent of metabolism and hence insensitive to metabolic inhibitors. Recent findings, to be discussed later, indicate that it is not possible to retain this notion of exchange diffusion. Classically, the definition of exchange diffusion rests on operational laboratory tests. For example, since a one-for-one exchange is involved, removal of the ions engaging in exchange diffusion from one side of the membrane should result in abolition of the process. In striated muscle cells Keynes (1954) and Keynes and Swan (1959) observed that sodium efflux was reduced by about 50 per cent when sodium ions were removed from the external bathing medium. As this component of efflux was insensitive to inhibitors of metabolism and since no other mechanism for the effect was obvious, it was suggested that exchange diffusion might be responsible.

It should be emphasized that exchange diffusion cannot be generally defined as the reduction in flux taking place when a certain ion is removed from the solution on one side of a cell membrane. For example, removal of external potassium ions reduces the efflux of potassium ions from striated muscle cells by about one half (Sjodin and Henderson, 1964). The effect is accompanied by a membrane hyperpolarization of some 20 mv, however, and this change in resting membrane potential is more than enough to account for the flux reduction without invoking the notion of exchange diffusion.

Exchange diffusion represents one type of coupling between ionic movements occurring in opposing directions across the cell membrane. It will be seen presently that active transport represents another type of coupling between ionic movements and that in some modes of coupling, the active transport process may exhibit properties that are strikingly similar to what has been described as exchange diffusion. In this regard, it is interesting that Keynes and Swan (1959) observed that the sodium-for-sodium exchange in muscle cells obeys the same sort of kinetic relation obeyed in the ionic-carrier hypothesis in active transport applications.

Active Transport and the Carrier Hypothesis

As early as 1940 it was known that striated muscle cells are capable of producing a net extrusion of sodium ions against an electrochemical gradient. Steinbach (1940, 1951, 1952) and later, Desmedt (1953) found that frog muscles gain sodium ions and lose potassium ions during storage in a potassium-free Ringer's solution. When such muscles were subsequently placed in Ringer's solution containing an adequate concentration of potassium ions, they were observed to extrude some of the sodium that was gained in exchange for potassium ions in the medium. Keynes (1954) observed that removal of potassium ions from Ringer's solution resulted in a diminution in the sodium efflux from frog muscle cells. Potassium-rich media increased the sodium efflux. A similar relation between sodium efflux and the external potassium ion concentration was observed in red blood cells by Harris and Maizels (1951) and by Glynn (1954). Hodgkin and Keynes (1955a) observed sodium efflux from invertebrate giant axons to behave in the same way. Removal of external potassium ions produced a rapid decline in sodium efflux while addition of potassium ions to a potassium-free medium brought about a rapid rise in sodium efflux. In giant axons from *Sepia*, the drop in sodium efflux was about equal to the magnitude of potassium influx in fibers previously subjected to electrical stimulation.

Experimentation on the effects of external potassium ions on sodium efflux suggests a linkage of a portion of the sodium efflux with the inward movement of potassium ions. The situation is reminiscent of the coupling of ionic movements discussed in the previous section. The obvious difference, of course, is that the ion producing an opposing and balancing flow is of a different species. Another important difference is that the coupling of a portion of sodium efflux with the influx of potassium ions is very sensitive to metabolic inhibitors and inhibitors of enzymatic reactions. The addition of strophanthin to the bathing medium was found to interfere with the extrusion of sodium ions and the net uptake of potassium ions in the sodium-enriched muscle system (Matchett and Johnson, 1954). Edwards and Harris (1957) observed that addition of g-strophanthin to Ringer's solution had about the same effect on tracer sodium loss from muscle cells as the removal of external potassium ions. The inhibitors dinitrophenol and cyanide were found by Hodgkin and Keynes (1955a) to markedly reduce both sodium efflux and potassium influx in *Sepia* giant axons. Since that time, many investigators have observed that application of cardiac glycosides reduces both sodium efflux and potassium influx in giant axons from the squid.

The observations on active transport discussed so far support the notion that sodium ions leave the cell and potassium ions enter the cell by first combining with a "carrier" molecule located in the cell membrane. Such a scheme was first proposed as a hypothesis to account for the apparently linked movements of sodium and potassium ions in red blood cells (Shaw, 1954; Glynn, 1956). This scheme has proven to be the embryonic form of a very general model that has since been applied to active ion transport in a wide variety of cell types including electrically excitable cells. The rationale for the scheme is as follows: in all the cell types studied, the movement of sodium ions out of the cell is a thermodynamically "uphill" process that requires an input of energy. A classical observation, for example, is that red blood cells washed free of glucose cannot continue to extrude sodium ions if no additional substrate is provided. A fundamental principle in biochemistry is that chemical free energy is transferred in cells by enzymatically catalyzed and controlled reactions. The "carrier hypothesis" results very naturally from the facts discussed above and from the interesting observation that the rate of active ion transport often follows the same kinetics that would be expected if the ion combined reversibly with an enzyme complex within the membrane.

The rate of an enzymatic reaction is determined by the concentration of an intermediate complex between the enzyme and its substrate. The velocity of such a reaction in its simplest form follows the kinetics:

$$v = v_{max} \frac{c_s}{c_s + K_m},$$ (3-12)

where v is the velocity of the reaction, v_{max} is the maximum velocity of the reaction obtained at very high substrate concentrations, c_s is the substrate concentration and K_m is a constant called the "Michaelis constant". K_m represents the substrate concentration at which the reaction velocity is one-half the maximum value. Derivations of this relation in biochemistry text-books are legion. If the ion to be transported must first combine with a membrane molecule having enzymatic properties, the ion can be looked upon as a substrate for the enzymatic part of the molecule. If the transport rate is determined by the concentration of the "Ion-Carrier Complex" within the membrane, Eq. (3-12) will be obeyed where c_s is equal to the ion concentration in the solution in contact with the membrane.

As an example, Glynn (1956) found that the amount of potassium-dependent sodium efflux from red cells can be fit by Eq. (3-12) with c_s being replaced by the external potassium ion concentration. The value observed for K_m was 1.8 mM. In this case the rate of the linked sodium and potassium transport is governed by the external potassium ion concentration. More recently, similar relations for sodium transport have been observed to be obeyed in nerve cells (Baker and Connelly, 1966; Rang and Ritchie, 1968) and in muscle cells (Sjodin, 1970).

Further details have been primarily elucidated by biochemical studies. By 1960, it was well established that, in red blood cells, the active transport of sodium and potassium ions depends ultimately upon energy in the form of the high-energy phosphate bonds of ATP (Glynn, 1957; Whittam, 1958; Hoffman, 1960). More recently, Mullins and Brinley (1967) demonstrated ATP to be the direct source of energy for sodium extrusion from giant axons of the squid. It is evident that the energy for ion transport must come from energy liberated during the hydrolysis of ATP to ADP.

As early as 1948, (Libet, 1948), it was known that the sheath from giant axons contains an enzyme (ATPase) that is activated by magnesium ions and hydrolyzes ATP to ADP, liberating inorganic phosphate. The appearance of a magnesium-activated ATPase in the particulate material associated with the sheath of rat peripheral nerve was observed by Abood and Gerard (1954). Skou (1957) made an extensive study of the properties of the magnesium-dependent ATPase activity residing in the particulate fraction of extracts of the sheaths of crab nerves. The most significant finding with regard to ion transport problems is that, in the presence of Mg^{++} and ATP, sodium and potassium ions have a synergistic action in activating the enzyme. The same has been observed for the ATPase activity associated with insoluble material extracted from red blood cells (Post, Merritt, Kinsolving and Albright, 1960; Dunham and Glynn, 1961). In these systems it has been shown that the sodium and potassium ion concentrations required for one-half maximal activation of the membrane-associated ATPases are similar to those ion concentrations required for one-half maximal stimulation of the rate of ion transport. In addition, cardiac glycosides inhibit the ATPase activity at the same concentration at which ion transport is strongly inhibited.

These facts have been assembled into a general "carrier" model for metabolically mediated active ion transport. The scheme is shown diagrammatically in Fig. 3-1. The membrane-located carrier molecule is denoted by X.

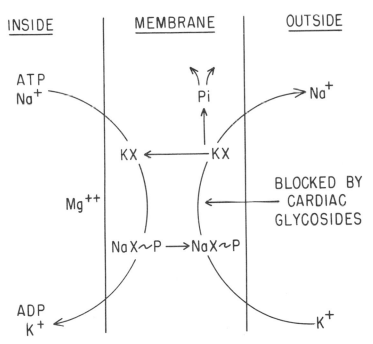

Fig. 3-1 *Diagram summarizing the carrier hypothesis of active sodium and potassium ion transport.*

The first step in the "transport cycle" is believed to be the phosphorylation of the carrier at the inner membrane surface accompanied by the hydrolysis of ATP. Sodium ions combine with the carrier in this form and are carried by diffusion of the carrier or passed along a chain of sites to the outside surface of the membrane where dephosphorylation of the carrier is believed to occur along with the liberation of sodium ions and the binding of potassium ions. The potassium is then either carried or passed along a chain of sites to the interior where potassium ions and the free carrier, X, are liberated.

In this particular scheme, two enzymatic activities are required, that of a phosphokinase and that of a phosphatase. An alternative is to envision a single enzyme with different catalytic conformations and activities depending upon the ionic environment. As the actual situation is unknown, either of these alternatives or as yet unexplored alternatives may suffice to account for the observations.

Returning to the observations on sodium and potassium transport in excitable cells described earlier, it is evident that the "carrier" model is able to account for many findings: 1) the observation that sodium outward transport is dependent upon the external potassium ion concentration and is apparently linked to inward potassium transport 2) the finding that ATP is the immediate energy source for ion transport and that the transport of sodium and potassium ions is accompanied by the hydrolysis of ATP 3) the finding that cardiac glycosides inhibit ion transport 4) the finding that the rate of ion transport can often be described by the kinetics of enzyme reactions and 5) the fact that membranes yield a particulate fraction showing ATPase activity that is dependent on the sodium and potassium ion concentrations.

A major difficulty with the model is that a high degree of discrimination between sodium and potassium ions is required for the carrier complex, depending upon the presence or absence of an ester phosphate group. No known organic compounds have been isolated that have this property. Also, the enzyme activity in the particulate extract from membranes has resisted attempts at purification. Nevertheless, the "carrier" model remains as the most attractive hypothesis at present.

AN ANALYSIS OF ION TRANSPORT IN EXCITABLE CELL MEMBRANES

Three basic processes underlying ion movements across excitable cell membranes have been briefly discussed. In any one measurement of ionic transmembrane flux, all three processes can contribute to the total measured flux. How the total flux can be subdivided into components is largely a matter of an analysis in terms of experimental operations. The subdivision of the ionic fluxes across excitable cell membranes into the three categories discussed is the subject of this section. The results discussed will be limited to two cell types: giant axons from the squid and cuttlefish, and striated muscle fibers.

Ion Fluxes in Invertebrate Giant Axons

Although there is some dispersion in the magnitudes of unidirectional ionic fluxes measured across excitable cell membranes, it is possible to define some average values that workers in the field would agree to be representative. The magnitudes of the unidirectional fluxes of potassium ions across the membranes of giant axons are K-influx, 22 pmole/cm^{2}sec (*Sepia*) and 23 pmole/cm^{2}sec (squid); K-efflux, 28 pmole/cm^{2}sec (*Sepia*) and 38 pmole/cm^{2}sec (squid). The ratio of efflux to influx is thus about 1.3 for *Sepia* giant axons and about 1.7 for squid giant axons (data from Hodgkin and Keynes, 1955a; Caldwell and Keynes, 1960; Sjodin and Beaugé, 1967). The internal concentrations of sodium and potassium in squid giant axons are: $c_{K_i} = 323$ mM and $c_{Na_i} = 46$ mM (Keynes and Lewis, 1951). The knowledge that the normal resting potential of the squid giant axon membrane is -60 mv permits the expected passive flux ratios to be calculated from Eq. (3-11).

The value calculated for the passive flux ratio for potassium ions when the external potassium ion concentration equals 10 mM is 3.0 at 20° C. This can be compared with the measured ratio of 1.7 for the total unidirectional potassium ion fluxes in squid giant axons. Actually, however, we know that this comparison is unrealistic because a very large part of the potassium influx must occur by a process dependent on metabolism. During maximal metabolic inhibition, potassium influx in giant axons is reduced to about 3 pmole/cm^{2}sec. The effect of inhibitors on potassium efflux is small and in the opposite direction. Inhibitors increase the efflux of potassium ions, on the average, by some 30 to 40 per cent. There is no evidence for an exchange diffusion of potassium ions in giant axons so that the fluxes in the presence of inhibitors should be due to passive diffusion. In the presence of inhibitors, the ratio of potassium efflux to potassium influx has a value in excess of 10, which is over 3 times the theoretically calculated value. There is no obvious explanation for this, but it is likely that the principle of independent ion movement underlying the derivation of Eq. (3-11) cannot be applied to passive potassium movement in giant axons (Hodgkin and Keynes, 1955b).

If potassium ions cross the membrane passively in narrow aqueous channels which do not readily permit passing of ions moving in opposing directions, the discrepancy can be explained. The situation would not be unlike that in a one-way tunnel with heavier traffic in one direction than the other. Traffic in the heavy direction will impede traffic from the light direction relatively more than traffic in the light direction will impede traffic from the heavy direction. It can be shown (Hodgkin and Keynes, 1955b; Sjodin, 1965) that ion movements based on such a model would be expected to obey a modified flux ratio equation which is Eq. (3-11) raised to the $(n + 1)$ power:

$$\frac{j_o}{j_i} = \left(\frac{c_i}{c_o}e^{EF/RT}\right)^{n+1}, \tag{3-13}$$

where n represents the effective number of positions at which passing of ions in opposite directions is not permitted. For the above data, $(n + 1)$ equals about 2 so that at least one ionic position must exist where passing is restricted. As no independent test of this model exists, it must remain as an interesting hypothesis.

To make a similar analysis of the resting sodium ionic fluxes in squid giant axons, the following data for he total measured unidirectional fluxes are considered: Na-influx = Na-efflux = 50 pmole/cm²sec. The data assume a steady state for the internal sodium ion concentration and are consistent with data obtained both on intact axons (Hodgkin and Keynes, 1955a) and on internally dialyzed axons (Brinley and Mullins, 1967). Hodgkin and Keynes (1955a) found metabolic inhibitors to reduce sodium efflux to about 10 per cent of the normal value and to reduce sodium influx only slightly. These findings suggest that passive sodium efflux has a value of about 5 pmole/cm²sec while almost all of the sodium influx is passive. The value of 5 pmole/cm²sec is now known to be too high for passive sodium efflux. The discrepancy is probably due to incomplete metabolic poisoning.

Mullins and Brinley (1967) were able to reduce sodium efflux to about 1 pmole/cm²sec in internally dialyzed squid axons by substrate deprivation and prolonged poisoning with CN to reduce endogenous energy sources. If we take the latter value as representing passive sodium efflux, the passive flux ratio (efflux/influx) for sodium ions has a nominal value of 0.02. For data given previously on the internal sodium concentration, the theoretical flux ratio calculated from Eq. (3-11) is 0.01. These figures have been arrived at by using average values of all parameters. Using values observed when ionic concentrations are accurately known, Mullins and Brinley (1967) in the same experiment, were able to show that the flux ratio observed for passive sodium fluxes agrees fairly well with the predictions of Eq. (3-11).

We can summarize these findings by stating that in giant nerve fibers sodium influx and potassium efflux have large passive components, while close to 90 per cent of potassium influx and nearly all of the sodium efflux is due to metabolically mediated active transport. Sodium efflux and potassium influx must be coupled to some extent but the coupling is neither complete nor on a one-for-one basis, since a large measurable component of active sodium efflux occurs in the absence of external potassium ions. We can also say that exchange diffusion of either potassium or sodium ions must play a relatively minor role in giant axons.

Ion Fluxes in Striated Muscle Fibers

The same kind of analysis of sodium and potassium movements can be made for striated muscle fibers. Most of the data have been obtained on sartorius muscle fibers from the frog. Let us first look at cases where both influx and efflux of potassium ions were measured on the same single muscle fiber (Hodgkin and Horowicz, 1959). For two cases the average values of potassium influx and the values of potassium efflux are as follows: K-influx = 3.7, K-efflux = 4.0; K-influx = 6.4, K-efflux = 6.5 (all values in units of pmole/ cm^{2}sec). The two resulting figures for the ratio K-efflux/K-influx are 1.08 and 1.02. The resting membrane potential of frog muscle fibers is -90 mv. With the additional knowledge that the average $c_{K_i} = 116$ mM and $c_{K_o} = 2.5$ mM, the flux ratio calculated from Eq. (3-11) is about 1.3. Though the small discrepancy between the calculated and experimental values may be significant, there is certainly approximate agreement between the measured ratio of total unidirectional potassium fluxes and the ratio calculated on the basis that the fluxes are passive. Using whole muscle data, the average ratio of potassium efflux to influx for 12 muscles was 1.14 when $c_{K_o} = 2.5$ mM (Sjodin and Henderson, 1964). The value calculated from Eq. (3-11) was 1.43 for the same conditions. The difference is significant and can be accounted for if a small percentage of the total measured potassium influx occurs via an active transport process.

It has already been stated that approximately one half of the total measured sodium fluxes are of the type that can be designated exchange diffusion. The remaining fraction of sodium efflux, after subtraction of the exchange diffusion component, must be due to active transport. This conclusion follows from thermodynamic considerations. Under normal conditions, the influx of sodium ions in muscle is unaffected by application of cardiac glycosides which have a marked inhibitory effect on the movements of cations that are actively transported (Horowicz and Gerber, 1965; Keynes and Steinhardt, 1968). One can conclude that sodium inward movement in muscle cells is normally a passive process.

Using Eq. (3-11), one can show that the passive component of sodium efflux must be less than 1 per cent of the magnitude of sodium influx. Thus, for steady state conditions, over 99 per cent of the sodium efflux must be due to an active transport mechanism. In muscle cells, this theoretical prediction has not as yet been experimentally verified. The verification necessary would be to reduce the efflux of sodium ions from muscle cells to less than one per cent of the normal value by application of inhibitors of active ion transport.

Horowicz and Gerber (1965) were able to abolish about 80 per cent of sodium efflux from muscle cells by application of 10^{-5} M strophanthidin. In other studies, about 10 per cent of the sodium efflux from muscle cells remained during inhibition with 10^{-5} M strophanthidin under conditions expected to abolish exchange diffusion components (Sjodin and Beaugé, 1968b; Beaugé and Sjodin, 1968).

Earlier, Keynes and Maisel (1954) had found sodium efflux from muscle cells to be insensitive to inhibitors such as dinitrophenol and cyanide. This is not surprising, however, as the energy source for active transport has been shown to be ATP. Muscle cells contain a storage form of high energy phosphate in the form of phosphocreatine which can resynthesize ATP from ADP in the presence of an appropriate enzyme called a phosphokinase. The fact that cardiac glycosides do not inhibit all of the active outward transport of sodium ions is interesting as it suggests that not all of the extrusion of sodium ions is due to a transport cycle such as that depicted in Fig. 3-1.

The literature is somewhat confusing as to the effects of inhibitors on the potassium fluxes in muscle cells. For example, Harris (1957) reports rather large effects of a cardiac glycoside on potassium fluxes in frog muscle whereas Sjodin (1965) found the potassium fluxes to be rather insensitive to strophanthidin and Keynes (1965) found that ouabain did not significantly affect potassium influx in muscle.

The situation has been somewhat clarified by the finding that the effect of glycosides on the potassium fluxes in muscle is a strong function of the internal sodium ion concentration (Sjodin and Beaugé, 1968b). When the internal sodium ion concentration is low, the potassium fluxes in muscle are insensitive to glycosides. When the internal sodium concentration becomes elevated, potassium influx is reduced by glycosides and potassium efflux is increased by glycosides. The potassium fluxes in muscles with an elevated sodium content thus tend to behave like the normal potassium fluxes in giant axons.

All of these findings become somewhat coherent if one takes into account the dependency of sodium extrusion in muscle cells on the internal sodium ion concentration. Keynes and Swan (1959) and Mullins and Frumento (1963) observed that the efflux of sodium ions into sodium-free solutions does not vary linearly with the internal sodium ion concentration. At low internal sodium ion concentrations, the sodium efflux rate varies approximately as the cube of the internal sodium ion concentration. The reason for this is not immediately obvious but some possible reasons will be examined later. The implication of the finding is clear enough, however. At low internal sodium concentrations, the rate of sodium ion extrusion will fall off very rapidly with further reduction in c_{Na_i}. As potassium inward transport is to some extent linked to the outward transport of sodium ions, we would expect, as is observed, that the metabolically dependent portion of potassium influx in muscle cells falls to very low values at low inside sodium ion concentrations. Also, as the internal sodium ion concentration is reduced, a greater and greater fraction of the total measured sodium efflux will be due to exchange diffusion as observed. The fractions of sodium efflux in muscle cells are plotted as a function of the internal sodium ion concentration in Fig. 3-2.

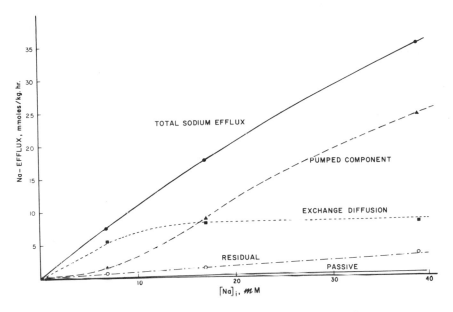

Fig. 3-2 *The components of sodium efflux from striated muscle cells.*

The line marked "residual" represents the efflux remaining after treatment with glycosides and sodium-free solutions to abolish exchange diffusion. The bottom line represents the passive component of sodium efflux calculated from Eq. (3-11). The difference between the bottom two lines must represent the amount of active sodium transport that occurs by a route which does not require full activity of the membrane-located ATPase.

In summary, at low sodium ion concentrations within the muscle fibers the potassium fluxes appear to be passive both by a theoretical flux ratio test and by an experimental test using cardiac glycoside inhibitors of transport. The sodium ion fluxes under these conditions show a large exchange diffusion component and a smaller component due to active transport. Sodium influx and potassium efflux appear to be passive, as in nerve, except for the component of sodium influx due to "exchange diffusion" which we have chosen to place in another category. There may be some arbitrariness in this choice, however, as this type of linked sodium movement is passive in the sense that it does not seem to be dependent on metabolism. Exchange diffusion of this sort might better be termed a tightly-linked passive movement. On this basis, passive fluxes would be categorized as to the "tightness" of the ionic linkage of the movement. The type of passive movement described by Eq. (3-8) would then be an example of a weakly linked passive movement. The types of passive movements that can be described by equations having the form of Eq. (3-12) would be examples of tightly linked passive movements, where a definite combination of the ion with a membrane site is required for transit to occur.

Potassium influx in muscle cells can be satisfactorily described by either formulation so that available experimental data does not force us to a choice (Sjodin, 1959, 1961). However, one of the characteristics which a tightly linked movement would be expected to show is that of competition for membrane sites. If an ion must combine with a membrane site before crossing the membrane, any other ionic species for which the membrane site has an affinity will competitively inhibit the movement of that ion. If the rate at which ions cross the membrane depends upon the number of ions that have combined with a site in the membrane, the rate in the presence of a competitive inhibitor will be given by:

$$v = v_{max} \frac{c_s}{c_s + \dfrac{K_m}{K_I} c_I + K_m}. \qquad (3\text{-}14)$$

The symbols here have the same significance as in Eq. (3-12). The concentration of the competitive inhibitor is c_I and K_I represents the "Michaelis constant" for the competitive inhibitor. In striated muscle cells, it is possible to describe inward potassium movement in the presence of foreign cations by equations having the form of Eq. (3-14) (Sjodin, 1961). The inference is that potassium ions cross the muscle cell membrane by first combining with a receptor site in the membrane.

PROPERTIES OF ACTIVE ION TRANSPORT

The carrier hypothesis provides a useful framework for planning experiments and interpreting results. To operate, the carrier model requires the presence of sodium ions on the inside of the membrane and potassium ions on the outside of the membrane. The question arises as to the significance of these ionic requirements. In red blood cell ghosts, it has been shown that the rate of hydrolysis of internal ATP increases as the internal sodium ion concentration rises and also as the external potassium ion concentration rises. As the hydrolysis of ATP is an enzymatic process, it seems evident that the total enzymatic activity involved in ATP hydrolysis depends upon the internal sodium ion concentration and the external potassium ion concentration.The activation of membrane-derived ATPase and of the ion transport mechanism itself has been the subject of much investigation.

Activation by Monovalent Cations

The activation of the magnesium-requiring ATPase from crab nerve by monovalent cations has been studied by Skou (1957, 1960). The effect of increasing sodium ion concentration on enzyme activity in the presence of ATP, Mg^{++} and different potassium ion concentrations is illustrated in Fig. 3-3. The influence of increasing potassium ion concentration on enzymatic activity at different fixed sodium ion concentrations is shown in Fig. 3-4. The manner in which sodium and potassium ions affect the enzyme activity can be surmised from these figures. Some features of the activation by K^+ and Na^+ can be pointed out. Each cation stimulates the

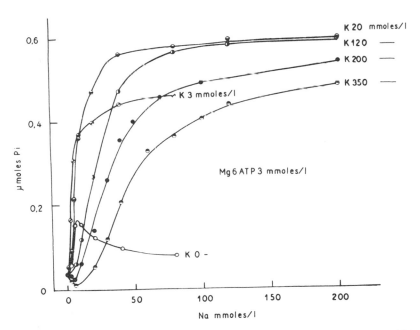

Fig. 3-3 *Effect of* Na^+ *on the enzyme activity in the presence of 6 mM/liter of* Mg^{++} *at different* K^+ *concentrations. (Taken from Skou (1957) with the kind permission of the author.)*

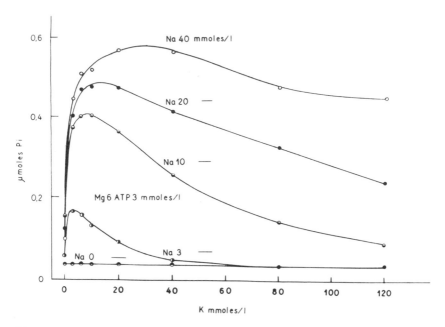

Fig. 3-4 *Effect of* K+ *on the enzyme activity in the presence of* 6 mM/liter *of* Mg++ *at different* Na+ *concentrations. (Taken from Skou [1957] with the kind permission of the author.)*

enzymatic activity considerably in the presence of the other cation if Mg^{++} is also present. Both cations also exert an inhibitory action at high concentrations. In these studies, an inhibitory action at high sodium ion concentrations was observed only in the absence of potassium ions. However, Post et al. (1960) found that high concentrations of sodium ions competitively inhibit the activation due to potassium ions of red blood cell membrane ATPase when the potassium concentration is varied from 0 to 10 mM.

It should be re-emphasized that the studies of the membrane-derived ATPases apply to disrupted and disoriented enzymatic sites. The studies on red blood cell ghosts indicate that the sites where sodium ions stimulate are located at the inner membrane surface whereas the sites at which potassium ions stimulate must be located at the outer membrane surface. We thus have the possibility for both activation and inhibition due to ions at two spatially separated regions of enzymatic sites. In the extracted system, it is clear that one measures the sum total of the actions at all sites due to the loss of spatial orientation. It is not likely, therefore, that one will find a one-to-one correspondence between studies on intact systems and studies on extracted systems. Even so, if the membrane-derived ATPases form a part of the molecular basis for ion transport, one should be able to detect some correspondence.

The first apparent similarity is that in studies of both ion transport rate and activation of membrane-ATPases, the activation often follows a sigmoidal curve when plotted against the concentration of the ionic activator. Measurements of the rate of sodium extrusion in striated muscle cells as a function of the internal sodium ion concentration demonstrate such a relationship (Fig. 3-2; see also Mullins and Frumento, 1963). Fig. 3-3 shows that membrane-derived ATPase is activated by sodium ions along a sigmoidal curve. Another example is the stimulation of sodium extrusion by external potassium ions in striated muscle cells (Sjodin and Beaugé, 1968b) and in squid giant axons (Baker, Blaustein, Hodgkin and Steinhardt, 1969a). In both cases, plotting the rate of sodium transport versus the external potassium ion concentration results in an S-shaped curve. Such sigmoidal curves are reminiscent of those obtained for the cooperative type of binding of oxygen to hemoglobin. Though their significance in transport studies is as yet undetermined, some possibilities will be discussed shortly.

It has already been mentioned that the ionic concentrations required to activate the membrane ATPases to half-maximal activity agree well with ionic concentrations required to produce one-half the maximal rate of ion transport. This is another point of correspondence. It has also been mentioned that cardiac glycosides like g-strophanthin (ouabain) and strophanthidin inhibit both active sodium and potassium transport and also inhibit the activity of membrane-derived magnesium-activated ATPases. It is noteworthy that the glycosides inhibit only that portion of the enzymatic activity that depends upon the presence of sodium and potassium ions. It should also be emphasized that the cardiac glycosides do not by any means inhibit all of the active cation transport in excitable cells. In striated muscle cells which are actively extruding sodium ions during recovery from the sodium-enriched state, about 90 per cent of sodium efflux not due to exchange diffusion is abolished by 10^{-5} M strophanthidin (Beaugé and Sjodin, 1968). When the internal sodium ion concentration falls toward the normal value, the percentage of sodium efflux that can be abolished by strophanthidin declines. In squid giant axons a considerable fraction of the total sodium efflux remains in the presence of cardiac glycosides (Caldwell and Keynes, 1959; Brinley and Mullins, 1967; Sjodin and Beaugé, 1969). Under normal conditions only about two-thirds of the sodium efflux from squid giant axons can be abolished by inhibition with cardiac glycosides. Nevertheless, the fact that glycosides inhibit the sodium and potassium ion dependent portions of both transport rate and enzymatic activity is impressive.

A final correspondence between transport and enzymatic studies lies in the area of ion selectivity, which has been studied by measuring the response of the enzyme activity to a series of monovalent cations. In the presence of Mg^{++} and Na^+, monovalent cations activated ATPase from crab nerve in the order $K^+ > Rb^+ > Cs^+ > Li^+$ (Skou, 1960). The same order has been observed for the activating effect of these cations on sodium extrusion in squid giant axons (Sjodin and Beaugé, 1968a). Furthermore, the cation concentrations required for attainment of one-half the maximal transport rate and the cation concentrations required for one-half maximal activation of membrane ATPase are comparable.

Nature of Sigmoidal Curves

In its simplest form, the "carrier" model assumes that one or more steps in the transport cycle requires a combination of the transported ionic species with a molecular complex within the membrane. By combining with sodium and potassium ions, it seems likely that the molecular complex becomes enzymatically active in the hydrolysis of ATP. It is clear that energy can be liberated in this way by an ionically regulated mechanism. It is not clear, however, how this energy is transduced into a net ionic movement. For example, is the cation that activates the enzymatic complex identical with the cation that is transported? Alternatively, are there separate "activator sites" which, when occupied by certain cations, induce the membrane molecular complex to become enzymatically active in transporting other cations via a different set of sites? Stated another way, are there separate "activator sites" and "transport sites" or are these sites one and the same? Available experimental evidence does not at present allow a decision to be made.

The simplest possible case would be that in which a single site at the inner membrane surface combines with a sodium ion and diffuses, via attachment to a carrier, to the outside membrane surface. There the site gives up the sodium ion and combines with a single potassium ion which then diffuses with the carrier back to the interior. In this case, the ion activating would be the same ion that is transported. Both enzymatic activation and transport rate would, in this instance, be expected to follow single site saturation kinetics described by Eq. (3-12). In addition, the ratio of the number of sodium ions transported to the number of potassium ions transported would be one-to-one. There is no experimentally studied case where all of these facts have been demonstrated. There are the cases discussed previously where the stimulation of sodium extrusion by external potassium ions follows Eq. (3-12). However, in red blood cells and in muscle cells the dependence of the transport rate on the internal sodium ion concentration is of the S-shaped type. Also, activation of membrane ATPases by cations generally follows a sigmoidal curve. Activation of sodium extrusion in muscle and nerve cells by external potassium ions in the presence of external sodium ions also follows a sigmoidal curve. In general, we have to conclude that a single site, univalent carrier model fails to explain the facts.

If the sites on the carrier are multivalent, however, the curve relating ion binding to ionic concentration in the solution will be of a sigmoidal character. A fairly simple case will be considered. The carrier molecule will be denoted by X and external potassium ions by K_o. It will be assumed that the carrier molecule can bind one, two or three potassium ions so that the following set of reactions can be visualized:

$$X + K_o \underset{k_2}{\overset{k_1}{\rightleftharpoons}} XK,$$

$$XK + K_o \underset{k_4}{\overset{k_3}{\rightleftharpoons}} XK_2,$$

$$XK_2 + K_o \underset{k_6}{\overset{k_5}{\rightleftharpoons}} XK_3.$$

We shall also assume that only the carrier form XK_3 is active in promoting active sodium extrusion. We might assume, for example, that only the form XK_3 can diffuse to the interior to react with internal sodium ions. The reaction with internal sodium ions can be represented by:

$$XK_3 + Na_i \underset{k_8}{\overset{k_7}{\rightleftharpoons}} XNaK_2 + K_i,$$

where the i subscript refers to ions inside of the cell membrane. If it is also assumed that the rate constant k_7 is low compared to all other rate constants, then the concentration of XK_3 can be computed by standard methods of chemical kinetics. The rate of sodium extrusion can then be set proportional to c_{XK_3} to give:

$$m_{Na} = \frac{Q_{Na}}{1 + \dfrac{a}{c_{Ko}} + \dfrac{b}{(c_{Ko})^2} + \dfrac{c}{(c_{Ko})^3}}, \tag{3-15}$$

where m_{Na} refers to the sodium extrusion rate, Q_{Na} to the maximal rate of sodium extrusion, $[c_{Ko}]$ to the external potassium ion concentration and a, b and c are constants which depend on the rate constants, k_1 through k_6. The data of Sjodin and Beaugé (1968b) on activation of the sodium pump in muscle cells by external potassium ions can be fit by Eq. (3-15). The results are compared with the predictions of Eq. (3-15) in Fig. 3-5. The fact that the agreement is good shows only that the model is a possible one. The assumptions underlying the derivation of Eq. (3-15) are reasonable but arbitrary and were made mainly as simplifications. There is no reason, for example, why other forms of the carrier should not react with internal sodium ions as well. Also, it appears likely that external sodium ions can react with the carrier to compete with external potassium ions for occupancy of carrier sites. External sodium ions are known to competitively inhibit both sodium active transport (Baker

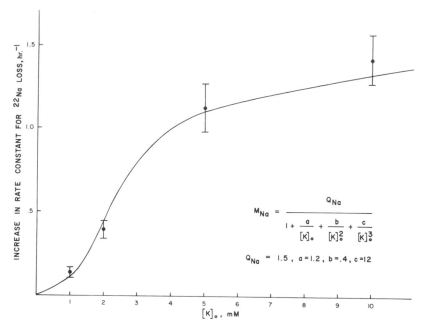

Fig. 3-5 *The potassium-sensitive portion of sodium efflux from striated muscle cells is plotted against the external potassium ion concentration. The curve drawn through the experimental points is a plot of Eq. (3–15).*

and Connelly, 1966) and activation of membrane ATPase by potassium ions (Post et al., 1960). These refinements can be incorporated into the carrier model with no great difficulty.

K-free and Na-free Effects and the Coupling of Ionic Movements

In excitable cells there is generally an effect on the sodium efflux when either potassium ions or sodium ions are removed from the external solution. Depending upon the ion involved, the effect is called a K-free or Na-free effect. The terminology is useful in helping to sort out the various types of ionic movements in excitable cells. In both muscle and nerve cells, the K-free effect is a considerable reduction in sodium efflux. A significant feature of the K-free effect is that it is abolished by the cardiac glycosides. Ouabain or strophanthidin reduce sodium efflux to at least the value observed in a potassium-free medium. Removal of external potassium ions in the presence of cardiac glycosides does not result in a significant further reduction in sodium efflux. There are at least two possible actions of external potassium ions on sodium efflux. Potassium ions may act as activators of sodium transport, probably by serving as co-factors for an enzyme or system of enzymes involved in cation transport. Potassium ions may also engage in an exchange with internal sodium ions during sodium extrusion to preserve the requirement of electrical neutrality. As mentioned previously, it is difficult to distinguish these functions and it has not yet been possible to clearly separate the two functions experimentally. Clearly, removal of external potassium ions will reduce sodium efflux by both mechanisms.

It is of interest to compare the magnitude of the reduction in sodium efflux in the K-free effect with the magnitude of the K influx itself. In invertebrate giant axons studied by a variety of experimental techniques, such a comparison indicates that the loss of sodium efflux is greater than the magnitude of potassium influx (Hodgkin and Keynes, 1955a; Sjodin and Beaugé, 1967, 1968a; Mullins and Brinley, 1967, 1969; Baker, Blaustein, Keynes, Manil, Shaw and Steinhardt, 1969b). In some instances, the loss in sodium efflux is up to twice the magnitude of potassium influx. It is evident that a good deal of the sodium efflux need not be coupled to inward potassium movement. In most cases the ionic movement that balances the residual sodium efflux is unidentified. At least in some cases, the balancing ionic movement is an inward sodium ion movement (Baker et al., 1969b). When the number of sodium ions actively transported outwardly does not equal the number of potassium ions actively transported inwardly, the ionic pump may be producing an electrical current which can contribute to the resting membrane potential. Such an ionic pump has been termed "electrogenic". Hodgkin and Keynes (1955a) were aware that the ionic pump in giant axons may be of this sort but pointed out that the contribution of the pump to the resting potential may be small and difficult to detect. In muscle cells, such electrogenic pump contributions to the resting membrane potential are detectable (Kernan, 1962; Mullins and Awad, 1965; Frumento, 1965; Adrian and Slayman, 1966).

In studying the effects of removing external sodium ions a problem arises that is not encountered in the study of K-free effects. Most of the osmotic pressure of the bathing medium is contributed by NaCl. A medium free of sodium ions must, therefore, contain an osmotic equivalent of some other solute. Ideally, the solute replacing NaCl should be physiologically inert. Unfortunately, the most inert solutes are nonelectrolytes such as dextrose. These substances do not preserve the normal ionic strength of the bathing medium. A commonly used alternative is replacement of NaCl with a salt containing a cation that is believed to be physiologically inert. A widely used salt has been lithium chloride. Using LiCl to replace NaCl, the Na-free effect in freshly dissected muscle cells is a reduction of approximately 50 per cent in the sodium efflux (Keynes and Swan, 1959).

In squid giant axons the Na-free effect is an increase in sodium efflux with LiCl replacement (Keynes, 1965; Baker et al., 1969a; Sjodin and Beaugé, 1969). In some cases, however, the Na-free effect in a lithium medium may be absent in squid giant axons or may be a decrease in the sodium efflux (Mullins and Brinley, 1967; Sjodin and Beaugé, 1969).

In frog muscle cells, the Na-free effect in a lithium medium reverses to an increase in sodium efflux when the internal sodium ion concentration is elevated (Keynes, 1965; Beaugé and Sjodin, 1968). When ouabain or strophanthidin is present in the solution, raising the internal sodium ion concentration does not bring about this reversal in the Na-free effect in muscle cells (Keynes, 1966; Beaugé and Sjodin, 1968). These observations can be explained as an action of lithium ions on the sodium pump. In muscle cells the stimulating effect of lithium ions on sodium extrusion is abolished by cardiac glycosides. In squid giant axons the lithium-stimulated sodium extrusion is not abolished by ouabain (Baker et al., 1969a).

The results obtained with lithium ions demonstrate that Li^+ is not a good choice as a substitute for Na^+ in attempting to demonstrate the influence of sodium-free conditions. In *Sepia* giant axons, Hodgkin and Keynes (1955a) used choline and dextrose as sodium substitutes. Using these solutes under K-free conditions, the Na-free effect was a large increase in sodium efflux. Since that time, the observations have been repeated and it has been shown that this type of Na-free effect is abolished by application of cardiac glycosides (Baker et al., 1969a; Sjodin and Beaugé, 1969). Using Tris as a sodium replacement, the presence of strophanthidin changes the Na-free effect in squid giant axons from an increase in sodium efflux to a decrease in sodium efflux. It is clear that, in general, Na-free effects are as metabolically dependent as are K-free effects. The only obvious exception is the Na-free effect observed in freshly dissected muscle cells.

Earlier in the chapter it was stated that, in excitable cells, one might expect an Na: Na exchange (exchange diffusion) to be revealed when sodium ions are removed from the external solution. Likewise it was stated that an active Na: K exchange will be revealed when potassium ions are removed from the external solution. It is possible to observe these responses under appropriate conditions in both nerve and muscle cells. In general, however, there must be a degree of interdependence between the two processes. Thus, there is a danger in attributing Na-free effects and K-free effects to two distinct mechanisms termed exchange diffusion and coupled active transport.

Any attempt to summarize the extent of coupling of the cation fluxes in muscle and nerve cell is made difficult by the fact that coupling ratios are not constant entities but vary with experimental conditions. In freshly dissected striated muscle cells about 50 per cent of the sodium fluxes is apparently coupled as an Na : Na exchange that is insensitive to metabolic inhibitors. Some 15 to 25 per cent of sodium efflux is coupled to potassium influx as an Na : K exchange. This leaves some 30 per cent of the sodium efflux unaccounted for. Such a fractionation of the sodium efflux may be deceptive, however. Application of cardiac glycosides along with Na-free conditions reduces sodium efflux to about 10 per cent of the normal value (Sjodin and Beaugé, 1968b; Keynes and Steinhardt, 1968). This must mean that most of the sodium efflux that cannot be accounted for on a coupled cation movement basis, as reckoned from K-free and Na-free effects, must be glycoside sensitive. Part of the explanation for this discrepancy may lie in the fact that, in the absence of external potassium ions, both the efflux and the influx of sodium ions are significantly reduced by ouabain (Keynes and Steinhardt, 1968). This means that a ouabain-sensitive Na : Na exchange occurs under K-free conditions. Hence the operational experimental test for deducing the extent of Na : K coupling seems to alter the mode of coupling of the pump.

Any attempt to deduce the extent of the flux coupling by the usual experimental operations is bound to meet with uncertainties if the coupling modes vary with the ionic composition of the medium. We might assume that the ouabain-sensitive Na : Na exchange in muscle cells is an abnormal coupling mode because ouabain is without effect on sodium influx if physiological concentrations of potassium ions are present in the external solution. Accordingly, we might assume that, in the presence of external potassium ions, K^+ is responsible for coupling to that part of the ouabain-sensitive sodium efflux that is coupled to Na^+ influx in the absence of external potassium ions. On this basis, in normal muscles in a physiological medium, 50 per cent of sodium efflux is involved in a glycoside-insensitive Na : Na exchange, 40 per cent is involved in a glycoside-sensitive Na : K coupling and 10 per cent is not coupled in any obvious way.

With regard to the potassium fluxes, a quantitative conclusion is made difficult by the strong dependence of potassium influx on the internal sodium ion concentration. The best estimate appears to be that, under normal conditions, over 80 per cent of the potassium movement in muscle is an electrically coupled K : K exchange. Though approximately 20 per cent of potassium influx is probably coupled to sodium efflux under normal conditions, the details of the coupling of the potassium fluxes are as yet unknown.

In giant axons, about two-thirds of the sodium efflux appears to be coupled to about 90 per cent of the potassium influx. The type of coupling engaged in by the remaining one-third of the sodium efflux is unknown at present as are the detailed natures of sodium influx and potassium efflux.

RECENT FINDINGS AND REMAINING PROBLEMS

This chapter has attempted to seek some order amongst a number of experimental findings. The general theoretical framework that has been discussed serves to partially explain some of the results. The main task that remains is to provide a more complete theoretical description that unifies the many findings and that will make predictions which can be subjected to experimental verification.

Some recent observations make it clear that the carrier model must be much more complex than initially supposed. For example, Baker et al. (1969a) have found that a part of the sodium efflux from squid giant axons occurs by way of carrier sites activated by lithium ions, requiring calcium ions for activity, and insensitive to the presence of cardiac glycosides. This sodium transport is definitely part of the total active transport complex because it is inhibitable by cyanide and also because we know that some 98 per cent of the sodium efflux in squid giant axons is ATP-dependent. The results emphasize an existing difficulty in the field of ion transport. Workers have often defined active transport in operational terms as that part of the total ionic movement that is abolished by cardiac glycosides. Clearly, it is not possible to do this. Mullins and Brinley (1969) have proposed that active transport be defined operationally as transport that requires ATP and, hence, ionic movement that disappears in the absence of ATP. This definition has the advantage of including the largest fraction of an active ionic movement that has been experimentally demonstrated so far. It also has the advantage of not missing ouabain-insensitive components such as those that have been recently observed.

Though cardiac glycosides have been extremely useful in sorting out different types of ionic movements in excitable cells, their usage suffers from the disadvantage that they only partially abolish active ion transport. Another difficulty is that, in some cases, they actually increase the rate of ionic movement. When the internal ATP concentration was very low in dialyzed squid axons, Brinley and Mullins (1968) found strophanthidin to increase the rate of outward sodium ion movement. It has also been observed that strophanthidin also increases potassium efflux from excitable cells (Harris, 1957; Sjodin and Beaugé, 1968b; Mullins and Brinley, 1969). The mechanisms for these effects are unknown.

Another difficulty encountered in using cardiac glycosides to detect active components of ion transport is that ATP-dependence and glycoside sensitivity do not always appear together. In normal squid giant axons, Mullins and Brinley (1969) found strophanthidin to be without effect on sodium influx. By a " glycoside test " for active transport, one would conclude that sodium influx occurs entirely by a passive process. On the basis of dependence on the presence of internal ATP, however, about one-half of the influx of sodium ions in squid giant axons appears to occur via a metabolically mediated mechanism, (Brinley and Mullins, 1968).

Any theoretical framework that accounts for the apparent multivalent nature of transport activation and successfully deals with these recent findings as well as classical observations may go far in providing a molecular basis for active membrane transport. The parallels between *in vitro* studies on " membrane ATPases " and studies on active cation transport in cells provide an encouraging beginning. From topics discussed in this chapter it should be clear that there exists no overwhelming theoretical or experimental justification for dividing total ion transport into the three separate categories: passive, exchange diffusion and active components. In many cases it will still prove useful to do so. In the final analysis, however, it may turn out that what was thought to represent these three processes really represents different aspects of the same basic membrane mechanism. Considerable overlap between the processes of exchange diffusion and active transport has already been observed. In some cases what is believed to be a passive ion movement obeys carrier type kinetics (Sjodin, 1961). It seems that the clearest distinction between passive and active fluxes occurs during action potentials where fluxes of thousands of pmoles/cm^{2}sec are insensitive to metabolic inhibitors. The highest fluxes observed in active transport processes are almost two orders of magnitude smaller than the fluxes occurring during membrane excitation. The implication is that we should look for separate and perhaps independent mechanisms here.

REFERENCES

ABOOD, L. G., and R. W. GERARD. 1954. Enzyme Distribution in Isolated Particulates of Rat Peripheral Nerve. *J. Cell. Comp. Physiol.* 43:379.

ADRIAN, R. H., and C. L. SLAYMAN. 1966. Membrane Potential and Conductance during Transport of Sodium, Potassium and Rubidium in Frog Muscle. *J. Physiol. (London).* 184:970.

BAKER, P. F., M. P. BLAUSTEIN, A. L. HODGKIN and R. A. STEINHARDT. 1969a. The Influence of Calcium on Sodium Efflux in Squid Axons. *J. Physiol. (London).* 200:431.

BAKER, P. F., M. P. BLAUSTEIN, R. D. KEYNES, J. MANIL, T. I. SHAW and R. A. STEINHARDT. 1969b. The Ouabain-sensitive Fluxes of Sodium and Potassium in Squid Giant Axons. *J. Physiol. (London).* 200:459.

BAKER, P. F., and C. M. CONNELLY. 1966. Some Properties of the External Activation Site of the Sodium Pump in Crab Nerve. *J. Physiol. (London).* 185:270.

BEAUGÉ, L. A., and R. A. SJODIN. 1968. The Dual Effect of Lithium Ions on Sodium Efflux in Skeletal Muscle. *J. Gen. Physiol.* 52:408.

BRINLEY, F. J., Jr., and L. J. MULLINS. 1967. Sodium Extrusion by Internally Dialyzed Squid Axons. *J. Gen. Physiol.* 50:2303.

———. 1968. Sodium Fluxes in Internally Dialyzed Squid Axons. *J. Gen. Physiol.* 52:181.

CALDWELL, P. C., and R. D. KEYNES. 1959. The Effect of Ouabain on the Efflux of Sodium from a Squid Giant Axon. *J. Physiol. (London).* 148:8P.

———. 1960. The Permeability of the Squid Giant Axon to Radioactive Potassium and Chloride Ions. *J. Physiol. (London).* 154:177.

DESMEDT, J. E. 1953. Electrical Activity and Intracellular Sodium Concentration in Frog Muscle. *J. Physiol. (London).* 121:191.

DUNHAM, E. T., and I. M. GLYNN. 1961. Adenosinetriphosphatase Activity and the Active Movements of Alkali Metal Ions. *J. Physiol. (London).* 156:274.

EDWARDS, C., and E. J. HARRIS. 1957. Factors Influencing the Sodium Movement in Frog Muscle with a Discussion of the Mechanism of Sodium Movement. *J. Physiol. (London).* 135:567.

FRUMENTO, A. S. 1965. Sodium Pump; Its Electrical Effects in Skeletal Muscle. *Science (Washington, D.C.)* 147:1442.

GLYNN, I. M. 1954. Linked Na and K Movements in Human Red Cells. *J. Physiol. (London).* 126:35P.

———. 1956. Sodium and Potassium Movements in Human Red Cells. *J. Physiol. (London).* 134:278.

———. 1957. The Ionic Permeability of the Red Cell Membrane. *Progr. Biophys. Biophys. Chem.* 8:241.

GOLDMAN, D. E. 1943. Potential, Impedance and Rectification in Membranes. *J. Gen. Physiol.* 27:37.

HARRIS, E. J. 1957. Permeation and Diffusion of Potassium Ions in Frog Muscle. *J. Gen. Physiol.* 41:169.

HARRIS, E. J., and M. MAIZELS. 1951. The Permeability of Human Erythrocytes to Sodium. *J. Physiol. (London).* 113:506.

HODGKIN, A. L., and P. HOROWICZ. 1959. Movements of Na and K in Single Muscle Fibres. *J. Physiol. (London).* 145:405.

HODGKIN, A. L., and B. KATZ. 1949. The Effects of Sodium Ions on the Electrical Activity of the Giant Axon of the Squid. *J. Physiol. (London).* 108:37.

HODGKIN, A. L., and R. D. KEYNES. 1955a. Active Transport of Cations in Giant Axons from *Sepia* and *Loligo*. *J. Physiol. (London).* 128:28.

———. 1955b. The Potassium Permeability of a Giant Nerve Fibre. *J. Physiol. (London).* 128:61.

HOFFMAN, J. F. 1960. The Link Between Metabolism and the Active Transport of Na in Human Red Cell Ghosts. *Federation Proc.* 19:127.

HOROWICZ, P., and C. J. GERBER. 1965. Effects of External Potassium and Strophanthidin on Sodium Fluxes in Frog Striated Muscle. *J. Gen. Physiol.* 48:489.

KERNAN, R. P. 1962. Membrane Potential Changes during Sodium Transport in Frog Sartorius Muscle. *Nature (London)*. 193:986.

KEYNES, R. D. 1954. The Ionic Fluxes in Frog Muscle. *Proc. Roy. Soc. (London) Ser. B.* 142:359.

———. 1965. Some Further Observations on the Sodium Efflux in Frog Muscle. *J. Physiol. (London)*. 178:305.

———. 1966. Exchange Diffusion of Sodium in Frog Muscle. *J. Physiol. (London)*. 184:31P.

KEYNES, R. D., and P. R. LEWIS. 1951. The Sodium and Potassium Content of Cephalopod Nerve Fibres. *J. Physiol. (London)*. 114:151.

KEYNES, R. D., and G. W. MAISEL. 1954. The Energy Requirement for Sodium Extrusion from a Frog Muscle. *Proc. Roy. Soc. (London) Ser. B.* 142:383.

KEYNES, R. D., and R. A. STEINHARDT. 1968. The Components of the Sodium Efflux in Frog Muscle. *J. Physiol. (London)*. 198:581.

KEYNES, R. D., and R. C. SWAN. 1959. The Effect of External Sodium Concentration on the Sodium Fluxes in Frog Skeletal Muscle. *J. Physiol. (London)*. 147:591.

LIBET, B. 1948. Adenosinetriphosphatase (ATPase) in Nerve. *Federation Proc.* 7:72.

MATCHETT, P. A., and J. A. JOHNSON. 1954. Inhibition of Sodium and Potassium Transport in Frog Sartorii in the Presence of Ouabain. *Federation Proc.* 13:384.

MULLINS, L. J., and M. Z. AWAD. 1965. The Control of the Membrane Potential of Muscle Fibers by the Sodium Pump. *J. Gen. Physiol.* 48:761.

MULLINS, L. J., and F. J. BRINLEY, Jr. 1967. Some Factors influencing Sodium Extrusion by Internally Dialyzed Squid Axons. *J. Gen. Physiol.* 50:2333.

———. 1969. Potassium Fluxes in Dialyzed Squid Axons. *J. Gen. Physiol.* 53:704.

MULLINS, L. J., and A. S. FRUMENTO. 1963. The Concentration Dependence of Sodium Efflux from Muscle. *J. Gen. Physiol.* 46:629.

POST, R. L., C. R. MERRITT, C. R. KINSOLVING and C. D. ALBRIGHT. 1960. Membrane Adenosinetriphosphatase as a Participant in the Active Transport of Sodium and Potassium in the Human Erythrocyte. *J. Biol. Chem.* 235:1796.

RANG, H. P., and J. M. RITCHIE. 1968. On the Electrogenic Sodium Pump in Mammalian Non-myelinated Nerve Fibres and its Activation by Various External Cations. *J. Physiol. (London)*. 196:183.

SHAW, T. I. 1954. Sodium and Potassium Movements in Red Cells. Cambridge University Ph.D. Thesis.

SJODIN, R. A. 1959. Rubidium and Cesium Fluxes in Muscle as Related to the Membrane Potential. *J. Gen. Physiol.* 42:983.

———. 1961. Some Cation Interactions in Muscle. *J. Gen. Physiol.* 44:929.

———. 1965. The Potassium Flux Ratio in Skeletal Muscle as a Test for Independent Ion Movement. *J. Gen. Physiol.* 48:777.

————. 1970. The Activation of Sodium Transport in Striated Muscle by Potassium Ions in the Absence of External Sodium Ions. *Biophys. Soc. Abstr.* (14th Annual Meeting) 222a.

SJODIN, R. A., and L. A. BEAUGÉ. 1967. The Ion Selectivity and Concentration Dependence of Cation Coupled Active Sodium Transport in Squid Giant Axons. *Currents Mod. Biol.* 1:105.

————. 1968a. Coupling and Selectivity of Sodium and Potassium Transport in Squid Giant Axons. *J. Gen. Physiol.* 51:152s.

————. 1968b. Strophanthidin-sensitive Components of Potassium and Sodium Movements in Skeletal Muscle as Influenced by the Internal Sodium Concentration. *J. Gen. Physiol.* 52:389.

————. 1969. The Influence of Potassium- and Sodium-free Solutions on Sodium Efflux from Squid Giant Axons. *J. Gen. Physiol.* 54:664.

SJODIN, R. A., and E. G. HENDERSON. 1964. Tracer and Non-tracer Potassium Fluxes in Frog Sartorius Muscle and the Kinetics of Net Potassium Movement. *J. Gen. Physiol.* 47:605.

SKOU, J. C. 1957. The Influence of Some Cations on an Adenosinetriphosphatase from Peripheral Nerves. *Biochim. Biophys. Acta.* 23:394.

————. 1960. Further Investigations on a Mg^{++}-Na^+ Activated Adenosinetriphosphatase Possibly Related to the Active, Linked Transport of Na^+ and K^+ across the Nerve Membrane. *Biochim. Biophys. Acta.* 42:6.

STEINBACH, H. B. 1940. Sodium and Potassium in Frog Muscle. *J. Biol. Chem.* 133:695.

————. 1951. Sodium Extrusion from Isolated Frog Muscle. *Am. J. Physiol.* 167:284.

————. 1952. On the Sodium and Potassium Balance of Isolated Frog Muscles. *Proc. Nat. Acad. Sci. U.S.A.* 38:451.

TEORELL, T. 1951. Zur Quantitativen Behandlung der Membranpermeabilität. *Z. Elektrochem.* 55:460.

USSING, H. H. 1949. Transport of Ions across Cellular Membrane. *Physiol. Rev.* 29:127.

WHITTAM, R. 1958. Potassium Movements and ATP in Human Red Cells. *J. Physiol. (London).* 140:479.

SOME ASPECTS OF ELECTRICAL STUDIES OF THE SQUID GIANT AXON MEMBRANE 4

K. S. COLE

Why all the fuss about the electrical behavior of nerves and such? Physiologists and electricians have been roundly criticized for wasting decades in puttering about with such stupid things as thresholds, resting potentials, spikes, farads, ohms, clamps and what not. This abusive attitude only shows a rather pitiful disdain for history (Cole, 1968). We have not waited for the experience and the vision to see the fundamental problems— we have only done what we could with what we had. Are we to be faulted because electrical techniques have long given the easiest and most powerful means for measurement and analysis of cells and tissues? Or because we used them without waiting for isotopes, for averaging and other electronic techniques or for optical, thermal, magnetic and transport analyses? Studies on such fundamental things as membrane conductances and potentials first required long cells. Should we have avoided these cells just because their usual function is to be excitable? Probably we can and should learn more about structures and ion permeabilities of passive membranes. But we have many more ingenious electrical measurements and intriguing analyses of them for excitable membranes than are available for any other cells or from any other observations. So we cannot but hope and expect that they will soon all fall together into a common interpretation of our often discordant information.

125

IN THE BEGINNING

Fricke (1923; 1925) produced the first quantitative evidence for the molecular dimensions of the red cell membrane in his measurement of its electrical capacity, C, where

$$I = C \, dV/dt$$

with a current I and potential V. Ideally we have

$$C = \varepsilon/4\pi\delta \simeq 1\mu f/cm^2$$

for a dielectric constant ε and thickness δ. This same capacity now seems to be a universal biological parameter—describing most membranes of cells and organelles. But neither ε or δ can yet be agreed upon for any membrane. Even worse, ε seemed to have some features of solid dielectrics which are yet to be explained.

Electrical resistances, R, or conductances, g, have come to be the most important single parameters in the description of membrane characteristics. The elementary operational definitions are, by Ohm's Law,

$$V = RI \text{ or } I = gV.$$

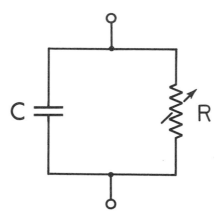

Fig. 4-1 *Electrical representation of living membranes with a passive capacity, C, and a highly variable resistance, R. (Reproduced, by permission, from Cole [1968].)*

Not until long after capacity measurements were well established was a conductance, g_M measured (Cole and Hodgkin, 1939) for the squid axon membrane. Here, formally,

$$g_M = \lambda/\delta \cong 1 \text{ mmho/cm}^2,$$

where the specific conductance, λ, in units of $ohm^{-1} cm^{-1}$, is not yet adequately explained.

This measurement of conductance had to be done because it had just been found (Cole and Curtis, 1939) that the conductance rose by 40 mmho/cm^2 during the passage of an impulse. But with it came the startling result that the capacity did not change by more than a few percent as the membrane conductance increased so drastically during excitation. As more and more attention has been devoted to these active ion conductances, the inert capacities have mostly been relegated to the role of one microfarad nuisances.

CAPACITY

Now with the vastly increased and highly gratifying realization that membrane structures and function may be the next most important, vulnerable frontier, the capacity regains its importance as one of the principal problems of membranes (Cole, 1970).

With analysis of various and sometimes complicated measurements, one has usually been able to dissect out this ubiquitous microfarad component which seems most likely to be a dielectric polarization. It seems to be largely independent of cell physiology, pathology and pharmacology, and probably of life itself. It does not depend significantly upon ion permeability or current flow; nor upon potential differences until an apparent breakdown at fields of at least 10^5 v/cm. The effects of pressure seem not to have been observed; the few temperature coefficients range from negligible up to about 1.5 per cent/$°C$ for squid axon membranes. Yet it must be pointed out that such sweeping generalizations may be only poor first approximations. Even at this stage they can serve to curb some ingenious speculations—such as perhaps some phase transition and depletion layer models. More detailed investigations, higher precision and new parameters may yet yield significant new information.

The interpretations of the capacity data are so far more matters of personal preference than of necessity. In spite of objections to almost every aspect, the phospholipid bilayer model has had increasingly convincing support from x-ray diffraction, electron microscope and optical data and from artificial models (Stockenius and Engelman, 1969).

Quite plausible phosphate and sterol arrangements have been worked out and there are strong indications of a considerable degree of ordering in hydrocarbon tails. But there seems not yet to be any calculation for the capacities of such structures. There are suggestions of changes of order by electric field, anesthetics, calcium and temperature. But how much should these affect the capacity? There must be proteins closely associated with living membranes and perhaps there are polypeptide paths for ions to follow across the membrane. Can the capacity tell us anything about them?

The recent artificial bilayer models made with various phospholipids (Mueller and Rudin, 1963, and others) have been quite encouraging in spite of some apparent differences from living membranes. In general the capacities have been rather low, 0.4–0.7 $\mu f/cm^2$, but it seems that a high proportion of low dielectric constant solvent may account for much of this. Small vesicles with solvent-free phospholipid membranes (Takashima, Redwood and Thompson, 1969) gave a capacity of about 2 $\mu f/cm^2$. This larger capacity and the apparently high effective dielectric constant are new challenges for explanation.

Each theoretical membrane model needs to be carefully examined for its ability to explain the dielectric capacity—first as to its magnitude of about a microfarad/cm^2 and its constancy in the face of spectacular biological and physical changes and then for other characteristics—perhaps smaller but no less important. Quite a few more and better experiments and analyses may be called for if theoretical efforts are to have the guidance and critique which they will need.

VOLTAGE CLAMP

The conductive properties of excitable membranes may be introduced by a brief summary of the development of voltage clamping.

The introduction of the squid giant axon for laboratory study first made intracellular potential recording possible (Hodgkin and Huxley, 1939; Curtis and Cole, 1940) which in turn focussed attention upon excitation. This process was difficult to investigate as it abruptly created an action potential and sped away, pulling the nerve impulse behind it. To hold it fixed in time and position soon became an obsession.

A cable analysis of a propagating impulse showed a wrong way current during the rising phase. Neither this nor the then-new increase of membrane conductance, nor both together made any sense. The only helpful hint, from a passive iron wire, was that the membrane might be made stable in the negative conductance region with a low resistance external circuit.

After World War II, the membrane potentials and current densities were made uniform by long low resistance electrodes placed inside and around the axon (Marmont, 1949). Although propagation was now eliminated in this clamp, the excitation and action potential remained much the same. It then seemed quite obvious that these phenomena were produced by the discharge of the membrane capacity. This could be prevented by keeping it at a constant potential—and besides an N shaped potential vs. current membrane characteristic might be stable with such a low resistance source.

Positive step changes of membrane potential quickly discharged the capacity, then gave an inward ion current flow which turned to a steady outward current (Cole, 1949). Nowhere in the time course was there any hint of an abrupt or threshold behavior. The outward currents were just what had already been thought of as the potassium current, shown with a slow ramp, Fig. 4-2, (Fishman, 1969), but the earlier inward flows were not

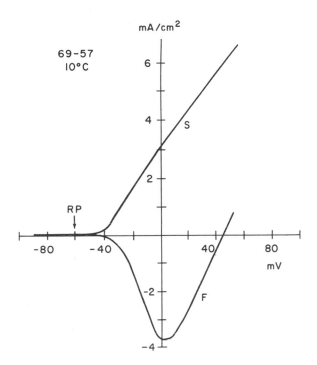

Fig. 4-2 *Conductive behavior of squid axon membrane shown with potentials increasing slowly, S, 2 V/sec and fast, F, 75 V/sec.*

then explained. The peaks of these transients were also a smooth function of potential which had a region of negative slope such as can be seen directly with a fast ramp of potential. Although it did not yet stand still in time, this region was an adequate basis for normal excitation and further the complete family of voltage clamp currents predicted a reasonably satisfactory action potential.

At this point Hodgkin, Huxley and Katz (1952) took over the concept, improved the technique and gave them to the world as the voltage clamp.

Criticisms of Voltage Clamping

There have been objections to voltage clamps. Some perhaps because they were new and quite different. Others may be because the basic simplicity was hidden in a maze of complicated and sophisticated electrodes and electronics. One of the first protests was that the clamp was unnatural, nonphysiological. The axon could not propagate an impulse nor even generate one. As clamping has spread this became almost irrelevant as well as superficial. The problems were reduced to the membrane level and now are clearly molecular, so that the threshold and spike are rather trivial consequences.

The squid clamps are brutal—if only by definition to those for whom control is brutality. The isolation, mounting, insertion of internal hardware and perfusion all contribute to the short survival of an axon. Then, by a single command, the membrane may be forced to pass ions up six times as fast as required for the passage of an impulse. Life was much more comfortable and easy in the animal. It is amazing and so very fortunate that axons are so tolerant and do as well as they do.

It is entirely proper to complain that a voltage clamp does not discriminate. In common with most other electrical measurements, a current is a current and flows of potassium and sodium ions look alike. Perhaps this will some day be answered by separate blue and yellow traces on a color screen. So far the separations have required difficult additional procedures. Specific drugs are much simpler to use than they are to justify to querulous purists.

Certainly the best known accusations are against the so-called clamps that do not clamp. But additional evidence suggested an indictment for negligence. There are good clamps and there are bad clamps and, if clamping is worth doing, it is worth doing well enough. It is the responsibility of the investigator to keep the bad results from bad clamps out of the literature—but how is he to know? The only rules suggested so far (Cole and Moore, 1960) are that, for the membrane measuring area:

1. The potential difference should be known and uniform within a few mv.
2. The delay in the measure and control of the potential should not be more than 10 μsec.
3. The current density should be uniform within a few per cent.

A continuing source of difficulty is the resistance of a few ohmcm^2 seen in series with the membrane capacity in the first current clamp measurements. So Hodgkin, Huxley and Katz introduced an electronic compensation but this is not simple to apply and has not come into general use. Resulting errors of 20 mv or more are not eliminated just because they are ignored.

The widespread use of the voltage clamp and the general utility of the results have certainly demonstrated the validity of the concept and the feasibility of its application to many preparations and problems. But as with other powerful approaches it has its limitations and hazards.

CONDUCTANCES

Conventional systems are said to be linear or ohmic and the R and g coefficients may be tacitly assumed to be constant over wide ranges of applied potential and current and of frequency or time.

Nonlinear and time dependent systems require more careful consideration and are more difficult to describe. As a point of departure, it is assumed that the systems are linear, ohmic, at sufficiently short times or high frequencies and for rather large potentials,

$$I = g_o V,$$

where g_o is the instantaneous or infinite frequency conductance. This assumption is not valid for some physical systems and appears not to apply exactly in some experiments on biological membranes. In systems containing an energy source, where $I = 0$ for $V = E_o$, this becomes

$$I = g_o (V - E_o).$$

From its representation on a V–I diagram, Fig. 4-3, g_o is usually called the chord conductance in biological literature and it is always positive. It is used exclusively in the Hodgkin-Huxley model axon. The tests for the region with this behavior are that g_o remain constant at shorter times or higher frequencies than limiting values and be independent of V in these ranges. The physical assumption and implication is that the state of the mechanism responsible for nonlinearity does not change instantaneously.

At the other extreme are the long time or low frequency characteristics in which the nonlinear components have reached steady state. Here the obvious and convenient parameter is

$$\partial I / \partial V = g_\infty,$$

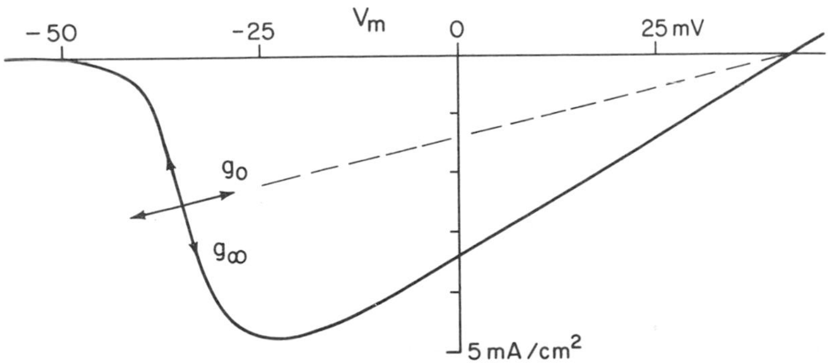

Fig. 4-3 *Quasi steady-state membrane behavior described in terms of the instantaneous, infinite frequency or chord conductance, g_o, and in terms of the steady state, zero frequency, or slope conductance, g_∞.*

which is often called the slope conductance and it may be either positive or negative.

Each of these two conductances, g_o or g_∞, can be converted to the other on a V–I diagram. It is usually easier to measure and manipulate g_o; g_∞ is more directly useful in stability problems and qualitative descriptions.

The transient or frequency dispersion behavior of the conductance between these two limits depends upon the nonlinear process and is an important characteristic of it. The state variable, S, which is responsible for non-linearity, is the all-important problem of membrane studies: in analogues it has been temperature, ionization, mobility or concentration of ions, oxide or conduction gates. In general, nonlinear behavior may be described by some functions p and q,

$$I = p(V,S); \, dS/dt = q(V,S),$$

and for small perturbations this reduces to the formal expression

$$dI/dt = g_o \, dV/dt + (I - g_\infty V)/\tau$$

with the time constant τ.

Probably the simplest and most obvious of analogues is the unusual thermistor with a positive temperature coefficient and a shunt, Fig. 4-8. This has an N characteristic on the V–I plane and temperature is the state variable.

THE HODGKIN-HUXLEY AXON

The Hodgkin-Huxley (1952) axon is a thorough, quantitative statement of the electrical properties of a normal squid axon membrane. These are derived from voltage clamp measurements of membrane current. Hodgkin and Huxley first separated this current into two principal components, a sodium current and a potassium current. This is a key operation that has been so well justified as to place it beyond reasonable criticism. The next step was to express each of these currents in terms of an instantaneous conductance and an electromotive force, Fig. 4-4 and 4-5. This was difficult to believe and appreciate but much subsequent work has strongly supported these para-meters as experimental facts.

Most amazingly these conductances, g_{Na}, g_K, are described by a family of empirical equations as functions of the membrane potential and the membrane potential alone, Fig. 4-6. They are expressed as smooth functions of membrane potential, time, and temperature. The functions are everywhere continuous with continuous derivatives. Their fundamental state variables are still unknown but even in empirical form these parameters are a sufficient basis for most of the normal and near-normal phenomena of the membrane and the axon.

Not only do these conductances represent the experimental voltage

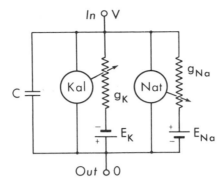

Fig. 4-4 *Hodgkin-Huxley description of squid membrane in which the conductances are controlled by the potential through the unknown mechanisms Kal and Nat. (Reproduced, by permission, from Cole* [1968].)

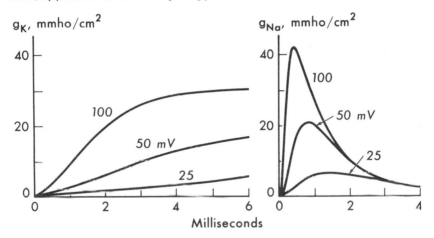

Fig. 4-5 *Behavior of the membrane conductances after the indicated changes from the resting potential. (Reproduced, by permission, from Cole* [1968].)

clamp data, but by computation they give everything, in excellent agreement with the experimental facts, from subthreshold linear behavior, through excitation, and on to include action potentials and propagation, as seen under usual physiological conditions.

There are, however, some restrictions. The conductance equations are based upon—and only upon—experiments. Only limited ranges of a limited number of variables were investigated, as summarized in Fig. 4-7. The equations could not, and should not, be expected to represent any other membrane or the squid membrane under any conditions for which there were no data. Increasingly broad explorations have shown extrapolations to be amazingly valid in several directions as well as deservedly wrong into other regions.

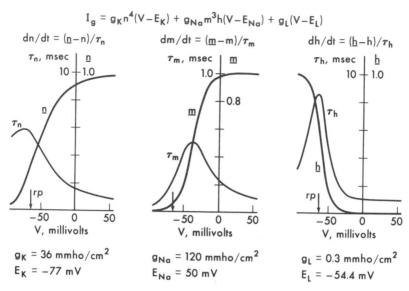

$$I_g = g_K n^4(V-E_K) + g_{Na}m^3h(V-E_{Na}) + g_L(V-E_L)$$

$$dn/dt = (\underline{n}-n)/\tau_n \qquad dm/dt = (\underline{m}-m)/\tau_m \qquad dh/dt = (\underline{h}-h)/\tau_h$$

$g_K = 36 \text{ mmho/cm}^2$ $g_{Na} = 120 \text{ mmho/cm}^2$ $g_L = 0.3 \text{ mmho/cm}^2$

$E_K = -77 \text{ mV}$ $E_{Na} = 50 \text{ mV}$ $E_L = -54.4 \text{ mV}$

Fig. 4-6 *Hodgkin-Huxley equations and constants with six of the variables shown graphically as functions of potentials. (Reproduced, by permission, from Cole [1968].)*

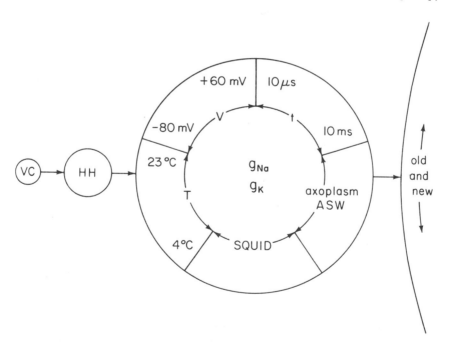

Fig. 4-7 *Hodgkin-Huxley country. The experimental conditions for which the membrane conductances were obtained from voltage clamp data and used to explain other membrane and axon behaviors.*

It would not be easy to find and catalogue either all of the good or all of the bad things that have been said about the Hodgkin and Huxley model axon. And such an opinion poll might not have any predictive value because this work may be established well enough to excite more voluble opposition than support. In this brief discussion no more than a minimum of personal, perhaps prejudiced, opinions are offered; only a few of the more annoying objections and insinuations are commented upon.

In support of Hodgkin and Huxley it seems entirely fair to insist, with but slight reservations, that their work is correct, although limited.

The Hodgkin and Huxley equations are, after all, but a presentation, in singularly popular form, of experimental facts. Although the original work was based on considerable ranges of the more common variables, Fig. 4-7, we must recognize that it may be limited to the ranges which were investigated. And it is quite amazing that there are not more extrapolations beyond these ranges which are significantly different from subsequent experiments. Yet it must be emphasized that any attempt to extend the range of any variable must not seriously modify any properties as given by Hodgkin and Huxley for their original coverage.

It has been said that only a measured current has reality. So a separation into Na and K components is fictitious because the components were not separately and directly measured. This unfortunate dictum would seem also to deny the application of many other useful concepts in many other fields. Aside from indirect confirmations there is increasing support from isotopes and blocking agents. Although the latter are literally out of Hodgkin and Huxley country, the interpretations are so consistent as to make them allowable extensions.

It was the natural consequence of the operational doctrine of inseparability that individual conductances had to be denied and on such a restrictive premise there is no defense. Although squid membrane current components are available, the point by point measurements of the corresponding conductances as given by Hodgkin and Huxley have not been widely used or substantially improved upon.

There is an arbitrary statement that membrane conductances, as measured in tens of μsec or at tens of kc, are steady state rather than instantaneous. This would be justified if the nonlinear processes were as fast as the ion redistributions are in free aqueous solution. The facts do not support such an assumption. With a few and perhaps uncertain exceptions, the measured conductances are reasonably linear and independent of time or frequency over amazing ranges.

The ion conductances are the key variables of the Hodgkin and Huxley model axon; they are conceptually clean and clear, they have been experimentally confirmed, but their fast and easy measurement is still a considerable challenge.

The labelling of the Hodgkin and Huxley axon as the sodium theory gives an appropriate acknowledgement to the genius of Hodgkin (1951) in the discovery and detailed description of the essential role of sodium in the normal functioning of the squid axon. This theory does not say and does not mean that sodium and only sodium can ever play this role. Any such implication is far away from the Hodgkin and Huxley country and into entirely foreign territory. But the careful investigations of old and new sodium substitutes have generalized the Hodgkin and Huxley axon and can be expected to be of great help in formulating a successful explanation.

A claim that the sodium and potassium current densities are not as calculated by Hodgkin and Huxley for a propagating impulse is difficult to test by isotopes. It would require the analysis of a small fraction of a pica mole net flux from a fraction of a cm of axon in a small fraction of a msec during each impulse. Neither the net flux nor such resolution have been achieved for a normal axon so there has not been a direct test. More practical are the measurements of each isotope flow into deficient media under voltage clamp where all requirements are much less stringent. The early and late components are then clearly identified as the predicted flows of Na^+ and K^+ with no abnormal leakages between pulses (Bezanilla, Rojas and Taylor, 1970). The unconvinced may have a long wait for a less sophisticated confirmation.

It is indeed true that the Hodgkin and Huxley axon and similar subsequent developments have required a high order of electrical and electronic background, imagination and invention. It makes no difference whether the results are expressed in numerical tables, by differential equations or as an equivalent electrical circuit. The circuit says so much with such elegance that references to the Hodgkin and Huxley axon as the equivalent circuit theory are far more flattering than may have been intended.

It is to be expected that some individuals are overly swayed by the criticisms of the Hodgkin and Huxley axon. At least a few authors of texts and reviews have ignored or slighted the work as being either trivial, controversial or wrong. This is a grave misfortune and the critics must accept some responsibility for probably retarding progress. The authors and the proponents of Hodgkin and Huxley may have seemed somewhat reluctant to promote polemics, so it is ironic that some of the more energetic attacks have turned out to be strong supports and extensions of the original work.

STABILITY AND THRESHOLDS

Certainly the best known and most striking property of a nerve fiber is that there is a threshold stimulus for a propagating impulse. Probably every possible adjective has been used to emphasize this all-or-none behavior. It is not surprising that the instability has been attributed to the axon membrane and that mechanisms have been proposed in tremendous variety just to explain it. So perhaps it is to be expected that most amateurs and even some professionals would not be interested in any other process that was comparatively dull, smooth and prosaic.

It was established more than twenty years ago that the squid axon membrane was not either inherently unstable or inherently stable. Its behavior was determined completely by an external boundary condition, the outside electrical conductance placed across the membrane. If there was no appreciable outside conductance in the current path the membrane had much the same threshold and action potential as for a propagating impulse. But if a near short circuit was placed across the membrane—the voltage clamp— the response was always a smooth function of time and potential.

This is even more obvious for a shunted positive thermistor, Fig. 4-8. As a controlled current from an outside zero conductance source is increased, the potential rises steadily until the current is above the top loop of the N. Then the potential increases as fast as the bead heats up until it reaches the upper leg of the N. Then that leg is followed down with decreasing current to the bottom loop. Here the potential goes back to the original lower limb as the bead cools. In this loop there are then two thresholds. For a controlled potential of nearly infinite conductance applied to this thermistor, every bit of the N, including the central negative conductance region, is accessible

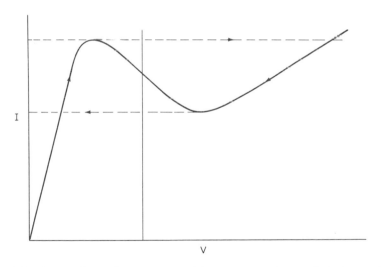

Fig. 4-8 *Assumed steady-state, potential vs. current characteristic for a shunted, positive thermistor. The horizontal dashed lines show the threshold transitions for constant current. For a constant applied potential, the vertical solid line makes a stable intersection.*

and a point of a stable steady state. So who can say that the thermistor, all by itself, is either unstable or stable? There is no answer until the outside conductance is known. Another example is the powder from a fire cracker heated inside an insulating cover. There comes a time when oxidation produces heat faster than it can escape and the reaction becomes explosive. On the other hand with a cooling system which can remove heat as fast as it is produced the burning will proceed smoothly at each and every temperature. So gunpowder is explosive or not according to whether its environment is adiabatic or isothermal.

The axon membrane is more complicated in that the negative conductance is only an approximately steady state for a fraction of a millisecond. This complicates the analysis but comes to the same conclusion (Chandler, FitzHugh and Cole, 1962). There is nothing in these characteristics to suggest that either instability or stability is an inherent internal property; the membrane can behave either way as allowed by external conditions.

It came as a rude shock, in 1959 (FitzHugh and Antosiewicz), to find that the Hodgkin and Huxley conductances predicted a graded response of the squid membrane to short stimuli at 6.3°C. But the response rose from 8 mv to 96 mv for an increase of stimulus by one part in 10^8. This seemed entirely academic, and faith in the all-or-none law as a physiological principle was not shaken. Since then there have been many threshold calculations and a usual criterion was a response of 50 mv. In the corresponding experiments the threshold was not sharp at higher temperatures. It was usually taken to be the stimulus for which the response rose most rapidly.

The Hodgkin and Huxley space clamp calculations showed that as much as a 5 per cent increase of stimulus should be necessary to increase the response from 10 mv to 80 mv—even at only 20°C. As the temperature was further increased the response became flatter until at 35°C it was increasing less than ten times as fast as the stimulus. Similarly for real axons there was a slight but obvious increase of response with increasing stimulus at 15°C and at 30°C and 35°C half a dozen or more well separated responses were obtained as the stimulus strength was raised.

Although quantitative differences are to be expected, the Hodgkin and Huxley equations and the real axons are in striking agreement (Cole, Guttman and Bezanilla, 1970). The membrane shows no basis to expect or assume any critical flip-flop, all-or-none, threshold behavior in a temperature range from probably 5°C up to perhaps 40°C (above 40°C the membrane is irreversibly damaged). This is a rather amazing extrapolation of the Hodgkin and Huxley equations which were limited to temperatures below 23°C. But the facts of real axons give reason to believe that the extrapolation is not grossly in error.

There is not yet a simple conceptual explanation for the increasingly graded membrane response at higher temperatures. The speed of the membrane potential change is limited by g_o/C which is relatively unaffected by temperature. It has been suggested that although the speed of the sodium-on process is increased by up to ten times, it cannot run ahead of the potential rise. On the other hand, as the recovery processes (sodium-off and potassium-on) speed up, they begin to overrun the excitation and so a stimulus becomes less effective.

The Bonhoeffer qualitative model showed such behavior, and the BVP approximation (FitzHugh, 1961) to the Hodgkin and Huxley model clearly gives the broader " No Man's Land " with increasing temperature. But it is quite disheartening to discover that the most interesting of the many other analogues for membrane behavior fail in this respect, although some can be patched up.

Where does this leave the time honored all-or-none law? Uncounted experiments and a few machine calculations (Cooley and Dodge, 1966) from the Hodgkin and Huxley conductances show that it is completely valid—but only under the conditions for which it was evolved—for the initiation of a propagating impulse. Another indication of the sharp distinction between membrane behavior and axon propagation is in the blockade of a conducting impulse by either an increase of temperature or of leakage (Huxley, 1959; Cole, 1968). Far from indicating a critical transition behavior these are but the consequences of the smooth membrane conductances placed in the more complicated, but more useful, configuration of an axon.

THEORIES

The electrical measurements on the squid axon membrane contain more accurate information than any other form of description. Any theoretical model of membrane structure and function is not adequately tested until it tries to explain the electrical properties. And any model which cannot explain these measurements is patently and grossly inadequate.

Chronic objectives of membrane theories have been to explain the threshold nature of the excitation process or to produce a propagating impulse. The sporadic developments of experiments, analyses and calculations have, over the past twenty years, made it increasingly clear that the normal squid axon membrane is almost certainly completely and smoothly continuous in its electrical behavior with respect to potential, time and temperature.

Thus any model is suspect if it produces critical phenomena in a normal membrane with spatially uniform potential difference and current density (Tasaki, 1968; Changeux, Thiéry, Tung and Kittel, 1967; Adam, 1968).

On the positive side, the measured sodium and potassium ion conductances, g_{Na} and g_K, and the analytical Hodgkin and Huxley approximations for them contain sufficient information to explain normal membrane and axon behavior. It is most pertinent to point out that the conductances are but an intermediate step, that they give, at most, only hints as to the mechanisms of ion permeability. But to dismiss them is to cast aside the simplest, best-defined and most descriptive data on ion permeabilities under normal conditions.

These conductances are continuous functions and have continuous derivatives with respect to potential, time and temperature. We are then confronted with the facts that neither the conductances nor the membrane responses show any evidence of critical behavior. If there must be such behavior, its effect on the conductances can only be small. Then it does not seem reasonable to expect that any abrupt transition plays a role of major importance in nerve membrane function. So one need not look for precipitous changes of membrane properties to explain normal behavior. These must all follow from an adequate explanation of the completely smooth conductance characteristics.

The one apparent exception to these generalizations is that the initiation of a propagating impulse is strictly all-or-none. The addition of the distance variable makes the description and the analysis far more complicated. (Cooley and Dodge, 1966). Yet the computations based on the smooth conductances do predict this threshold as well as the failure of conduction at high temperatures. Thus there is no need to work out a theory of impulse initiation or propagation. In fact it should be obvious that it is a manifest, not to say stupid, waste of effort to attempt to concoct *ab initio* any theory of any aspect of nerve impulse transmission. This is not only because of the enormously increased difficulty of solving such problems as compared with the primitive, basic challenge of the membrane conductances for spatial uniformity of potential and current. It is even more ridiculous to produce a theory of propagation which does not contain, at least implicitly, an explanation for the conductances because it will likely be obscure, almost certainly inadequate and probably quite wrong.

A model which is a good explanation of the membrane conductances will automatically give good descriptions of propagation as well as most of the other classical nerve phenomena.

REFERENCES

ADAM, G. 1968. Ionenstrom nach einem Depolarisierenden Sprung im Membran-potential. *Z. Naturforsch.* 23b:181–197.

BEZANILLA, F., E. ROJAS and R. E. TAYLOR. 1970. Time Course of the Sodium Influx in Squid Giant Axon during a Single Voltage Clamp Pulse. *J. Physiol. (London).* 207:151–164.

CHANDLER, W. K., R. FITZHUGH and K. S. COLE. 1962. Theoretical Stability Properties of a Space-clamped Axon. *Biophys. J.* 2:106–127.

CHANGEUX, J. P., J. THIÉRY, Y. TUNG and C. KITTEL. 1967. On the Cooperativity of Biological Membranes. *Proc. Nat. Acad. Sci. USA.* 57:336–341.

COLE, K. S. 1949. Dynamic Electrical Characteristics of the Squid Axon Membrane. *Arch. Sci. Physiol.* 3:253–258.

———. 1968. *Membranes, Ions and Impulses.* Univ. of California Press, Berkeley.

———. 1970. Dielectric Properties of Living Membranes, pp. 1–17. *In* F. Snell, J. Wolken, G. J. Iverson and J. Lam [eds.] *Physical Principles of Biological Membranes.* Gordon and Breach, New York.

COLE, K. S., and H. J. CURTIS. 1939. Electric Impedance of the Squid Giant Axon during Activity. *J. Gen. Physiol.* 22:649–670.

COLE, K. S., and A. L. HODGKIN. 1939. Membrane and Protoplasm Resistance in the Squid Giant Axon. *J. Gen. Physiol.* 22:671–687.

COLE, K. S., and J. W. MOORE. 1960. Ionic Current Measurements in the Squid Giant Axon. *J. Gen. Physiol.* 44:123–167.

COLE, K. S., R. GUTTMAN and F. BEZANILLA. 1970. Nerve Membrane Excitation without Threshold. *Proc. Nat. Acad. Sci. USA.* 65:884–889.

COOLEY, J. W., and F. A. DODGE, Jr. 1966. Digital Computer Solutions for Excitation and Propagation of the Nerve Impulse. *Biophys. J.* 6:583–599.

CURTIS, H. J., and K. S. COLE. 1940. Membrane Action Potentials from the Squid Giant Axon. *J. Cell. Comp. Physiol.* 15:147–157.

FISHMAN, H. M. 1969. Direct Recording of K and Na Current-Potential Characteristics of Squid Axon Membrane. *Nature.* 224:1116–1118.

FITZHUGH, R. 1961. Impulses and Physiological States in Theoretical Models of Nerve Membrane. *Biophys. J.* 1:445–466.

FITZHUGH, R., and H. A. ANTOSIEWICZ. 1959. Automatic Computation of Nerve Excitation. Detailed Corrections and Additions. *J. Soc. Ind. Appl. Math.* 7:447–458.

FRICKE, H. 1923. The Electric Capacity of Cell Suspensions. *Physic. Rev.* 21:708–709.

———. 1925. The Electric Capacity of Suspensions with Special Reference to Blood. *J. Gen. Physiol.* 9:137–152.

HODGKIN, A. L. 1951. The Ionic Basis of Electrical Activity in Nerve and Muscle. *Biol. Rev.* 26:339–409.

HODGKIN, A. L., and A. F. HUXLEY. 1939. Action Potentials Recorded from Inside a Nerve Fibre. *Nature.* 144:710–711.

———. 1952. A Quantitative Description of Membrane Current and its Application to Conduction and Excitation in Nerve. *J. Physiol. (London).* 117:500–544.

HODGKIN, A. L., A. F. HUXLEY and B. KATZ. 1952. Measurement of Current-Voltage Relations in the Membrane of the Giant Axon of *Loligo*. *J. Physiol. (London)*. 116:424–448.

HUXLEY, A. F. 1959. Ion Movements during Nerve Activity. *Ann. N. Y. Acad. Sci.* 81:221–246.

MARMONT, G. 1949. Studies on the Axon Membrane, I. A New Method. *J. Cell. Comp. Physiol.* 34:341–382.

MUELLER, P., and D. O. RUDIN. 1963. Induced Excitability in Reconstituted Cell Membrane Structure. *J. Theo. Biol.* 4:268–280.

STOCKENIUS, W., and D. M. ENGELMAN. 1969. Current Models for the Structure of Biological Membranes. *J. Cell Biol.* 42:613–646.

TAKASHIMA, S., W. R. REDWOOD and T. E. THOMPSON. 1969. Dielectric Properties of Aqueous Suspensions of Homogeneous Phospholipid Vesicles. *Abstracts, III Int. Biophys. Cong.* IUPAB, Cambridge, I. M8.

TASAKI, I. 1968. *Nerve excitation.* C. C. Thomas, Springfield, Ill.

VOLTAGE CLAMP METHODS 5

J. W. MOORE

In the previous chapter, Cole has set out the strategy of the voltage clamp method. In this chapter, we will go into more detail as to the tactics and logistics of voltage clamping. We will review the general concepts, spend some time on both potential and current measurement methods and show how to put them together into a feedback circuit which will control the potential and measure the corresponding current in the membrane of a single nerve fiber. We will also review some of the types of arrangements which have been used for clamping various excitable membranes. Voltage clamp techniques have been written up in some detail in the past and four illustrative references have been selected from the work of Cole and Moore (Cole and Moore, 1960; Moore and Cole, 1963; Moore, 1959; Moore, 1963).

CONTROL OF MEMBRANE POTENTIAL

General Concepts

A generalized circuit for voltage control is shown in Fig. 5-1. The voltage, V, between the membrane and ground is measured by a potentiometric or electrometer amplifier* with a gain of +1. Its output is compared with a set of input command signals at the input of a differential amplifier.† Any

---

* An amplifier which draws negligible current from the potential source being measured.

† Implementation of a particular circuit, using operation amplifiers, will be described later. Note that the convention in Fig. 5-1 is to indicate the sign of a potential change at each input which gives rise to a positive change in the output.

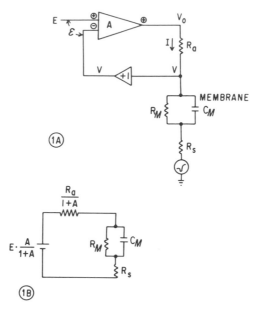

Fig. 5-1 A *A simplified schematic diagram of a generalized voltage clamp of a membrane.* B *The effective equivalent circuit of a voltage clamp.*

small potential difference, ε (between the command signal, E, and the potential, V), is amplified by a large gain, A, to give an output for potential control and labeled V_o. The polarities are so arranged that, if E is larger than V, the output will be increased and, conversely, decreased if E is less than V. This output voltage, V_o, is applied to the membrane (via an access resistance) with a polarity such as to make the potential across the membrane and the control potential the same. The current through the membrane is measured between ground and the small resistance that invariably (at least for the squid axon) seems to be in series with the membrane. The characteristics of the current measuring device will be considered later and separately. There is always a resistance between the output of the control amplifier and the interior surface of the membrane (labeled R_a). This may take the form of a microelectrode or the resistance of a surface of a metal to electrolyte junction. It will also include axoplasm or cytoplasm resistance. The relative contributions of electrodes and cytoplasms will vary from arrangement to arrangement and with cell type and size, but will always have to be considered because they are important aspects in determining the speed and accuracy of the control circuit.

The equations which describe the operation of this control circuit may be developed very simply. The output of the control amplifier is equal to the difference in the input potentials, ε, times the gain of the amplifier*:

$$V_o = \varepsilon \times A = A(E - V). \tag{5-1}$$

Next we can also say that this output potential is distributed between the drop across the access resistance and the membrane potential or:

$$V_o = V_M + R_a I. \tag{5-2}$$

If we solve these two equations for the potential across the membrane, we have:

$$V_M = \frac{EA}{1 + A} - \frac{R_a I}{1 + A}. \tag{5-3}$$

This description applies to the simple circuit shown in Fig. 5-1B, which is spoken of as an "equivalent circuit" for the voltage clamp. This figure shows that the voltage control circuit is equivalent to a battery, in series with a resistance, driving the element to be controlled, in this case an active membrane. Here there is a potential source equal to $E \times A/(1 + A)$. For large values of A, the battery potential approaches E. However, if A is not large (for example, at high frequencies), there is a difference between the command potential and the effective battery driving the membrane. The equivalent resistance in series with the battery is the access resistance R_a divided by the factor $(1 + A)$. A large gain for the amplifier reduces the effect of the access resistance to a very small value. Again reduced performance (less gain) degrades the overall performance of the circuit from ideal.

Thus the element to be controlled is effectively put across the terminals of a battery with a low internal resistance. As the gain of the amplifier increases, the value of the battery approaches the command potential and the effect of the series access resistance becomes minimized. In other words, the potential across the load (the membrane plus R_s) very nearly approaches the command potential and an ideal voltage control is obtained. The small resistance, R_s, in series with the membrane is not affected by the feedback. R_s is associated with the membrane, and the feedback potential is measured across it and the membrane. The access resistance, R_a, is said to be "inside" the negative feedback circuit, while the series resistance, R_s, is said to be "outside" the feedback loop. For the squid membrane, R_s is so closely associated with the membrane that it is extremely difficult to penetrate between it and the membrane even with the finest microelectrode. This

---

* Note that this is true because ε is connected to the + input and V to the − input.

resistance is included for purposes of generality even though it does not seem to be important in the case of some excitable membranes. Where a potential probe can be placed between R_s and the membrane, the spurious $R_s I$ voltage drop may be subtracted from V and a better membrane voltage clamp obtained.

Upon inspection of Fig, 5-1B, one might well raise the question as to why batteries could not be used directly instead of the electronic circuit of Fig. 5-1A. There are several reasons as to why this would not be satisfactory. Among them is the fact that potentials must be switched from one level to another very rapidly in order to introduce step changes to the membrane potential. Secondly, the access resistance, R_a, is frequently of the same order as the magnitude of the membrane resistance, R_M, at least when the membrane is depolarized. This would mean that the potential control would be very poor. By making use of an electronic circuit in which a control amplifier with a very large gain, A, is employed, the access resistance is drastically reduced, as we have shown above.

Potential Measurement

There are several difficulties associated with the measurement of potential across the membrane of a single cell. They become particularly important when a measurement with high absolute accuracy and very high speed is required for voltage clamping. Since it is innately impossible for the voltage control system to be any more accurate than the accuracy of the potential measurement, it is worth considering briefly the pitfalls and the compromises that have to be made.

Frequently the potential must be measured via a high resistance pathway whether it be a 5 to 50 megohm electrolyte-filled micropipette (or microelectrode) or a cytoplasmic resistance, as is the case usually employed in myelinated nerve fibers. This potential access resistance appears to be a part of the source resistance. This puts very stringent requirements on the electronic device which is used to measure the potential. In order to prevent attenuation through a very high input source resistance such as might be encountered in a micropipette, the potential detector input resistance should be many times higher than the 50 megohms of the microelectrode. This clearly rules out any possibility of connecting a microelectrode directly to a cathode ray oscilloscope, which usually has an internal resistance in the order of 1 megohm. Such a low resistance for the detector would inordinately attenuate the potential signal to be measured.

Furthermore, the device used to measure the potential must not inject or withdraw a significant current through the microelectrode and cell. It is apparent that any current through a large resistance microelectrode would insert an error in the potential measurement equal to the product of the microelectrode's resistance and the current flowing through it. If it is desired to keep this error to a reasonable value (e.g., 0.1 mv), the upper limit of current would be 2×10^{-12} amps* because microelectrode resistances may range as high as 50 megohms. Furthermore, one must also ascertain that this current does not disturb the membrane potential of the cell under measurement. For most large cells this is not significant, but it may be for very small cells.

Because the microelectrode is composed of a glass capillary pulled out to a fine point and filled with an electrolyte solution, the thin glass wall near the tip may act as an insulating material between two plates of a capacitor, one plate being the external electrolyte bath and the other being the electrolyte within the capillary. This capacitance amounts to about 1 picofarad ($10^{-12}f$) per mm of immersion of the micropipette. Although this appears to be small compared to many circuit capacitors, it must be remembered that it is in parallel with a tip resistance of many megohms. The product of this capacitance and resistance results in a time constant of response at the input to the amplifier which may be long compared to that needed even for measuring action potentials. For example, a 20 megohm microelectrode in parallel with a 10 picofarad capacitor (1 cm immersion) would have a 200 μsec time constant.

It is possible to compensate for this input capacitance in a potentiometric amplifier by means of positive feedback. A capacitor (C_{fb}) between the input and the output as shown in Fig. 5-2A provides the small current which flows through the wall of the micropipette. For this scheme to be effective, a very high frequency response in the amplifier itself is necessary. Fig. 5-2B shows the equivalent circuit for measurement of a transmembrane potential via a micropipette. The membrane variables are subscripted with an "M", the micropipette tip characteristic with "t" and the amplifier input capacitance is designated by C_a.

These requirements of a very low input current and high frequency response represent a rather formidable set of specifications for the design of an amplifier. However, such specifications have been considered and met and several designs have been published. Furthermore, a number of such potentiometric (or "electrometer") amplifiers have been described (Moore and Gebhart, 1962) and several are now available commercially.

Because considerable confusion has arisen with respect to methods of calibration of the tip resistance and the speed of response of the measuring system via the microelectrode, a brief summary is in order. In Fig. 5-3, three systems are shown which have been proposed and used for such calibrations,

---

* Fortunately special purpose vacuum tubes, called electrometer tubes, and certain types of field effect transistors, called FET's, which fulfil this requirement are available.

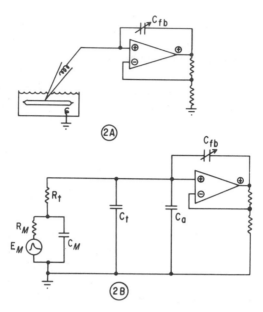

Fig. 5-2 A *Schematic diagram of a membrane potential measurement via a micro-pipette and high impedance electrometer amplifier.* **B** *The equivalent circuit showing the membrane equivalent circuit, the tip resistance R_t, capacitance C_t and the effective input capacitance to the amplifier C_a.*

each amplifier being of the potentiometric type. Some workers have attempted to estimate the ability of the system to measure fast changes in the trans-membrane potential by the method shown in Fig. 5-3A, where the bath is driven with respect to ground by a potential pulse. The equivalent circuit shows that the placement of the micropipette tip capacitor is different from that in the circuit of Fig. 5-2 in which the measurement is made from inside the cell. The configuration of Fig. 5-3A gives a deceptively fast response at the amplifier output because the source adds a capacitive current, proportional to the derivative of the input signal, to that through the resistance.

In their early work, Nastuk and Hodgkin (1950) recognized this potential difficulty and interposed a small grounded loop of wire around the micro-pipette (as in Fig. 5-3B). The loop was placed just high enough above the surface of the bath so that a drop of electrolyte, held (by surface tension) between the micropipette and loop, did not touch the meniscus of the bath. This method offers some improvement and is approximately represented by the equivalent circuit in Fig. 5-3B. However, this is a tedious and time-consuming procedure.

A much more accurate and convenient test of the tip resistance and potential measurement dynamics is afforded by a step current into the input node

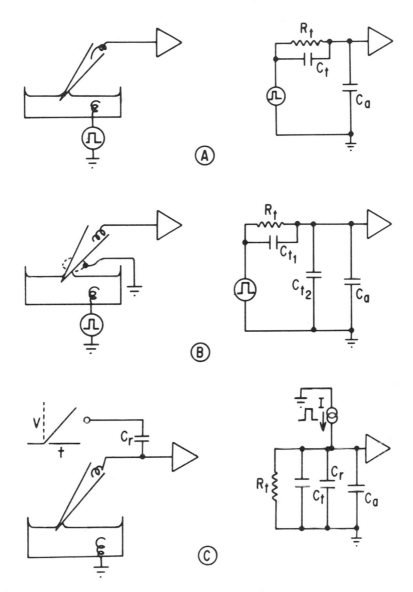

Fig. 5-3 *Methods of testing the response of a micropipette electrode recording system. The method in Fig. 5–3A gives a deceptively fast response because the source adds a capacative current. The method of Nastuk and Hodgkin is shown in 5–3B where a small ground loop of wire surrounds the micropipette. 5–3C then gives a most accurate and convenient test of the tip response by injecting a step current into the amplifier via the capacitor C_r.*

of the amplifier. A number of persons have tried this with various types of current devices. The most successful and generally used has been that of applying a ramp (or linearily rising voltage) to a very small capacitor which is connected to the input, as shown in Fig. 5-3C. As long as the voltage change on the ramp side is large compared to that at the input of the amplifier (this is easily achieved), the current through the capacitor is essentially a step function. This method of Lettvin, Howland and Gesteland (1958) is applicable before, during and after penetration. The ramp may be initiated early during an oscilloscope sweep so that the calibration of the tip resistance and system response appears with a membrane potential record on each sweep.

Current Measurement

Currents are most often measured by being passed through a standard resistor, transforming them to a voltage as is shown in Fig. 5-4A. The

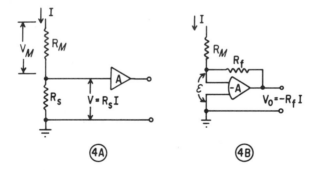

Fig. 5-4 *Two methods of measurement of current. A conventional method is shown in 5–4A. The operational amplifier method in 5–4B approaches the ideal because no significant potential drop appears at the input of the amplifier.*

standard series resistor R_s is made small enough so that the voltage generated across it is small compared to the voltage across the load (in our case, the membrane). A compromise must always be made between having the signal proportional to current large enough to be sufficient to drive an oscilloscope and small enough not to introduce an appreciable error into the measurement of the V_M. The compromise may be affected by using either a high sensitivity oscilloscope or an amplifier A between the series resistance and the oscilloscope. However, there is a better way to arrange these elements, namely by putting the resistor in the feedback path of an operational amplifier as shown in the configuration in Fig. 5-4B. The operational amplifier is chosen to have a current at its input terminal which is very much smaller than any signal current to be measured. Our convention for operational amplifier circuits will be that the lower input terminal is connected to ground and the upper is signal input. A positive signal input at the upper input terminal causes the output to go negative. Thus the amplifier symbol is shown with a gain of " $-A$ ". Input current cannot enter the amplifier but goes on through the feedback resistance R_f. Thus the input current is approximately equal to the current through R_f. Furthermore, the output voltage of the amplifier is simply equal to the voltage across the feedback resistor, $-R_fI$, plus a very small signal ε existing at the input of the amplifier which causes the output signal. That is to say the amplifier output is

$$V_o = -R_fI + \varepsilon, \qquad (5\text{-}4)$$

and it is also

$$V_o = -\varepsilon A. \qquad (5\text{-}5)$$

Because amplifiers with gains in excess of 10^6 are readily available, ε can be made negligible in comparison with V_o* and we have

$$V_o \simeq -R_fI. \qquad (5\text{-}6)$$

There are several advantages to this system which, incidentally, I have used for many years now.

1. It does not introduce an appreciable error into the measurement of V_M.
2. Large output signals with low resistance are available to drive cathode ray oscilloscopes, recorders, plotters or other instruments.
3. An area variation from axon to axon can be easily accommodated because a potentiometer in the feedback network, as shown in Fig. 5-5, can be set proportional to the area, giving a voltage proportional to current density at the output of the amplifier.
4. The potentiometer output may be automatically set to be proportional to the membrane area.†

---

* To satisfy ourselves that the input ε is in fact negligible in Eq. (5-4), let us calculate its value for a typical situation. Generally, operational amplifiers today are limited to a ± 10 v output range. Assume the maximum product R_fI of 10 v and the gain of the amplifier at 10^6. In this case ε is equal to 10 μv. This is also quite small even compared to ± 200 mv, the normal operating range of excitable biological membranes. The maximum value of ε, only 10 μv, is also small compared to the error introduced by having in series a standard resistor which is large enough to give reasonable oscilloscope deflections.

† For example, in the sucrose gap to be described later, the area may fluctuate a bit with the rate of flow of solutions, but measurement of the effective area may be accomplished between test pulses, and a servomotor may be used to automatically adjust the area potentiometer.

5. The speed of measurement of the current can easily be made in the fractional μsec range by using the very high speed operational amplifiers available today.

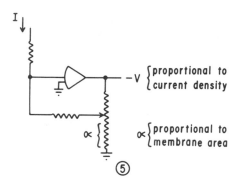

Fig. 5-5 *Method of measurement of membrane current density where the wiper on a potentiometer is set at a resistance (from ground) proportional to membrane area. This results in a voltage proportional to current density at the output of the amplifier.*

Control Amplifier

Operational amplifiers may also be used most conveniently and effectively in the control of the membrane potential. Fig. 5-6 shows a conventional operational amplifier circuit for giving an output signal proportional to the sum of several input signals. Choosing an amplifier which draws no appreciable current at the input, we may write an expression for Kirchhoff's Law

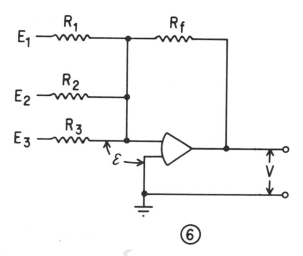

Fig. 5-6 *Method of summation of potentials with an operational amplifier.*

which states that the sum of the input currents equals the current through the feedback resistor.

$$I_{in} = \frac{E_1 - \varepsilon}{R_1} + \frac{E_2 - \varepsilon}{R_2} + \frac{E_3 - \varepsilon}{R_3} = \frac{\varepsilon - V}{R_f} . \tag{5-7}$$

Again, the configuration of the circuit is such that the potential at the upper input is held very near ground so that the small potential difference between the two inputs, ε, is held very close to zero. This simplifies the above equation to the following form:

$$\frac{E_1}{R_1} + \frac{E_2}{R_2} + \frac{E_3}{R_3} = \frac{-V}{R_f} . \tag{5-8}$$

Alternatively expressed,

$$V = -E_1\left(\frac{R_f}{R_1}\right) - E_2\left(\frac{R_f}{R_2}\right) - E_3\left(\frac{R_f}{R_3}\right). \tag{5-9}$$

Thus it is clear such a control circuit may sum a number of independent and simultaneous inputs without interaction. This, of course, is most convenient for the voltage control circuit. Instead of tying the feedback resistor to the output of its operational amplifier it may be connected to a voltage which is a precise representation of the potential across the membrane.

The complete control circuit, using the operational amplifier concept is shown in Fig, 5-7. Various potential commands are entered through weighting

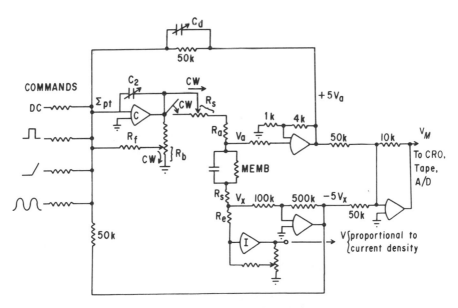

Fig. 5-7 *Schematic diagram of a complete voltage clamp circuit for an axon membrane.*

resistances to the summing point of the control amplifier. The membrane potential is measured from the difference between the potential just inside the membrane and that just outside. This is implemented by using a high input impedance amplifier to measure the inside potential V_a and a normal operational amplifier configuration (which inverts the polarity) to measure the potential outside V_x through a relatively low resistance electrode. The sum of these two potentials may then be taken (with appropriate scaling) by another operational amplifier which generates a signal V_M for display on a cathode ray oscilloscope, for tape recording, for plotting or for other data collection devices. These two signals, internal and external membrane potentials, are also sent back through weighting resistances to the summing point ($\sum$ pt) of the control amplifier. I have normally used a conservative scheme to avoid difficulties in oscillations and unintended commands by starting with a low gain on the control amplifier and by having the amplifier disconnected from the membrane. Starting with a rotary switch in the extreme counterclockwise position, it is turned clockwise (cw), closing the switch, and then reducing R_s and R_b simultaneously. When the rotary switch is fully clockwise, there is no resistance between the control amplifier output and the current electrode. Furthermore, the feedback resistance R_f is now returned to ground potential, resulting in no current flow through it. Thus it no longer affects the gain of the amplifier. The only remaining feedback signals are through the resistances from the internal and external potential (and the capacitors C_d and C_2 whose insertion will be considered later).

The current through the membrane is measured by amplifier I. The feedback potentiometer is set proportionally to the area of the membrane as observed under a microscope. This provides a voltage at output of the amplifier which is directly proportional to current density and ready for display on an oscilloscope or for recording, etc.

The fact that voltages proportional to both current density and membrane potential are obtained from operational amplifiers means that these signals are available from sources of low resistance and may be used to drive a number of recording instruments without any affect on the control or measuring circuits.

Controlling the Dynamic Characteristics of the Voltage Clamp

Dynamic problems such as oscillations (or the tendency towards this) following a step change in input potential are common in any negative feedback circuit which has more than one time constant in the feedback path. In this control circuit (Fig. 5-7) there are at least two significant time constants which tend to make the system oscillate. Special attention must be paid to compensation for these in order to achieve a reasonably smooth voltage step across the membrane.

The problems may be illustrated rather simply if we consider the open loop situation; that is, when the potential across the membrane is measured but not connected back to the control amplifier. In general, it can be assumed that the control amplifier responds quite rapidly. Let us consider that a step at the input results in a step of voltage at the output. However, because of the capacitance across the membrane and the access resistance in series, the response of voltage across the membrane is a first order lag, as shown in Fig. 5-8 where the responses and time are normalized. In addition, there is usually a lag in the measurement of membrane potential because of the capacitance at the input of the amplifier. When the potential is measured via a microelectrode with a capacitance shunting the tip resistance (which was previously discussed), the resulting measured membrane potential will have a second order lag response; that is to say that the response is delayed or has a sigmoid or "s" shape.

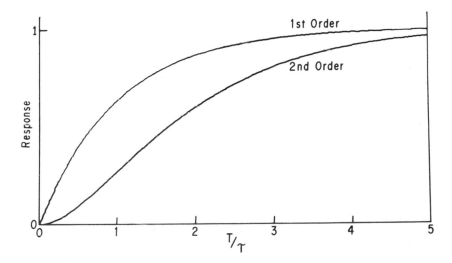

Fig. 5-8 *Open loop dynamic responses of a voltage clamp system. Without frequency compensation, the open loop response of the system appears as a sigmoid such as the second order curve shown. Compensation must be introduced to make it approach the response of the first order curve in order to have stability from oscillations when feedback is applied.*

When such a signal is fed back to the control amplifier with the feedback loop closed, this system will tend to oscillate. The amplitude and frequency of oscillation will depend on the delay or shape of the sigmoid curve. In order to make the system follow a step function input smoothly, the open loop response measured for the membrane potential must closely approximate a first order lag. This can be achieved by taking a derivative of the membrane potential signal and adding it to the measured membrane potential itself. In Fig. 5-7, a capacitor, C_d, supplies current to the control amplifier proportional to the derivative of 5 V_a. Such a compensation for the delay in measurement of the membrane potential is generally sufficient for a second order (or two time constants) linear system. However, in practice there are usually more than two time constants in such a control circuit as shown in Fig. 5-7. For example, there are actually two time constants in the preamplifier and one to two in the control amplifier. The membrane capacity and access resistance introduce a major time constant and stray capacitances within the circuit may add additional ones. These different time constants may range over several orders of magnitude. In general, there always appear to be more than two time constants, but they may be dealt with in a very pragmatic way; the system is connected and one adjusts the amount of derivative added to the membrane potential feedback signal and the band width of the control amplifier by varying a capacity, C_2, from output to feedback point on it until one achieves the optimum or a best compromise in response.

TYPES OF VOLTAGE CLAMP ARRANGEMENTS

Axial Wire for Large Nonmyelinated Axons

The squid giant axon is large enough to allow placement of more than one electrode along the axis. This system was first used by Marmont (1949) and Cole (1949), exploited later by Hodgkin, Huxley and Katz (1952) and further developed by Cole and Moore (Fig. 5-9 shows our schematic arrangement). This method depends on producing a space clamp for a large area of membrane which is maintained at uniform potential by means of an internal short circuit achieved through the axial wire electrode. Cole devoted a great deal of effort to find out just how to prepare the surface of an axial wire so as to achieve a low surface impedance as well as setting the limits for what this surface impedance could be in order to meet the criteria for a space clamp. Cole had originally used a single axial electrode to measure the potential as well as to pass the current. This had bad polarization problems associated with the passage of the rather large currents available from a membrane. Hodgkin, Huxley and Katz improved on this situation by using

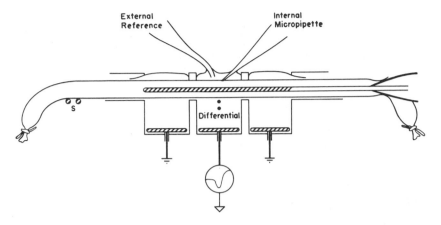

Fig. 5-9 *Schematic diagram of the voltage clamp of Moore and Cole (1963) using an axial wire for current and a micropipette just inside and a minipipette just outside the membrane for potential measurement.*

a second electrode (wrapped around current electrode) for measuring the potential. The potential electrode did not pass current and therefore was not subject to the RI drop associated with Cole's original electrode. Later Cole and Moore used a microelectrode to just penetrate the cell membrane and a reference "millielectrode" just outside the membrane. Thus, the voltage drops across the external sea water and the internal axoplasmic resistances, as well as potential drops across the current carrying electrodes, were eliminated. This work has been described in great detail (Cole and Moore, 1960).

Myelinated Axons

A single myelinated nerve fiber in a voltage clamp arrangement adapted from Dodge and Frankenhaeuser (1958) is schematically represented and an equivalent circuit is shown in Fig. 5-10. The axoplasmic internodal resistance of about 40 megohms offers access (on the right) for potential measurement and (on the left) for current injection. The internodal axoplasmic resistance is approximately the same as the resistance of the exposed membrane at a node. In order for the circuit shown to measure the potential across the central node, the external leakage resistance along the outside surface of the myelin must be made very much larger than the sum of the resistances of the node and internodal axoplasm (a total of about 80 megohms). Frankenhaeuser and Dodge employed two vaseline seals between the central and right nodes. Although the resistance of these vaseline seals was only 5 or 10 megohms each, a negative feedback circuit was used to effectively multiply the resist-

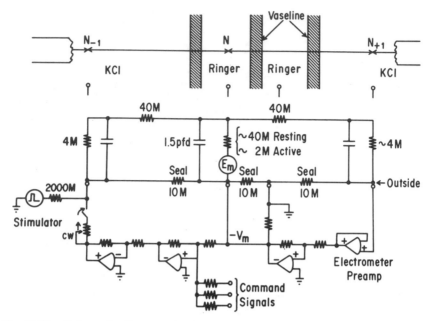

Fig. 5-10 *Schematic diagram of a single myelinated nerve fiber in a voltage clamp arrangement adapted from Dodge and Frankenhaeuser.*

ance across one seal by such a large factor that a high precision measurement of the potential node was possible. This can be achieved if the input resistance of the potentiometric preamplifier is also very much larger than 80 megohms. An amplifier suitable for recording potentials through a micropipette will also be effective here.

Tasaki and Bak (1958) were able to achieve a reasonable potential measurement with an air gap between the pool for the central node and the pool for an adjacent node to which his potentiometric amplifier was connected. Very careful control of the humidity was required.

The happy advantage of using a preparation of this sort lies in the fact that one has a natural access path of the axoplasm through which to measure potential and inject current rather than having to use the microelectrodes which must be physically forced through the membrane. It is usually difficult to penetrate a single node with a micropipette without injury or, at least, relatively rapid degradation of activity. This arrangement allows one to use large or macroelectrodes in the three pools and avoids the difficulties and "witchcraft" associated with making, filling and inserting micropipettes. Furthermore, it also avoids problems of space clamping (in which it has been pointed out that great care must be used to produce an axial wire whose surface has a uniformly low resistance). Because the node is so short (never more than a few microns), the membrane potential is relatively uniform in spite of the fact that there is a longitudinal voltage gradient (dv/dx) resulting from the entrance of all of the membrane current from one end. A very rough estimate of the maximum potential difference between the ends of a node lies in the neighborhood of 5 or 10 mv.

In summary, use of this preparation allows one to exchange the problems of making electrodes (micropipettes and axial wires with low surface resistance) for the problems associated with a very high access resistance. For this case, where the access resistance equals the resting nodal membrane, only one-half of the control amplifier output appears across the nodal membrane (refer to Fig. 5-1). Furthermore, the time necessary to charge the nodal membrane capacitor through a very large resistance is long. This means that the voltage across the nodal membrane is difficult to control because of the attenuation and delay of the control amplifier output. This difficulty is exacerbated when the membrane resistance is drastically reduced upon depolarization.

Sucrose Gap for Nonmyelinated Axons

The experimental convenience of a sucrose gap appeared so attractive that Julian, Moore and Goldman (1962) worked out this methodology for the intermediate sized lobster axon, using a sucrose solution to simulate a myelin sheath. At the left of Fig. 5-11 is a photograph of a lobster nerve threaded through two holes in two lucite walls and, on the right, sea water flows from top to bottom and sucrose is flowing through small holes in the walls and out along the axon. The boundaries seen are the changes of index of refraction at the junction of the solutions. The width of this artificial node may be adjusted by controlling the flow rate for the sucrose and seawater. Thus we were able to make a non-myelinated fiber simulate a myelinated one by means of a flowing sucrose solution of high resistance.*

Fig. 5-12 shows a schematic diagram of the solution flow pattern (above) and the electrical equivalent circuit (below). The membrane potential, dominated by the potassium equilibrium potential, is reduced to nearly zero in the right pool by flowing isosmotic KCl. This treatment also increases the potassium conductance so effectively that the membrane as indicated is as a "short circuit" in the KCl pool. As long as the sucrose insulation is high and the resistance of the potentiometric amplifier is high compared to both the axial resistance and the resistance across the sucrose stream, the

---

* Sucrose solutions are made of deionized water and are passed through a filter and a deionizer column before use.

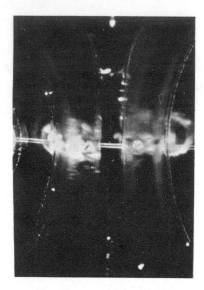

Fig. 5-11 *Photograph of a lobster nonmyelinated axon threaded through two lucite partitions. On the left, no sucrose is flowing. On the right, sucrose is flowing, approaching the axon from below and through the lucite partitions. Note the open exposed segment or artificial node on the axon in the center. (From Julian, Moore and Goldman, 1962.)*

amplifier output will be a faithful measure of the potential at the inside of the membrane at the center " node." It has been routinely possible to achieve a resistance of 10 to 30 megohms between the central and right pools. Thus, for a "nodal" membrane resistance of the order of 50 K- to 500 Kohms, there is essentially no attenuation of the potential measured between the right pool and ground. This potential very closely approximates that across the "nodal" membrane because the central pool is held at a virtual ground, being connected to the input of a current-measuring operational amplifier (as previously described). The membrane area may be estimated by microscopic observation, the feedback point of the current amplifier adjusted to be proportional, giving an output voltage from this amplifier proportional to membrane current density. This node may be stimulated to give an action potential when current is injected through a high resistance shown at the lower left. It may be voltage clamped by closing the switch and short-circuiting the resistor connected to the output of the "V clamp" amplifier. The voltage measured across the node is forced to match a set of command potentials as we have indicated before.

The sucrose gap voltage clamp offers several advantages for large non-myelinated fibers. Notice that the access resistance is much less than the forty megohm resistance of the myelinated fiber. Because of this, the ratio

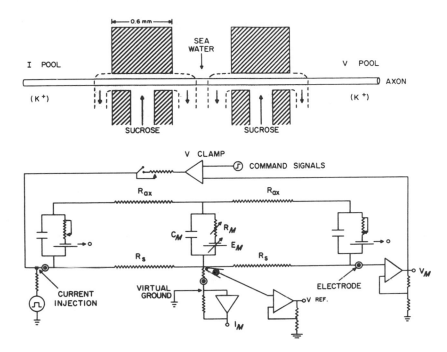

Fig. 5-12 *Schematic diagram of the sucrose gap system showing the flow pattern above and an electrical equivalent circuit below. (From Julian, Moore and Goldman, 1962.)*

of access to controlled resistance may be quite low. For a squid axon, this ratio may be as low as 0.02 (10KΩ/500KΩ); this allows very good clamping. Very short lengths of axon are exposed to the electrolyte solution in the central pool, usually 1/4 to 1/2 of a diameter, so as to minimize voltage difference from one edge of a node to the other. Although certain experimental conditions may destroy a single "node," the balance of the axon is not affected.

When experimental treatments, time, or errors have degraded the performance of a "node," a new one may be obtained simply by pulling the axon through the streams. The search for a new patch with good activity may be accomplished very rapidly. A squid axon of several centimeters may provide up to 8 to 10 fresh nodes while there are usually three or four of these in a lobster axon. Usually the axon is moved enough to avoid making a "node" from an area which had been surrounded by flowing sucrose because of the possibility of altered internal ion content in that area. That is, any ions which left the axoplasm, going through the membrane into the sucrose, would simply have been washed away.

The major inconvenience with the sucrose gap results from this process which leaches out the axoplasmic ions and increases the axoplasm resistance so that, after some time, the clamp is no longer really effective. For lobster axons this may be 20 minutes to one-half hour. For squid axons with larger diameters, the surface-to-volume ratio is much more favorable, and experiments of an hour or two present no significant problems. This process of leaching out ions with the sucrose stream also means that ions have to move in from other areas to replace them. This causes a minute current to flow in the central node, which results in membrane hyperpolarization. The amount of hyperpolarization usually depends on the length of the artificial node; the narrower it is, the more the hyperpolarization. For voltage clamping experiments the hyperpolarization generally presents no problem, because it is standard procedure to hyperpolarize the membrane so as to completely "activate" the sodium conductance system. We have seen no appreciable difference in axon performance between those hyperpolarized electrically or by this method. Thus, while one would not want to use the sucrose gap method to measure resting potentials, it is quite appropriate for voltage clamping.

Another inconvenience of the sucrose gap method is that the flow of the streams is not completely controllable. Tiny bits of dirt can cause fluctuations in the flow pattern, changing the area of the node. We have recently circumvented that problem by measuring the area as proportional to the effective capacity of the nodal membrane, during the time when a test pulse is not being applied. That is to say, we are "time-sharing" the axon for the two measurements (capacitance, during the period between pulses [1–2 sec] and voltage clamp currents, during the few milliseconds when the pulse is on). The nodal membrane capacity is measured by its response to a 10–20 KC sine wave and used to automatically set the area potentiometer so that the output is always proportional to current density, regardless of stream fluctuations.

Voltage Clamping Other Systems

Let us now briefly consider the problems associated with applying a voltage clamp to other systems. Cells of a few microns in diameter and of nearly spherical shape may be nearly isopotential throughout. It is possible to measure the internal potential with one microelectrode, and to pass enough current through a second microelectrode to completely change the membrane potential throughout the cell and thus effect a voltage clamp. If the object to be clamped is the soma of a neuron, there are additional problems associated with the additional membrane of the axon and the dendritic tree. Not only is it difficult to supply the additional current and also voltage clamp loosely coupled membranes, but if the extraneous membrane is excitable, it may introduce spurious current patterns when activity generated in these regions propagate back into the soma.

Although it would not be suitable to clamp a cylindrical structure such as nonmyelinated axon or muscle with a point source of current because currents would spread along the cable in a nonuniform manner and give rise to a nonuniform potential, it is possible to use this method to examine the currents in a muscle motor end-plate region or a synaptic region of a nerve. In such cases, where the current of interest is known to be limited to a very small patch, it is both possible and useful to clamp the membrane at a single point of interest; it has been accomplished both at motor end-plates (Gage and Armstrong, 1968; Takeuchi and Takeuchi, 1960), and at the synaptic region in the squid giant axon (Hagiwara and Tasaki, 1958; Takeuchi and Takeuchi, 1962). This method will work admirably for the study of synaptic currents at the resting level. However, if it is desirable to change the potential of the post-synaptic fiber to note the effect of hyperpolarization or depolarization on the synaptic current, one must have a much more powerful source of current than a microelectrode. It is simply impossible to force enough current through a multimegohm micropipette to supply the amount of current needed to go down the giant axon cable just to maintain the potential in the synaptic region at a displaced level. Gage and I (1969) have used a wire, insulated except for a small area at the tip, which was pushed down the axis of the post-synaptic fiber. This had much lower resistance than a micropipette and it was possible to supply the amount of current required to control the potential at the post synaptic region. We were then able to examine the variations of the synaptic current as a function of the polarization of the post synaptic membrane.

CHARACTERISTIC CURRENT PATTERNS IN A VOLTAGE CLAMPED MEMBRANE

Dr. Cole has presented (Chapter 4) examples of the current patterns which he originally observed. One of the fundamental advantages of his voltage clamp method was that all discontinuities in current were eliminated. Every current curve was completely continuous with time and small voltage changes made only small changes in the current. A family of axon membrane currents in response to voltage clamp steps is shown in Fig. 5-13. The different components of a current pattern are as follows:

1. First there is a fast current transient through the membrane capacitance. This component is not faithfully reproduced for large voltage steps because of speed limitations in the data acquisition system.
2. This is immediately followed by a component which has been called a leakage current, carried by as yet unidentified ions. It is normally small compared to the other components.

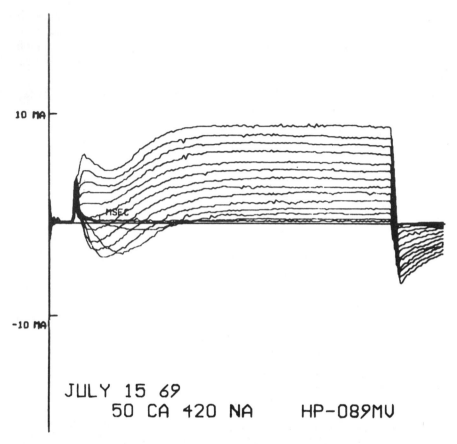

Fig. 5-13 *A typical family of axon membrane currents in response to voltage clamp steps.*

3. This is followed by a transient phase of current normally carried by sodium ions. The direction of this transient current flow reverses at, or near, the sodium equilibrium potential.
4. This is followed, in turn, by a potassium current which builds slowly to a maintained level.

Thus the voltage clamp step spreads out, in a temporal sequence, the four different current components. Three appear to be related to an emf resulting from concentration gradients, each with its own internal conductance or resistance. In the case of the potassium and sodium current, the conductances change with both time and voltage.

Notice the differences between the current patterns in Fig. 5-13 and those of Cole's original experiment where there was polarization of the electrodes.

When electrode polarization is eliminated, the potassium current is maintained nearly constant as long as the potential is maintained. It was found that the early transient current reversed in direction at, or close to, the equilibrium potential for sodium. Moore and Adelman (1961) have found this to be so predictable that the procedure could be turned around so as to use this methodology to measure the internal sodium concentration. In contrast, the potential at which the last steady potassium current reverses (once it has been allowed to be fully "turned-on") is quite variable. This appears to be a consequence of the piling up of potassium in the space between the Schwann cell and the axon itself, as first noted by Frankenhaeuser and Hodgkin (1956).

These current patterns are usefully and frequently summarized by plotting both the early peak current and the steady state current against the potential during a pulse as shown in Fig. 5-14. If the leakage current is subtracted

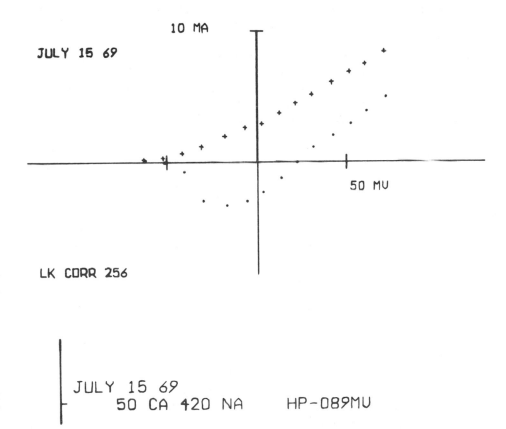

Fig. 5-14 *Membrane current-voltage relations. The peak and steady-state currents, with leakage subtracted, are plotted as a function of the potential during the step.*

from the total ionic current, the intersection of the peak transient current with the abscissa will be at the sodium equilibrium potential. Drug-, ion-, or toxin-induced changes in the slope or intercepts of such current voltage characteristic curves are easily detected. This accounts for the popularity of this convenient summary form of plotting. However it does not show kinetic changes. They must be presented in other forms.

REFERENCES

COLE, K. S. 1949. Dynamic Electrical Characteristics of the Squid Axon Membrane. *Arch. Sci. Physiol.* 3:253.

COLE, K. S., and J. W. MOORE. 1960. Ionic Current Measurement in the Squid Giant Axon Membrane. *J. Gen. Physiol.* 44:123.

DODGE, F A., and B. FRANKENHAEUSER. 1958. Membrane Currents in Isolated Frog Nerve Fibre under Voltage Clamp Conditions. *J. Physiol. (London).* 143:76.

FRANKENHAEUSER, B., and A. L. HODGKIN. 1956. The After-effects of Impulses in the Giant Nerve Fibres of *Loligo. J. Physiol. (London).* 131:341.

GAGE, P. W., and C. M. ARMSTRONG. 1968. Miniature End-plate Currents in Voltage Clamped Muscle Fibres. *Nature.* 218:363.

GAGE, P. W., and J. W. MOORE. 1969. Synaptic Current at the Squid Giant Synapse. *Amer. Assoc. Adv. Sci.* 166:510.

HAGIWARA, S., and I. TASAKI. 1958. A Study of the Mechanism of Impulse Transmission across the Giant Synapse of the Squid. *J. Physiol. (London).* 143:114.

HODGKIN, A. L., A. F. HUXLEY and B. KATZ. 1952. Measurement of Current-Voltage Relations in the Membrane of the Giant Axon of *Loligo. J. Physiol. (London).* 116:424.

JULIAN, F. J., J. W. MOORE and D. E. GOLDMAN. 1962. Current-Voltage Relations in the Lobster Giant Axon Membrane under Voltage Clamp Conditions. *J. Gen. Physiol.* 45:1217.

LETTVIN, J. Y., B. HOWLAND and R. C. GESTELAND. 1958. Footnotes on a Headstage. *IRE Trans. Med. Electronics PGME.* 10:26.

MARMONT, G. 1949. Studies on the Axon Membrane. *J. Cell. Comp. Physiol.* 34:351.

MOORE, J. W. 1959. Electronic Control of Some Active Bioelectric Membranes. *Proc. IRE.* 47:1869.

_____. 1963. Operational Amplifiers. *In* W. L. Nastuk [ed.] *Physical Techniques in Biological Research*, Vol. 6. Academic Press, New York.

MOORE, J. W. and W. J. ADELMAN, Jr. 1961. Electronic Measurement of the Intracellular Concentration and Net Flux of Sodium in the Squid Axon. *J. Gen. Physiol.* 45:77.

MOORE, J. W., and K. S. COLE. 1963. Voltage Clamp Techniques, p. 263. *In* W. L. Nastuk [ed.] *Physical Techniques in Biological Research,* Vol. 6. Academic Press, New York.

MOORE, J. W., and J. GEBHART. 1962. Stabilized Wide-Band Potentiometric Pre-amplifiers. *Proc. IRE.* 50:1928.

NASTUK, W. L., and A. L. HODGKIN. 1950. The Electrical Activity of Single Muscle Fibers. *J. Cell. Comp. Physiol.* 35:39.

TAKEUCHI, A., and N. TAKEUCHI. 1960. Further Analysis of Relationship between End-plate Potential and End-plate Current. *J. Neurophysiol.* 154:52.

————. 1962. Electrical Charges in the Pre- and Postsynaptic Axons of the Giant Synapse of *Loligo. J. Gen. Physiol.* 45:1181.

TASAKI, I., and A. F. BAK. 1958. Current-Voltage Relations of Single Nodes of Ranvier as Examined by Voltage-Clamp Technique. *J. Neurophysiol.* 21:124.

DESCRIPTION OF AXON MEMBRANE IONIC CONDUCTANCES AND CURRENTS

6

Y. PALTI

THE VOLTAGE CLAMP CONCEPT

The generation of the nerve impulse and its propagation along the nerve fiber are associated with the simultaneous variation of numerous parameters. Furthermore, each of these parameters is usually a function of more than one variable. For example, the membrane current is determined by the membrane potential and conductance which are themselves a function of time and position along the length of the fiber. Because of these complexities, an analysis of the excitation process and the basic mechanisms underlying it has progressed slowly for many years.

As in any complex system of this type, analysis can be greatly facilitated by eliminating one or more of the variables. This can be done by forcing some of the variables to remain constant, while others change. This concept was first applied to neurophysiology by Marmont (1949) and Cole (1949). Initially, membrane current, I_M, was held constant. However, it soon became apparent that analysis of the currents generated when the membrane potential, E_M, was clamped to a constant value is more profitable. This technique, the voltage clamp technique, rapidly developed and has become one of the most valuable tools for studying membrane mechanisms and ionic currents associated with the generation and propagation of the nerve impulse.

Basically, the voltage clamp technique involves temporal and spatial fixation of membrane potential. The temporal aspect of the voltage fixation is achieved by means of a feedback circuit (see chapter 5). The spatial fixation or space clamp is achieved by keeping the external electrodes at ground potential and introducing a long platinum black electrode (axial wire) through the whole length of the axon. In this way, while the whole external surface of the axon is kept at practically ground potential, the axoplasm is short-circuited by the axial wire. Potential gradients along the outer and inner axon surfaces must therefore be negligible. Under these conditions, any variation in membrane potential occurs simultaneously along the whole length of axon under investigation, making E_M independent of position along the fiber. As a result, the membrane current density is independent of position along the fiber and the electric field is homogeneous.

In a voltage clamp experiment, E_M is held constant while I_M is measured by means of an operational amplifier. From the known value of E_M and the experimental values of I_M one can define the changes, with time, in membrane conductance, g_M. By stepping E_M to different values, one can determine g_M as a function of both E_M and time.

MEMBRANE CONDUCTANCE AND CURRENT

On the basis of such experiments in the squid giant axon, Hodgkin and Huxley (1952a, b) concluded that in a general sense (see below), the axon membrane does indeed obey Ohm's law:

$$I_M = g_M \Delta E_M, \tag{6-1}$$

where ΔE_M is the net driving force acting on the ions, and g_M the membrane chord conductance (see page 174).

Since the bulk of the axon membrane consists of a dielectric material, any change in E_M* is associated with a capacitive current in addition to the ionic currents (Cole, 1968). The total membrane current, I_M, is thus the sum of the currents carried by the individual ions and the displacement (capacitive) current, I_C. Let C_M be the membrane capacity. The capacitative current is given by (Hodgkin, Huxley and Katz, 1952):

$$I_C = C_M \cdot dE_M/dt. \tag{6-2}$$

From Eq. (6-1) and (6-2) the total membrane current would be:

$$I_M = (C_M \cdot dE_M/dt) + \sum_{n=1}^{i} g_i \Delta E_i, \tag{6-3}$$

* Note that E_M is defined here in absolute terms in contrast to the original definition of Hodgkin and Huxley, who defined the resting potential, V_M, as zero: i.e.

$$E_M = -V_M - 60.$$

where i refers to a permeable ion. For any permeable ion, g_i is greater than zero. Therefore, any specific ionic current will be zero only when ΔE_i equals zero, that is, when the force acting on the ion is zero. The measured membrane potential at which a specific ionic current equals zero is usually referred to as the reversal potential, E_{rev}, of the ion, i^*. It follows that when $E_M = E_{rev}$, the ionic current is zero. Consequently, the force driving I_i will be $E_M - E_{rev}$. Eq. (6-1) can now take the form

$$I_i = g_i(E_M - E_{rev}). \tag{6-4}$$

Using this relationship, Hodgkin and Huxley derived the following relationship for the major membrane current components:

$$I_M = I_C + I_{Na} + I_K + I_L$$
$$= C_M \cdot dE_M/dt + g_{Na}(E_M - E_{Na}) + g_K(E_M - E_K) + g_L(E_M - E_L), \tag{6-5}$$

where g_L refers to the lumped conductance of all ions other than sodium and potassium. E_{Na} and E_K are the sodium and potassium reversal potentials and E_L is the potential at which the net current of all the other ions equals zero.

Of the three conductances specified in Eq. (6-5), Hodgkin and Huxley found only g_L to be, to a good approximation, a simple ohmic conductance. That is, g_L was found to be independent of E_M and time, while g_K and g_{Na} were dependent on both.[†]

POTASSIUM CONDUCTANCE AND CURRENTS

Let us examine the experimental currents generated by the axon membrane under voltage clamp when sodium currents have been eliminated, for example, by TTX (Moore, Blaustein, Anderson and Narahashi, 1967; Hille, 1968).[‡] A typical membrane current generated under such conditions upon stepping E_M from a resting value of -60 mv to about zero is illustrated

---

[*] In most biological systems the reversal potential of an ion is close but not equal to its equilibrium potential. The equilibrium potential, E_{eq}, is given by (Cole, 1968):

$$E_{eq} = \frac{RT}{zF} \ln \frac{a_1}{a_2}$$

where a_1 and a_2 are the ion activities at the two sides of the membrane.

[†] It was later demonstrated that g_L was voltage dependent. Such a dependency was predicted by Goldman in 1943.

[‡] Since I_L is so much smaller than I_K, for the moment we can ignore I_L.

in Fig. 6-1B. The capacitive current, which lasts a few microseconds (Hodgkin and Huxley, 1952a), can hardly be seen with the given time scale. On the other hand, the onset of the ionic current is relatively slow. After an initial delay (Cole and Moore, 1960), I_K gradually increases to reach a plateau within five msec. Upon the return of E_M to its original value, the potassium current decays exponentially. Stepping the membrane to different potentials results in somewhat different current turn-on and turn-off rates and very different steady-state values (Fig. 6-1A, C). Since under these conditions E_M is held constant, the changes in I_M reflect changes in g_K.

To describe the potential and time dependencies of ionic conductances, g_i, Hodgkin and Huxley (1952c) introduced a constant, $\bar{g}_i$, which is the maximal possible value of g_i (usually in mmho/cm^2). This definition of $\bar{g}_i$ is based on the assumption that there is a finite number of channels (carriers or sites, depending on the membrane model one choses) in the membrane, through which ions pass at some specific rate and when all these channels are conducting simultaneously, one gets the maximal conductance: $\bar{g}_i$. The

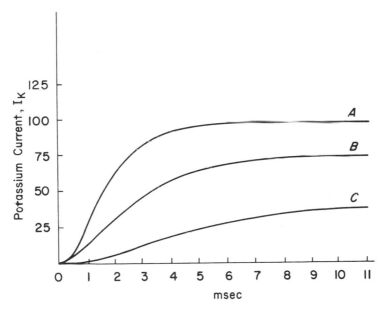

Fig. 6-1 *Plot of potassium current (expressed as the percentage of the maximal obtainable I_K) as a function of time. The curve begins after onset of depolarization in the voltage clamped squid axon. Data was obtained from axons bathed in choline sea water (sodium free) at 6–7° C; prior to the depolarizing step, E_M was held at resting level. In A the depolarizing step was 109 mv, in B 63 mv and in C 32 mv (From Hodgkin and Huxley, 1952c.)*

temporal and potential dependencies of any g_i are described by means of time and potential dependent parameters, m, n and h, which vary between 0 and 1. The parameters are so defined that, when introduced properly and raised to appropriate powers in the current equations, they represent the fraction of active channels. Thus, by multiplying $\bar{g}_i$ by the appropriate parameters, one gets the actual membrane conductance at any time and at any potential.*

To describe the experimental data obtained for potassium currents, a parameter n is introduced and raised to the fourth power. Since n^4 represents the fraction of active membrane ion channels passing potassium, it is defined experimentally as the ratio between I_K at any given set of conditions, to the maximal value of I_K obtainable: I_K/I_{Kmax}. Hodgkin and Huxley's equation is as follows:

$$I_K = \bar{g}_K n^4 (E_M - E_K). \tag{6-6}$$

At the start of depolarization n is a small fraction. When this fraction is raised to the fourth power, its value is considerably reduced. As n approaches unity, elevating it to the fourth power has very little effect on its magnitude. In this way, Eq. (6-6) accounts for the delay in the turn-on of I_K and lack of delay in the turn-off of I_K. It is important to note that the fourth power gives a correct delay only under normal conditions; after prolonged membrane hyperpolarization, the power has to be increased almost tenfold to fit the experimental results (Cole and Moore, 1960).

The behavior of n as a function of voltage and time can best be described by the following differential equation:

$$dn/dt = \alpha_n(1 - n) - \beta_n n, \tag{6-7}$$

where α_n and β_n are voltage dependent rate constants. The voltage dependency of α_n and β_n, as determined by Hodgkin and Huxley from their experimental curves, is given by:

$$\alpha_n = -0.01(E_M + 50)/\left[\exp\left(\frac{-(E_M + 50)}{10}\right) - 1 \right]; \tag{6-8}$$

$$\beta_n = 0.125 \ \exp -\left(\frac{E_m + 60}{80}\right). \tag{6-9}$$

By means of Eq. (6-8) and (6-9), the α's and β's can be calculated for any potential. Rearranging Eq. (6-7) gives:

$$dn/dt = \alpha_n - (\alpha_n + \beta_n)n. \tag{6-10}$$

* For typical values of g_i and reversal potentials, see chapter 7.

Let

$$y = \alpha_n - (\alpha_n + \beta_n)n. \tag{6-11}$$

Then

$$dy = -(\alpha_n + \beta_n)dn. \tag{6-12}$$

Introducing dn from Eq. (6-12) to Eq. (6-10) and rearranging:

$$dy/y = -(\alpha_n + \beta_n)dt. \tag{6-13}$$

In the case of a step voltage clamp, the membrane potential is known and constant at one value before the step and at another value after the step in E_M. Under these conditions, α_n and β_n are constant so that Eq. (6-13) can be readily integrated:

$$n_{(t)} = n_\infty - (n_\infty - n_0)e^{-t/\tau_n}, \tag{6-14}$$

where

$$n_0 = \alpha_n/(\alpha_n + \beta_n), \tag{6-15}$$

and the values of α_n and β_n are those calculated from E_M before the step in potential and

$$n_\infty = \alpha'_n/(\alpha'_n + \beta'_n), \tag{6-16}$$

where α'_n and β'_n are values of the rate constants after the step in potential. The time constant, τ_n, is defined by:

$$\tau_n = 1/(\alpha'_n + \beta_n'). \tag{6-17}$$

The voltage dependency of n can analytically be solved for the steady-state condition, that is, n_∞ or n_0, by means of Eq. (6-8), (6-9) and (6-10). The resulting curve is given in Fig. (6-4). In the neighborhood of the resting potential, $n \cong 0.3$. It tends toward zero for hyperpolarization potentials and towards 1.0 for depolarizations. According to present theories, the changes in n_∞ should probably be interpreted as changes in the fraction of sites, or channels, available at any potential or membrane field strength.

The potassium currents computed on the basis of Eq. (6-5) through (6-14) closely correspond to the experimental results.

If one plots the steady-state or plateau values of I_K (also called the steady-state current, I_{ss}) as a function of the membrane potential at which it was obtained, one obtains the so-called steady-state, or potassium, current-voltage (I-V) relationship. As seen in Fig. (6-2); the curve which crosses the I = 0 point at E_K (-80) shows considerable current rectification. The rectification is due to the fact that, in general, the dependency of g_K on E_M is much stronger for depolarizing than hyperpolarizing potentials.

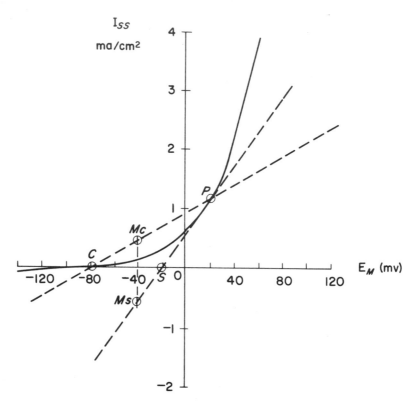

Fig. 6-2 *Plot of steady state potassium current, I_{ss}, as a function of membrane potential, E_M. Outward current is upward.* Mc *and* Ms *are hypothetical ionic current values obtained immediately after stepping E_M from +20 mv (point P) to −40 mv. Point* Mc *represents the current that would have been obtained were membrane conductance a chord conductance, while point* Ms *corresponds to that of a slope conductance. The point obtained in actual experiments is* Mc. *Experimental I_{ss} values from Adelman and Senft, 1966.*

DEFINITION OF MEMBRANE CHORD CONDUCTANCE

Figure (6-2) also gives experimental evidence which illustrates the logic behind the definition of g_M as a chord conductance. In some systems, the conductance can be defined as a slope conductance, that is: $g = dV/dI$. Following this definition in Fig. (6-2), g would be given by the reciprocal of the slope of line $\overline{PS}$. However, if one uses the chord conductance definition, g would be given by the reciprocal of the slope of line $\overline{PC}$. To distinguish between these two definitions, Hodgkin and Huxley (1952*b*) chose a set of conditions which brought the axon I-V relationship to point P. They then stepped E_M, for example, from +20 mv to −40 mv. Since the n parameter changes relatively slowly, the current step associated with the potential step is determined only by g_M or, in this specific case, the g_K which the system had reached before the step [Eq. (6-6)].* Therefore, the new potential and current define the same g_M as that defined at P. A point on either line $\overline{PS}$ (point Ms) or $\overline{PC}$ (point Mc) should be obtained with the potential step. As the experimental points fall on the $\overline{PC}$ line (point Mc) rather than on $\overline{PS}$ (point Ms), the proper definition for the axon membrane conductance, g_M, is the chord conductance.

THE POTASSIUM "TIME CONSTANT," τ_n

As can be seen in Fig. (6-1), the higher the value of depolarization, the higher the steady-state level of I_K. However, for different values of E_M, the rate of change of n and therefore the changes in I_K, with time, are not a linear function of E_M. These rates of change are usually expressed by the time constant τ_n of Eq. (6-14). For any given E_M, τ_n can be calculated from Eq. (6-17). The resulting τ_n vs. E_M curve has a peak around the membrane resting potential. Fig. (6-5) shows that at potentials typical of the resting state, the rates of change of both potassium and sodium parameters are minimal. These rates increase both in the depolarizing and hyperpolarizing directions. From the values of I_K obtained upon stepping E_M to any specific value, τ_n can be determined experimentally by plotting log I_K vs. time. From the slope of this linear relationship τ_n can be obtained. The I_K values can be best obtained from the decay of the current at the end of a long depolarizing pulse (see Fig. 7-2) as under these conditions there is practically no sodium current.

If we assume that the membrane permeability to potassium ions depends on the movement of a charged or polar carrier, then, for fields of resting strength, the configuration or position of the carrier is such that it is at some energetic minimum. Any change in the membrane field strength, resulting from an E_M change, may force the carrier away from this energy well and thus increase the probability that it will carry an ion upon a step in membrane potential.

SODIUM CONDUCTANCE AND CURRENTS

The time and potential dependencies of the sodium conductance are more complex than those of potassium. Fig. (6-3) illustrates a typical sodium current elicited by a depolarizing potential step. Sodium current turns on rapidly,

---

* As a first approximation, leakage current I_L and I_{Na} can be neglected.

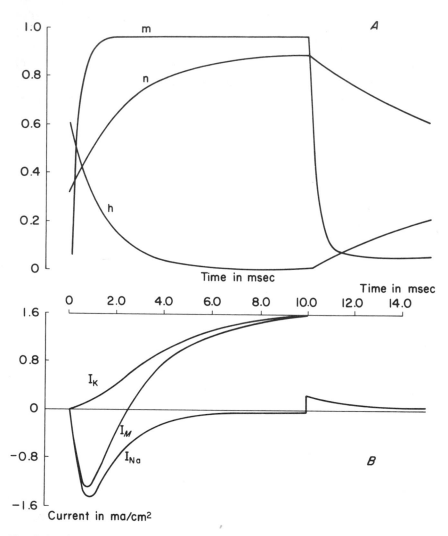

Fig. 6-3 *Computed values of m, n, h, I_{Na}, I_M, I_K and I_C as a function of time after stepping E_M from $E_H = -60$ mv to 0 mv at $t = 0$ and back to -60 mv at $t = 10$ msec. Outward current is up. A. Values of parameters m, n, and h. Note exponential changes in parameters and faster time constant of m. B. Ionic Currents. Since $dE_M/dt = 0$, $I_C = 0$. Values of I_L are so small that they are indistinguishable from zero. Note transient behavior of I_{Na}, the tendency of I_K towards a steady-state and the tail-current elicited upon stepping E_M to its original resting value. For details of computer program, see Chapter 7.*

as compared with I_K, but it is quickly turned off while the depolarization continues (Hodgkin and Huxley, 1952b). It is obvious that the membrane depolarization may be interpreted as having a dual effect on g_{Na}. These effects can be referred to as sodium activation and sodium inactivation. To describe experimental results, two parameters, m and h were introduced in the following way:

$$I_{Na} = \bar{g}_{Na} m^3 h(E_M - E_{Na}). \tag{6-18}$$

The conductance parameter m is defined similarly to n; it is time and potential dependent and varies between zero and 1.0. At any E_M for any fixed value of h, m^3 is defined as the ratio between the peak value of I_{Na} obtained at this E_M and the maximal obtainable value of I_{Na} (at any E_M): I_{Na}/I_{Namax}. The time and voltage dependency of m and h are described by equations similar to those for n:

$$dm/dt = \alpha_m(1 - m) - \beta_m m, \tag{6-19}$$

$$m_{(t)} = m_\infty - (m_\infty - m_0)e^{-t/\tau_m}, \tag{6-20a}$$

where

$$m_0 = \alpha_m/(\alpha_m + \beta_m); \ m_\infty = \alpha'_m/(\alpha'_m + \beta'_m); \text{ and } \tau_m = 1/(\alpha'_m + \beta'_m).$$
$$\tag{6-20b,c,d}$$

The equations which give the potential dependency of α_m, β_m, α_h, and β_h are listed in chapter 7.

The potential dependency of m_∞, i.e. the steady-state values of m, is similar to that of n. With depolarization, m_∞ increases, and it decreases with hyperpolarization (see Fig. 6-4). However, under resting conditions ($E_M = -60$ mv), m is close to zero (0.05) while n has a value of 0.3. This is the main reason why g_K is greater than g_{Na} under normal resting conditions. Raising m to the third power introduces a delay in the turn-on of I_{Na} which fits the experimental results. As illustrated in Fig. 6-5, the τ_m vs. E_M curve is similar in its general shape to the τ_n or τ_h curves. However, the values of τ_m are smaller in amplitude by a factor of about ten. Therefore, upon depolarization m increases faster than n and the rate of turn-on of I_{Na} is much faster than that of I_K. Like τ_n, τ_m can be determined experimentally for any E_M from the I_{Na} values obtained upon stepping E_M to the desired value. As predicted by Eq. (6-20a), the plot of log I_{Na} vs. time gives a linear relationship, the slope of which is used to determine τ_m. It should be pointed out here that since $\tau_n > \tau_m < \tau_h$, under normal conditions, the m process is the fastest and the initial ionic current component recorded from voltage clamped axons is practically pure I_{Na} so that τ_m can be directly determined from the initial current accurately.

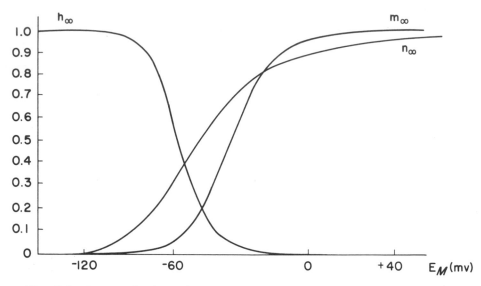

Fig. 6-4 *Computed values of steady-state values of m, n and h as a function of membrane potential, E_M. Note that when the axon membrane is depolarized, h_∞ decreases to zero while m_∞ and n_∞ increase towards one. The data is computed from Eq. (6–16), (6–20b) and (6–22b).*

SODIUM INACTIVATION, 1–h

Like m and n, h is defined as a fraction. For any particular set of experimental conditions where m is constant, h can be defined as the ratio between I_{Na} obtained and the maximal I_{Na} value obtainable under any of these conditions.

$$d\mathrm{h}/dt = \alpha_\mathrm{h}(1 - \mathrm{h}) - \beta_\mathrm{h}\,\mathrm{h}, \qquad (6\text{-}21)$$

$$\mathrm{h}_{(t)} = \mathrm{h}_\infty - (\mathrm{h}_\infty - \mathrm{h}_0)e^{-t/\tau_\mathrm{h}}, \qquad (6\text{-}22a)$$

where

$$\mathrm{h}_0 = \alpha_\mathrm{h}/(\alpha_\mathrm{h} + \beta_\mathrm{h}); \; \mathrm{h}_\infty = \alpha'_\mathrm{h}/(\alpha'_\mathrm{h} + \beta'_\mathrm{h}); \; \text{and } \tau_\mathrm{h} = 1/(\alpha'_\mathrm{h} + \beta'_\mathrm{h}).$$
$$(6\text{-}22b,c,d)$$

The potential dependency of the h parameter is the inverse of the potential dependency of m and n. Upon depolarization h decreases and during hyperpolarization it increases. The steady-state value, h_∞, falls from its resting value of about 0.6 towards zero at about $E_M = -20$ mv (see Fig. 6-4). Since the rate of change of h, as given by τ_h, is considerably slower than that of m, with a reasonable membrane depolarization (i.e. $E_M \geq -30$ mv), m reaches a value close to 1.0 before h decreases significantly. The values of

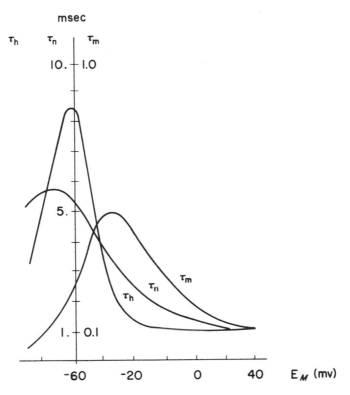

Fig. 6-5 *Computed values of conductance parameters time constants τ_m τ_n and τ_h as a function of membrane potential, E_M. Since all time constants peak close to the resting potential (-60 mv in the model axon) the rates of conductance or current changes are minimal near that value. Note that the scale of τ_m is ten fold larger than that of τ_n and τ_h. The rate constant values were computed from Eq. (6–17), (6–20c) and (6–22c). (From Cole, 1968.)*

I_{Na} thus first increase with m. However, as h slowly decreases, sodium conductance is inactivated and I_{Na} is attenuated (Fig. 6-3). This sodium inactivation together with the increase in outward potassium current are responsible for the decay of the initial phase of the total inward membrane current, I_M (Fig. 6-3). Therefore, the decay of the initial transient inward I_M can not be generally used to determine the time course of change of h. For this purpose, a different type of experiment has been designed (Hodgkin and Huxley, 1952b).

Fig. 6-6 gives a schematic representation of such an experiment. The experiment involves clamping an axon to three different potentials. First E_M is held at the resting level, in this case -60 mv. Then it is prepulsed to the specific potential at which the behavior of h is investigated (-120 mv)

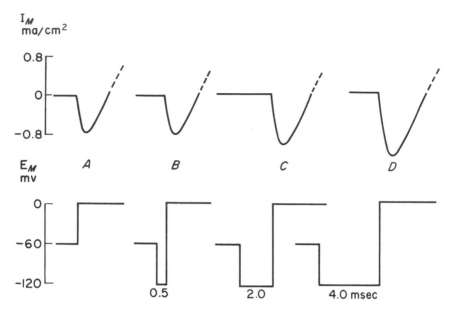

Fig. 6-6 *Schematic representation of an experiment illustrating the method for determining the changes in values of the h parameter has a function of time. The inward current is plotted downward. Upper traces: membrane currents, I_M, are generated by the changes in E_M given in lower traces. A. I_M corresponds to a change in E_M from its resting value of $E_H = -60$ mv to zero mv. This transient, inward current, like the following ones, is carried mostly by sodium ions. B, C, D. The following transient currents increase as the duration of the hyperpolarization, which precedes the depolarization step, increases. The changes in the peak value of the inward current correspond to the changes in the conductance parameter h with hyperpolarization.*

and finally pulsed to a strongly depolarized value. Upon the onset of the hyperpolarizing prepulse, m and n tend towards zero so that I_M becomes negligible and consists mostly of leakage. At the same time, h increases gradually from its resting value ($h_0 = 0.6$) to its final value, h_∞ (at this hyperpolarized potential, close to 1.0). One-half msec after the beginning of the hyperpolarizing prepulse, E_M is stepped to zero (Fig. 6-6B). During this short period, h has barely enough time to change. The I_{Na} generated upon this depolarization is practically the same as that elicited when E_M is stepped directly from resting value to zero with no prepulse (Fig. 6-6A). However, when the duration of the hyperpolarizing prepulse is increased to 2 msec (Fig. 6-6C) and then 4 msec (Fig. 6-6D), h has sufficient time to increase and with it, I_{Na}. Assuming that, in all these cases, when I_{Na} reaches its peak, I_K is still negligible and m has reached its steady-state value, the changes of the peak values of I_M correspond to the changes in h with time during the selected hyperpolarization. A plot of the log of this peak of I_M vs. time will again give a linear relationship [see Eq. (6-22)]. From the slope of the line, one can find τ_h at the selected E_M. For the sodium I-V relationship, see chapter 10.

ANALYSIS OF AXON VOLTAGE CLAMP RECORDS

On the basis of the parametric analysis given above, we can now analyze a typical example of the behavior of a Hodgkin-Huxley model axon under voltage clamp. Fig. 6-3 plots the computed values of m, n and h and the membrane currents generated upon stepping E_M from a resting value of -60 mv to zero. With the exception of the instant of the potential step, throughout the step clamp $I_C = 0$ ($\partial V/\partial t = 0$). The small τ_m and the large depolarization result in a rapid increase of m from its resting level of 0.05 to its maximal value of unity. Since $\tau_n > \tau_m$, I_{Na} is the first current to turn on. For the potential step in question, $E_{Na} > E_M$. Since E_{Na} is directed inward (sodium concentration is higher on the outside of the membrane), the sodium current is inward. The potassium current turns on slower corresponding to the relatively slow increase of n (Fig. 6-3). As n tends towards n_∞, I_K tends to plateau. At about the same rate that n increases, h decreases from its resting value of 0.6 to practically zero,[*] and the sodium current turns off at a rate similar to the turn-on of I_K. Note that under the above experimental conditions, I_L is usually small as compared with I_{Na} and I_K.

As E_M is stepped back to its resting value, I_M does not return immediately to its zero resting level. The outward current, carried by potassium ions, persists for a while. This persistent current is usually termed the tail current. Since sodium inactivation is maximal at and just after the end of the depolarization and since $g_K \gg g_L$, the tail current is carried mostly by potassium ions. From the plot of the log of the tail currents vs. time, one can therefore determine τ_n. By stepping E_M to different values (at the end of the depolarizing pulse), one can obtain the τ_n vs. E_M curve. The tail current is generated because under resting conditions E_M does not equal E_K ($E_M = -60$ mv, $E_K = -80$ mv), and just after the end of the depolarization, n still retains its high value (close to unity). As n decays to its resting value (with a time constant, τ_n), I_K also decays and eventually reaches a value equal to the sum of all the other ionic currents under resting conditions. At this moment, $I_M = 0$ and the system reaches its stable steady-state condition.

---

[*] Note that the level of sodium inactivation is measured by $1 - h$ rather than h itself. The parameter h is sometimes referred to as sodium facilitation (Cole, 1968).

REFERENCES

ADELMAN, W. J., Jr., and J. P. SENFT. 1966. Voltage Clamp Studies on the Effect of Internal Cesium Ion on Potassium Currents in the Squid Giant Axon. *J. Gen. Physiol.* 50:279–293.

COLE, K. S. 1949. Dynamic Electrical Characteristics of the Squid Axon Membrane. *Arch. Sci. Physiol.* 3:253–258.

————. 1968. *Membranes, ions and impulses.* Univ. of California Press, Berkeley.

COLE, K. S., and W. J. MOORE. 1960. Potassium Ion Current in the Squid Giant Axon; Dynamic Characteristics. *Biophys. J.* 1:1–14.

GOLDMAN, D. E. 1943. Potential, Impedance and Rectification in Membranes. *J. Gen. Physiol.* 27:37–60.

HILLE, B. 1968. Pharmacological Modification of the Sodium Channels of Frog Nerve. *J. Gen. Physiol.* 51:199–220.

HODGKIN, A. L., and A. F. HUXLEY. 1952a. Currents Carried by Sodium and Potassium Ions through the Membrane of the Giant Axon of *Loligo*. *J. Physiol. (London)*. 116:449–472.

————. 1952b. The Dual Effect of Membrane Potential of Sodium Conductance in the Giant Axon of *Loligo*. *J. Physiol. (London)*. 116:497–506.

————. 1952c. A Quantitative Description of Membrane Current and its Application to Conduction and Excitation in Nerve. *J. Physiol. (London)*. 117:500–544.

HODGKIN, A. L., A. F. HUXLEY and B. KATZ. 1952. Measurement of Current Voltage Relations in the Membrane of the Giant Axon of *Loligo*. *J. Physiol. (London)*. 116:424–448.

MARMONT, G. 1949. Studies on the Axon Membrane. I. A New Method. *J. Cell. Comp. Physiol.* 34:351–382.

MOORE, J. W., M. P. BLAUSTEIN, N. C. ANDERSON and T. NARAHASHI. 1967. Basis of Tetrodotoxin Selectivity in Blockage of Squid Axons. *J. Gen. Physiol.* 50:1401–1411.

DIGITAL COMPUTER SOLUTIONS OF MEMBRANE CURRENTS IN THE VOLTAGE CLAMPED GIANT AXON

7

Y. PALTI

The following computer program is written in Fortran IV. It is designed to solve, as a function of time, the changes in squid giant axon membrane parameters (I_C, I_{Na}, I_K, I_L, m, n, h,) when membrane potential, E_M, is clamped to various potentials. The program is not limited to step changes in potential and can be used for sine waves, linear potential changes or other functions. However, any potential change investigated must take place simultaneously along the whole length of membrane under investigation, i.e. the membrane has to be space-clamped (see chapter 6).

For a step membrane potential change, when E_M is constant with time, the program calculates the membrane currents and other variables by means of the analytical solution of Eq. (6-3), (6-7), etc. For other types of potential changes when the differential equations cannot be solved analytically, the solution is made by numerical methods. Two examples of such solutions are given for:

1. Sine waves:

$$E_M = E_H + E_o \sin \omega t,$$

where E_o is the membrane potential measured from the initial holding potential, E_H, to the peak of the wave; $\omega = 2\pi f$, and f is the frequency.

2. Triangular waves:

$$E_M = E_H + At$$

where A is the coefficient determining the slope of the linear potential change; the sign of A changes in pace with $\sin \omega t$.

Other potential functions can also be easily used by addition of an appropriate potential function statement.

The program uses the following relations formulated by Hodgkin and Huxley (1952) to describe the membrane currents:

$$I_M = C_M dE_M/dt + \bar{g}_{Na}\, m^3 h(E_M - E_{Na}) + \\ \bar{g}_K n^4(E_M - E_K) + g_L(E_M - E_L), \tag{7-1}$$

$$dm/dt = \alpha_m(1 - m) - \beta_m m, \tag{7-2}$$

$$dn/dt = \alpha_n(1 - n) - \beta_n n, \tag{7-3}$$

$$dh/dt = \alpha_h(1 - h) - \beta_h h, \tag{7-4}$$

$$\alpha_m = -0.1(E_M + 35)/(\exp\left[\frac{-(E_M + 35)}{10}\right] - 1), \tag{7-5}$$

$$\beta_m = 4 \exp\left[\frac{-(E_M + 60)}{18}\right], \tag{7-6}$$

$$\alpha_n = -0.01(E_M + 50)/(\exp\left[\frac{-(E_M + 50)}{10}\right] - 1), \tag{7-7}$$

$$\beta_n = 0.125 \exp\left[\frac{-(E_M + 60)}{80}\right], \tag{7-8}$$

$$\alpha_h = 0.07 \exp\left[\frac{-(E_M + 60)}{20}\right], \tag{7-9}$$

$$\beta_h = 1/(\exp\left[\frac{-(E_M + 30)}{10}\right] + 1), \tag{7-10}$$

$$m_\infty = \frac{\alpha_m}{\alpha_m + \beta_m}. \tag{7-11}$$

For definition of terms and parameters, see chapter 6.

For the convenience of the reader, a typical set of constants of a normal axon is listed below (FitzHugh, 1960):

$E_{Na} = 55.0$ mv	$\bar{g}_{Na} = 120.0$ mmho/cm^2
$E_K = -72.0$ mv	$\bar{g}_K = 36.0$ mmho/cm^2
$E_L = -49.4011$ mv	$g_L = 0.3$ mmho/cm^2.

The solution follows the flow chart presented in Fig. 7-1. To solve Eq. (7-1), the program first reads from an input card or (given as an option) calculates the initial values of parameters m, n and h: m_0, n_0 and h_0 (for $t = t_0$). Calculation is carried out by means of Eq. (7-5), (7-6) and (7-11) for m_0 and

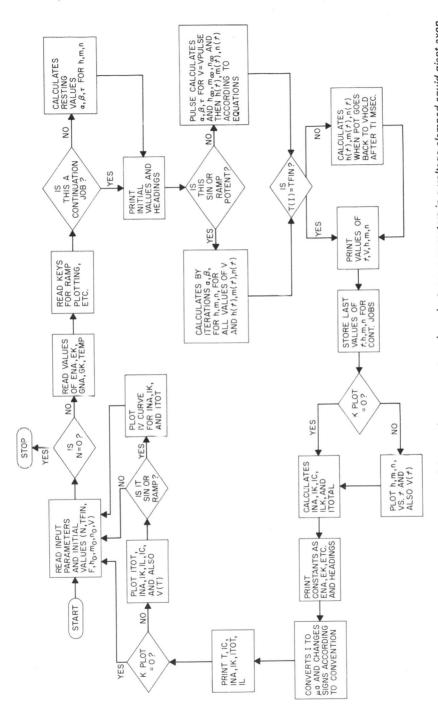

Fig. 7-1 Flow chart of program for solution of the membrane currents and conductance parameters in a voltage clamped squid giant axon.

by means of analogous equations for n_0 and h_0. On the basis of the initial parameter values at $t = t_0$, their values after a small increment of time $(t_0 + \Delta t)$ are calculated by means of the Taylor Series. These series give the value of a function at $(t + \Delta t)$ when its value at Δt is known (Courant, 1959). For the m parameter, the relation would be:

$$m_{(t_0 + \Delta t)} = m_{t_0} + \Delta t \, dm/dt + \frac{\Delta t^2}{2!} d^2m/dt^2 + \cdots \qquad (7\text{-}12)$$

For the continuous functions given in Eq. (7-1), (7-2), (7-3) and (7-4), and for a sufficiently small element of time, Δt, one can ignore the terms subsequent to $\Delta t \cdot dm/dt$ without introducing a significant error.* For the solution of our problem, using Eq. (7-2), Eq. (7-12) can be rewritten (for m):

$$m_{(t_0 + \Delta t)} = m_{t_0} + \Delta t[\alpha_m(1 - m) - \beta_m m]. \qquad (7\text{-}13)$$

As the α's and β's are known for any E_M[Eq. (7-5), (7-6)], Eq. (7-13) can now be solved. Using equations similar to (7-13), Eq. (7-3) and (7-4) are solved for $n_{(t_0 + \Delta t)}$ and $h_{(t_0 + \Delta t)}$. The calculated parameters at $(t + \Delta t)$, together with the given membrane constants, are now substituted in Eq. (7-1). Since:

$$I_C = C_M dE_M/dt \qquad\qquad I_{Na} = \bar{g}_{Na}m^3h(E_M - E_{Na})$$
$$I_K = \bar{g}_K n^4(E_M - E_K) \qquad\qquad I_L = g_L(E_M - E_L),$$

the individual currents as well as I_M can be easily calculated for time $t + \Delta t$.

The above calculations are now repeated by means of a "do-loop" (see Fig. 7-1). At each new iteration, the time is increased by an additional Δt and the last calculated m, n and h values are substituted for m_0, n_0 and h_0.

The calculated values of membrane potential, m, n and h parameters, the α's and β's (for any E_M), the total membrane current, as well as the individual ionic and capacity currents, are thus calculated and tabulated as a function of time. The values of m, n and h, E_M and the membrane currents can also be plotted as a function of time. Such plots are given in Fig. 8-1, 8-5, 8-7, etc. However, this last option which is included in the flow chart is not given in this listing.

The following is a listing of the program Fortran statements. The larger typeface includes mainly the logic and mathematics of the solution while the smaller typeface consists mainly of input, output and other statements which are not essential for the understanding of the program. For definition of any additional parameters used, see the dictionary at the end of the program.

_____

* Using the Runga-Kutta method (Ralston and Wilf, 1960), a sufficiently small element of time, $\triangle t$, is automatically determined for each step in the solution of the equations so that the error is kept below a prefixed criterion.

FORTRAN PROGRAM

```
      REAL IK, INA, IC, ITØT, IL
      DØUBLE PRECISIØN DELTAN, TEMP, ALFAH, BETAH, ALFAEM
      DØUBLE PRECISIØN DELTAT, EX, DELTAH, DELTAM
      DØUBLE PRECISIØN BETAEM, ALFAEN, BETAEN, PIHALF
      DØUBLE PRECISIØN PI, PI3HLF, PI2, UAI, VCØNST, VHØLD
      DIMENSIØN IL(1000)
      DIMENSIØN T(1000), EX(1000), H(1000), EM(1000), EN(1000)
      DIMENSIØN INA (1000), IK(1000), IC(1000), ITØT(1000)
      EQUIVALENCE (EN(1), INA(1)), (EM(1), IC(1)), (H(1), IK(1))
      EQUIVALENCE (EX(1), ITØT(1))
      PI = 3.1415926535897932
      GLK = 0.3
      EL = 10.5989
      CM = 1.0
C     READ INITIAL VALUES
31    READ (5, 101) N, TFIN, F, HIN, EMIN, ENIN, VØ, VHØLD,
     1VPULSE, T1, NDEL
101   FORMAT (I4, 9F8.3, I4)
      IF (N.EQ.0) STØP
      READ (5, 201) ENA, EK, GNA, GK, TEMPFI, KRAMP
201   FORMAT(5F10.5, I2)
      FIMNH = 3.0 ** ((TEMPFI-6.3)/10.0)
      IF (KRAMP.EQ.0) GØ TØ 750
      PI = 0.5/ F
      TANALF = VØ/(PI * 0.5)
750   PIHALF = 0.5 * PI
      PI3HLF = 1.5 * PI
      PI2 = 2.0 * PI
      READ (5,200) KHØLDH, KHØLDM, KHØLDN, KCØNT
200   FØRMAT (4I2)
      FN = N
      IF (KCØNT.EQ.0) T(1) = 0.0
      IF (T(1).GT.0.0)GØ TØ 32
      T(1) = 0.0
      EX(1) = VHØLD
      ALFAH1 = 7.0D - 2 * DEXP (VHØLD/2.OD1)
      UAI = 1.OD -1 * VHØLD
      BETAH1 = 1.0/(DEXP(3.0D0 + UAI) + 1.0)
      ALFAM1 = (2.5D0 + UAI)/DEXP(2.5D0 + UAI) - 1.0)
      BETAM1 = 4.0 * DEXP(VHØLD/1.8D1)
```

```
      ALFAN1 = (1.0D − 1 + 1.0D − 2 * VHØLD)/(DEXP(1.0D0
     1+UAI) − 1.0)
      BETAN1 = 1.25D − 1 * DEXP(VHØLD/8.0D1)
      TAUH1 = 1./(ALFAH1 + BETAH1)
      H(1) = ALFAH1 * TAUH1
      H1 = H(1)
      IF (KHØLDH.EQ.0) H(1) = HIN
      TAUM1 = 1.0/(ALFAM1 + BETAM1)
      EM(1) = ALFAM1 * TAUM1
      EM1 = EM(1)
      IF (KHØLDM.EQ.0) EM(1) = EMIN
      TAUN1 = 1.0/(ALFAN1 + BETAN1)
      EN(1) = ALFAN1 * TAUN1
      EN1 = EN(1)
      IF (KHØLDN.EQ.0) EN(1) = ENIN
   32 DELTAT = (TFIN − T(1))/FN
      N1 = T1/DELTAT
C PRINT INITIAL VALUES AND CØLUMN HEADINGS
      WRITE (6,163) VPULSE, VHØLD, KHØLDH, KHØLDM, KHØLDN, KCØNT,
     1KRAMP
  163 FØRMAT (1H1, 8HVPULSE = , F8.3, 3X, 7HVHØLD = , F8.3,
     13X, 8HKHØLDH = , I1, 3X, 8HKHØLDM = , I1, 3X, 8HKHØLDN = ,
     2I1, 3X, 8HKCØNT = , I1, 3X, 7HKRAMP = , I1)
      WRITE (6,102) N, F, VØ, HIN, EMIN, ENIN, T1
  102 FØRMAT (1H0, 3HN = , I5, 8X, 11HFREQUENCY = , F10.5, 3X,
     14HCPMS, 8X, 4HV0 = ,
     21F10.5, 3X, 2HMV//5H HIN = , F10.5, 10X, 6HEMIN = , F10.5, 10X,
     36HENIN = , F10.5, 4X, 4HT1 = , F5.1,
     4///2X, 4HT(I), 14X, 4HX(I), 6X, 4HH(I), 5X, 5HEM(I), 4X, 5HEN(I))
      W = F * PI2
      DØ 11 I = 2,N1
      IF (VPULSE.NE.0.0) GØ TØ 712
      T(I) = T(I − 1) + DELTAT
      IF (KRAMP) 700, 700, 701
  700 EX(I) = W * T(I)
      GØ TØ 702
  701 EX(I) = T(I)
  702 EX(I) = DMØD (EX(I), PI2)
      IF (EX(I).LT.PI2.AND.EX(I).GE.PI3HLF) NQUART = 4
      IF (EX(I).LT.PI3HLF.AND.EX(I).GE.PI) NQUART = 3
      IF (EX(I).LT.PI.AND.EX(I).GE.PIHALF) NQUART = 2
      IF (EX(I).LT.PIHALF.AND.EX(I).GE.0.0) NQUART = 1
```

```
       SIGN = 1.0
       IF (NQUART − 2) 12, 13, 14
12  UAI = EX(I)
       GØ TØ 15
13  UAI = PI − EX(I)
       GØ TØ 15
14  SIGN = − 1.0
       IF (NQUART.EQ.3) GØ TØ 16
       IF (NQUART.EQ.4) GØ TØ 17
16  UAI = EX(I) − PI
       GØ TØ 15
17  UAI = PI2 − EX(I)
15  IF (KRAMP) 710, 710, 720
720 EX(I) = UAI * SIGN * TANALF + VHØLD
       GØ TØ 711
710 EX(I) = SIGN * VØ * DSIN(UAI) + VHØLD
       GØ TØ 711
712 I = N
       EX(I) = VPULSE
711 TEMP = DEXP(EX(I)/20.0)
       ALFAH = 7.0D − 2 * TEMP
       BETAH = 1.0/(DEXP(3.0D0 + 0.1EX(I)) + 1.0)
       DELTAH = (ALFAH − (ALFAH + BETAH) * H(I − 1))*
      1DELTAT
       DELTAH = DELTAH * FIMNH
       H(I) = H(I − 1) + DELTAH
       S = EX(I)
       IF(ABS(S + 25.0). * LT.1.E − 5) GØ TØ 20
       TEMP = 1.0D − 1 * (EX(I) + 25.D0)
       ALFAEM = TEMP * (1./DEXP(TEMP) − 1.0)
       GØ TØ 18
20  ALFAEM = 1.0
18  BETAEM = 4.0 * DEXP(EX(I)/18.D0)
       DELTAM = (ALFAEM − (ALFAEM + BETAEM) * EM(I − 1)) *
      1DELTAT
       DELTAM = DELTAM * FIMNH
       EM(I) = EM(I − 1) + DELTAM
21  IF(EM(I).LT.0.0) EM(I) = 0.0
       IF (ABS(S + 10.0).LT.1.0E − 5) GØ TØ 22
       TEMP = EX(I) + 1.0D1
       ALFAEN = 1.0D − 2 * TEMP * (1.0/(DEXP(TEMP/1.0D1) − 1.0))
       GØ TØ 19
```

```
22  ALFAEN = 0.1
19  BETAEN = 0.125 * DEXP(EX(I)/8.0D1)
    DELTAN = (ALFAEN - (ALFAEN + BETAEN) * EN(I - 1)) *
    1DELTAT
    DELTAN = DELTAN * FIMNH
    EN(I) = EN(I - 1) + DELTAN
11  CØNTINUE
    IF (VPULSE.EQ.0.0) GØ TØ 717
    DØ 716 I = 2, N1
716 EX(I) = VPULSE
    TAUH = 1.0/(ALFAH + BETAH)
    TAUEM = 1.0/(ALFAEM + BETAEM)
    TAUEN = 1.0/(ALFAEN + BETAEN)
    HINF = ALFAH * TAUH
    EMINF = ALFAEM * TAUEM
    ENINF = ALFAEN * TAUEN
    DO 714 I = 2, N1
    T(I) = T(I - 1) + DELTAT
    H(I) = HINF + (H(1) - HINF) * EXP(- T(I) * FIMNH/TAUH)
    EM(I) = EMINF + (EM(1) - EMINF) * EXP(- T(I) * FIMNH/
    1TAUEM)
714 EM(I) = ENINF + (EN(1) - ENINF) * EXP(- T(I) * FIMNH/
    1TAUEN)
717 N2 = N1 + 1
    IF (TFIN.EQ.T1) GØ TØ 715
    DØ 713 I = N2, N
    EX(I) = VHØLD
    T(I) = T(I - 1) + DELTAT
    TN2 = T(I) - T(N1)
    H(I) = H1 + (H(N1) - H1) * EXP(- TN2 * FIMNH/TAUH)
    EM(I) = EM1 + (EM(N1) - EM1) * EXP(- TN2 * FIMNH/
    1TAUEM)
713 EN(I) = EN1 + (EN(N1) - EN1) * EXP(- TN2 * FIMNH/TAUEN)
715 DØ 23 I = 1,N,NDEL
    WRITE (6,103) T(I),EX(I),H(I),EM(I),EN(I)
103 FORMAT (1H,1PE12.5,D14.5,0P3F9.5)
23  CØNTINUE
    TMPCT1 = T(N)
    TMPCT2 = H(N)
    TMPCT3 = EM(N)
    TMPCT4 = EN(N)
    TMPCNT = (EX(N) - EX(N - 1))/DELTAT
```

C   CØMPUTE CURRENT
30 RDT = (1.0/DELTAT) * CM
    TEMPNA = GNA * (EX(1) − ENA) * EM(1) ** 3 * H(1)
    TEMPK = GK * (EX(1) − EK) * EN(1) ** 4
    TEMP = EX(1) − EL
    TEMPX = EX(1)
    INA(1) = TEMPNA
    IK(1) = TEMPK
    IC(1) = CM * VØ
    IF (VPULSE.NE.0.0) IC(1) = 0.0
    IF (T(1).GT.0.0) IC(1) = CM * TMPCNT
    ITØT(1) = TEMP * GLK + INA(1) + IK(1) + IC(1)
    DØ 44 I = 2,N
    TEMPNA = GNA * (EX(I) − ENA) * EM(I) ** 3 * H(I)
    TEMPK = GK * (EX(I) − EK) * EN(I) ** 4
    TEMP = EX(I) − EL
    IC(I) = (EX(I) − TEMPX) * RDT
    TEMPX = EX(I)
    INA(I) = TEMPNA
    IK(I) = TEMPK
    ITØT(I) = TEMP * GLK + IK(I) + IC(I) + INA(I)
44 CØNTINUE
    IF (VPULSE.EQ.0.0) GØ TØ 42
    ITØT(2) = ITØT(2) − IC(2)
    IC(2) = 0.0
    IC(N1 + 1) = 0.0
C PRINT VALUES ØF CØNSTANTS AND HEADINGS
42 WRITE (6,165) ENA,EK,TAUH,TAUEM,TAUEN
165 FORMAT (1H1, 5HENA = , F8.3, 8X, 4HEK = , F8.3, 8X, 6HTAUH = ,
    1F8.3, 8X, 6HTAUEM = , F8.3, 8X, 6HTAUEN = , F8.3, 14HIN MV
    2AND MSEC)
    WRITE (6,107)GNA, GK, GLK
107 FORMAT(1H0, 5HGNA = , F10.5, 8X, 4HGK = , F10.5, 8X, 5HGLK = ,
    1F10.5, 8X,
    219HALL IN MMHØ PER SCM//5X, 4HT(I), 8X, 5HIC(I), 7X, 6HINA(I), 6X,
    35HIK(I), 7X, 7HITØT(I), 5X, 6HILK(I), 6X, 18HALL IN MA PER SCM///)
    DØ 43 I = 1, N
    INA(I) = INA(I) * 0.001
    IK(I) = IK(I) * 0.001
    IC(I) = IC(I) * 0.001
43 ITØT(I) = ITØT(1) * 0.001
    DØ 45 I = 1, N

```
      TEMP = ITØT(I) − IC(I) − INA(I) − IK(I)
45   IL(I) = TEMP
      DØ 41 = 1, N, NDEL
      WRITE (6,108)T(I), IC(I), INA(I), IK(I), ITØT(I), TEMP
108 FØRMAT (1H , 1PE12.5, 0P5F12.5)
41   CØNTINUE
      EX(1) = EX(N)
      T(1) = TMPCT1
      H(1) = TMPCT2
      EM(1) = TMPCT3
      EN(1) = TMPCT4
      GØ TØ 31
      END
```

DICTIONARY

E—potential. In the equations the potential is defined in Hodgkin Huxley terms, i.e. resting potential is zero mv.

TEMPFI—Temperature coefficient.

N—Number of calculated points or iterations.

TFIN—the duration of the experiment.

HIN, EMIN, ENIN—h_0, m_0, n_0.

VØ—amplitude of sine or triangular wave measured from E_H to peak of wave.

VPULSE—the membrane potential during the step in a step voltage clamp experiment.

NDEL—of all the computed points, the program prints or plots every NDELth point.

KRAMP—if 0, the potential the membrane is clamped to is a sine wave; if 1, it is a triangular wave.

KCONT—if 0, $t_0 = 0$; if 1, t_0, n, m and h are to be substituted from a previous calculation so that the new calculation becomes a continuation of that program.

T1—the time the potential change is terminated.

VHOLD—holding potential, i.e. E_M before the membrane potential is changed.

KHØLDH—if 0, the initial value of h(h(1)) is HIN; if 1, (h(1)) is calculated from the α's and β's.

KHØLDM—if 0, the initial value of m(m(1)) is EMIN; if 1, m(1) is calculated from the α's and β's.

KHØLDN—if 0, the initial value of n(n(1)) is ENIN; if 1, n(1) is calculated from the α's and β's.

IK—I_K.

INA—I_{Na}.

IC—I_C.

IL—I_L.

ITØT—total membrane current, m_a.

ENA—E_{Na}. EK—E_K. EL—E_L.

ALFAEN—α_n.

BETAEN—β_n, and similarly for m and h.

GK—$\bar{g}_K$.

GNA—$\bar{g}_{Na}$.

GL—g_L.

CM—membrane capacity.

REFERENCES

COURANT, R. 1959. *Differential and Integral Calculus,* Second Ed., Vol. 1. Blakie & Son, London.

FITZHUGH, R. 1960. Thresholds and Plateaus in the Hodgkin-Huxley Nerve Equations. *J. Gen. Physiol.* 43: 867–896.

HODGKIN, A. L., and A. F. HUXLEY. 1952. A Quantitative Description of Membrane Current and its Application to Conduction and Excitation in Nerve. *J. Physiol. (London).* 117: 500–544.

RALSTON, A., and H. S. WILF. 1960. *Mathematical Methods for Digital Computers.* Wiley, New York.

VARYING POTENTIAL CONTROL VOLTAGE CLAMP OF AXONS

8

Y. PALTI

The voltage clamp technique has become the main tool for studying the electric behavior of axon membranes. Since the early 1950's the technique has been used in its original form in axons, i.e., membrane potential E_M was stepped from one value to another and the currents generated were measured. As discussed in Chapter 6, the main virtue of this technique is that E_M is known and constant during the course of the experiment. This eliminated a major variable from the multivariable system and thus simplified the analysis of the axon behavior.

Recently, a modification of the technique was introduced (Palti and Adelman, 1969, 1970 and Fishman and Cole, 1969). In the new method, the axon membrane is clamped to varying rather than constant potentials. These potentials may be sine waves, triangular waves, ramps or others. While this procedure seems to contradict the basic concept of the voltage clamp technique, in numerous cases it not only retains the advantages of the step voltage clamp, but improves on them.

PROSPECT ADVANTAGES OF A VARYING POTENTIAL CONTROL

In the Hodgkin-Huxley axon model, the voltage and time dependency of the membrane conductance parameters, n, m, and h, are described by a set of differential equations [Eq. (6-7), (6-19), and (6-20)]. These equations can be solved analytically for a constant potential when the α's and β's are constant.

194

When E_M is a function of time, as it is in this new method, the α's and β's vary with time so that generally the equations cannot be solved analytically. However, in these cases, the differential equations can still be solved numerically by means of digital computers (see chapter 7). As such computers are readily accessible and easy to use today, the treatment of axon behavior with a varying membrane potential is simple. The question is now what advantages may the new method provide over the step voltage clamp.

IONIC CURRENT SEPARATION

Sodium current is known to follow voltage changes much faster than potassium does. This is due to the faster response of the m conductance parameter as compared with the n parameter. For a step potential, these rates of response are measured by the time constants τ_m and τ_n (chapter 6). As can be seen in Fig. 6-6, for almost any E_M, τ_n is ten times larger than τ_m, indicating a factor of ten in the response time of sodium vs. potassium current. Consequently, sodium is able to follow E_M changes of a much higher frequency than potassium. Therefore, one may expect that using a rapidly changing potential wave, for example, a high frequency sine wave or a steep ramp, one would obtain $I_{Na} \gg I_K$. On the other hand, for depolarizing potential changes which are slow with respect to τ_h, sodium inactivation $(1 - h)$ would increase at a rate similar to the increases in m. Under these conditions, g_{Na}, which depends on m^3h, would decrease and so would I_{Na}.* While I_{Na} becomes negligible at this slow depolarization, in accordance with the increase of n, I_K increases. Therefore, at these frequencies, $I_K \gg I_{Na}$.

By choosing the proper frequencies or rates of E_M changes one can effectively separate sodium and potassium currents without need for chemical substitutions or drugs. Use of these chemicals is not only tedious but also involves numerous assumptions as to their chemical and pharmacological properties.

MEMBRANE CAPACITY CURRENT

In the step clamp, $dE_M/dt = 0$ (except at the very instant of step). As the membrane capacity current is:

$$I_C = C_M \, dE_M/dt. \tag{8-1}$$

---

* Note that m tends towards one while h towards zero. Their product will therefore tend towards zero.

I_C should be zero. Consequently, it is impossible to study membrane capacity by means of the step clamp method. However, upon the potential step, a capacitative transient is generated by the voltage clamp system itself. This transient frequently interferes with the study of the initial segment of sodium current.

If one were to use as a command potential a triangular wave, $|dE_M/dt|$ would equal a constant, so that I_C would become directly proportional to C_M. If the wave frequency is sufficiently high (wavelength short with respect to τ_m and τ_n), the ionic currents do not have time to turn on before the cycle reverses and never develop beyond a negligible value. Under these conditions, the measured membrane current, I_M, is practically a pure I_C. As $|dE_M/dt| =$ constant, this current is a square wave. Let the peak to peak amplitude of the square current wave be I_0 (in $\mu a/cm^2$), the peak to peak amplitude of the triangular voltage wave be V_0 (in mv) and λ the wavelength (in msec); then, from Eq. (8-1), the membrane capacity, C_M, is given in $\mu f/cm^2$ by:

$$C_M = I_0\lambda/4V_0. \tag{8-2}$$

Since I_0 and V_0 are easily measurable, one can expect an accurate C_M determination using the variable potential clamp.

In contrast to a step voltage change where the capacitative current is an unpredictable transient, when one uses a sine wave to command a voltage clamp system, I_C is a predictable cosine wave. Such a current is less likely to interfere with the study of the early sodium current.

I-V CURVES

A third distinct advantage of the varying potential voltage clamp method is the possibility to directly display the membrane current-voltage relationships (I-V curves) rather than to reconstruct them from numerous step clamp measurements. By applying the varying potential to the time base (vertical plates) of an oscilloscope and I_M to the vertical gain amplifier (horizontal plates), one gets an "on line" direct display of I_M vs. E_M, i.e., the membrane I-V characteristics.

Additional advantages as well as limitations of the varying potential voltage clamp will become more evident as the method and results of further experiments carried out using it are described in detail.

The basic difference between the varying command potential and the classical step voltage clamp is in the command voltage. The standard square command voltages are replaced by varying potentials and the voltage clamp system forces E_M to follow its command potential.

COMPUTED AND EXPERIMENTAL RESULTS

In this chapter we shall deal with the application of two types of varying potentials to the voltage clamp: sine waves and triangular waves (Palti and Adelman, 1969, 1970). A preliminary report on an additional type of varying potential, a ramp, has been given by Fishman and Cole (1969). The ramp is very convenient for direct display of membrane I-V characteristics. Its special properties will be discussed at the end of this section.

Low Frequencies

Let us consider first a very slow sine wave. The potential at any time, is given by:

$$E_M = E_H + E_o \sin \omega t, \tag{8-3}$$

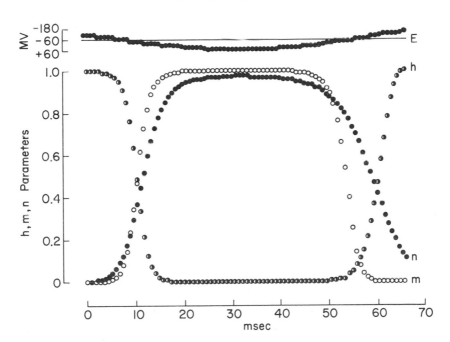

Fig. 8-1 *Computer readout of Hodgkin-Huxley m, n and h parameters plotted as a function of time during a sine wave voltage clamp of f = 10 Hz and amplitude of ±120 mv, symmetric with respect to $E_H = -60$ mv. Ordinate: upper trace, E_M change from just before beginning of depolarization. Downward inflection depolarization. Lower trace, m, n and h values: minimal value is 0.0, maximal value is 1.0. Abscissa: time in msec. Note initial saturation of h at maximal value of 1.0 and initial saturation of both m and n at minimal values of 0.0. At peak of depolarization, h is at zero while n is close to maximal. (From Palti and Adelman, 1969.)*

where E_H is the holding potential, E_o is the potential measured from E_H to the peak of the wave, $\omega = 2\pi f$, and f is the frequency of the wave.

Let f = 10 Hz and E_o = 120 mv. Fig. 8-1 plots the values of m, n, and h as a function of time for the above conditions. These values were calculated by means of a digital computer from the Hodgkin-Huxley equations (see chapter 7 and Palti and Adelman, 1970). As the rate of E_M change is slow with respect to τ_m, τ_n, and τ_h, the three parameters follow this low frequency wave, each being saturated either at zero or at unity when E_M approaches peak values. During the hyperpolarizing phase, m approaches zero, while h approaches zero as the depolarization develops. This explains why sodium current,

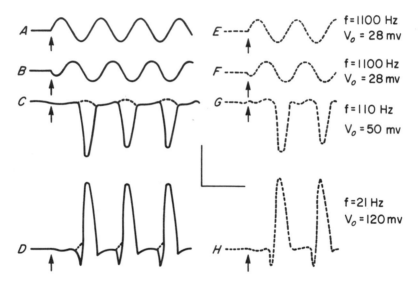

Fig. 8-2 *Typical experimental and calculated current and voltage traces of axons voltage clamped to sinusoidally varying potentials. V_o is the holding potential to peak amplitudes. Holding potential was to resting potential in living axon and −60 for the computed model axon.*
A. *An experimental E_M trace from an axon voltage clamped to a sine wave of 1100 Hz and ±28 mv.*
B. *The experimental trace of I_M obtained upon clamping an axon to the varying E_M illustrated in Fig. 8-2A.*
C and D. *The experimental traces of I_M obtained upon clamping axons to sine waves of 110 Hz and 21 Hz respectively. In C, V_o = 50 mv and in D, V_o = 120 mv. Interrupted line is I_M trace obtained in ASW in which all Na^+ was substituted by $Tris^+$.*
E. *Values of E_M calculated for the same conditions as the experimental trace of Fig. 8-2A.*
F, G and H. *Values of I_M calculated for the same conditions as the experimental traces of Fig. 8–2B, C, D, respectively.*
Arrows indicate beginning of sine oscillation. Horizontal bar = 1 msec in A, B, E and F, 10 msec in C and G and 50 msec in D and H.
Vertical bar = 100 mv in A and E, 1 ma/cm² in B, C, F and G, and 2 ma/cm² in D and H. For further details, see text. (From Palti and Adelman, 1969.)

which is determined by the product m^3h, is negligible during most of the cycle (see Fig. 8-2). There is only a short inward sodium transient which rapidly disappears as h tends to zero. Since n tends towards zero for hyperpolarization and towards unity for depolarization (see Fig. 6-5 and 8-1), the potassium current is almost completely rectified and consequently, so is the total membrane current (Fig. 8-2D, H). For this low frequency, dE_M/dt is so small that I_C becomes negligible compared to I_K [Eq. (6-5)]. Since g_L is so much smaller than g_K, throughout the depolarization $I_K \gg I_L$ (see chapter 7). Therefore, except when $E_M = 0$ and excluding the short initial transient sodium current, the measured I_M is predicted to be practically identical with the outwardly directed I_K (Fig. 8-2H). Fig. 8-2D illustrates that indeed the experimentally measured current closely resembles the predicted, predominating I_K current. For these conditions (depolarizing potentials beyond −40 mv), I_K is within 1–2 per cent of I_M.

The membrane currents generated reflect not only the time dependency of the various membrane properties, but also the membrane potassium I-V characteristics. An I-V curve calculated for the above conditions is presented in Fig. 8-3. All currents are negligible during the hyperpolarizing half cycle. Upon depolarization* an inward sodium current appears first. Sodium conductance is, however, soon inactivated and I_{Na} decays towards zero. For

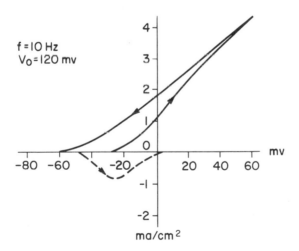

Fig. 8-3 *Computed potassium (continuous line) and sodium (interrupted line) current voltage characteristics of an axon voltage clamped to a sine wave of 10 Hz of* ±120 mv *in amplitude. Holding potential:* −60 mv. *Arrows indicate direction of potential and current changes.*
Ordinate: membrane current in ma/cm², *outward current up.*
Abcissa: membrane potential, E_M, *in* mv. *(From Palti and Adelman, 1969.)*

* The resting potential of the axon serving as a model for computations is −60 mv.

depolarization beyond $E_M = 5$ mv and upon membrane repolarization all the way from $E_M = 60$ mv to $E_M = -80$ mv, $I_{Na} = 0$. Throughout this range, the potassium I-V curve, I_K-V, superposes the measurable I_M-V curve. As the E_M changes are slow with respect to τ_n, the I_K-V relationship of Fig. 8-3 represents the so-called steady-state I-V (I_{ss}-V) curve. Indeed, the shape of the I_K-V curve in the specified range (see part of curve with arrow pointing to the left) closely resembles the I_{ss}-V curves obtained with the step clamp. Fig. 8-4

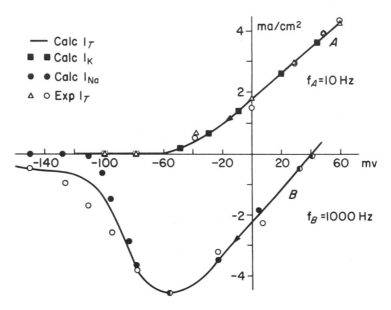

Fig. 8-4 *Typical experimental and computed membrane currents as a function of membrane potential (I — V curves). Axon voltage clamped to a sine wave of 10 Hz in A and 1000 Hz in B. $E_H = -60$ mv. Continuous lines are the calculated total currents (I_T) for each frequency. Open symbols experimentally determined total currents of two nerves. Filled symbols, calculated ionic currents; squares, potassium currents; circles, sodium currents. (From Palti and Adelman, 1969.)*

demonstrates that the experimental I-V relationship is in good agreement with the computed values.

The experimental I-V relationships can be directly displayed on an oscilloscope screen by feeding E_M to the horizontal amplifier time base and I_M to the vertical gain amplifier. Very similar results are obtained if one uses triangular waves with slopes corresponding to the above sine waves.

Since the conductance parameters can follow E_M at these relatively slow potential changes, the system is at a quasi steady-state. Therefore, the current generated by the first cycle is practically identical with that of consecutive cycles. Clamping the axon membrane potential for more than one cycle is necessary only when the preparation generates small currents and averaging is required to increase the signal-to-noise ratio.

Medium Frequencies

When the rates of E_M change are about ten times higher, i.e. corresponding to a frequency of about 100–500 Hz, potassium conductance (n) becomes too slow to respond to E_M. At the same time, the sodium conductance turn-on rate (τ_m) is sufficiently fast (see Fig. 6-6) so that it approaches unity upon depolarization (see Fig. 8-5). Depolarization may thus result in a twenty fold

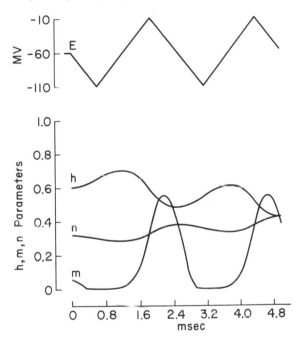

Fig. 8-5 *Computer readout of Hodgkin and Huxley m, n and h parameters plotted as a function of time during a triangular wave of f = 400 Hz and amplitude of ±50 mv. Ordinate: upper trace, membrane potential changes in mv. Lower trace, m, n and h values. Abscissa: time in msec. (From Palti and Adelman, 1969.)*

increase in m and a very small change in n and h. The result of this is that $I_{Na} \gg I_K$ [Eq. (6-6) and (6-18)]. Since dE/dt is still relatively small at these frequencies, I_C is also small; since $\bar{g}_{Na} \gg g_L$ during most of the depolarizing cycle, the total membrane current is almost pure I_{Na}. Fig. 8-6 illustrates the currents generated when f = 100 Hz, E_o = 50 mv for a sine wave when the nerve is slightly hyperpolarized, i.e., held at $E_H = -80$ mv. Under these conditions, when the potential change begins, the initial value of h is relatively high and those of m and n are low. However, while m recovers

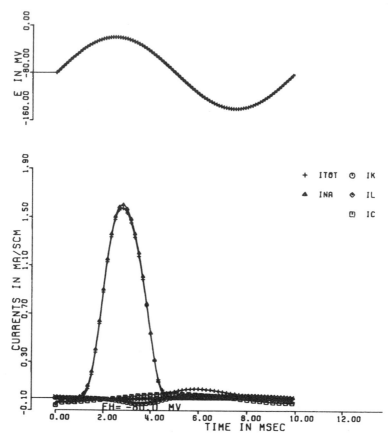

Fig. 8-6 *Computed values of membrane potential and currents as a function of time. Axon voltage clamped to a sine wave of 100 Hz ± 50 mv. $E_H = -80$ mv. Note similarity between the I_{Na} and total membrane current curves. Curve taken from computer Calcomp plotter.*

immediately to approach unity, n has a slow recovery and it only reaches a maximum of 0.4. The total current generated during the depolarization is therefore practically pure I_{Na}.

As in the case of lower frequencies, by feeding E_M to the x-axis and I_M to the y-axis of an oscilloscope, one may obtain a direct display of membrane sodium, I_{Na}-V, characteristics.

Fig. 8-2C and 8-2G illustrate that the actual currents measured under such conditions are in agreement with the predictions. Fig. 8-4B illustrates that the experimental I-V characteristics also conform with the computations. Note that in Fig. 8-2 sodium is frequently replaced by Tris to demonstrate that the current separation obtained by the different frequency response of the sodium and potassium conductances is in agreement with that obtained by chemical substitutions.

In this frequency range of 100–400 Hz, some of the conductance rate constants (τ_n and τ_h) are of the same order as the rate of E_M change. It therefore takes a few cycles until the system reaches a steady-state. The sodium current generated by the first cycle is usually different from that generated by the second cycle. However, as can be seen in Fig. 8-2C, the system reaches a steady-state within a few cycles.

High Frequencies

When the rate of E_M changes is increased by a factor of ten to reach frequencies of 1–10 KHz, I_{Na} completely predominates the ionic currents. However, since dE_M/dt is appreciable, capacity current no longer can be neglected. In fact if one uses a sine wave, a triangular wave or a fraction thereof (a ramp), of 1 KHz or more, and amplitudes of only 10–50 mv, the ionic currents become negligible with respect to I_C.

As shown earlier, if one uses a triangular wave, the capacity current is a square wave. From the amplitude of this wave, C_M can be easily calculated. Any ionic current superposed on the capacity current is easily detectable as it results in a distortion of the square wave. Conditions can therefore be sought where the current is purely capacitive. Moreover, as C_M is determined from the amplitude of the step in I_M, ionic currents which are relatively slow to develop do not change this jump but rather modify the plateau of the square wave sometime after the jump.

Fig. 8-7 illustrates that the amplitude of the computed jump in I_M is identical with that of I_C. Therefore the step in I_M may be used to determine C_M. Fig. 8-8 illustrates a typical membrane current generated by a 2 KHz triangular potential wave. The current is seen to be a square wave. From the amplitude of such waves, membrane capacity can be determined [see Eq. (8-2)] as a function of membrane potential, medium composition, temperature or any other set of conditions. The value of C_M in the squid axon obtained by means of the above method under normal conditions is $1.09 \pm 0.1^* \mu f/cm^2$ at 1 KHz. In contrast to C_M determinations using external electrodes, membrane capacity C_M was shown to be practically independent of frequency in the 200–2000 Hz range. Membrane capacity was also shown to be very little affected, if at all, by medium composition.

The varying potential control voltage clamp method enables nonchemical current separation, or line display of membrane current-voltage characteristics and simple and accurate determination of membrane capacity.

* Standard deviation.

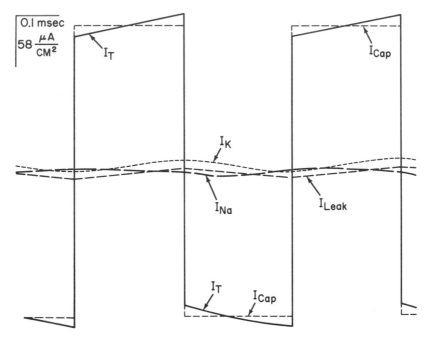

Fig. 8-7 *Computer readout of membrane currents generated by an axon voltage clamped to a triangular wave of* 2000 Hz *and amplitude of* ± 10 mv, *symmetrical with respect to the holding potential,* $E_H = -60$ mv. *Ordinate: membrane currents. Maximal* $I_C = \pm 160$ *μa/cm^2. Abscissa: time in msec. Note that the jumps in* I_T *equal those of* I_C. (*From Palti and Adelman,* 1969.)

Kinetics studies of the effect of rapidly acting drugs, TTX, for example, on membrane I-V characteristics, are practically impossible to carry out by means of the step clamp, as each I-V curve takes a few minutes to construct. The direct display of the whole sodium I-V relationship within less than 10 msec makes such studies simple and accurate with the varying potential control.

As I-V curves and separated currents can be obtained using sine wave of 10 Hz to 500 Hz. The frequency response bandwidth required from the electronic system is only about 1 to 10^4 Hz as compared with D.C. to 1 MHz for the step clamp. This fact makes it possible to clamp preparations which, because of the high impedances associated with the electrodes which penetrate them and the resulting large RC, cannot be clamped by the step clamp. Preparations which generate very small currents and thus have a very poor signal to noise ratio can also be studied by averaging numerous consecutive cycles.

The major difficulties encountered using the varying potential control

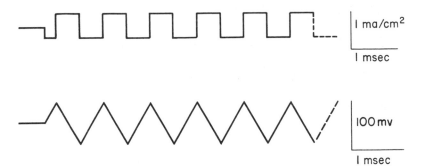

Fig. 8-8 *Typical capacity current wave (upper trace) generated by an axon voltage clamped to a triangular potential wave (lower trace) of f = 900 Hz and amplitude of ±40 mv. Note the square wave shape of the membrane current. (From Palti and Adelman, 1970.)*

voltage clamp are its inability to give simple determinations of membrane time constants (τ_m, τ_h and τ_n) as defined by Hodgkin and Huxley (unless the wave is interrupted and current decay measured) and the difficulty in determining the inactivation parameters in Hodgkin-Huxley terms.

REFERENCES

FISHMAN, N. J., and K. S. COLE. 1969. On Line Measurement of Squid Axon Current Potential Characteristics. *Federation Proc.* 28: 333.

PALTI, Y., and W. J. ADELMAN, Jr. 1969. Voltage Clamp Measurements of Membrane Conductance and Capacity by Oscillating Potential Control. *Federation Proc.* 28: 333.

———. 1969. Measurement of Axonal Membrane Conductances and Capacity by means of a Varying Potential Control Voltage Clamp. *J. Memb. Biol.* 1: 431–458.

ANALYSIS AND RECONSTRUCTION OF AXON MEMBRANE ACTION POTENTIAL

9

Y. PALTI

The Hodgkin-Huxley equations (see chapter 6) describe the axon membrane ionic conductances as a function of membrane potential, E_M, and time. They can be used to predict E_M and membrane currents under a variety of conditions including the active state. Making use of these equations we shall try to analyze the generation of the nerve impulse.

In steady-state resting axon the net membrane current, I_M, is zero and E_M a constant. To activate such an axon, i.e. to elicit an action potential, one may depolarize its membrane by driving a stimulatory current of proper direction through the membrane. To understand how such stimulation results in generation of an action potential, let us limit the discussion here to the processes taking place in a space-clamped axon.* In such an axon, potential differences cannot prevail along the membrane so that any change in E_M is simultaneous along the whole length of the axon under investigation. This experimental condition is extremely useful for analyzing nerve behavior as it simplifies the major variables† and thus simplifies analysis. Furthermore, as there is no current flow along the membrane, there are no local circuit currents.

When an axon is stimulated, membrane current, I_M, equals the stimulating current, I_p. Ionic currents take time to turn on, and are negligible at the

---

* In the space-clamped axon, a metal wire inserted along the axon axis short circuits the axon internal medium while a set of external electrodes shorts the medium just outside the membrane.

† In an axon which is not space clamped, membrane potential, conductances and currents are complex functions of location along the fiber.

beginning of the stimulating period; therefore the main component of I_M is the capacity current, I_C. This current either charges, or, for depolarizations, discharges the membrane capacity.

When E_M is displaced from its resting value of about -60 mv towards zero by means of a stimulating current, the membrane permeabilities for sodium, potassium, and perhaps other ions, increase. The resulting changes in ionic membrane conductance parameters m and h (for sodium) and n (for potassium) are given as a function of potential and time by the following equations (see chapter 6):

$$dm/dt = \alpha_m(1 - m) - \beta_m m \qquad (9\text{-}1)$$

$$dh/dt = \alpha_h(1 - h) - \beta_h h \qquad (9\text{-}2)$$

$$dn/dt = \alpha_n(1 - n) - \beta_n n, \qquad (9\text{-}3)$$

where the α's and β's are voltage dependent rate constants. The rates of change of n and h are relatively slow with their time constants, τ_n and τ_h in the order of 3–10 msec (see chapter 6), while m varies faster, $\tau_m \le 1$ msec. Fig. (9-1) illustrates typical changes in m, n and h during and after a square current pulse.

The membrane ionic conductances, g_i, are related to the conductance parameters, m, n and h by the following relationships:

$$g_{Na} = \bar{g}_{Na} m^3 h \qquad (9\text{-}4)$$

$$g_K = \bar{g}_K n^4, \qquad (9\text{-}5)$$

where $\bar{g}_{Na}$ and $\bar{g}_K$ are constants. The membrane ionic currents, I_i, are related to the ionic conductances by relationships of the following general nature (Hodgkin and Huxley, 1952):

$$I_i = g_i(E_M - E_i), \qquad (9\text{-}6)$$

where E_i is the ionic current reversal potential. Stimulatory currents thus change both E_M and membrane conductances and result in changes in the amplitudes of the individual ionic currents. These changes do not end with the stimulation but persist after this current is shut off; I_{Na} decays with time constant τ_M and I_K with τ_n (see chapter 6 and Fig. 9-1). However, since there are no local circuit currents in the space-clamped axon, just after termination of the stimulus, when I_p will equal zero, $I_M = 0.0$.

Since membrane current, I_M, is given at any time by:

$$I_M = C_M dE_M/dt + \sum_1^i I_i, \qquad (9\text{-}7)$$

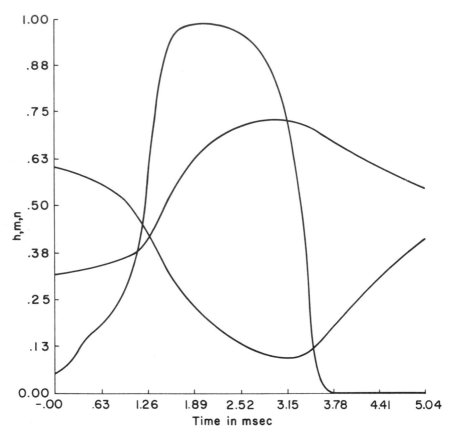

Fig. 9-1 *Computer print-out of computed axon conductances parameters m, n and h. Values are computed during and after a 1 msec, 30 μa depolarizing current pulse.*

where C_M = membrane capacity, the changes in E_M are given by (see chapter 10):

$$dE_M/dt = (1/C_M) \times (I_M - \sum_{1}^{i} I_i) \qquad (9\text{-}8)$$

$$dE_M/dt = (1/C_M) \times [I_M - \bar{g}_{Na}m^3h(E_M - E_{Na}) \\ - \bar{g}_K n^4(E_M - E_K) - g_L(E_M - E_L)]. \qquad (9\text{-}9)$$

At the end of the stimulus, I_i is different from that in the resting state. Therefore, $\sum I_i$ which equalled zero at rest, now has a finite value. The difference between I_M and $\sum I_i$ [Eq. (9-8)] is made up by the capacity current such that

$I_M = 0$. The fact that the term $(I_M - \sum I_i)$ in Eq. (9-8) now has a finite value requires that $dE_M/dt \neq 0$, i.e. that E_M continues to change (because of the I_i and not only I_{C_p} changes) even after the end of the stimulus. These E_M changes can be calculated by integrating Eq. (9-9) after substituting proper values for the conductance parameters at any E_M and t (see chapter 10). The direction of E_M change is determined by the direction (sign) of the net ionic current of Eq. (9-8). (Note that I_M is now zero.) Thus the initial change in E_M [Eq. (9-4)] is membrane depolarization.* The immediate result of this additional depolarization is a further increase in m (see Fig. 9-1). Since the decrease in h is much smaller ($\tau_h \gg \tau_m$), g_{Na} is increased and with it inward sodium current. As a consequence, E_M is further depolarized. The process of depolarization thus tends to maintain itself. Once the conductances were changed enough to result in a net inward ionic current, depolarization progresses further whether or not the stimulus is still present.

This increasing rate of growth of inward I_{Na} together with the membrane depolarization is eventually checked by three factors:

1. As depolarization progresses, E_M approaches E_{Na} ($E_{Na} \approx +40$ mv) so that the sodium ion driving force ($E_M - E_{Na}$) [see Eq. (9-4)] approaches zero. This decrease of the driving force soon overrides the effect of increasing g_{Na}. Obviously E_{Na} is the limit beyond which E_M cannot be driven by the above mechanisms.

2. While depolarization progresses, both g_K and ($E_M - E_K$) increase so that the outward† I_K is increased. The increase in g_K follows the increase of n^4 with depolarization. As seen in Fig. (9-1), the turn-on rate of n lags considerably as compared with that of m. However, its rate is comparable with that of h. The increasing outward I_K tends to balance out the inward I_{Na} such that $\sum I_i$ approaches zero and so does the rate of change of E_M [dE_M/dt of Eq. (9-8)].

3. As time progresses, sodium inactivation develops at a rate comparable to that of n. The decrease in sodium conductance parameter, h, is

---

*A different approach to explain the E_M changes which are coupled to the g_i changes is to use the constant field equation:

$$E_M = \frac{p_{Na}a_{Na_o} + p_K a_{K_o} + \cdots}{p_{Na}a_{Na_{in}} + p_K a_{K_{in}} + \cdots}$$

The stimulus results in an initial increase in g_{Na} which tends to shift E_M towards E_{Na}, i.e., depolarization. Later when the relative change in g_K predominates, E_M tends back towards E_K and the resting potential.

† Under the above conditions I_{Na} is inward while I_K is outward as ($E_M - E_{Na}$) is negative while ($E_M - E_K$) has a positive value ($E_K \approx -90$ mv).

thus the third factor checking the depolarization due to inward I_{Na}. Eventually as outward I_K grows and inward I_{Na} attenuates, outward current becomes larger than inward current (see Fig. 9-2).

Following Eq. (9-8), dE_M/dt reverses its sign, E_M repolarizes and tends back to its original value (see Fig. 9-3). As repolarization progresses, the conductance parameters (m, n and h) reverse their direction of change (see Fig. 9-1); while h increases, m and n decrease. However in spite of the rapid decrease in m (Fig. 9-1), I_{Na} is seen to increase (Fig. 9-2). This increase is only slightly affected by the increase in h. The main factor responsible for

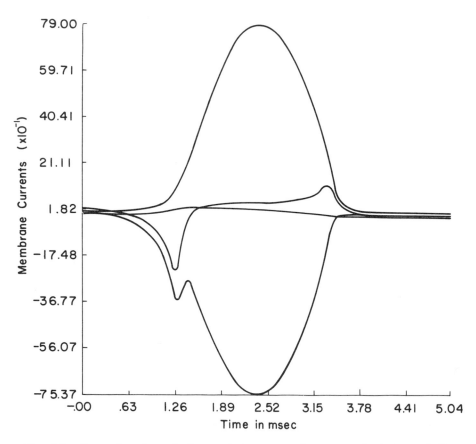

Fig. 9-2 *Computer print-out of computed axon membrane currents during and after 1 msec 30 μa depolarizing current pulse. Outward currents are upward. Ordinate: Currents in μa/cm². Max. $I_{Na} = 780$ μa/cm², Max. $I_K = 840$ μa/cm². Other details as in Fig. 9-1.*

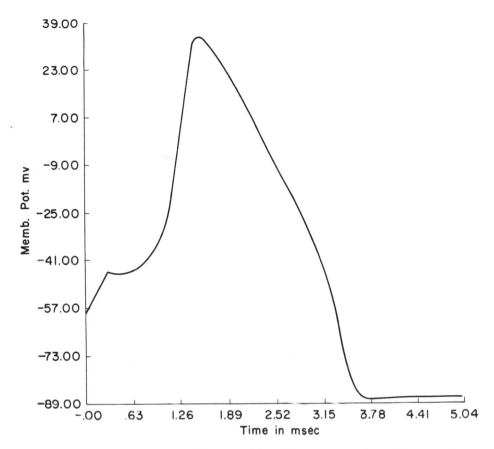

Fig. 9-3 *Computer print-out of computed E_M values corresponding to the parameter and current changes of Fig.* 9-1 *and* 9-2. *Initial E_M value,* -60 *mv. Other details as in Fig.* 9-1.

it is the fact that the value of $(E_M - E_{Na})$ which was close to zero at the peak of the action potential, now enlarges rapidly. However, as E_M approaches resting level, m tends to its very small resting value of 0.05 and I_{Na} decays to its negligible resting value.

I_K also continues to increase beyond the peak of E_M since n is slow to respond to E_M; I_K reaches a peak well after the E_M peak (Fig. 9-1). Since E_M tends to its resting level both $(E_M - E_K)$ and n decrease so that I_K also decays towards its resting level.

At the end of the action potential, when E_M has reached its resting value, I_{Na} is almost completely inactivated. This inactivation results from the fact that m, which follows E_M with no delay, reaches its resting value while h is still very small (Fig. 9-1). However, at the same time, n, which is also slow to respond to E_M, has a value above its resting value. Therefore, when E_M first reaches its resting value, I_{Na} is smaller than its resting value, while I_K is larger. Thus, at this point there is a net outward current.* From Eq. (9-8) it follows that dE_M/dt is not zero and E_M is hyperpolarized. This positive after potential reaches a peak of about 10 mv and then decays as n and h tend toward their resting level.

The ionic current associated with the action potential can also be obtained from the solution of Eq. (9-9). Such currents are given in Fig. 9-2. The value of the capacity current, I_C, is given at any time by the derivatives of E_M;

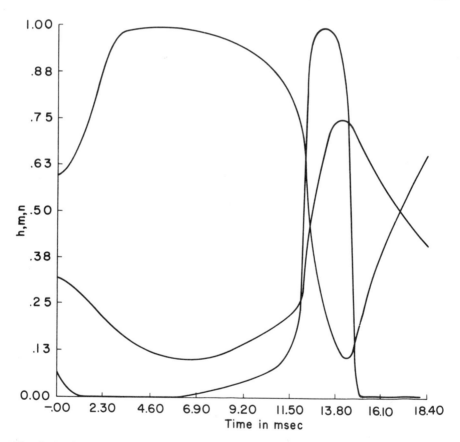

Fig. 9-4 Computer print-out of computed axon conductance parameters, m, n and h. Values computed during and after a 3 msec, 15 μa/cm² hyperpolarizing pulse. Other details as in Fig. 9-1.

* Under these conditions, leakage current, I_L, is inward but is relativaly small.

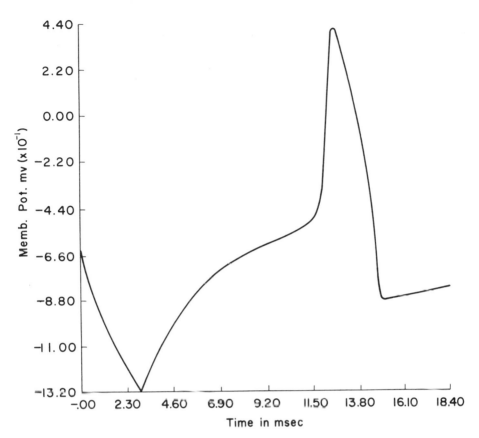

Fig. 9-5 *Computer print-out of computed E_M values during and after the 3 msec hyperpolarizing current pulse which induced the conductance parameter changes of Fig. 9-4. Initial E_M value −60 mv. Note the initial charging of membrane capacity and long delay between end of stimulus and beginning of action potential.*

$I_C = C_M \cdot dE_M/dt$. Its locus in time is therefore biphasic, crossing the zero line when E_M peaks (see Fig. 9-2).

Fig. 9-4 illustrates changes in the conductance parameters associated with a 15 μa hyperpolarizing current. The changes are opposite to those of Fig. 9-1. These changes continue after the current is turned off for the same reasons they continue after a depolarizing pulse is shut off. Since $\tau_n > \tau_m < \tau_h$, m recovers sooner than n at the end of the hyperpolarizing current. Since at the same time h has a value higher than normal, g_{Na} has a considerably higher value than that of resting while the value of g_K is still below that of resting. These abnormal conductances result in a net inward depolarizing membrane current. This result is identical with that obtained in the case of a depolarizing stimulating current. The depolarization obtained results in a further increase in inward I_{Na} and further depolarization. The process thus progresses into the regular action potential illustrated in Fig. 9-5. This process, which takes place after the end of a hyperpolarizing stimulus is generally referred to as an anodic* break response.

The shapes of action potentials computed as described above (Fig. 9-3 and 9-5) are in excellent agreement with the experimental ones. By methods similar to the above, membrane action potentials can be reconstructed for a variety of other conditions. In all cases, the results are in good agreement with experiments.

REFERENCES

HODGKIN, A. L., and A. F. HUXLEY. 1952. A Quantitative Description of Membrane Current and its Application to Conduction and Excitation in Nerve. *J. Physiol. (London)*. 117:500–544.

* The term "anodic" comes from stimulation by means of a pair of external electrodes. In such a case depolarization occurs under the cathode and hyperpolarization under the anode.

DIGITAL COMPUTER RECONSTRUCTION OF AXON MEMBRANE ACTION POTENTIAL

10

Y. PALTI

The following computer program is written in Fortran IV.* The program solves for changes in the membrane potential and conductance parameters when the membrane current is stepped from one constant value to another. The computation is carried out for a space-clamped axon in which any change in membrane potential, E_M, occurs simultaneously along the whole length of nerve under investigation (see chapter 6). Obviously under this set of conditions the action potential does not propagate. This type of activity is termed membrane action potential.

The computer program uses the same dictionary as the program in chapter 7, together with the additions given below and follows the flowchart given in Fig. 10-1. The program solves for changes produced in E_M as well as other membrane parameters by the stimulating current on the basis of Eq. (7-1). Simple rearrangement of that equation gives:

$$dE_M/dt = 1/C_M[I_M - \bar{g}_{Na}m^3h(E_M - E_{Na}) - \bar{g}_K n^4(E_M - E_K) - g_L(E_M - E_L)]. \tag{10-1}$$

For the condition described in chapter 7 and for a sufficiently small increment of time, Δt:

$$\Delta E_M = \Delta t \cdot dE_M/dt. \tag{10-2}$$

Under the steady-state resting conditions, $I_M = 0$, $dE_M/dt = 0$ and E_M has the value E_{M_0}. By means of Eq. (6-15), (6-20b) and (6-22b), the corresponding

* Note that the input-output statements are for IBM 360 series and may differ in other machines.

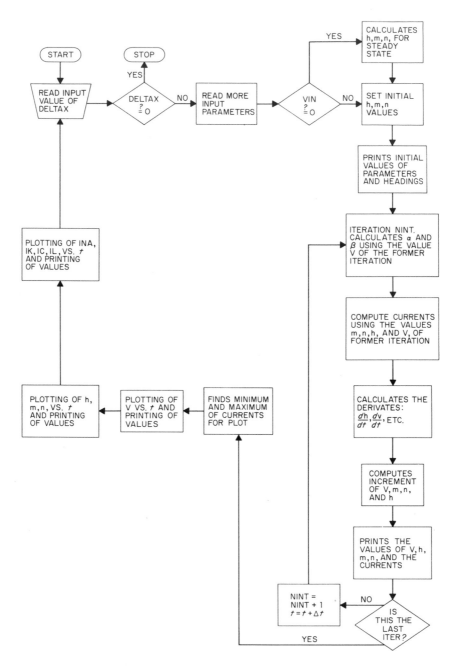

Fig. 10-1 *Flowchart of computer program for reconstruction of axon membrane action potential.*

values of m, n and h (m_0, n_0 and h_0) are calculated for E_{M_0}. Let I_M be stepped now to some value, I_M' at $t = 0$. Introducing the calculated values of m_0, n_0, h_0, the proper constants ($\bar{g}_{Na}$, E_{Na}, $\bar{g}_K$, etc.), Δt and the value of I_M' into Eq. (10-1), one can solve for dE_M/dt. The value of dE_M/dt thus obtained is introduced into Eq. (10-2) and ΔE_M solved. Membrane potential, E_M at time $t_0 + \Delta t$ now has a new value, $E_{M_0} + \Delta E_M$.

As E_M changes, the membrane conductance parameters, n, m and h, will change from their steady-state values, n_0, m_0 and h_0, towards a new steady-state as described by differential equations (6-10), (6-19) and (6-21). A solution of such an equation is given in Eq. (7-12). Using the new value of E_M, α_n and β_n can now be recalculated by means of Eq. (6-8) and (6-9). Similarly, α_m, β_m, α_h and β_h can be calculated by means of analogous equations. Using these new values of the α's and β's, the program now solves Eq. (6-14), (6-20a) and (6-22a) for the values of m, n and h at time $t_0 + \Delta t$. These new values are now set as initial conditions and Eq. (10-1) and (10-2) are solved again, this time for time $t_0 + 2\Delta t$. By repeating this process over and over, E_M, m, n and h are obtained as a function of time. From these values, the membrane currents are also calculated. At any desired time (designated t_1 in the program), I_M can be stepped back to zero or to a new value and the solution process continued.

For example, to reconstruct a typical excitation process, I_M is stepped from its initial value of zero to about 20 μa for a duration of close to a msec. As illustrated in Fig. 9-1, this short current suffices to disturb the values of m, n and h enough so that a membrane action potential is elicited. Note that at the end of the stimulus, while I_M is zero, the ionic currents are not. This is due to the fact that the capacity current is equal in amplitude but opposite in direction to the sum of all the ionic currents [see Eq. (6-5)].

The given program lists the values of E_M, I_{Na}, I_K, I_L, m, n and h during the stimulating period and for any desired period thereafter. These values are plotted by the computer print out. They can be also plotted by a Calcomp or a similar plotter program. A set of statements for the print out plotting is included in the program.

Typical plots of the data computed by this program are given in Fig. 9-1, 9-2 and 9-3. Fig. 9-1 plots E_M as a function of time and illustrates a membrane action potential elicited by a 20 μa, one msec-long depolarizing current. Fig. 9-2 gives the values of the conductance parameters (m, n and h) during the stimulus and action potential, while Fig. 9-3 gives the separated ionic currents during the same periods.

It is obvious that this computer program can be used for the investigation of axon electrical behavior other than action potential reconstruction. For example, it can be used for the reconstruction of local responses, or to predict membrane behavior under the influence of a hyperpolarizing current.

FORTRAN PROGRAM

```
C     NUMERICAL SØLUTIØN FØR H-H MNHV FUNCTIØNS IN CURRENT CLAMP
      REAL INA, IK, ILK, IC
      INTEGER DØT, FNA, FMA, HA, FK, C, FL, VA, EX, BLANK, ALINE
      DATA DØT, FNA, FMA, HA, FK, C, FL, VA/'.', 'N', 'M', 'H', 'K', 'C', 'L',
     1'V'
      DATA EX, BLANK/'X', '   '/
      DIMENSIØN V(1500), EM(1500), EN(1500), H(1500), ALINE(101)
      DIMENSIØN FINA(1500), FIK(1500), FIL(1500), FIC(1500)
      DIMENSIØN GAMMA(7), Y(4), DERY(4)
      EQUIVALENCE (EN(1), FINA(1)), (EM(1), FIC(1)), (H(1), FIK(1))
      EQUIVALENCE (V(1), FIL(1))
      CM = 1.0
      RCPCM = 1.0/CM
      EL = -10.5989
C     READ PARAMETERS AND INITIAL VALUES
99    READ (5,100) DELTAX
100   FØRMAT (F10.6)
      IF (DELTAX.EQ.0.) STØP
      READ (5,101)CØNSTI, VIN, HIN, EMIN, ENIN, T0, T1, TFIN
      READ (5,110) ENA, EK, GNA, GK, GL, TEMPFI, FI2, NDEL
110   FØRMAT (7F10.5, I5)
101   FØRMAT (8F10.5)
      FIMNH = 3.0 ** ((TEMPFI - 6.3)/10.0)
C     TEST FØR STEADY STATE
      IF (VIN.EQ.0.0) GØ TØ 2
      GØ TØ 3
2     ALFAH = 0.07
      BETAH = 1.0/(EXP(3.0) + 1.0)
      HIN = ALFAH/(ALFAH + BETAH)
      ALFAEM = 2.5/(EXP(2.5) - 1.0)
      BETAEM = 4.0
      EMIN = ALFAEM/(ALFAEM + BETAEM)
      ALFAEN = 0.1/(EXP(1.0) - 1.0)
      BETAEN = 0.125
      ENIN = ALFAEN/(ALFAEN + BETAEN)
3     Y(1) = HIN
      Y(2) = EMIN
      Y(3) = ENIN
      Y(4) = VIN
      X = T0
```

```
      FITØT = CØNSTI
      NINT = 1
C     PRINT INITIAL VALUES AND CØLUMN HEADINGS
      WRITE (6,109) ENA, EK, GNA, GK, GL, TEMPFI
109   FØRMAT (1H1, 4HENA = , F10.5, 6X, 3HEK = , F10.5, 6X, 4HGNA = ,
      1F10.5, 6X, 3HGK = , 1F10.5, 6X, 3HGL = , F10.5, 6X,
      212HTEMPERATURE = , F10.5)
      WRITE (6,105) T0, T1, TFIN, CØNSTI, VIN, HIN, EMIN, ENIN
105   FØRMAT (1H0, 3HT0 = , F10.5, 10X, 3HT1 = , F10.5, 10X, 5HTFIN = ,
      1F10.5, 10X, 3HIT = , F10.5,
      210X, 36HTIME IS IN MSEC, AND I IS IN MICRØA.//4HVIN = , F10.5,
      33X, 3HMV., 4X, 4HHIN =,
      4F10.5, 10X, 5HEMIN =, F10.5, 10X, 5HENIN =, F10.5///1H0, 4X, 4HT(I),
      58X, 1HH, 8X, 1HM, 8X, 1HN, 8X, 5HV MV., 7X, 3HINA, 9X, 2HIK, 9X,
      63HILK, 10X, 2HIC///)
C     CØMPUTE ALFA AND BETA FØR EVERY V
10    TEMP = 0.1 * (Y(4) + 25.0)
      BETAEM = 4.0 * EXP(Y(4)/18.0)
      IF (ABS(Y(4) + 25.0).LT.1.0E − 5) GØ TØ 5
      ALFAEM = TEMP/(EXP(TEMP) − 1.0)
      GØ TØ 4
5     ALFAEM = 1.0
4     ALFAH = 0.07 * EXP(Y(4)/20.0)
      BETAH = 1.0/(EXP(0.1 * (Y(4) + 30.0)) + 1.0)
      TEMP = 0.1 * (Y(4) + 10.0)
      BETAEN = 0.125 * EXP(Y(4)/80.0)
      IF (ABS(Y(4) + 10.0).LT.1.0E − 5) GØ TØ 6
      ALFAEN = (0.1 * TEMP)/(EXP(TEMP) − 1.0)
      GØ TØ 7
6     ALFAEN = 0.1
C     CØMPUTE IØNIC CURRENTS
7     INA = GNA * Y(2) ** 3 * Y(1) * (Y(4) − ENA)
      IK = GK * Y(3) ** 4 * (Y(4) − EK)
      ILK = GL * (Y(4) − EL)
      IF (X.GE.T1) CØNSTI = FI2
      IC = CØNSTI − INA − IK − ILK
      RCPCM = 1.0/CM
      GAMMA(1) = ALFAH
      GAMMA(2) = ALFAH + BETAH
      GAMMA(3) = ALFAEM
      GAMMA(4) = ALFAEM + BETAEM
      GAMMA(5) = ALFAEN
```

```
        GAMMA(6) = ALFAEN + BETAEN
        GAMMA(7) = IC * RCPCM
        DERY(1) = (GAMMA(1) − GAMMA(2) * Y(1)) * FIMNH
        DERY(2) = (GAMMA(3) − GAMMA(4) * Y(2)) * FIMNH
        DERY(3) = (GAMMA(5) − GAMMA(6) * Y(3)) * FIMNH
        DERY(4) = GAMMA(7)
        DØ 8 I = 1, 4
8       Y(I) = Y(I) + DERY(I) * DELTAX
        X = X + DELTAX
        DØ 9 J = 1, 3
        IF (Y(J).GT.1.0) Y(J) = 1.0
9       IF (Y(J).LT.0.0) Y(J) = 0.0
        H(NINT) = Y(1)
        EM(NINT) = Y(2)
        EN(NINT) = Y(3)
        V(NINT) = Y(4)
        NINT = NINT + 1
        WRITE (6,130)X, (Y(I), I = 1, 4), INA, IK, ILK, IC
130     FØRMAT (1HO, 1PE12.5, 0P3F9.5, F11.4, 4F12.5)
        IF (NINT.GT.1500) GØ TØ 20
        IF (X.GE.TFIN) GØ TØ 20
        GØ TØ 10
C       THIS PART IS FØR THE PLØTTING ØF THE GRAPHS
20      STEPT = NDEL * DELTAX
        JM = 1
        JN = 1
        JH = 1
        CØNSTI = FITØT
        TLIMT = 0.0
C       NG = 1 IS THE PLØT ØF V(T),
C       NG = 2 IS THE PLØT ØF MNH
C       NG = 3 IS THE PLØT ØF THE CURRENTS
        DØ 30 NG = 1, 3
        YMIN = 1.0E5
        YMAX = −1.0E5
        IF (NG − 3) 31, 32, 31
C       FIND MINIMUM AND MAXIMUM CURRENTS
        NINT1 = NINT − 1
32      DØ 29 I = 1, NINT1, NDEL
        Y(1) = H(I)
        Y(2) = EM(I)
        Y(3) = EN(I)
```

```
      Y(4) = V(I)
      FINA(I) = GNA * Y(2) ** 3 * Y(1) * (Y(4) − ENA)
      FIK(I) = GK * Y(3) ** 4 * (Y(4) − EK)
      FIL(I) = GL * (Y(4) − EL)
      TLIMT = TLIMT + STEPT
      IF (TLIMT.GE.T1) FITØT = FI2
      FIC(I) = FITØT − FINA(I) − FIK(I) − FIL(I)
      YMIN = AMIN1(YMIN, FINA(I), FIK(I), FIC(I), FIL(I))
29    YMAX = AMAX1(YMAX, FINA(I), FIK(I), FIC(I), FIL(I))
31    DØ 22 J = 7,101
22    ALINE(J) = DØT
C     PRINT A LINE ØF DØTS AS Y AXIS
      WRITE (6,103) ALINE
103   FØRMAT (1H1, 101A1)
      DØ 23 J = 1, 101
23    ALINE(J) = BLANK
      ALINE(1) = DØT
      TLIMT = 0.
      IF (NG − 2) 50, 51, 52
50    DØ 25 IV = 1, NINT1
      YMAX = AMAX1 (V(IV), YMAX)
25    YMIN = AMIN1 (V(IV), YMIN)
52    STEP = (YMAX − YMIN)/100.0
      GØ TØ 53
51    STEP = 0.01
      YMIN = 0.0
      YMAX = 1.0
53    DØ 24 I = 1, NINT1, NDEL
      TLIMT = TLIMT + STEPT
      TLIMT2 = TLIMT + STEPT
      IF (NG − 2) 60, 61, 62
60    JV = 1.5 + (V(I)-YMIN)/STEP
      ALINE(JV) = VA
      GØ TØ 26
61    JN = 1.5 + EN(I)/STEP
      JH = 1.5 + H(I)/STEP
      JM = 1.5 + EM(I)/STEP
      ALINE(JN) = FNA
      ALINE(JH) = HA
      ALINE(JM) = FMA
C     N IS NA CURRENT, C IS FØR IC,
C     K IS K CURRENT, L FØR IL
```

```
      GØ TØ 26
62    JN = 1.5 + (FINA(I) − YMIN)/STEP
      JM = 1.5 + (FIC(I) − YMIN)/STEP
      JH = 1.5 + (FIK(I) − YMIN)/STEP
      JV = 1.5 + (FIL(I) − YMIN)/STEP
      ALINE(JN) = FNA
      ALINE(JH) = FK
      ALINE(JM) = C
      ALINE(JV) = FL
C     THE PØINT MARKED BY X IS THE TIME WHEN THE EXTERNAL CURRENT IS
C     STØPPED
26    IF (TLIMT.LE.T1.AND.TLIMT2.GE.T1)ALINE(1) = EX
      WRITE (6,104) ALINE
104   FØRMAT (1H, 101A1)
      ALINE(JN) = BLANK
      ALINE(JM) = BLANK
      ALINE(JH) = BLANK
      ALINE(JV) = BLANK
24    ALINE(1) = DØT
      WRITE (6,102) YMIN, YMAX, STEPT, STEP, T0, T1, TFIN, CØNSTI
102   FØRMAT (1H0, 5HYMIN =, F10.5, 3X, 5HYMAX =, F10.5, 3X, 6HSTEPT =,
     1F10.5, 3X, 5HSTEP =, F10.5, 3X, 3HT0 =, F7.3, 3X, 3HT1 = , F7.3, 3X,
      25HTFIN =, F7.3, 3X, 3HIC =, F8.4)
30    CØNTINUE
      GØ TØ 99
      END
```

DICTIONARY

(See the dictionary at the end of chapter 7 for additional definitions.)
CONSTI—stimulating current intensity in $\mu a/cm^2$
TFIN—time of end of computation
T0—time of beginning of first step-change in I_M
T1—time of second step-change in I_M
DELTAX—Δt

ANALYSIS AND RECONSTRUCTION OF PROPAGATED IMPULSE IN THE SQUID GIANT AXON

11

Y. PALTI

In contrast to the membrane action potential described in chapter 9, the natural action potential is a propagated response. It is associated with the generation of potential differences and ionic currents along the nerve fiber.

The analysis of the generation and propagation of a nerve action potential is based on both the voltage and time dependent ionic conductances as formulated by Hodgkin and Huxley (1952*a, b, c*) and the so-called axon cable properties which describe the spread of potential and current along a fiber. We will limit the discussion here to nonmyelinated fibers. (For myelinated nerves see Goldman and Albus, 1968). Let us consider first the cable properties of an axon. In this connection, one pictures the axon membrane as being composed of a series of elements each made out of a resistance and capacity connected in parallel. These elements are interconnected at one end by resistances R_2, representing the axoplasm resistance and at the other end by resistance R_1, representing the external medium resistance (see Fig. 11-1).

When the potential at any point ($x = 0$) along the axon is changed by ΔE_o, its value at any point x at a distance from x_0 is given by:

$$E_x = E_o\,e^{-x/\lambda}, \tag{11-1}$$

where λ is the space or length constant of the axon. For a constant ΔE_o, λ gives the distance at which E equals E_o/e. In typical axons $\lambda = 1$–3 mm, depending on the values of internal, external and membrane resistances. The potential differences along the axon obviously generate currents. Such currents are generally called local circuit currents and for an impulse travelling

223

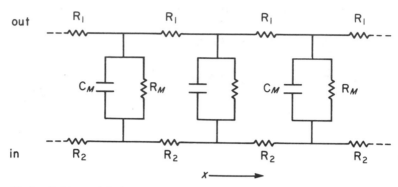

Fig. 11-1 *Cable model of an axon in a conductive medium.*

along an axon they have the general form given in Fig. 11-3. The cable theory assumes that these currents flow either along the membrane or normal to it.

On the basis of this assumption the so-called one-dimensional electric models (Fig. 11-1) of axons placed in their natural conductive medium were constructed. It should be noted that the one-dimensional cable model is only an approximation of the true three-dimensional axon in its surrounding medium. As Clark and Plonsey (1966) have shown, under a variety of conditions this approximation may introduce a significant error.

To compute the currents in an axon let us analyse the currents at a point P inside an infinitely long axon when P is sufficiently removed from any electrode generating currents or potential differences. Let the internal longitudinal current at P be I_2 (Fig. 11-2) and the axoplasm resistance be R_2 (in ohms/unit length). Let I_M be membrane current at P (in amp/unit axon length) and I_1 and R_1 the corresponding external current and unity length resistance.

As no current is injected near P, to maintain a closed circuit, internal and external currents must be equal:

$$I_1 = I_2. \tag{11-2}$$

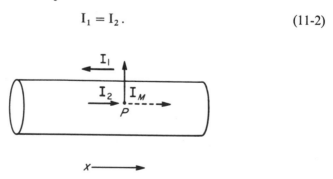

Fig. 11-2 *Schematic representation of currents around a point P inside an axon.*

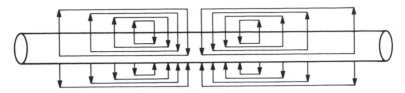

Fig. 11-3 *Schematic representation of local circuit currents associated with a propagated action potential in a nonmyelinated axon. Internal currents are represented only for upper side of the membrane.*

By Ohm's law ($\Delta V = IR$) along the external membrane surface:*

$$\partial E_1/\partial x = -R_1 I_1, \tag{11-3}$$

and the internal surface

$$\partial E_2/\partial x = R_2 I_2. \tag{11-4}$$

From Eq. (11-2), (11-3) and (11-4) we get for the membrane potential difference at P:

$$\partial E_M/\partial x = I_2(R_1 + R_2) \tag{11-5}$$

Any change in axial current along a small distance along x, $\partial I_2/\partial x$, is obviously due to I_M, i.e.

$$\partial I_2/\partial x = I_M \tag{11-6}$$

Differentiation of Eq. (11-5) gives:

$$\frac{\partial^2 E_M}{\partial x^2} = (R_1 + R_2)\frac{\partial I_2}{\partial x}. \tag{11-7}$$

From Eq. (11-6) and (11-7) we get:

$$\frac{\partial^2 E_M}{\partial x^2} = (R_1 + R_2)I_M,$$

or   (11-8)

$$I_M = \frac{1}{R_1 + R_2}\frac{\partial^2 E_M}{\partial x^2}$$

Eq. (11-8) shows that under these conditions membrane current is proportional to the second derivative of membrane potential.

As illustrated in Fig. 11-4 (see also Fig. 9-3) the action potential is, to a

* The signs in the system are defined positive along the x-axis.

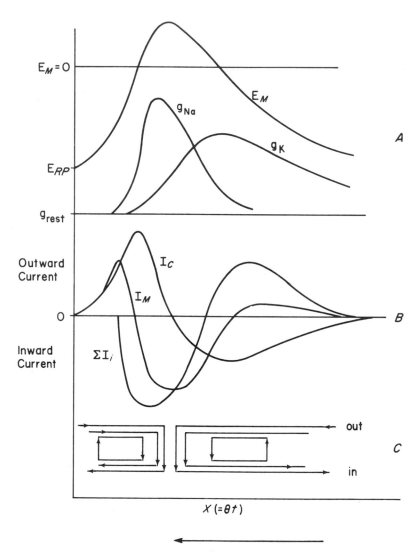

Fig. 11-4 *Schematic representation of events associated with a propagated action potential plotted as a function of distance (x) along the axon.*
A.   Relative changes of membrane potential; E_M; sodium conductance, g_{Na}; and potassium conductance, g_K; E_{RP} is the membrane resting potential; g_{rest} is the normalized resting conductance levels. Ordinate: relative values of potential or conductance.
B.   Total membrane current I_M; membrane capacity current, I_C; and total membrane ionic current ΣI_i in relative units.
C.   Local circuit currents associated with action potential. Arrow indicates direction of propagation. For further details see text.

good approximation, a monophasic phenomenon. Its first derivative is therefore biphasic and the second derivative triphasic (Lorente de No, 1947).

In chapter 6, we have seen that in the squid giant axon, membrane current is given by:

$$I_M = C_M \frac{\partial E_M}{\partial t} + \sum_1^i I_i, \qquad (11\text{-}9)$$

where I_i refers to any specific ionic membrane current. In an axon segment where there is no stimulating current, one gets from Eq. (11-8) and (11-9):

$$\frac{1}{R_1 + R_2} \frac{\partial^2 E_M}{\partial x^2} = C_M \frac{\partial E_M}{\partial t} + \sum_1^i I_i. \qquad (11\text{-}10)$$

To solve Eq. (11-10) we must have E_M as a derivative of either x or t in both sides of the equation. As is well known, the action potential travels along the axon in a wave like fashion. Therefore, from wave theory, for a propagated potential wave (Stratton, 1941):

$$\frac{\partial^2 V}{\partial x^2} = \frac{1}{\theta^2} \frac{\partial^2 V}{\partial t^2}, \qquad (11\text{-}11)$$

where θ is the wave propagation velocity.*

Assuming that the nerve impulse propagation velocity is constant, we can substitute Eq. (11-11) in Eq. (11-10):

$$\frac{1}{R_1 + R_2} \cdot \frac{1}{\theta^2} \cdot \frac{\partial^2 E_M}{\partial t^2} = C_M \frac{\partial E_M}{\partial t} + \sum_1^i I_i \qquad (11\text{-}12)$$

When E_M is determined experimentally, Eq. (11-12) can be easily solved. However, as Eq. (11-12) is an ordinary differential equation it can be solved numerically for E_M or the ionic currents. The method of solution is similar to the solutions given in chapters 7 and 9. One has only to introduce the proper value of θ. If this value is not known, a value can be guessed and the solution carried out. If the guessed value is too large, E_M tends towards infinity and if it is too small, towards minus infinity. The guess can thus be corrected in the proper direction until E_M returns to the resting level at the end of the action potential.

---

* The derivation of Eq. (11-11) for a specific case of a sine potential wave is simple. It will be given here as an illustration. For such a wave:

$E = \sin (x + \theta t)$
$\partial E/\partial x = \cos (x + \theta t)$ $\partial E/\partial t = \theta (x + \theta t)$
$\partial^2 E/\partial x^2 = -\sin (x + \theta t)$ $\partial^2 E/\partial t^2 = -\theta^2 \sin (x + \theta t)$

It therefore follows that $\partial^2 E/\partial x^2 = \partial^2 E/\partial t^2 \cdot 1/\theta^2$

Note also that the dimensions of $\partial^2 V/\partial x^2$ are volts/cm^2, while those of $\partial^2 v/\partial t^2$ are volts/sec^2. As the dimensions of $1/\theta^2$ are sec^2/cm^2, it is the proper factor in Eq. (11-11).

Fig. 11-4 is a schematic illustration of E_M, g_{Na}, g_K and the membrane currents as a function of location along an active axon. The values traced are based on solution of Eq. (11-12). Note the three-phasic nature of I_M. Note also the faster turn-on rate of g_{Na} as compared with g_K.

Fig. 11-5 illustrates the separated membrane ionic currents as a function of

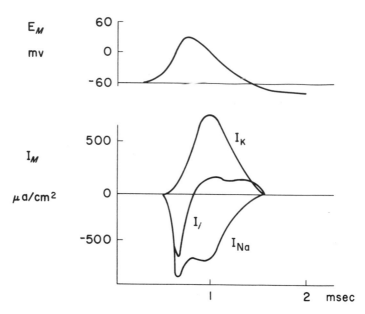

Fig. 11-5 *Schematic representation of membrane potential and membrane currents as a function of time at any specific point along an active axon. Upper trace: membrane potential, E_M; lower traces: total membrane ionic current, I_i; sodium current, I_{Na}; and potassium current, I_K. Outward current is up. (From Hodgkin and Huxley, 1952c.) For further details see text.*

time at a single point along an axon along which an action potential propagates. Note that initially I_M is mostly I_{Na}. Later I_{Na} and I_K almost cancel each other. These currents are obviously different from those of a membrane action potential (see chapter 9) where $I_M = 0$. One of the most significant differences is the much higher initial rise of I_{Na} in the propagated case (compare with Fig. 9-2). However, in both cases the "hump" in I_{Na} is due to the same cause.

Action potentials reconstructed as shown above have been found to be in excellent agreement with the ones measured experimentally. The agreement holds true for action potentials generated under a large variety of conditions. This fact illustrates the amazing success of the Hodgkin-Huxley model to predict nerve behavior.

REFERENCES

CLARK, J., and R. PLONSEY. 1966. A Mathematical Evaluation of the Core Conductor Model. *Biophys. J.* 6:95–112.

GOLDMAN, L., and J. S. ALBUS. 1968. Computations of Impulse Conduction in Myelinated Fibers: Theoretical Bases of the Velocity Diameter Relation. *Biophys. J.* 8:596–607.

HODGKIN, A. L., and A. F. HUXLEY, 1952a. Currents Carried by Sodium and Potassium Ions through the Membrane of the Giant Axon of *Loligo. J. Physiol. (London)*. 116:449–472.

_____. 1952b. The Dual Effect of Membrane Potential on Sodium Conductance in the Giant Axon of *Loligo. J. Physiol. (London)*. 116:497–506.

_____. 1952c. A Quantitative Description of Membrane Current and its Application to Conduction and Excitation in Nerve. *J. Physiol. (London)*. 117:500–544.

LORENTE DE NO, R. 1947. *A Study of Nerve Physiology*, Part I. The Rockefeller Inst., New York.

STRATTON, J. A. 1941. *Electromagnetic theory*. McGraw Hill, New York.

VOLTAGE CLAMP STUDIES ON MYELINATED NERVE FIBERS 12

B. HILLE

Nearly all the properties of nerve impulse conduction known before 1935 were discovered in experiments on myelinated nerves. Action currents, an external negativity traveling at constant velocity, core-conductor cable properties, propagation by local circuits, a constant charge requirement for excitation, the all-or-nothing principle and refractoriness were described in myelinated nerves. Then the era of the squid giant axon began. In rapid succession the impedance change, the overshoot, the sodium hypothesis and finally the complete voltage clamp analysis of sodium and potassium permeability changes appeared. These fundamental discoveries ensure a primary position for squid giant axons in the history of excitable membranes. Nevertheless, vertebrate myelinated nerve fibers have continued to be interesting and useful experimental objects from a fundamental, comparative and medical point of view. Indeed, myelinated fibers are the only other axons to have been analyzed by the quantitative method developed by Hodgkin and Huxley (1952b) for squid giant axons. This chapter concerns the properties of the nodal membrane of myelinated nerve fibers as revealed by voltage clamp experiments.

The largest myelinated nerve fibers in *Rana pipiens* are axons of ca. 12 μ in diameter covered over most of their lengths by a layer of myelin several microns thick. Myelin is a low-capacity, high-resistance condensation of Schwann cell membranes, which insulate much of the axon. At roughly 2 mm intervals there are 1 μ wide gaps in the myelin where one Schwann cell ends and a new one takes over. These gaps, the nodes of Ranvier, form a conducting path of extracellular fluid to the excitable axon membrane of the node.

Because of the insulation covering the internodal axon, stimulating currents and local circuit currents enter and leave the axons mainly at the nodes (Tasaki, 1953; Huxley and Stämpfli, 1949). Only the nodal membrane generates the ionic currents needed for propagation of impulses. It is not known whether or not the internodal membrane is excitable in the absence of myelin. Functional conduction is lost in demyelinating diseases and returns after remyelinization.

THE POTENTIOMETRIC RECORDING METHOD

Historically the problem in studying single myelinated fibers has been to achieve a faithful recording method with such a small structure. Axial wires are impossible and glass microelectrodes are too unreliable in these fibers. The successful methods are all based on Tasaki's observation (Kato, 1936; Tasaki, 1953) that the node already has two useful connections attached to it, the insulated internodes. The internodes on each side of a node can serve as pipets to record the potential inside a node and to deliver current to the node. The major problem is that the high resistance internodal "pipets" are totally immersed in the extracellular medium, a short circuit which effectively shunts out most of the signal. Hence, in any practical recording system the internodes must traverse an insulating gap of air, oil, Vaseline or sucrose to increase the extracellular shunting resistance. Even under these conditions, an appreciable short circuit current flows and the recorded potentials are attenuated. Tasaki and Mizuguchi (1949) were able to measure the impedance change during the action potential using this method. A relatively unstable and slow voltage clamp can also be achieved with a double air gap of this type (del Castillo, Lettvin, McCulloch and Pitts, 1957; Tasaki and Bak, 1958).

By an ingenious nulling procedure, Huxley and Stämpfli (1951a) circumvented the attenuating effect of the extracellular shunt. They adapted the potentiometric measuring principle of slide-wire recording devices to the myelinated nerve fiber. The idea of the method is shown in Fig. 12-1. The upper diagram represents a node of Ranvier and two internodes of a fiber. The left hand internode passes through two insulating gaps (stippled regions) separating saline pools A, B and C. Pools A and B contain frog Ringer's solution and C contains isotonic KCl to depolarize the end of the fiber. The resistances associated with the node, the axoplasm and the two gaps are drawn to show the path $DCBAD$ of short circuit current flow. To reduce this current, Huxley and Stämpfli placed a battery between the electrodes in pools A and B while monitoring the short circuit current by measuring the voltage drop across the resistance R_{CB} of the gap between C and B

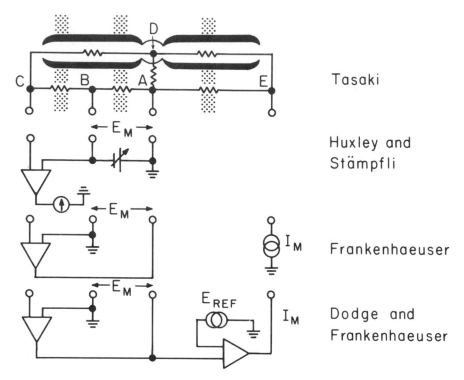

Fig. 12-1 *Evolution of the Frankenhaeuser-Dodge voltage clamp. A single myelinated fiber lies across four fluid filled compartments C, B, A, and E separated by insulating gaps (stippling). The distance from pool C to pool E is typically 800 μm. The node under investigation is in pool A. The ends of the fiber in pools C and E may be cut across in the internode or treated with cocaine, formalydehyde, or KCl to eliminate unwanted responses. If the figure were drawn in correct proportion, the nodal gap would be ten times shorter and the internodal length twenty times longer than drawn here. The D.C. equivalent circuit of the preparation is superimposed on the drawing of the fiber. The circuits below represent the progression from static potentiometric recording, to dynamic recording with known membrane current, and finally to the voltage clamp.*

(Fig. 12-1). The potential of the battery is adjusted to eliminate the measured current. At this point the applied voltage E_{BA} must be equal to the emf which drives the short circuit current. Hence E_{BA} is equal to the unattenuated resting potential of the nodal membrane. The mean resting potential for nodes of *Rana esculenta* bathed in a Ringer's solution with 1 mM $CaCl_2$ is found to be -71 mv (Huxley and Stämpfli, 1951a).

If the potentiometric system is in balance at the resting potential, it is out of balance during an action potential or during any other deviation from the resting potential. However, the height of the action potential can be measured by adding a brief adjustable pulse to the steady voltage of the battery during the action potential. Again the pulse is adjusted to cancel the short circuit current at the time of the peak, and the height of the pulse is equal to the height of the action potential. The peak of the action potential overshoots zero by 45 mv (Huxley and Stämpfli, 1951a).

Huxley and Stämpfli's method served well to establish that the resting nodal membrane behaves roughly as a potassium electrode and the active membrane as a sodium electrode when the external concentrations of potassium and sodium are changed (Huxley and Stämpfli, 1951b). However, because it was cumbersome and basically static, the method was not used again.

The next major step was Frankenhaeuser's (1957) conversion of the static potentiometric method to a dynamic measurement. He observed that the nulling process of adjusting the balancing voltage E_{BA} to eliminate short circuit current could be accomplished very rapidly by substituting a feedback amplifier for the human hand (Fig. 12-1). A high gain, broad bandwidth amplifier reduces E_{CB} by applying the amplified error signal to pool A. As before, E_{BA} is equal to the unattenuated membrane potential, but now the balancing voltage follows the action potential faithfully and automatically. Because the system always stays in balance, no current is drawn from the node by the measuring circuit. In effect the input impedance of the measuring circuit is increased by a factor equal to the gain of the feedback loop.

Frankenhaeuser's method has two further advantages. By grounding pool B it creates a virtual ground of the high impedance point, pool C. Hence the attenuation of high frequencies arising from stray capacitance between pool C and ground is minimized without introducing a possibly erroneous correction from a "capacity neutralization" circuit. Indeed the same principle has been used to reduce the effect of stray capacitance in recording from other cells with conventional glass microelectrodes (Eisenberg and Gage, 1969). Frankenhaeuser's method also permits a simple measurement of the nodal membrane current. The nodal membrane current is equal to any current flowing to the node in the path ED, because no current flows in the path DC from the node. Point D, like point C, is held at ground potential by the feedback amplifier, so the current in ED is equal to the voltage in pool E divided by the resistance R_{ED}. After suitable treatment of the nodes in pool E, the resistance R_{ED} remains relatively constant and the current is considered to be proportional to the voltage in pool E. Although most of the current from pool E flows in the low extracellular resistance R_{EA} (typically 5 megohms) of the Vaseline gap rather than in the higher intracellular resistance R_{ED} (typically 40 megohms) of the axon, this extra current does not affect the measurement of nodal current by the above criterion. In brief, Frankenhaeuser's method offers a reliable measurement of the response of single nodes of Ranvier to known stimulating currents.

Dodge and Frankenhaeuser (1958) developed a voltage clamp from the dynamic potentiometric method (Fig. 12-1). The original feedback amplifier measures the membrane potential and a second feedback amplifier, the voltage clamp amplifier, applies the amplified difference between the membrane potential and the reference waveform of the clamp to the current injecting point, pool E. The results compare favorably with the voltage clamp of the squid giant axon. The errors and pitfalls are somewhat different. Most of the difficulties arise from the stability and frequency response of two high gain feedback loops operating on a high impedance tissue rather than from series resistance and non-uniformity of electrodes and membranes (Dodge and Frankenhaeuser, 1958, 1959; Dodge, 1963; Nonner, 1969). Another significant difficulty is a greater uncertainty in the D.C. level arising from the many liquid junctions and electrodes coupled with a shunted high impedance system (Hille, 1967c). A recent modification of the method using only one amplifier achieves a good clamp of small mammalian fibers even at $37°C$ (Nonner, 1969; Horackova, Nonner and Stämpfli, 1968).

THE MATHEMATICAL MODEL FOR VOLTAGE CLAMP CURRENTS

As is described in other chapters of this book, the voltage clamp currents of the squid giant axon can be separated into different independent components by ionic substitution experiments. The time course of these components can be described mathematically as the product of some dimensionless parameters h, m, and n, taken to appropriate powers, multiplied by the maximum possible currents $\bar{I}_{Na}$, $\bar{I}_K$, and $\bar{I}_L$ at that clamp voltage. The parameters, h, m, and n, change in time with rate constants α and β which are functions of the membrane potential (Hodgkin and Huxley, 1952b). One simple interpretation of the Hodgkin-Huxley equations is that there are a fixed number of available ionic channels which if fully open could carry a current $\bar{I}_{Na}$, $\bar{I}_K$, and $\bar{I}_L$. At any time a fraction m^3h of the sodium channels and a fraction n^4 of the potassium channels are open.

The voltage clamp currents of amphibian nerves (Fig. 12-2) closely resemble those of squid giant axons. Ionic substitution experiments reveal sodium, potassium and leakage components of ionic current in *Rana* and *Xenopus* (Dodge and Frankenhaeuser, 1959; Frankenhaeuser, 1960; Dodge, 1961, 1963). Leakage seems to be carried primarily by K ions. In *Xenopus* there is an additional late, time-dependent current carried primarily by Na ions. This minor component has been called I_p (Frankenhaeuser, 1962b). The permeabilities of the nodal membrane seem to be specially adapted to permit the tiny node to depolarize the relatively enormous expanse of internode leading to the next node. The resting resistance is very low, about 10 ohm cm^2, and the peak inward I_{Na} exceeds 50 ma cm^{-2}.

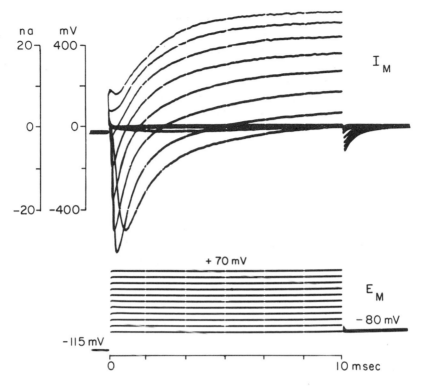

Fig. 12-2 *Voltage clamp currents and voltages on the oscilloscope screen. Ten superimposed traces with 10 msec test pulses ranging from −80 mv to +70 mv. The voltage trace E_M is the negative of the voltage in pool A. The current trace I_M is the voltage in pool E. I_M is converted to units of current by assuming that the node has a resting resistance of 40 megohm (which in this example is equivalent to assuming a value of 20 megohm for R_{EU}). Both E_M and I_M are recorded through a 5 kHz low pass filter. The brief tail of inward current after the larger test pulses is not present in all fibers. It may correspond to Frankenhaeuser's I_p. Rana pipiens fiber at 13°C; ends cut in KCl.*

The kinetic properties of the ionic currents can be described by equations like those in the Hodgkin-Huxley (1952) model. For the ionic currents of *Xenopus laevis*, Frankenhaeuser (1959, 1960a, 1962a,b,c, 1963a) has used the equations:

$$I_{Na} = m^2h\bar{I}_{Na} \tag{12-1}$$

$$I_K = n^2k\bar{I}_K \tag{12-2}$$

$$I_p = p^2\bar{I}_p \tag{12-3}$$

$$I_L = \bar{I}_L, \tag{12-4}$$

where k gives a slow inactivation of I_K for long depolarizations and the other symbols have the same significance as in the Hodgkin-Huxley model. Koppenhöfer and Schmidt (1968a,b; Koppenhöfer, 1967) have preferred to use n^4 instead of n^2 to describe the activation of I_K. A complete set of average values of the rate constants α and β for h, m, and n in many experiments with *Xenopus* are summarized by Frankenhaeuser and Huxley (1964).

For *Rana pipiens* Dodge (1961, 1963) has used the equations

$$I_{Na} = m^3 h \bar{I}_{Na} \tag{12-5}$$

$$I_K = n^4 \bar{I}_K \tag{12-6}$$

$$I_L = \bar{I}_L. \tag{12-7}$$

A factor k is needed here too for slow inactivation of I_K (Hille, 1967c). The steady state values h, m and n and time constants τ_h, τ_m and τ_n for one of the nodes studied by Dodge are given in Fig. 12-3. The empirical equations used to generate these curves are given in the Appendix to this chapter.

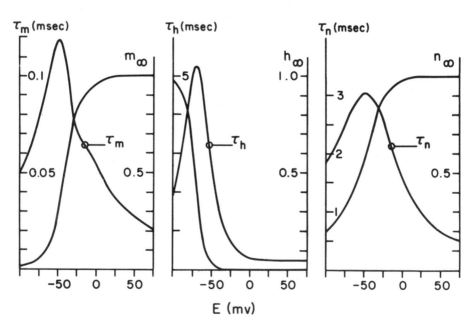

STANDARD PARAMETERS AT 22°C

Fig. 12-3 *The steady-state values and time constants for m, h, and n as a function of membrane potential. The curves are computed from the empirical equations for rate constants given in the Appendix. The empirical equations closely approximate experiments with a node of Ranvier from Rana pipiens in 1.8 mM Ca Ringer's at 22°C.*

Dodge (1963) also has developed complete empirical equations for four other nodes.

The functions $\bar{I}_i$ are frequently called the "instantaneous" I-V relations, but a preferable term is Noble and Tsien's (1969) "fully activated" I-V relations which explicitly implies that they are the I-V relations expected if all the ionic channels could be activated. For many models of ionic mechanisms, the fully activated I-V relations would have the same shape as the I-V relations of a single (fully) open ionic channel. In the squid giant axon bathed in sea water, the fully activated I-V relations $\bar{I}_{Na}$, $\bar{I}_K$, and $\bar{I}_L$ are linear. They are said to be "ohmic" and are conveniently represented in the mathematical model as the product of a maximum conductance times an electrical driving force. For *Xenopus* nodes $\bar{I}_L$ is ohmic and may be represented as in the squid axon model. However, $\bar{I}_{Na}$, $\bar{I}_K$, and $\bar{I}_p$ are nonlinear and must be represented in some other way. Happily the curvature of these functions is like that given by the Goldman (1943) equation for ionic currents in simple barriers, so, for example, the maximum sodium current in nodes can be written (Dodge and Frankenhaeuser, 1959):

$$\bar{I}_{Na} = \bar{P}_{Na}F\omega \, \frac{c_{Na_o} - c_{Na_i} \exp \omega}{1 - \exp \omega} \, , \quad \omega = \frac{EF}{RT} \tag{12-8}$$

where $\bar{P}_{Na}$ is the maximum Na permeability, c_{Na_o} and c_{Na_i} the outside and inside Na concentrations, E the membrane potential, and F, R, and T have their usual meanings. The same type of expression can be used for $\bar{I}_K$ and $\bar{I}_p$. For *Rana* nodes $\bar{I}_L$ is ohmic, $\bar{I}_K$ is close to ohmic, and $\bar{I}_{Na}$ is non-linear and fits the Goldman equation. The appropriate fully activated I-V relations inserted into the kinetic expressions 1-4 or 5-7 give the dynamic I-V relations of the node of Ranvier.

When all the equations of the mathematical model of voltage clamp currents are assembled (see, for example, the Appendix to this chapter), they may be used to calculate the responses of myelinated nerve fibers to various stimuli. Nodal action potentials can be computed with very close correspondence between experimental and theoretical curves (Dodge, 1961, 1963; Frankenhaeuser and Huxley, 1964; Frankenhaeuser, 1965; Schoepfle, Johns and Molnar, 1969). Even the characteristically different nodal action potentials of the five nodes described by Dodge (1963) are specifically matched by the calculations. Propagated action potentials with reasonable shapes, currents and conduction velocities can be obtained if the equations for nodes are suitably combined with the cable equations for internodes (Goldman and Albus, 1968; Hardy, 1969; see also FitzHugh, 1962). Thus the quantitative description of propagation in uniform myelinated fibers is nearly complete.

There still remain many interesting problems regarding propagation at branch points and at the transitions between myelinated axons and the un-myelinated parts of cell bodies or nerve terminals. In the strictest sense the experiments described in this chapter relate only to the largest fibers in leg nerves of toads, frogs and rats, but it may be supposed that all mye-linated fibers are at least broadly similar. There is little specific information regarding possible differences among fibers of different type and function. The late currents, I_K and I_p, differ between sensory and motor fibers (Bergman and Stämpfli, 1966). The late currents of mammalian myelinated fibers are much smaller than those of amphibian fibers (Horackova et al., 1968). The importance of these trends is not known, but it is likely that they relate in some way to function.

THE COMPARATIVE LESSON

Probably the single most interesting result of the analysis of the membrane of the node of Ranvier is the astounding similarity to the membrane of the squid giant axon. For the 0.5×10^9 years since they may have had a common ancestor, the two great superphyla of the animal kingdom have diverged, producing such contrasting forms as frogs and squids. Still the action potential mechanism, which apparently was fully developed by that early time, has been passed along with only trivial change.

THE MECHANISMS OF CURRENT GENERATION

Inasmuch as the mathematical properties of the currents of the nodal mem-brane closely resemble those of squid giant axons, the empirical models of *Xenopus* or *Rana* nodes cannot suggest many new mechanisms not already implicit in the Hodgkin-Huxley model. Only the areas of dissimilarity and new types of experiments offer additional clues. Many of these have been reviewed elsewhere (Hille, 1970) and are mentioned only briefly here.

Current-Voltage Relations and Free Diffusion

In inquiring how ions cross membranes it is instructive to review a property of the free diffusion of neutral molecules. In his random walk model Einstein (1908) showed that free diffusion and Fick's Law are accounted for by the random thermal agitations of *individual* molecules. A molecule does not need to know what its neighbors are doing, it does not need to know of the existence of a gradient; by moving to the right and to the left with equal probability it will make its contribution to the diffusion flux.

The diffusion of ions is more complicated because the motions of ions are biased by electric fields. Part of the field may be applied externally but part of it arises locally from the neighboring ions. Hence, except for very dilute solutions and for a few other special cases, the diffusion of ions does not occur by motions which are independent of neighboring ions.

Three criteria have been used to test how closely the ionic movements in membranes resemble free diffusion fluxes: (1) The independence principle (Hodgkin and Huxley, 1952a) assumes simply that the movement of an ion is independent of any other ion. It can be used to predict the net flux of an ion in one set of ionic concentrations from the net flux at the same membrane potential but in another set of ionic concentrations. Thus the independence relation predicts the changes in the fully activated I-V relations upon changing the bathing solution. (2) The Goldman (1943) theory assumes free diffusion of ions in a homogeneous membrane with a constant electric field from one side to the other and predicts I-V curves [see Eq. (12-8)] for any ionic conditions. The Goldman equation fits the independence relation because it assumes that the field remains constant and thereby removes the electrical coupling that otherwise exists between ions. The same equation can be derived for some other initial assumptions (Cole, 1968). (3) The Ussing (1949) flux ratio assumes free diffusion without restriction on the amount of electrical coupling between ions but with no other coupling such as might be caused by saturatable carrier system or by a long narrow pore in which ions must pass in single file. This theory predicts that the simultaneous influx and efflux of an ion at one potential have the same ratio as the outside and inside electrochemical activities of the ion. The independence principle also gives this result because it is a special case of free diffusion (Hodgkin and Huxley, 1952a).

The fully activated I-V relations of *Xenopus* nodes fit the Goldman equation and satisfy the independence principle when c_{Na_o} and c_{K_o} are varied (Frankenhaeuser, 1960a,b, 1962a,b,c). Hence the ionic movements have much in common with free diffusion. The Ussing flux ratio test would be extremely difficult to apply with present methods to myelinated nerves.

The agreement with simple theories in amphibians highlights disagreements in squid giant axons. Why are the fully activated I-V relations linear in squid? As $\bar{I}_{Na}$ in squid does fit independence when c_{Na_o} is varied, free diffusion constant field theories might still apply. Two variants with asymmetrical membranes substituted for Goldman's homogeneous membrane have been shown to work (Frankenhaeuser, 1960b; Woodbury, 1969). They postulate an asymmetrical distribution of diffusion barriers in the membrane which exactly compensates for the asymmetrical distribution of sodium ions. In addition, K flux ratios in squid do not satisfy the Ussing flux ratio (Hodgkin and Keynes, 1955). The flux ratios deviate in the direction expected for single file diffusion. Independence has not been tested. Although independence should in theory not be satisfied, perhaps it is not a very sensitive test. If it is not, then the K channels of frogs may also have some single-file character. The answer is not known. In any case, the known I-V and flux properties of axons may be fit by models based on free diffusion of ions across the membrane.

Selectivity

The characteristic discrimination between Na and K ions in axon membranes is a property not ordinarily seen in free diffusion. The selectivity of sodium channels for Na and K ions was first demonstrated to be imperfect in myelinated axons. Frankenhaeuser and Moore (1963b) voltage clamped a fiber in a Na-free, hypertonic K Ringer's solution. An appreciable inward K current, with the same time course as the original Na current, showed that the permeability ratio, $P_K:P_{Na}$, is about 1:20 for the sodium channel of *Xenopus* nodes. Since then the selectivity of ionic channels has been studied in several different axons. Recently it has been found that the cation content of the axoplasm of myelinated fibers can easily be varied (Koppenhöfer and Vogel, 1969, 1970), so many more selectivity experiments are now feasible. The method is to cut across the internodes in pools C and E allowing the cations in these pools to equilibrate with those of the axoplasm. It is not known whether the anions equilibrate as well. In these experiments, the rate of diffusion of Na ions in the axoplasms is about 50 per cent of the rate in free solution. The exchange is nearly complete after 30 minutes.

Pharmacology

The node is especially useful for pharmacological studies of externally applied agents because new solutions can be washed in and out in one or two seconds. Many drugs selectively alter one or a few parameters of the sodium permeability change. The poisons tetrodotoxin and saxitoxin specifically block $\bar{P}_{Na}$ in nanomolar concentrations (Hille, 1966, 1967c, 1968a). The poisons veratrine, veratridine and DDT specifically impede the reduction of the m and h parameters after the sodium permeability has been activated by a depolarization (Ulbricht, 1965, 1969; Hille, 1967c, 1968a). Scorpion venom specifically lengthens τ_h and prevents h from ever falling to zero (Koppenhöfer and Schmidt, 1968a,b). Changes of calcium concentration shift the curves of m and h parameters vs. voltage (Fig. 12-3) along the voltage axis without affecting n (Hille, 1967c, 1968b).

One drug, tetraethylammonium ion, acts selectively on the potassium permeability changes. It specifically blocks $\bar{g}_K$ or $\bar{P}_K$ (Koppenhöfer, 1967;

Koppenhöfer and Vogel, 1969, 1970; Hille, 1967a, b, c). Numerous agents have some effects on both sodium and potassium permeability changes. The local anesthetics procaine and Xylocaine, phenothiazine tranquilizers and nitrous oxide depress $\bar{P}_{Na}$ strongly and $\bar{g}_K$ only slightly (Hille, 1966, 1967c). Changes of pH shift the curves of m and h equally along the voltage axis (Hille, 1968b). The curve of n is also shifted, but by a different amount. Reducing the pH below pH 5 reversibly blocks $\bar{P}_{Na}$ as well as shifting the curves of m, h and n.

Together all of these effects suggest, as has been assumed in this chapter, that sodium channels and potassium channels are physically independent specializations of the membrane (see review Hille, 1970). The shifts with changes of calcium concentration and of pH suggest areas of negative fixed charge near the ionic channels. The block of P_{Na} at low pH suggests the presence of an essential ionized acid group at the sodium channel with a pK_a of 5.2.

Density of Sodium Channels

The node is by far the smallest patch of membrane (30 μ^2) which can be reliably studied today and offers a unique possibility of observing quantal effects of single ionic channels. Existing studies report quantal steps of membrane potential attributable to steps of Na current (del Castillo and Suckling, 1957; Lüttgau, 1958; Verveen, Derksen and Schick, 1967; Verveen and Derksen, 1968, 1969). Near the resting potential the steps are roughly 0.5 to 1.0 mv, suggesting increments of 12 to 24 pa in I_{Na} or increments of 0.1 to 0.2 nmho in g_{Na}. It remains to be determined if these steps represent the opening and closing of single channels. If they are, there may be 10^4 Na channels on a large node with a maximum sodium conductance of 750 nmho. If evenly spaced, the channels would be about 60 nm apart.

Although the conclusions reached so far are helpful, they have been insufficient to explain the fundamental properties of ionic selectivity and voltage-dependent permeability in membranes. Many new types of experiments are needed before detailed hypotheses can be profitably discussed. Undoubtedly both myelinated and giant unmyelinated axons will be valuable in these investigations.

ACKNOWLEDGMENT

Original work supported by NIH grant number NS08174.

APPENDIX

An empirical model for the ionic currents of a node of Ranvier from *Rana pipiens*.

The following equations are suitable for calculating the responses of the nodal membrane of a normal nerve fiber at 22°C. The values of the constants are taken with small modification from Dodge's (1961, 1963) "node 7."

$$I_M = m^3 h \bar{I}_{Na} + n^4 \bar{I}_K + \bar{I}_L + C \frac{\partial E}{\partial t} \tag{12-9}$$

$$\bar{I}_{Na} = \bar{P}_{Na} c_{Nao} \frac{F^2 E}{RT} \frac{\exp[(E - E_{Na})F/RT] - 1}{\exp[EF/RT] - 1} \tag{12-10}$$

$$\bar{I}_K = \bar{g}_K(E - E_K) \tag{12-11}$$

$$\bar{I}_L = \bar{g}_L(E - E_L) \tag{12-12}$$

$$\frac{dm}{dt} = \alpha_m(1 - m) - \beta_m m \tag{12-13}$$

$$\frac{dh}{dt} = \alpha_h(1 - h) - \beta_h h \tag{12-14}$$

$$\frac{dn}{dt} = \alpha_n(1 - n) - \beta_n n. \tag{12-15}$$

For computational purposes the following equation is a simple approximation of Eq. (12-10) in external solutions containing 110 mM Na.

$$\bar{I}_{Na} = \bar{g}_{Na}\left(1 - \frac{E}{0.183}\right)(E - E_{Na}). \tag{12-16}$$

Eq. (12-10) and (12-16) are written with units of volts. The constants in the equations are:

$\bar{P}_{Na} = 3.9 \times 10^{-9}$ cm³ sec⁻¹ $E_{Na} = 48$ mv

$c_{Nao} = 110$ mM $= 1.1 \times 10^{-4}$ moles cm⁻³

$\bar{g}_{Na} = 0.750$ μmho

$\bar{g}_K = 0.130$ μmho $E_K = -75$ mv

$\bar{g}_L = 0.025$ μmho $E_L = -75$ mv

$C = 2$ pf.

The empirical formulae for the rate constants in units of msec⁻¹ as a function of absolute membrane potential in units of mv are:

$$\alpha_m = \frac{1.6\phi}{\exp\phi - 1} + \frac{6.0}{1 + \left(\dfrac{E + 25}{18}\right)^2} \qquad \phi = \frac{-37.5 - E}{3.8}$$

$$\beta_m = \frac{4.05\phi}{\exp\phi - 1} \qquad\qquad \phi = \frac{37.5 + E}{13.2}$$

$$\alpha_h = \frac{0.14\phi}{\exp(0.8\phi) - 1} \qquad\qquad \phi = \frac{79 + E}{5.9}$$

$$\beta_h = \frac{4.0}{\exp\phi + 1} \qquad\qquad \phi = \frac{-17 - E}{14.5}$$

$$\alpha_n = \frac{0.34\phi}{\exp\phi - 1} + 0.1 \qquad\qquad \phi = \frac{-15 - E}{15.9}$$

$$\beta_n = \frac{0.085\phi}{\exp\phi - 1} \qquad\qquad \phi = \frac{37.5 + E}{12.2}$$

Working with *Xenopus laevis*, Frankenhaeuser and Moore (1963a) found a Q_{10} of about 1.8 for α_m and β_m and a Q_{10} of about 2.9 for the other rate constants. The Q_{10} of $\bar{P}_{Na}$ is 1.3 and $\bar{P}_K$, 1.2.

REFERENCES

BERGMAN, C., and R. STÄMPFLI. 1966. Différence de Perméabilité des Fibres Nerveuses Myélinisées Sensorielles et Motrices à l'ion Potassium. *Helv. Physiol. Pharmacol. Acta.* 24:247–258.

CASTILLO, J. del, and E. E. SUCKLING. 1957. Possible Quantal Nature of Subthreshold Responses at Single Nodes of Ranvier. *Federation Proc.* 16:29.

CASTILLO, J. del, J. Y. LETTVIN, W. S. McCULLOCH and W. PITTS. 1957. Membrane Currents in Clamped Vertebrate Nerve. *Nature.* 180:1290–1291.

COLE, K. S. 1968. *Membranes, Ions and Impulses.* University of California Press, Berkeley.

DODGE, F. A. 1961. Ionic Permeability Changes Underlying Nerve Excitation. *In Biophysics of Physiological and Pharmacological Actions.* AAAS, Washington, D.C.

———. 1963. *A Study of Ionic Permeability Changes Underlying Excitation in Myelinated Nerve Fibres of the Frog.* Thesis. The Rockefeller University. Univ. Microfilms. Ann Arbor, Mich. (No. 64–7333).

DODGE, F. A., and B. FRANKENHAEUSER. 1958. Membrane Currents in Isolated Frog Nerve Fibre under Voltage Clamp Conditions. *J. Physiol. (London).* 143:76–90.

———. 1959. Sodium Currents in the Myelinated Nerve Fibre of *Xenopus laevis* Investigated with the Voltage Clamp Technique. *J. Physiol. (London).* 148:188–200.

EINSTEIN, A. 1908. The Elementary Theory of the Brownian Motion. Reprinted in *Investigations on the Theory of Brownian Movement,* 1956, Dover Publ., Inc., New York.

EISENBERG, R. S., and P. W. GAGE. 1969. Ionic Conductances of the Surface and Transverse Tubular Membranes of Frog Sartorius Fibers. *J. Gen. Physiol.* 53:279–297.

FITZHUGH, R. 1962. Computation of Impulse Initiation and Saltatory Conduction in a Myelinated Nerve Fiber. *Biophys. J.* 2:11–21.

FRANKENHAEUSER, B. 1957. A Method for Recording Resting and Action Potentials in the Isolated Myelinated Nerve Fibre of the Frog. *J. Physiol. (London).* 135:550–559.

_____. 1959. Steady State Inactivation of Sodium Permeability in Myelinated Nerve Fibres of *Xenopus laevis. J. Physiol. (London).* 148:671–676.

_____. 1960a. Quantitative Description of Sodium Currents in Myelinated Nerve Fibres of *Xenopus laevis. J. Physiol. (London).* 151:491–501.

_____. 1960b. Sodium Permeability in Toad Nerve and in Squid Nerve. *J. Physiol. (London).* 152:159–166.

_____. 1962a. Delayed Currents in Myelinated Nerve Fibres of *Xenopus laevis* Investigated with Voltage Clamp Technique. *J. Physiol. (London).* 160:40–45.

_____. 1962b. Instantaneous Potassium Currents in Myelinated Nerve Fibres of *Xenopus laevis. J. Physiol. (London).* 160:46–53.

_____. 1962c. Potassium Permeability in Myelinated Nerve Fibres of *Xenopus laevis. J. Physiol. (London).* 160:5–61.

_____. 1963a. A Quantitative Description of Potassium Currents in Myelinated Nerve Fibres of *Xenopus laevis. J. Physiol. (London).* 169:424–430.

_____. 1963b. Inactivation of the Sodium-Carrying Mechanism in Myelinated Nerve Fibres of *Xenopus laevis. J. Physiol. (London).* 169:445–457.

_____. 1965. Computed Action Potential in Nerve from *Xenopus laevis. J. Physiol. (London).* 180:780–787.

FRANKENHAEUSER, B., and A. F. HUXLEY. 1964. The Action Potential in the Myelinated Nerve Fibre of *Xenopus laevis* as Computed on the Basis of Voltage Clamp Data. *J. Physiol. (London).* 171:302–315.

FRANKENHAEUSER, B., and L. E. MOORE. 1963a. The Effect of Temperature on the Sodium and Potassium Permeability Changes in Myelinated Nerve Fibres of *Xenopus laevis. J. Physiol. (London).* 169:431–437.

_____. 1963b. The Specificity of the Initial Current in Myelinated Nerve Fibres of *Xenopus laevis. J. Physiol. (London).* 169:438–444.

GOLDMAN, D. E. 1943. Potential, Impedance, and Rectification in Membranes. *J. Gen. Physiol.* 27:37–60.

GOLDMAN, L., and J. S. ALBUS. 1968. Computation of Impulse Conduction in Myelinated Fibers; Theoretical Basis of the Velocity-Diameter Relation. *Biophys. J.* 8:596–607.

HARDY, W. L. 1969. *Propagation in Myelinated Nerve: Dependence on External Sodium.* Thesis. University of Washington.

HILLE, B. 1966. The Common Mode of Action of Three Agents that Decrease the Transient Change in Sodium Permeability in Nerves. *Nature.* 210:1220–1222.

_____. 1967a. Quaternary Ammonium Ions that Block the Potassium Channel of Nerves. *Abstr. 11th Meet. Biophys. Soc.,* 19.

_____. 1967b. The Selective Inhibition of Delayed Potassium Currents in Nerve by Tetraethylammonium Ion. *J. Gen. Physiol.* 50:1287–1302.

_____. 1967c. *A Pharmacological Analysis of the Ionic Channels of Nerve.* Thesis. The Rockefeller University. Univ. Microfilms, Inc., Ann Arbor, Mich. (No. 68–9584).

_____. 1968a. Pharmacological Modifications of the Sodium Channels of Frog Nerve. *J. Gen. Physiol.* 51:199–219.

_____. 1968b. Charges and Potentials at the Nerve Surface: Divalent Ions and pH. *J. Gen. Physiol.* 51:221–236.

_____. 1970. Ionic Channels in Nerve Membranes. *Progr. Biophys. Mol. Biol.* 21:1–32.

HODGKIN, A. L., and A. F. HUXLEY. 1952a. Currents Carried by Sodium and Potassium Ions through the Membranes of the Giant Squid *Loligo. J. Physiol. (London).* 116:449–472.

_____. 1952b. A Quantitative Description of Membrane Current and its application to Conduction and Excitation in Nerve. *J. Physiol. (London).* 117:500–544.

HODGKIN, A. L., and R. D. KEYNES. 1955. The Potassium Permeability of a Giant Nerve Fibre. *J. Physiol. (London).* 128:61–88.

HORACKOVA, M., W. NONNER and R. STÄMPFLI. 1968. Action Potentials and Voltage Clamp Currents of Single Rat Ranvier Nodes. *Proc. Int. Union Physiol. Sci.* 7:198.

HUXLEY, A. F., and R. STÄMPFLI. 1949. Evidence for Saltatory Conduction in Peripheral Myelinated Nerve Fibres. *J. Physiol. (London).* 108:315–339.

_____. 1951a. Direct Determination of Membrane Resting Potential and Action Potential in Single Myelinated Nerve Fibres. *J. Physiol. (London).* 112:476–495.

_____. 1951b. Effect of Potassium and Sodium on Resting and Action Potentials of Single Myelinated Nerve Fibres. *J. Physiol. (London).* 112:496–508.

KATO, G. 1936. On the Excitation, Conduction, and Narcotisation of Single Nerve Fibres. *Cold Spring Harbor Symp. Quant. Biol.* 4:202–213.

KOPPENHÖFER, E. 1967. Die Wirkung von Tetraäthylammoniumchlorid auf die Membranströme Ranvierscher Schnürringe von *Xenopus laevis. Arch. Ges. Physiol.* 293:34–55.

KOPPENHÖFER, E., and H. SCHMIDT. 1968a. Incomplete Sodium Inactivation in Nodes of Ranvier Treated with Scorpion Venom. *Experientia.* 24:41–42.

_____. 1968b. Die Wirkung von Skorpiongift auf die Ionenströme des Ranvierschen Schnürrings. I. Die Permeabilitäten P_{Na} and P_K. II. Unvollständig Natrium-Inaktivierung. *Arch. Ges. Physiol.* 303:133–149, 150–161.

KOPPENHÖFER, E., and W. VOGEL. 1969. Intraaxonal Application of Tetraethylammonium Chloride (TEA) at the Node of Ranvier. *Arch. Ges. Physiol.* 312:R101.

————. 1969. Effects of Tetrodotoxin and Tetraethylammonium Chloride on the Inside of the Nodal Membrane of *Xenopus laevis*. *Arch. Ges. Physiol.* 313:361–380.

LÜTTGAU, H. C. 1958. Sprunghafte Schwankungen unterschwelliger Potentiale an markhaltigen Nervenfasern. *Z. Naturforsch.* 13*b*: 692–693.

NOBLE, D., and R. W. TSIEN. 1969. Outward Membrane Currents Activated in the Plateau Range of Potentials in Cardiac Purkinje Fibres. *J. Physiol. (London).* 200: 205–231.

NONNER, W. 1969. A New Voltage Clamp Method for Ranvier Nodes. *Arch. Ges. Physiol.* 309:176–192.

SCHOEPFLE, G. M., G. C. JOHNS and C. E. MOLNAR. 1969. Simulated Responses of Depressed and Hyperpolarized Medullated Nerve Fibers. *Am. J. Physiol.* 216:932–938.

TASAKI, I. 1953. *Nervous transmission.* C. C. Thomas, Springfield, Ill.

TASAKI, I., and A. F. BAK. 1958. Current-Voltage Relations of Single Nodes of Ranvier as Examined by Voltage-Clamp Technique. *J. Neurophysiol.* 21: 124–137.

TASAKI, I., and K. MIZUGUCHI. 1949. The Changes in the Electric Impedance during Activity and the Effect of Alkaloids and Polarization upon the Bioelectric Processes in the Myelinated Nerve Fiber. *Biochim. Biophys. Acta.* 3: 484–493.

ULBRICHT, W. 1965. Voltage Clamp Studies of Veratrinized Frog Nodes. *J. Cell Comp. Physiol.* 66 (Suppl. 2):91–98.

————. 1969. The Effect of Veratridine on Excitable Membranes of Nerve and Muscle. *Ergeb. Physiol.* 61: 18–71.

USSING, H. H. 1949. The Distinction by Means of Tracers between Active Transport and Diffusion. *Acta Physiol. Scand.* 19:43–56.

VERVEEN, A. A., and H. E. DERKSEN. 1968. Fluctuation Phenomena in Nerve Membrane. *Proc. IEEE.* 56:906–916.

————. 1969. Amplitude Distribution of Axon Membrane Noise Voltage. *Acta Physiol. Neerl.* 15: 353–379.

VERVEEN, A. A., H. E. DERKSEN and K. L. SCHICK. 1967. Voltage Fluctuations of Neural Membrane. *Nature.* 216:588–589.

WOODBURY, J. W. 1969. Linear Current-Voltage Relation for Na$^+$ Channel from Eyring Rate Theory. *Biophys. J.* 9. Abstracts Biophys. Soc. Meeting, A250.

OPTICAL STUDIES OF ACTION POTENTIALS

13

L. B. COHEN and D. LANDOWNE

This chapter considers the question of how the structure of an axon changes during the action potential. Included in this question is the more interesting one of how the structure of the membrane changes during the increases in conductance which underlie the action potential.

These questions have attracted considerable attention among physiologists, especially in the period following the quantitative description of the ionic basis of the action potential (Cole, 1949; Hodgkin, Huxley and Katz, 1952; Hodgkin and Huxley, 1952a ; Keynes, 1951). There are many difficulties involved in studying the molecular basis of the nerve action potential. There is only a small amount of active material available, and it is intimately entangled in structures which must be considered as impurities. In the largest single axon preparation, less than 0.002 per cent of the material is responsible for the action potential. The action potential is a transient, nonequilibrium phenomenon which occurs fairly rapidly. To be useful, a method of observation must be able to resolve time differences of about 0.1 msec. Finally, an axon is a delicate structure and cannot withstand many kinds of manipulations.

Two methods usually used to study changes in structure are chemical analysis and measuring the interaction of various electromagnetic waves with the material. But for nerves, ordinary chemical analyses are generally too slow and often lethal (Ungar, Aschheim, Psychoyos and Romano, 1957), and methods using high-energy waves, including x-ray diffraction (Schmitt, Bear and Clark, 1935) and electron microscopy, have not been applied successfully.

Using visible light, we have been able to measure optical changes in the axon during the action potential. This approach was originated by Hill and Keynes (1949, Hill, 1950; Bryant and Tobias, 1952), who found a long-lasting decrease in light scattering during and following a train of action potentials in the crab nerve. We have extended this finding and now are able to measure changes in birefringence and light scattering in single axons with a' time resolution better than 0.1 msec (Cohen, Keynes and Hille, 1968; Cohen, Hille and Keynes, 1969; Cohen and Keynes, 1969; Cohen and Landowne, 1970; Landowne and Cohen, 1970). The analysis of the changes constitutes the major portion of what follows. These findings have inspired efforts to measure other optical changes and to study changes in indicator molecules added to the axon. In chapter 19, Tasaki and Carnay describe the changes in fluorescence discovered in their laboratory (Tasaki, Carnay, Sandlin and Watanabe, 1969; Tasaki, Watanabe, Sandlin and Carnay, 1968). A related area, presently being explored, is the electron spin and nuclear magnetic resonance of nerves (Commoner, Woolum and Larsson, 1969; Hubbell and McConnell, 1968). Addition of the spin label tetramethyl-pipiridine-1-oxl to rabbit vagus nerves shifted the spectrum of one hyperfine component (Hubbell and McConnell, 1968). Presumably there will be studies of how the action potential affects this shift.

BIREFRINGENCE

When light interacts with matter its velocity of propagation is reduced. The ratio of the velocity *in vacuo* to the velocity in the material is the refractive index. Many materials have different indices of refraction to light polarized in different planes and are thus called birefringent. For an object with a single optic axis, the amount of birefringence is the difference between the refractive indices for light polarized parallel and perpendicular to that axis. The birefringence times the thickness equals the retardation, i.e., the amount one plane-polarized wave front is retarded with respect to the other. The birefringence of biological materials is related to the extent of structural order; e.g., a myelinated nerve is more birefringent than a nonmyelinated one. In our experiments the direction of propagation was perpendicular to the long axis of the axon. With respect to this axis the birefringence of the squid giant axon averaged 6.6×10^{-5}, giving a retardation of 56 nm.

When a birefringent object is placed between perpendicular polarizers, with its optic axis at 45° to the plane of polarization of the incident light, the intensity of light passing through the second polarizer is directly related to the retardation. Several unsuccessful attempts had been made to find changes in the retardation of axons during the action potential (Schmitt and Schmitt, 1940), most of which have not been reported.

The left side of Fig. 13-1 illustrates measurements in a crab nerve, and in each trace there is an intensity decrease after the stimulus. In most experiments the signal-to-noise ratio was smaller than that in Fig. 13-1, and a number of sweeps were averaged. Since the noise is random, while the signal is the same in each sweep, averaging a number of sweeps increases the signal-to-noise ratio. This is illustrated in the right side of Fig. 13-1, where, in the average

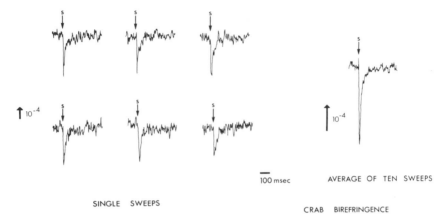

100 msec   AVERAGE OF TEN SWEEPS

SINGLE   SWEEPS

CRAB   BIREFRINGENCE

Fig. 13-1   Intensity changes in the light passing through perpendicular polarizers with the nerve between the polarizers at 45° to the plane of polarization of the incident light. Averaging 10 sweeps increased the signal-to-noise ratio by about threefold. Nerve from the walking leg from the crab Maia squinado. The arrows labeled s indicate the times of stimuli. In all figures the vertical arrows indicate the direction of an increase in intensity; the length of each represents the stated value of the change in intensity divided by the resting intensity per sweep. All records are traced from the original. Temperature, 13°C. (From L. B. Cohen and R. D. Keynes, unpublished.)

of ten sweeps, the signal-to-noise ratio was about three times larger. Further experiments showed that the intensity decrease during the action potential represented a retardation decrease.

Once a change was found, we wished to answer two questions. First: Where were the changes occurring, in the axoplasm, in the axon membrane, or in the Schwann cell and connective tissue? This is a real problem because each of these components has its own retardation. Second: Which of the many events that occur during the action potential give rise to the retardation change? Is it one of the ionic currents, one of the conductance changes, or the potential change itself?

It seemed impossible to answer these questions by doing experiments on a bundle of axons like the crab nerve, so we tried to measure a retardation change in the squid giant axon. But the giant axon has about 200 times less membrane per unit volume than the crab nerve, and if the retardation change occurred only in the membrane, then the change would be difficult to measure. This turned out to be so, and Fig. 13-2 shows the average of 27,500 sweeps

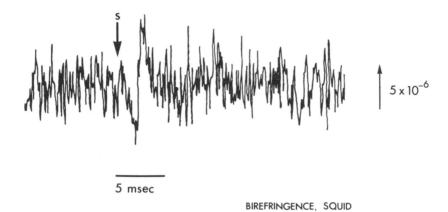

5 msec

BIREFRINGENCE, SQUID

Fig. 13-2 *First retardation measurement on a giant axon, from the squid* Loligo forbesi. *The record is noisy but a signal was present. Optical arrangement as in Fig.* 13-1. *27,500 sweeps averaged. Temperature,* 17°C. *(From L. B. Cohen and R. D. Keynes, unpublished.)*

in our first attempt to measure a retardation change in the giant axon. The record is noisy but there does appear to be a change during the 5-msec period following the stimulus. After improving the signal-to-noise ratio (Cohen, Hille and Keynes, 1969), we could measure the retardation decrease during the action potential with better sensitivity and time resolution. Fig. 13-3, compares the retardation decrease with the action potential; the intensity decrease has been inverted to facilitate the comparison of the two time courses. The time course of the retardation change was very similar to that of the potential and therefore not like the time courses of ionic currents or conductance increases that occur during the action potential (Hodgkin and Huxley, 1952a). Thus our initial hypothesis was that the retardation change depended upon potential. This hypothesis could be tested by doing combined retardation and voltage-clamp experiments.

We were concerned that the current-passing electrode might introduce artifacts; so, in all of the voltage-clamp experiments an image of the axon and electrode was formed and then the image of the electrode was blocked so that no light from the electrode was measured. Also, the results from voltage-clamp experiments were quantitatively consistent with the results from action potential experiments. So we feel that the electrode did not cause artifacts.

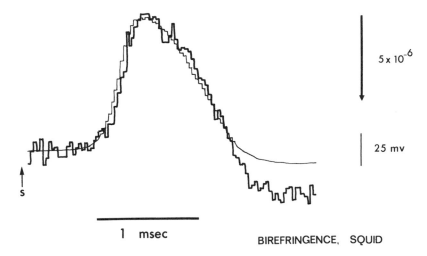

5×10^{-6}

25 mv

1 msec

BIREFRINGENCE, SQUID

Fig. 13-3 *Retardation measurement on a giant axon from* L. forbesi. *The retardation change and the potential had similar time courses. The tip of the potential-measuring electrode was at the center of the region where the retardation was measured, so that action potential and retardation could be recorded at the same time and place. The thick, noisy trace is the retardation, inverted;* 10,000 *sweeps averaged. The thin trace is the action potential. Optical arrangement as in Fig. 13-1. Temperature,* 11°C. (*From L. B. Cohen, B. Hille, and R. D. Keynes, unpublished.*)

Fig. 13-4 illustrates a voltage-clamp and retardation experiment. During the hyperpolarizing step, on the left, there was very little ionic current and no conductance change, but there was an appreciable retardation increase having a time course similar to that of the potential. Thus some of the retardation change depended upon the potential. During the depolarizing step, on the right, when there were large currents and conductance changes there was a retardation decrease. It was smaller than the increase during the hyperpolarizing step and it differed somewhat from the rectilinear shape of the potential. When compensation for the series resistance (Hodgkin, Huxley and Katz, 1952) was included, the retardation change had the same shape as the potential. Since the retardation change always looked like the potential and was not affected by ionic currents or permeability changes, we concluded that at least 90 per cent of the retardation change depended upon the potential. Because the potential occurs mainly across the axon membrane, we tentatively concluded that the retardation change was occurring in the membrane. Now to test this suggestion, we removed the axoplasm and perfused the axon. The retardation decrease was still present and had the same characteristics; therefore, the change did not occur in the axoplasm. A similar retardation change was found during the action potential in the electric organ of the eel (Cohen, Hille and Keynes, 1969), and because this

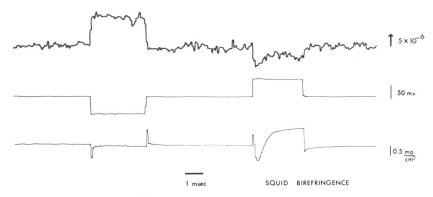

$\uparrow$ 5×10^{-6}

$|$ 50 mv

$|$ 0.5 $\frac{ma}{cm^2}$

—— 1 msec   SQUID BIREFRINGENCE

Fig. 13-4 *Combined retardation and voltage clamp experiment on a giant axon from L. Forbesi. During the hyperpolarizing voltage step there were small currents and no conductance changes; the retardation change must depend on the potential. Top trace: intensity; middle trace: potential between internal and external electrodes (downward is hyperpolarizing); bottom trace: current density (downward is inward current). Experiment performed on Reichert Zetopan polarizing microscope. Optical arrangement as in Fig. 13-1. A stop placed in the objective image plane blocked the image of the voltage clamp electrode. 13,100 sweeps averaged. Temperature, 16.5°C. (From L. B. Cohen, B. Hille, and R. D. Keynes, unpublished.)*

tissue lacks Schwann cells it was highly improbable that the retardation change in the squid axon occurred in the Schwann cells. The only likely remaining structure is the axon membrane.

The membrane makes only a small contribution to the resting retardation, so the percentage change in membrane retardation is considerably greater than the percentage change in intensity, which was 0.001 per cent. Assuming that the axon membrane has a retardation similar to an erythrocyte ghost (Mitchison, 1953), we estimated the decrease in membrane retardation during the action potential to be about 0.2 per cent.

Some preliminary experiments were done to find out what molecules might be responsible for the retardation change. Several pharmacologically active substances were added to the axon to see if they would affect the retardation change. Neither 10 mM procaine nor 100 mM butanol affected the retardation change in either magnitude or time course. Both substances have marked effects on lipid model systems (Skou, 1954; Skou, 1958; Johnson and Bangham, 1969); therefore, we decided provisionally that the retardation change did not originate in the lipid portion of the membrane.

Two other substances, tetrodotoxin and saxitoxin, had marked effects on the retardation change. Added to the bathing medium in low concentrations (110–300 nM), they increased the size of the retardation change by a factor of two to four and slowed the time course of the change. The actions of these poisons on the retardation change was apparently not directly related to their effects on the early conductance increases, because when they were removed the early conductance increase returned but the retardation change remained enlarged and slowed. Tetrodotoxin injected into the axoplasm did not affect the retardation change, suggesting that the retardation change originates in some nonlipid molecules in the outer portion of the axon membrane.

LIGHT SCATTERING

The second optical property of axons that we studied was light scattering. Light scattering is often used to measure the size of molecules and particles (Tanford, 1961) and the time course of volume changes in cells and cell organelles (Ørskov, 1935). Dr. B. B. Shrivastav has measured the light scattering of intact erythrocytes, hemolysed erythrocytes and hemocyanin as a function of scattering angle. The results in Fig. 13-5 show that for objects as large as erythrocytes the scattering angle was very important; the forward scattering (10°) was about 10^4 times larger than the 90° scattering. When intact and hemolysed erythrocytes were compared, elimination of the difference in refractive indices between the inside and the outside of the cell reduced the 90° scattering by about 10-fold and the 40° scattering by about 50-fold. Thus, both 90° and forward scattering were influenced by the difference in refractive indices, but the forward scattering was more sensitive. For particles small compared to the wavelength (hemocyanin), the angular dependence and the total scattering were much reduced. Light scattering thus depends upon angle, size, shape and refractive index.

Since the scattering angle is an important variable, we tried to measure changes in scattering at several angles during the action potential. Fig. 13-6 illustrates compiled results from three giant axons. Clearly there were changes in light scattering, ranging in size from about 2×10^{-6} to 10×10^{-6}. (The small size of the changes contradicts the results of Ludkovskaya, Emel'yanov and Lemazhikin, (1965), who claimed a change in scattering about 10^5 times larger. We assume that they were measuring artifacts.) The scattering changes during the action potential were of two distinct types. At 10° to 45° there was a decrease during the action potential, while at 70° to 120° there was an increase. At 60° the sign of the scattering change was different with different axons. Further experiments indicated that the most representative angle for the low-angle scattering was 10° and that 90° adequately represented the higher angle change. To determine what events during the action potential were giving rise to the light scattering change, combined light scattering and voltage clamp experiments were carried out.

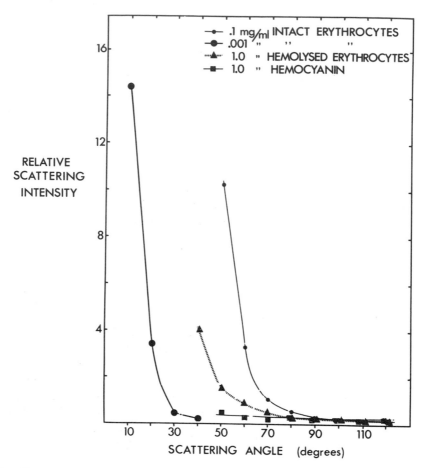

Fig. 13-5 *Light scattering by intact and hemolysed erythrocytes and by hemocyanin, as a function of scattering angle. Angle and refractive index differences were important in determining the amount of scattering. Erythrocytes from* Homo sapiens; *hemocyanin from the lobster* Homarus americanus. *Temperature,* 14°–17°C. (*From B. B. Shrivastav, unpublished.*)

10° Light Scattering

Fig. 13-7 illustrates an experiment where both hyperpolarizing and depolarizing voltage steps were given. In contrast to the birefringence findings (Fig. 13-4), there was a large light scattering change during the depolarizing step and a relatively small increase, barely visible in Fig. 13-7, during the hyperpolarizing step. This increase during the hyperpolarizing step was clearly seen when ten times as many sweeps were averaged. Then it had the same shape as the potential, and, therefore, this part of the light scattering

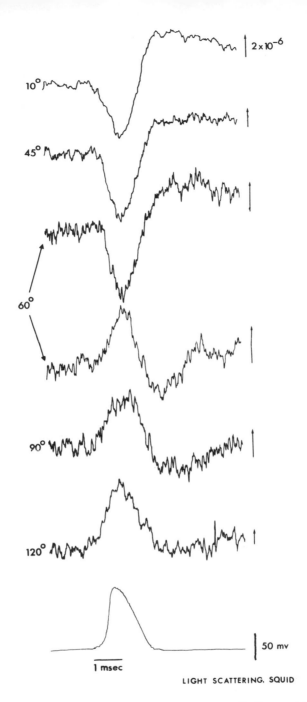

10°

45°

60°

90°

120°

2×10^{-6}

50 mv

1 msec

LIGHT SCATTERING, SQUID

Fig. 13-6 *Changes in light scattering from giant axons* (Loligo pealei) *as a function of angle. The changes during the action potential were of two types, one noticed at low angles, the other near right angles. Top six traces: light scattering; bottom record: representative action potential. Between 5,000 and 16,000 sweeps were averaged. Temperature, 12°–15°C. (From D. Landowne and L. B. Cohen, unpublished.)*

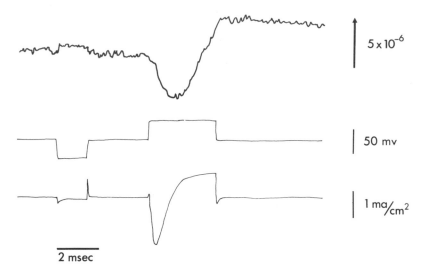

LOW ANGLE LIGHT SCATTERING, SQUID

Fig. 13-7 *Combined low-angle light scattering and voltage clamp experiment on giant axon (L. forbesi). The large scattering change occurred during the depolarizing step when there were large currents and conductance increases. Top trace: light scattering; middle trace: potential; bottom trace: current density. 800 sweeps averaged. Temperature, 15°C. (From L. B. Cohen and R. D. Keynes, unpublished.)*

change was apparently potential dependent. This 10° potential-dependent light scattering change will be called $I_{10}V$.

The large decrease occurring with the depolarizing step presumably resulted from ionic currents or conductance changes. To decide which, the light scattering during a brief 50 mv depolarizing step was measured in ordinary sea water and in sea water in which all of the sodium was replaced by choline. The change in sodium conductance occurred in both media, but in ordinary sea water the currents were much larger, as was the change in light scattering. In the choline sea water the currents were small and there was no detectable light scattering change. The light scattering appeared to ignore the conductance change but to depend upon the ionic current ($I_{10}C$). Thus the 10° light scattering depended partly on potential and partly on current.

When the time courses of $I_{10}C$ and the current were compared, that of $I_{10}C$ was the same as the time integral of the current; all of the change occurred during the flow of current. This is true of the change shown in Fig. 13-7, and is easily demonstrated when the currents are in one direction. An inward current gives rise to a decrease in scattering, while an outward current gives rise to an increase in scattering. $I_{10}C$ was a persistent change; the scattering returned to the baseline with a time constant greater than 100 msec.

The experiment with erythrocytes indicated that refractive index was an important variable; we tried to modify $I_{10}C$ by raising the external refractive index. When 7 grams of bovine albumin were added to 100 ml of bathing medium, the light scattering change reversed in direction and was less persistent. Now, with an outward current there was a decrease in scattering which returned to the resting level with a time constant of about 20 msec.

Possible mechanisms that might account for $I_{10}C$ will be proposed in the discussion section. The time course and the fact that inward and outward currents give rise to light scattering changes opposite in direction are important factors to be considered.

90° Light Scattering

At 90° the scattering changes during the action potential were different from those at 10°, suggesting that they probably had different structural or physiological origins. When experiments similar to those shown in Fig. 13-7 were carried out at 90° there was a light scattering change dependent upon potential ($I_{90}V$). Again, the largest scattering change occurred during the depolarizing voltage steps and appeared to depend on either current or conductance. To decide which was causal, we compared the light scattering changes associated with 40 mv and 100 mv depolarizing voltage steps (Fig. 13-8). During the 40 mv step there was a large increase in scattering which

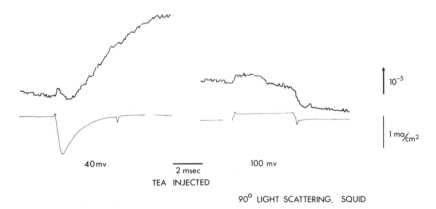

40mv   2 msec   100 mv

TEA INJECTED

90° LIGHT SCATTERING, SQUID

Fig. 13-8 *Combined 90° light scattering and voltage clamp experiment on axons (L. forbesi) injected with tetraethylammonium chloride (TEA). The light scattering change depended on the amount and direction of the current. Left: 40 mv depolarizing voltage step used; right: 100 mv depolarizing step used. Top records: light scattering; bottom records: current density. Records from two axons. 200 sweeps averaged. Temperature, 12°C and 15°C. (From L. B. Cohen and R. D. Keynes, unpublished.)*

accompanied the inward current and the conductance increase. However, the 100 mv step reached the sodium equilibrium potential so that there was less current during the step but a similar conductance change. (Hodgkin and Huxley, 1952*b*). Whereas the inward current gave rise to an increase in scattering, the small outward current during the 100 mv step gave rise to a small decrease in scattering. The increase in conductance in both situations did not seem to affect the light scattering changes. Thus the changes depended upon current ($I_{90}C$). Clearly, the time course of $I_{90}C$ was different from that of $I_{10}C$ since the largest component of $I_{90}C$ occurred after the current when the duration of the current was less than 10 msec. As seen in Fig. 13-8, outward currents gave rise to scattering changes opposite in direction from those arising from inward currents. A large outward current (not illustrated) resulted in a large, persistent decrease in scattering that occurred after the current (for short pulses) and returned to the baseline with a time constant of 50 to 100 msec. In addition to this large component there was also a small increase in scattering during the outward current. Note also the small decrease in scattering during the inward current in Fig. 13-8. There was no change in $I_{90}C$ when the refractive index of the external solution was raised with bovine albumin.

DISCUSSION

Potential-Dependent Changes

The retardation change and the two potential-dependent light scattering changes apparently had different structural origins. The addition of 150 nM tetrodotoxin to the bathing medium had no effect on $I_{90}V$ or $I_{10}V$ but did modify the retardation change. When $I_{10}V$ and $I_{90}V$ were compared following steps in potential, the time course of $I_{90}V$ followed the potential with a time constant of about 70 μsec while that of $I_{10}V$ followed with a time constant of about 200 μsec. This suggested that $I_{10}V$ and $I_{90}V$ reflected different structural changes. The structural events producing all three potential-dependent effects presumably occurred in the axon membrane.

Two mechanisms that might account for potential-dependent optical changes were considered. First, a change in membrane potential causes membrane dipoles to rotate. If the dipolar molecules are optically anisotropic, their rotation will lead to a change in retardation called a Kerr effect (LeFevre and LeFevre, 1960) or to a transient electric light scattering change (Stoylov and Sokerov, 1967). A simple calculation, using the Kerr constant of possible membrane constituents with an estimation of the potential gradient, showed that a Kerr effect could be large enough to account for the retardation change during the action potential. A second possible mechanism is a change in thickness due to electrostrictive forces. Potential-dependent increases in capacity which may be the result of thinning of the membrane have been reported for nerve membranes (Cole and Curtis, 1939) and black lipid films (Babakov, Ermishkin and Liberman, 1966; Rosen and Sutton, 1968). A thickness change alone or a concomitant orientation or refractive index change could lead to the optical changes we have observed. Unfortunately, we have not done experiments that would decide whether molecular orientation or membrane thickness change was responsible for any of the three potential-dependent optical effects.

Although it seems simplest to assume that the potential-dependent changes are not directly related to changes in membrane conductance, possibly one or more might be in the causal chain between the changes in membrane potential and membrane conductance.

Current-Dependent Changes

The amplitude of both current-dependent light scattering changes was proportional to the integral of the ionic current; i.e., the amplitudes were determined by the amount of charge that crossed the membrane rather than by the rate of crossing. The direction of both $I_{10}C$ and $I_{90}C$ reversed when the direction of the current was reversed. Since the amplitude of the change in scattering per coulomb was about the same for inward and outward currents, it seemed unlikely that the scattering changes resulted from specific effects of sodium or potassium ions. Ouabain did not seem to affect any of the light scattering changes; presumably they are not related to recovery processes.

The mechanisms considered most seriously were based on the fact that cations carry the current across the membrane while both cations and anions carry current elsewhere in the current path between electrodes. In an "outward potassium current", only the current across the membrane is carried by outward-going potassium ions, while in the sea water just outside the axon membrane more than half of the current is carried by inward-going chloride ions. This transport number effect (Barry and Hope, 1969; Girardier, Reuben, Brandt and Grundfest, 1963) results in an increased KCl concentration just outside the membrane and a decreased salt concentration just inside the membrane. The KCl concentration just outside the membrane remains high for some milliseconds because of the presence of the diffusion barrier presented by the Schwann cell structures (Geren and Schmitt, 1954; Villegas and Villegas, 1968; Frankenhaeuser and Hodgkin, 1956). Frankenhaeuser and Hodgkin (1956) showed that excess potassium would disappear with a time constant that we found to average 40 msec. There is no barrier on the inside of the axon membrane, and the diminished salt concentration just inside the membrane would disappear with a time constant of less than 1 msec. However, a slightly diminished salt concentration would persist for longer periods.

If a light scattering change were directly related to the increase in KCl concentration in the space between the axon membrane and Schwann cell (periaxonal space), then it should occur as soon as the KCl arrives in the space and it should return to the baseline with a time constant of 40 msec. $I_{10}C$ did occur as soon as the KCl arrived in the space, but it lasted longer than 40 msec. When the external refractive index was increased, the modified time course of $I_{10}C$ sometimes resembled the time course expected if the scattering change were directly related to the excess KCl. It thus seemed possible that $I_{10}C$ might result in part from the increased salt concentration in the periaxonal space. This explanation was unlikely to account for $I_{90}C$, because this change was not complete for some time after the arrival of the excess KCl. However, the excess KCl generates an osmotic gradient, and water flows into the space to neutralize this gradient. If most of the water flowed through the same pathway by which potassium diffuses away, then the resulting volume change would occur with a time constant of about 5 msec and the return to the original size would have a time constant greater than 40 msec. These two time constants describe a time course similar to the time course of $I_{90}C$, suggesting that this scattering change was the result of volume changes in the periaxonal space.

Several experiments were carried out in the spring of 1970 to test this hypothesis by changing the major anion in the bathing medium from chloride to isethionate. The equivalent conductivity of isethionate was estimated to be about 42 per cent of the chloride conductivity, and a transport number effect should be reduced by 36 per cent when isethionate replaces chloride. In our best experiment, five measurements of the scattering change per μcoul in chloride sea waters bracketed four measurements in isethionate sea waters. In isethionate the intensity change per μcoul was reduced by an average of 37 per cent. The correspondence between the measured and the predicted values supports the hypothesis that $I_{90}C$ does result from volume changes in the periaxonal space. Thus, we feel that the structural origins of this optical change have been positively identified.

CONCLUSION

Five separable optical changes in giant axons have been described: a retardation change and four light scattering changes. It was relatively simple to determine what aspects of the action potential gave rise to each change.

The retardation change and two of the light scattering changes depended upon potential. The remaining two scattering changes depended upon the integrated current. It was also possible to localize the structures in the axon in which the optical changes happened. The three potential-dependent changes probably occurred in the axon membrane. The two current-dependent changes probably occurred in structures just adjacent to the membrane. Our speculative attempts at providing more detailed structural explanations were given in the body of the paper.

Although none of the five changes was directly related to the increases in conductance, we expect that within a few years some hundreds of optical or spectroscopic effects will be reported (already the total is above 20) and that some of these will be related to conductance. Detailed studies of these effects should provide important information concerning the structural alterations occurring during changes in membrane conductance.

ACKNOWLEDGMENTS

These experiments were carried out at the Laboratory of the Marine Biological Association, Plymouth, England, and at the Marine Biological Laboratory, Woods Hole, U.S.A. The authors are grateful to the directors and staffs at these laboratories for their assistance. Supported in part by Public Health Service grant number NB08437 and NB08304.

Note added in proof. Further experiments have indicated that the current dependent scattering which we called $I_{10}C$ should be separated into two different changes, one occurring at 5° to 15°, and a second at 15° to 30°.

REFERENCES

BABAKOV, A. V., L. N. ERMISHKIN and E. A. LIBERMAN. 1966. Influence of Electric Field on the Capacity of Phospholipid Membranes. *Nature.* 210:953.

BAKER, P. F., A. L. HODGKIN and T. I. SHAW. 1962. Replacement of the Axoplasm of Giant Nerve Fibres with Artificial Solutions. *J. Physiol. (London).* 164:330.

BARRY, P. H., and A. B. HOPE. 1969. Electroosmosis in Membranes: Effects of Unstirred Layers and Transport Numbers. I. Theory. *Biophys. J.* 9:700.

BRYANT, S. H., and J. M. TOBIAS. 1952. Changes in Light Scattering Accompanying Activity in Nerve. *J. Cell. Comp. Physiol.* 40:199.

COHEN, L. B., B. HILLE and R. D. KEYNES. 1969. Light Scattering and Birefringence Changes during Activity in the Electric Organ of *Electrophorus electricus.* *J. Physiol. (London).* 203:489.

COHEN, L. B., and R. D. KEYNES. 1969. Optical Changes in the Voltage-Clamped Squid Axon. *J. Physiol. (London).* 204:100. (Abstract).

COHEN, L. B., R. D. KEYNES and B. HILLE. 1968. Light Scattering and Birefringence Changes During Nerve Activity. *Nature.* 218:438.

COHEN, L. B., and D. LANDOWNE. 1970. Light Scattering Changes in Voltage-Clamped Squid Giant Axon. *J. Gen. Physiol.* 55: 144 (Abstract).

COLE, K. S. 1949. Dynamic Electrical Characteristics of the Squid Axon Membrane. *Arch. Sci. Physiol.* 3:253.

COLE, K. S., and H. J. CURTIS. 1939. Electrical Impedance of Squid Giant Axon during Activity. *J. Gen. Physiol.* 22:649.

COMMONER, B., J. C. WOOLUM and E. LARSSON. 1969. Electron Spin Resonance Signals in Injured Nerve. *Science.* 165:703.

FRANKENHAEUSER, B., and A. L. HODGKIN. 1956. The After-Effects of Impulses in the Giant Nerve Fibres of *Loligo. J. Physiol. (London).* 131:341.

GEREN, B. B., and F. O. SCHMITT. 1954. The Structure of the Schwann Cell and its Relation to the Axon in Certain Invertebrate Nerve Fibers. *Proc. Nat. Acad. Sci. USA.* 40:863.

GIRARDIER, L., J. P. REUBEN, P. W. BRANDT and H. GRUNDFEST. 1963. Evidence for Anion-Permselective Membrane in Crayfish Muscle Fibers and its Possible Role in Excitation Contraction Coupling. *J. Gen. Physiol.* 47:189.

HILL, D. K. 1950. The Effect of Stimulation on the Opacity of a Crustacean Nerve Trunk and its Relation to Fibre Diameter. *J. Physiol. (London).* 111:283.

HILL, D. K., and R. D. KEYNES. 1949. Opacity Changes in Stimulated Nerve. *J. Physiol. (London).* 108:278.

HODGKIN, A. L., and A. F. HUXLEY. 1952a. A Quantitative Description of Membrane Current and its Application to Conduction and Excitation in Nerve. *J. Physiol. (London).* 117:500.

————. 1952b. Currents Carried by Sodium and Potassium Ions through the Membrane of the Giant Axon of *Loligo. J. Physiol. (London).* 116:449.

HODGKIN, A. L., A. F. HUXLEY and B. KATZ. 1952. Measurement of Current-Voltage Relations in the Membrane of the Giant Axon of *Loligo. J. Physiol. (London).* 116:424.

HUBBELL, W. L., and H. M. McCONNELL. 1968. Spin Label Studies of the Excitable Membranes of Nerve and Muscle. *Proc. Nat. Acad. Sci. USA.* 61:12.

JOHNSON, S. M., and A. D. BANGHAM. 1969. The Action of Anaesthetics on Phospholipid Membranes. *Biochim. Biophys. Acta* 193:92.

KEYNES, R. D. 1951. The Ionic Movements During Nervous Activity. *J. Physiol (London).* 114:119.

LANDOWNE, D., and L. B. COHEN. 1970. Light Scattering Changes in Squid Giant Axons that Depend upon Current. *Abstracts Biophys. Soc. 14th Ann. Meet. Biophysical J.* 10:13a.

LEFEVRE, C. G., and R. J. W. LEFEVRE. 1960. The Kerr effect. *In* A. Weissberger [ed.] *Technique of Organic Chemistry*, Vol. I, Part 3, Chapter 36. Interscience.

LUDKOVSKAYA, R. G., V. B. EMEL'YANOV and B. K. LEMAZHIKIN. 1965. Investigation of Optical Properties of the Giant Axon of the Squid at Rest and in Various Phases of Excitation. *Tsitologiya.* 7:520.

MITCHISON, J. M. 1953. A Polarized Light Analysis of the Human Red Cell Ghost. *J. Exp. Biol.* 30: 397.

ØRSKOV, S. L. 1935. Untersuchungen über den Einfluss von Kohlensäure und Blei auf die Permeabilität der Blut Körperchen für Kalium und Rubidium. *Biochem. Z.* 279: 250.

ROSEN, D., and A. M. SUTTON. 1968. The Effects of a Direct Current Potential Bias on the Electrical Properties of Bimolecular Lipid Membranes. *Biochim. Biophys. Acta.* 163: 226.

SCHMITT, F. O., R. S. BEAR and G. L. CLARK. 1935. X-ray Diffraction Studies on Nerve. *Radiology.* 25: 131.

SCHMITT, F. O., and O. H. SCHMITT. 1940. Partial Excitation and Variable Conduction in the Squid Giant Axon. *J. Physiol. (London).* 98: 26.

SKOU, J. C. 1954. Local Anaesthetics. VI. Relation between Blocking Potency and Penetration of a Monomolecular Layer of Lipoids from Nerve. *Acta Pharmacol. Toxicol.* 10:325.

———. 1958. Relation between the Ability of Various Compounds to Block Nervous Conduction and their Penetration into a Monomolecular Layer of Nerve-Tissue Lipoids. *Biochim. Biophys. Acta.* 30: 625.

STOYLOV, S. P., and S. SOKEROV. 1967. Transient Electric Light Scattering. I. A Method for Determination of the Rotational Diffusion Constant. *J. Colloid Interface Sci.* 24: 235.

TANFORD, C. 1961. *Physical Chemistry of Macromolecules.* John Wiley & Sons, Inc., New York. 278.

TASAKI, I., L. CARNAY, R. SANDLIN and A. WATANABE. 1969. Fluorescence Changes During Conduction in Nerves Stained with Acridine Orange. *Science.* 163: 683.

TASAKI, I., A. WATANABE, R. SANDLIN and L. CARNAY. 1968. Changes in Fluorescence, Turbidity, and Birefringence Associated with Nerve Excitation. *Proc. Nat. Acad. Sci., U.S.A.* 61: 883.

UNGAR, G., E. ASCHHEIM, S. PSYCHOYOS and D. ROMANO. 1957. Reversible Changes of Protein Configuration in Stimulated Nerve Structures. *J. Gen. Physiol.* 40: 635.

VILLEGAS, G. M., and R. VILLEGAS. 1968. Ultrastructural Studies of the Squid Nerve Fibers. *J. Gen. Physiol.* 51:44s.

INTERNAL PERFUSION OF SQUID AXONS: TECHNICAL CONSIDERATIONS 14

D. L. GILBERT

The history of perfusing the squid axon goes back at least to 1938. Cole (1968) states: "As he (Hodgkin] was resting his eyes during the cleaning of an axon at Woods Hole [Massachusetts] in 1938 he had prophesied that he would someday perfuse the inside of an axon." Hodgkin (1964) wrote that he and Katz tried unsuccessfully to perfuse an axon in 1948.

Baker, Hodgkin, and Shaw (1961) reported in 1961 a successful technique. Their method is based upon applying external pressure to the axon and rolling out the axoplasm with a small roller. This method can be designated as the toothpaste-squeezing technique or the lawn-roller technique. This technique has been described in detail by Baker, Hodgkin, and Shaw (1962).

Oikawa, Spyropoulous, Tasaki, and Teorell (1961) also reported in 1961 another successful technique. This method is performed by inserting a glass capillary longitudinally through the axon. The axoplasm is taken up in the capillary and subsequently forced out. In this capillary technique either one or two capillaries are used. When two capillaries are used, one contains the perfusion fluid and the other is the outflow capillary. The capillary method has been described in detail by Tasaki, Watanabe, and Takenaka (1962).

MODIFICATIONS OF THE BASIC PERFUSION TECHNIQUES

The lawn-roller technique has proved successful only by workers in England and by Narahashi (1963) in Woods Hole. The axons of *Loligo pealei* available in Woods Hole average about 350 to 450 μ in diameter, whereas the axons

264

of *Loligo forbesi*, available in Plymouth, England, are about twice this size. Needless to say, it is more difficult to perfuse small axons.

One of the greatest difficulties in perfusion is removing the axoplasm. If the pressure is increased too much, the axon can be irreversibly damaged. With the capillary method, the perfusion and flow become clogged with axoplasm more than with the lawn-roller technique. Suction has also been applied to remove the axoplasm. In giant axons of the large squid off the coast of Chile (*Dosidicus gigas*), it was found that the largest axons were sometimes impossible to perfuse due to the viscosity of the axoplasm. Adelman and Gilbert (1964) used a single capillary which eroded away the axoplasm as shown in Fig. 14-1. A constant volume pump can be used, but

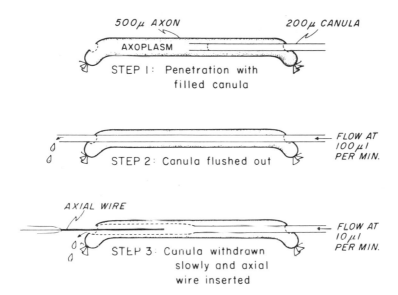

Fig. 14-1 *Perfusion technique showing how infusion pipette and current axial wire is inserted. (From Adelman and Gilbert, 1964.)*

when an axoplasm plug clogs the capillary, then pressure can increase, damaging the axon. Hydrostatic pressure can also be used. However, if the perfusion flow ceases, the axoplasm is totally replaced with a massive injection of the internal fluid.

Brinley and Mullins (1967) used a dialysis method, which has the advantage of avoiding clogging the axoplasm within the tubing. This method presents a combination of the perfusion and injection methods.

Tasaki, Watanabe and Singer (1966) have used pronase (a protease) to remove axoplasm for approximately 1.5 minutes. A pronase concentration of

0.05–0.1 mg/ml and glycerol were also added to remove axoplasm. Huneeus-Cox, Fernandez, and Smith (1966) used a hyperosmotic cysteine solution to remove axoplasm. The solution consisted of 400 mM potassium cysteinate, 400 mM KCl, and 200 mM KF. Perhaps the observed effect on the axoplasm is due to the osmolarity influence alone.

TYPES OF PERFUSION FLUID

Potassium is the most predominately used cation. Tasaki, Singer and Takenaka (1965) claim that the beneficial effectiveness of cations in decreasing order are: Cs, Rb, K, NH_4, Na, and Li. The fluoride ion was discovered by Tasaki and Takenaka (1964) to be extremely effective. Its beneficial effect may be to remove calcium ions, which are detrimental to the internal environment. Tasaki, Singer and Takenaka (1965) claim that the beneficial effective anions in decreasing order are: F, HPO_4, glutamate or aspartate, citrate, tartrate, propionate or butyrate, SO_4, acetate, Cl, NO_3, Br, I, and SCN.

The fluoride anion is a poison for enzymatic systems and cannot be present in active transport studies. Brinley and Mullins (1967) have, therefore, used in their active transport studies, potassium with some sodium and negligible concentration of magnesium for cations; taurine, then equal amounts of isethionate and aspartate, and some chloride were used for anions.

VOLTAGE CLAMPING OF PERFUSED AXONS

Moore, Narahashi and coworkers (Moore, Narahashi and Ulbricht, 1964; Narahashi, Anderson and Moore, 1967) have utilized the sucrose-gap technique in conjunction with the voltage clamp technique.

Chandler and Meves (1965) used a differential measurement for obtaining currents. The advantage is that the length of the axon to be perfused is decreased. However, calibration can be a problem. If the solution resistance is altered, then the calibration has to be changed.

Adelman and Gilbert (1964), using a microelectrode for measuring the electrical potential, were the first to voltage clamp a perfused axon. Fig. 14-2 is a diagram of this voltage clamp perfusion technique. The axial current wire can seriously impede the flow of the perfusion fluid. Voltage clamping with this technique requires guard electrodes, current measuring electrode, current (axial) electrode and the internal and external potential probing electrodes. A swage-lock at the end of the infusion pump was utilized to change solutions by Adelman and Senft (1966).

The resistance of a microtip electrode is high; therefore an alternative method is to use a capillary inserted longitudinally inside the axon for sensing the internal electrical potential. The advantage here is that the capillary has

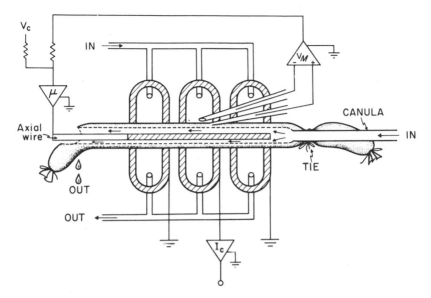

Fig. 14-2 *Diagram showing a combination perfusion, voltage clamped axon. (From Adelman and Gilbert, 1964.)*

a much lower resistance than the microtip, but the disadvantage is that the internal perfusion flow is impeded.

Armstrong and Binstock (1964) have developed an electrode with a hole in it some distance from the end. The outside of this glass capillary is coated with black platinum and serves as the current electrode. The inside of the capillary is filled with potassium chloride solution and makes electrical contact with the internal environment through the hole and serves as the internal electrical potential sensing electrode. When this electrode is inserted longitudinally into the axon, it serves as both the current and internal electrical potential electrode. It is, however, difficult to make.

Rojas and Ehrenstein (1965) placed the current electrode next to the internal electrical potential electrode and carefully adhered the two together using dental sticky wax. This piggyback arrangement works very well, but the wax can possibly disturb the homogeneity of the platinum wire surface.

TECHNIQUE USED IN PRESENCE OF REDOX SUBSTANCES

Fig. 14-3 shows the experimental set-up of a voltage clamped axon in the presence of redox substances in which the axoplasm is replaced with an internal solution. This technique of Gilbert and Robbins (1969) is modified

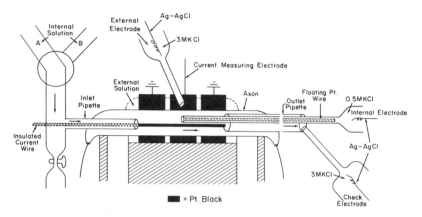

Fig. 14-3 *Side view diagram of voltage clamped axon.*

from the method used by Rojas and Ehrenstein (1965). The outlet pipette is pushed through the axon, using suction, or sometimes just capillary action, to remove the axoplasm. Axoplasm in the outlet pipette is then removed by washing out the outlet pipette with the internal solution. Then the current wire and inlet pipette is inserted into the outlet pipette and the internal solution flow is begun. The inlet pipette and the outlet pipette are both slowly moved to the right until they are in the position shown in Fig. 14-3. An internal electrode is then inserted into the outlet pipette and placed in position. The current is measured with a black platinum electrode. The difference in electrical potential between the internal and external electrodes (after correcting for junction potentials) is the membrane potential. An additional internal electrode (check electrode) is also used. It senses the potential at the end of the outlet pipette instead of at the center of the axon. The check electrode is not used for voltage clamping, and therefore does not include a floating platinum wire, which is used for high frequency responses by acting as a capacitive shunt (Baker, Hodgkin and Meves, 1964; Chandler and Hodgkin 1965).

It is possible with this procedure to check junction potentials with the check electrode in the middle of an experiment. Baker, Hodgkin and Meves (1964) also took care in measuring junction potentials. Under ordinary conditions, they are small. However, if redox chemicals are present, they can affect the potential at the platinum surface and affect the membrane electrical potential measurement. The error involved in making electrical potential measurements with platinum surfaces has been mentioned by Tasaki and Singer (1968).

Error in Measurement of Electrical Potential Difference

Fig. 14-4 illustrates how we measure in our experimental set-up the electrical

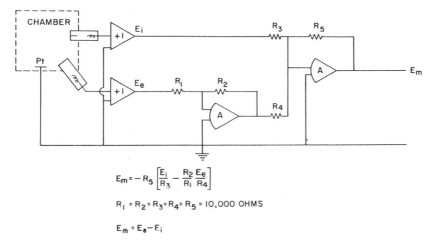

$$E_m = -R_5 \left[\frac{E_i}{R_3} - \frac{R_2}{R_1} \frac{E_e}{R_4} \right]$$

$R_1 = R_2 = R_3 = R_4 = R_5 = 10,000 \text{ OHMS}$

$E_m = E_e - E_i$

Fig. 14-4 *Measurement of electrical potential.*

potential difference between the internal and external electrodes. Both the internal and external electrodes are filled with potassium chloride solutions and are connected to the amplifiers by silver-silver chloride electrodes. E_i is the sum of the potentials due to the internal probe and platinum surface of the electrode which leads to ground. Likewise, E_e is the sum of the potentials due to the external probe and the platinum surface of the electrode which goes to ground.

If we examine the external probe potential, it follows that:

$$(E_e/R_1) = -(E_{ref}/R_2), \qquad (14\text{-}1)$$

where R_1 and R_2 are the resistances in Fig. 14-4 and E_{ref} is the electrical potential of the first operational amplifier.

Next, let us examine the effect of the second operational amplifier:

$$(E_i/R_3) + (E_{ref}/R_4) = -(E_m/R_5), \qquad (14\text{-}2)$$

where R_3, R_4, and R_5 are the resistances in Fig. 14-4, and E_m is the measured potential.

Combining Eq. (14-1) and (14-2) gives

$$E_m = -R_5 \left[\frac{E_i}{R_3} - \left(\frac{R_2}{R_1} \right) \left(\frac{E_e}{R_4} \right) \right]. \qquad (14\text{-}3)$$

The actual membrane potential, which also includes junction potentials is:

$$E_M = E_i - E_e, \qquad (14\text{-}4)$$

where E_M is the actual membrane potential.

Substituting E_i from eq. (14-4) into Eq. (14-3) yields:

$$E_m = -\left[\frac{R_5}{R_3}\right]E_M + \left[\frac{R_2}{R_1} \cdot \frac{R_5}{R_4} - \frac{R_5}{R_3}\right]E_e. \qquad (14\text{-}5)$$

If R_3 equals R_5 and if the product of R_2 times R_5 equals the product of R_1 times R_4, then E_m, the measured potential, equals the negative value of the actual membrane potential. Experimentally, the five resistors in Eq. (14-5) are set equal to one another, so that E_m equals $-E_M$. However, if there is a small error in the designated resistors, then the measured potential will also be a function of E_e. Table 14-1 gives some examples showing this discrepancy

Table 14-1   Error in measurement of Electrical potential difference

R_5/R_3	R_2R_5/R_1R_4	E_e (mv)	E_M (mv)	E_m (mv)
1.01	1.00	−100	−60.0	61.6
1.01	1.00	−200	−60.0	62.6
1.01	1.00	−500	−60.0	65.6
1.00	1.01	−100	−60.0	59.0
1.00	1.01	−200	−60.0	58.0
1.00	1.01	−500	−60.0	55.0

for a value of -60.0 mv for the actual membrane potential.

The value of E_e is dependent upon the unstable electrical potential platinum surface as shown in Fig. 14-4. In the presence of redox substances, the electrical potential at the platinum surface can be greatly altered. Thus, addition of redox substances can possibly significantly alter the measured potential, without really affecting the actual membrane potential. One way to avoid this error is to substitute for the platinum surface a stable silver-silver chloride electrode with a 3 M KCl agar plug next to the silver-silver chloride electrode. Thus, changes in E_e will be minimized, and so a difference in E_m will reflect differences in E_M.

Currents in the Internal Electrode

Fig. 14-5 illustrates in an oversimplified way of how the internal electrode can be sensitive to a redox reaction at point *B*. In this model, it is assumed

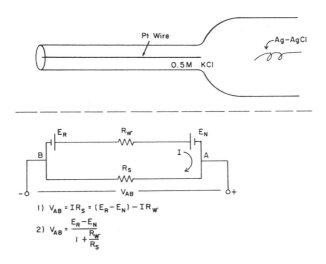

1) $V_{AB} = IR_s = (E_R - E_N) - IR_w$

2) $V_{AB} = \dfrac{E_R - E_N}{1 + \dfrac{R_w}{R_s}}$

Fig. 14-5 *Internal electrode.*

that the platinum wire is insulated except at points A and B. Actually the platinum wire used is not insulated. The difference in electrical potential between A and B is

$$V_{AB} = IR_s = (E_R - E_N) - IR_w,\qquad (14\text{-}6)$$

where V_{AB} = potential at A minus potential at B

I = current

R_s = resistance of solution surrounding the platinum wire

R_w = resistance of the platinum wire

E_R = potential at the platinum surface at point A

E_N = potential at the platinum surface at point B.

Deleting the current term in Eq. (14-6) yields

$$V_{AB} = \frac{E_R - E_N}{1 + (R_w/R_s)}.\qquad (14\text{-}7)$$

If the environment is the same at points A and B, then E_R will equal E_N, and V_{AB} will equal zero. However, a redox substance in contact with the external tip of the electrode can possibly cause a change in E_R. If the solution resistance is low or if the wire resistance is high, then V_{AB} will be small. However, if the solution resistance is high or if the wire resistance is low, then V_{AB} will be equal to the difference between E_R and E_N. Hence, it is desirable to decrease the resistance in the solution as much as possible; i.e., by increasing the ionic concentration in the electrode. This error can be avoided by using the check electrode which does not contain a wire and is hence insensitive to redox reagents.

Errors in Electrodes

By measuring junction potentials between electrodes, one can usually determine which electrode is faulty when an error in measuring the potential is noted. Bubbles are quite a common annoyance; their presence can be noted by a lot of amplifier noise and often by an unusual potential not predicted by known junction potentials.

Other electrical paths to ground can be detected by removing the platinum surface which is connected to ground and replacing it with a stable electrode, such as a calomel or silver-silver chloride electrode in agar, and measuring E_{ref} [see Eq. (14-1)]. The measured potential should be stable and predicted from any calculated junction potential. If the potential deviates from the calculated value, then there is an alternate pathway to ground.

SUMMARY

Both the lawn-roller and capillary methods of performing internal perfusion in the giant axons of squid have proved to be successful. Perfused axons have been successfully voltage clamped with both of these techniques.

Redox chemicals can influence the electrical potential on metal surfaces, and can produce errors in some of the standard measurements made. A description on the avoidance of these errors has been given.

REFERENCES

ADELMAN, W. J., Jr., and D. L. GILBERT. 1964. Internally Perfused Squid Axons Studied under Voltage Clamp Conditions. I. Method. *J. Cell. Comp. Physiol.* 64:423.

ADELMAN, W. J., Jr., and J. P. SENFT. 1966. Voltage Clamp Studies on the Effect of Internal Cesium Ion on Sodium and Potassium Currents in the Squid Giant Axon. *J. Gen. Physiol.* 50:279.

ARMSTRONG, C. M., and L. BINSTOCK. 1964. The Effects of Several Alcohols on the Properties of the Squid Giant Axon. *J. Gen. Physiol.* 48:265.

BAKER, P. F., A. L. HODGKIN and H. MEVES. 1964. The Effect of Diluting the Internal Solution on the Electrical Properties of a Perfused Giant Axon. *J. Physiol. (London).* 170:541.

BAKER, P. F., A. L. HODGKIN and T. I. SHAW. 1961. Replacement of the Protoplasm of a Giant Nerve Fibre with Artificial Solutions. *Nature.* 190:885.

_____. 1962. Replacement of the Axoplasm of Giant Nerve Fibres with Artificial Solutions. *J. Physiol. (London).* 164:330.

BRINLEY, F. J., Jr., and L. J. MULLINS. 1967. Sodium Extrusion by Internally Dialyzed Squid Axons. *J. Gen. Physiol.* 50:2303.

CHANDLER, W. K., and A. L. HODGKIN. 1965. The Effect of Internal Sodium on the Action Potential in the Presence of Different Internal Anions. *J. Physiol. (London).* 181:594.

CHANDLER, W. K., and H. MEVES. 1965. Voltage Clamp Experiments on Internally Perfused Giant Axons. *J. Physiol. (London).* 180:788.

COLE, K. S. 1968. *Membranes, Ions and Impulses. A Chapter of Classical Biophysics.* Univ. of Calif. Press, Berkeley, p. 465.

GILBERT, D. L., and M. ROBBINS. 1969. Effect of Potassium Ferricyanide on Sodium Currents in the Squid Giant Axon. *Biophys. J. Abst.* 9: A-249.

HODGKIN, A. L. 1964. *The Conduction of the Nervous Impulse.* Charles C. Thomas, Pub., Springfield, Illinois, p. 36.

HUNEEUS-COX, F., H. L. FERNANDEZ and B. H. SMITH. 1966. Effects of Redox and Sulfhydryl Reagents on the Bioelectric Properties of the Giant Axon of the Squid. *Biophys. J.* 6:675.

MOORE, J. W., T. NARAHASHI and W. ULBRICHT. 1964. Sodium Conductance Shift in an Axon Internally Perfused with a Sucrose and Low-Potassium Solution. *J. Physiol. (London).* 172:163.

NARAHASHI, T. 1963. Dependence of Resting and Action Potentials on Internal Potassium in Perfused Squid Giant Axons. *J. Physiol. (London).* 169:91.

NARAHASHI, T., N. C. ANDERSON and J. W. MOORE. 1967. Comparison of Tetrodotoxin and Procaine in Internally Perfused Squid Giant Axons. *J. Gen. Physiol.* 50:1413.

OIKAWA, T., C. S. SPYROPOULOS, I. TASAKI and T. TEORELL. 1961. Methods for Perfusing the Giant Axon of *Loligo pealei. Acta Physiol. Scand.* 52:195.

ROJAS, E., and G. EHRENSTEIN. 1965. Voltage Clamp Experiments on Axons with Potassium as the only Internal and External Cation. *J. Cell Comp. Physiol. (Suppl. 2)* 66:71.

TASAKI, I., and I. SINGER. 1968. Some Problems Involved in Electric Measurements of Biological Systems. *Ann. N.Y. Acad. Sci.* 148:36.

TASAKI, I., I. SINGER, and T. TAKENAKA. 1965. Effects of Internal and External Ionic Environment on Excitability of Squid Giant Axon. *J. Gen. Physiol.* 48:1095.

TASAKI, I., and T. TAKENAKA. 1964. Effects of Various Potassium Salts and Proteases upon Excitability of Intracellularly Perfused Squid Giant Axons. *Proc. Nat. Acad. Sci. USA.* 52:804.

TASAKI, I., A. WATANABE and I. SINGER. 1966. Excitability of Squid Giant Axons in the Absence of Univalent Cations in the External Medium. *Proc. Nat. Acad. Sci. USA.* 56:1116.

TASAKI, I., A. WATANABE, and T. TAKENAKA. 1962. Resting and Action Potential of Intracellularly Perfused Squid Giant Axon. *Proc. Nat. Acad. Sci. USA.* 48:1177.

ELECTRICAL STUDIES OF INTERNALLY PERFUSED SQUID AXONS

15

W. J. ADELMAN, Jr.

"The giant axon of the squid *Loligo* has made possible a spectacular advance toward an understanding of excitable membranes. It allowed the first measurements of membrane potential, membrane conductance, and axoplasm composition. The potential of the membrane was then brought under control and a few years ago, after a quarter century of increasingly powerful insults to its autonomy, the membrane was made to submit to a control of its internal environment by perfusion.

"Each of the earlier membrane experiments were steps that confirmed old concepts or inspired new ones and each led to at least similar results for other membranes. With complete electrical and chemical control to within at most a few micra of the squid membrane, the number of combinations of experimental parameters was vastly increased while hypotheses and theories had to become more complete and could be tested more rigorously."

With this introduction, Kenneth S. Cole (1965) described the status of squid axon work in this generation. It was approximately in 1960 that the first internal perfusion experiments were performed (Oikawa, Spyropoulos, Tasaki and Teorell, 1961; Baker, Hodgkin and Shaw, 1961, 1962a). Since then, many experiments have been performed using this preparation in laboratories, primarily in Woods Hole, Massachusetts and Plymouth, England. For details of the experimental methods used by various laboratories, see chapter 14. These experiments have been of two general types: 1. Membrane potential measurements and voltage clamp studies of membrane currents, and 2. Chemical studies involving ion substitutions, blocking agents, metabolites, and metabolic inhibitors in which chemical control of both sides of the membrane was achieved.

It is now generally accepted that the mechanisms responsible for nervous excitation and for the maintenance of the electrolyte distributions in cells are found in the 100 Å thick cell phase generally referred to as the plasma membrane. While it would be of great interest to isolate this thin structure for exact chemical study and analysis of its functional units, extraneous external coats and its extreme sensitivity to insult have precluded this possibility for the present. Most work in the natural membrane field has either been to identify structural units through such techniques as electronmicroscopy and X-Ray diffraction (see chapter 1), or to deduce the membrane properties by means of electrical or indirect chemical studies. The direct approach has often involved the death of the cell whereas the indirect approach has maintained the physiological integrity within some limits. The internally perfused squid giant axon is one neural membrane system which has permitted close scrutiny of its physiological structure *in vitro* while maintaining a semblance of the *in vivo* structure. As has been indicated in the opening quote from Dr. Cole, this preparation has allowed the degree of chemical and electrical control over the properties of the nerve membrane to approach within a few micra of its ultimate structure. This chapter describes attempts to use this preparation in combination with sophisticated electrical techniques in order to achieve a combined chemical and electrical approach to the mechanisms of the nerve impulse.

INTERNAL PERFUSION WITH INORGANIC SALT SOLUTIONS OF HIGH IONIC STRENGTH

Resting Potentials

Variations in c_{Na} and c_K. Some of the earliest experiments with internally perfused axons sought to determine the effects of simple changes in the internal ionic environment on membrane potentials (Baker, Hodgkin and Shaw, 1961; Tasaki, Watanabe and Takenaka, 1962). Baker, Hodgkin and Shaw (1962b) measured resting potentials recorded from squid giant axons internally perfused with solutions having various $c_{K_i}:c_{Na_i}$ ratios. These internal solutions were made up from isotonic KCl and NaCl solutions, and thus were similar in osmolarity and ionic strength to artificial and/or natural sea water used for external perfusion.

Fig. 15-1 shows the experimental relations between internal potassium concentration, c_{K_i}, and the internally recorded resting potential plotted for three different external potassium concentrations, c_{K_o}. Baker, Hodgkin and

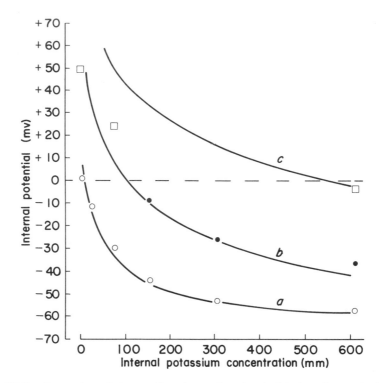

Fig. 15-1 *Comparison between the observed and calculated resting potentials of perfused fibres. The external potassium concentrations are indicated thus:* ○, 10 mM ●, 100 mM □, 540 mM.
    The theoretical curves were calculated on the assumption that the ionic currents vary with membrane potential according to the equations of Hodgkin and Huxley (1952b) and that for a given resting potential the currents vary with concentration as for ions moving independently (Hodgkin and Huxley, 1952a). The "leakage" current has been attributed to chloride and the internal ionic concentrations of intact axons have been selected to be consistent with the respective equilibrium potentials. The resting potential of intact axons has been taken as 62 mv. Curves a, b and c are for external potassium concentrations of 10, 100 and 540 mM, respectively. (From Baker, Hodgkin and Shaw, 1962b.)

Shaw (1962b) concluded that this relation gave strong support for the consideration that the resting potential of perfused axons is a function of the potassium ion concentration gradient across the membrane.

From the Hodgkin and Huxley hypothesis (1952b), it can be shown that the ratio of sodium to potassium conductance, $g_{Na} : g_K$, (for discussion and definitions, see chapter 6) is a function of membrane potential, E_M, while the leakage conductance, g_L, is a constant. For more depolarized potentials than $E_M = -30$ mv, g_{K_∞} increases while g_{Na_∞} declines and approaches zero. If we assume that P_K, P_{Na} and P_{Cl} are proportional to g_{K_∞}, g_{Na_∞} and g_L respectively, then any relation of the Hodgkin and Katz (1949) form of the Goldman equation (1943):

$$E_M = \frac{RT}{F} \ln \frac{P_K c_{K_o} + P_{Na} c_{Na_o} + P_{Cl} c_{Cl_i}}{P_K c_{K_i} + P_{Na} c_{Na_i} + P_{Cl} c_{Cl_o}} \tag{15-1}$$

used to describe E_M would contain variable permeability coefficients for potassium (P_K) and sodium (P_{Na}) ions and constant permeability coefficients for those ions contributing to the leakage current; i.e., P_{Cl}, the chloride permeability coefficient, is a constant. See chapter 2 for a general derivation of Eq. (15-1).

Therefore, Baker, Hodgkin and Shaw attempted to fit their data with relations for E_M derived from the 1952 Hodgkin and Huxley hypothesis. The correlation between theory and experimental data shown in Fig. 15-1 was reasonably good except at high external c_K. In addition, Baker et al. found that membrane selectivity decreased with perfusion time and that the decrease was more rapid when the $c_{Na_i}:c_{K_i}$ ratio was high. When axons are kept depolarized over a period of time, there is a gradual loss of selectivity which accounts for the deviations from Hodgkin-Huxley theory. Despite these difficulties, Baker, Hodgkin and Shaw claimed that for axons internally perfused with mixtures of isotonic K and Na solutions, the resting potential was roughly determined by the potassium ion concentration gradient across the membrane.

It is possible to compare the effect of variations in the internal potassium concentration with the effects of variations in external c_K (Curtis and Cole, 1942) on the resting potential of squid axons. Fig. 15-2 was prepared by Adelman and Gilbert for this comparison by plotting the internal potential against the log of c_{K_i}/c_{K_o}. Data were taken from Baker, Hodgkin and Shaw (1961); Tasaki, Watanabe and Takenaka (1962, Fig. 2); and Adelman and Gilbert (1964) for internal perfusion with various mixtures of isotonic potassium and sodium salts (Cl^- and $SO^=$). Curtis and Cole's data are plotted assuming a c_{K_i} of 400 mm.

Notice that the points are roughly fit by a single nonlinear E_M vs. log (c_{K_i}/c_{K_o}) curve (continuous line) and that deviations from a simple Nernst concentration potential for $10°C$ (dashed line) are most severe when the concentration gradients across the membrane are large. Implicit in the figure are the following: 1. The functional relation between resting potential and potassium concentrations is the same for internal and external K^+ and 2. The functional relation between resting potential and the K^+ gradient is the same for internally perfused and intact axons.

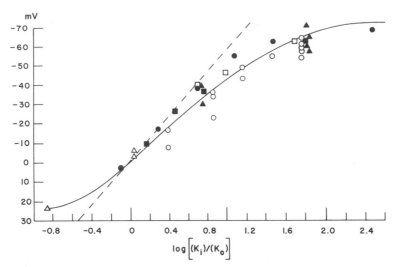

Fig. 15-2 *Resting potential, E_{RP} (ordinate) plotted against the $\log c_{K_i}/c_{K_o}$ (abscissa). Comparison of some data from internally perfused squid giant axons with data from intact squid axons (Curtis and Cole, 1942). Filled circles: Curtis and Cole data. Open circles: 10 mM K_o. Filled squares: 100 mM K_o. Open triangles: 540 mM K_o (data from Baker, Hodgkin and Shaw, 1961). Open squares: 10 mM K_o (data from Tasaki, Watanabe and Takenaka, 1962). Filled triangles: 10 mM K_o (data from Adelman and Gilbert, 1964.)*

Adelman and Senft (1966*a* and unpublished) have measured the membrane resting potential, E_{RP}, during internal perfusion with isotonic fluoride solutions. Table 15-1 gives values of resting potentials recorded from axons internally perfused with mixtures of KF and NaF. By adding dextrose internal solutions were made isosmotic to artificial sea water (ASW). Activities (a) and activity coefficients (γ) for ASW and internal perfusion solutions were estimated by Dr. Leon A. Cuervo in a manner similar to that given in Baker, Hodgkin and Shaw (1962*a*) and in Baker, Hodgkin and Meves (1964). Electrodes sensitive to sodium, total cation or chloride were used. Electrode potentials of experimental solutions were compared with those potentials obtained for a concentration series of known simple salt solutions, some solutions containing dextrose. Calibration curves for the specific ion electrodes were prepared using the Debye-Huckel limiting law for dilute solutions. The activities cited throughout this chapter are only considered estimates as no complete treatment is available for ion activities in complex solutions (cf. Robinson and Stokes, 1959).

Resting potentials were recorded between an internal glass micropipette electrode (3 M KCl salt bridge connected to a calomel half-cell) inserted just through the membrane and an external reference electrode (3 M KCl agar salt bridge connected to a calomel half-cell). Any potential difference between

Table 15-1   Effects of variation in internal and external sodium and potassium ion activities, a, on the resting potential, E_{RP}, of internally perfused squid giant axons. High ionic strength solutions. Temperature $= 5\,°C$

Axon	a_{K_o}	a_{K_I}	a_{Na_o}	a_{Na_I}	E_{RP}	P_{Na}/P_K
	mM	mM	mM	mM	mv	
65–82	0	175	240	0	−61	0.041
65–87	0	175	240	0	−70	0.038
65–91	0	175	240	0	−64	0.049
65–92	0	175	240	0	−52	0.081
65–66	5.5	175	234	0	−67	0.020
65–68	5.5	175	234	0	−67	0.020
65–93	5.5	175	234	0	−57	0.044
65–94	5.5	175	234	0	−59	0.038
66–59	5.5	169	234	0	−62	0.022
66–59	0	169	240	0	−73	0.031
65–95	0	150	240	19	−51	0.073
65–94	5.5	150	234	19	−50	0.055
65–96	5.5	150	234	19	−51	0.051
65–92	0	75	240	103	−32	0.090
65–93	5.5	75	234	103	−39	0.042

average $= 0.046$

the two electrodes was balanced to zero prior to penetrating the axon with the microtip. Additional corrections for junction potentials were not made as there is no adequate way of measuring the junction potentials. Addition of another 3 M KCl salt bridge between test pools of internal and external solutions seems to compound the problem.

Calculations of resting $P_{Na} : P_K$ ratios were made from resting potential data using the exponential form of Eq. (15-1) for $a_i = \gamma c_i$:

$$\frac{P_{Na}}{P_K} = \frac{a_{K_o} - a_{K_i}[\exp(E_M F/RT)]}{a_{Na_i}[\exp(E_M F/RT)] - a_{Na_o}}. \tag{15-2}$$

These calculated values are given in the last column in Table 15-1. Unlike Baker, Hodgkin and Shaw (1962b), we have assumed that the leakage current component was not significantly carried by Cl⁻ ions, and in addition, that at high concentrations of internal F⁻, g_L is small over the range from $E_M = -70$ to $E_M = -30$ mv. The average value of $P_{Na} : P_K$ given in Table 15-1 is 0.046, which is somewhat less that the value of 0.08 obtained by Baker, Hodgkin and Shaw.

The difference between the data of Adelman and Senft (1966*a*) and that of Baker, Hodgkin and Shaw (1962*b*) may lie in the difference in the anion used in the internal perfusion solution. Adelman, Dyro and Senft (unpublished) have shown that ionic selectivity is lower for axons perfused with KCl than with KF. Baker, Hodgkin and Shaw question their own choice of Cl^- as the charge carrier for the leakage current. They point out that resting chloride fluxes are generally low (Caldwell and Keynes, 1960) and that measurements of resting potentials for internal chloride vs. internal sulfate salts indicate that Cl^- contributes only slightly to the resting ionic permeability.

If internal Cl^- raises the leakage conductance in some nonspecific way, the leakage permeabilities become even more complex. Some of the difficulties in estimating leakage permeability in *Myxicola* axons have been raised by L. Goldman and Binstock (1969). However, this problem may not be as complex as it seems, as both Moore, Narahashi, Poston and Arispe (1970) and Fishman (1970) have determined linear current-voltage relations in squid axons. They applied TTX externally and TEA internally (TTX is expected to block sodium permeability and TEA potassium permeability) and found that the leakage component appears either as a fixed ohmic resistance in series with a battery (Fishman, 1970) or as a shunt, independent of ionic species, similar to that found in an imperfect capacitor (Moore *et al.*, 1970).

Variation in Other Alkali Metal Cations as well as Sodium and Potassium

Baker, Hodgkin and Shaw (1962*b*) give the following order for the electrogenic actions of internal alkali metal ions (periodic table group IA) on the resting potential of internally perfused squid axons: K > Rb > Cs > Na > Li. The E_{RP}'s were measured with sea water externally and isotonic sulphate salt solutions of the alkali metals internally. Average resting potential values in the presence of the ions were: K = -61.8 mv, Rb = -51.5 mv, Cs = -18.5 mv, Na = -12.5 mv and Li = -11 mv.

Similar results were obtained by Adelman and Senft (unpublished) for internal perfusion with alkali metal (Na, K, Rb, Cs) fluoride solutions. Roughly the same E_{RP} values were obtained by Adelman and Senft as by Baker, Hodgkin and Shaw (1962*b*). Adelman and Senft also varied the composition of the external sea water solutions. The effects on resting potential of 440 mM Na–, Li–, K–, Rb–, and Cs– sea water solutions, in combination with each of the four internal alkali metal cation solutions, were studied. These 20 combinations showed that for any internal solution, the external E_{RP} series was Li $\geq$ Na > Cs > Rb > K, such that for any internal solution, the E_{RP} was most negative when the external solution was LiCl. Thus, the combination of NaF internally and KCl externally produced the most positive internal E_{RP} values. These experiments support the contention that the resting permeabilities of the internally perfused squid axon to alkali metal ions are in the following order: $P_K > P_{Rb} > P_{Cs} > P_{Na} \geq P_{Li}$. This series agrees well with the series obtained by Adelman and Senft (1968) for the resting permeabilities of intact, non-perfused squid axons.

Action Potentials

Variations in the c_{Na_i}/c_{Na_o} Ratio. One of the most extraordinary findings coming from internal perfusion studies has been the discovery that squid axons remain excitable and conduct thousands of impulses when internally perfused with certain inorganic salt solutions (Oikawa *et al.*, 1961; Tasaki, Watanabe and Takenaka, 1962; Baker, Hodgkin and Shaw, 1961, 1962*a*).

Baker, Hodgkin and Shaw (1962*b*) varied internal sodium concentrations by internally perfusing squid giant axons with mixtures of isotonic K_2SO_4 and isotonic Na_2SO_4. Inasmuch as resting potential values varied between -59 and -64 mv (sodium inactivation changed only slightly) for internal solutions having a range of sodium and potassium activities from $a_{Na} = 0$, $a_K = 440$ mM to $a_{Na} = 220$ mM, $a_K = 220$ mM, action potentials could be elicited and the amplitudes of these compared with the sodium concentration ratio across the membrane. Roughly, the amplitude of the action potential decreased as the c_{Na_o}/c_{Na_i} ratio decreased. Under these conditions, the membrane potential at the peak of the action potential, E_{AP}, was predicted by the relation:

$$E_{AP} = \frac{RT}{F} \ln \frac{a_{K_o} + ba_{Na_o}}{a_{K_i} + ba_{Na_i}}, \qquad (15\text{-}3)$$

where the a's are the ion activities and b is the permeability ratio, P_{Na}/P_K, which was given a value of 7.

Baker, Hodgkin and Shaw (1962*b*) has some misgivings about the use of Eq. (15-3) for predicting the peak of the action potential because the equation neglects the effects of inward current and anion leakage. However, the general conclusion reached was that the membrane underwent an increase in sodium ion permeability during the rising phase of the action potential. For axons giving large amplitude spikes, b was 0.06 in the resting state, 14 in the active state (peak of the overshoot) and 0.03 in the refractory state (at the peak of the undershoot following the spike). These findings are in general agreement with the ionic theory (Hodgkin and Huxley, 1952*b*, Hodgkin, 1964).

Fig. 15-3 shows a membrane action potential recorded from a squid giant axon internally perfused with a 695 mM K, 11 mM Na sulfate solution, isosmotic with sea water (Adelman and Gilbert, 1964). This action potential

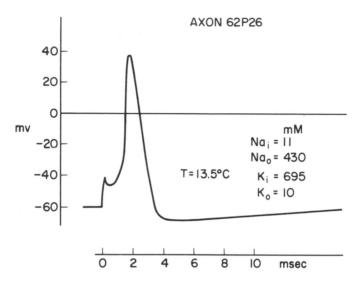

Fig. 15-3 *Nonpropagated membrane action potential of an axon perfused with a solution containing low internal sodium. (From Adelman and Gilbert, 1964.)*

was initiated by passing a brief depolarizing current between an axial wire platinum/platinum black electrode and external platinum/platinum black electrode. The potential at the peak of the spike was $+38$ mv. Using Eq. (15-3) with activities: $a_{K_i} = 358.2$ mм, $a_{Na_i} = 5.9$ mм, $a_{K_o} = 6.0$ mм and $a_{Na_o} = 234.0$ mм, at $E_M = +38$ mv, b was calculated as 8.04. The values of b in the resting state and in the refractory state were calculated to be 0.10 and 0.068 respectively. These results confirmed the finding of Baker, Hodgkin and Shaw (1962*b*).

Tasaki and Luxoro (1964) recorded large amplitude action potentials when internally perfusing *Docidicus gigas* axons with 0.6 м sodium aspartate (or glutamate) solutions. Tasaki and Luxoro described these action potentials as gradually increasing in amplitude upon changing the internal solution from 0.6 м potassium aspartate (or glutamate) to 0.6 м sodium aspartate solutions. They state, "The increase in the amplitude was characterized by an increased sharpness at the peak of the action potential." As recorded resting potentials were about -40 mv internally, the overshoot was between 30 and 55 mv. Tasaki and Luxoro concluded that the overshoot could not be determined by the Nernst equation applied to sodium ions.

Adelman and Senft (1966*a*) recorded action potentials with various internal sodium concentrations and never found an action potential whose overshoot exceeded the Nernst potential for the sodium concentration gradient. On the contrary, the potential at the peak of the overshoot was always less than the calculated sodium concentration potential.

Chandler and Hodgkin (1965) presented an explanation for the Tasaki and Luxoro finding by considering that the increased sharpness of the action potential peak derived from the use of a long high resistance axial salt bridge (microelectrode) to record the internal potential (Tasaki and Takenaka, 1963). Chandler and Hodgkin presented theoretical and experimental evidence for the expectation that the use of a long high resistance electrode may lead to a canula artifact which increases the spike amplitude beyond its true value. According to Baker, Hodgkin and Meves (1964), this artifact could be considered to be the result of an increase in the potential of the column of perfusion fluid inside the canula when the nerve action potential reaches the tip of the canula. The rapid change in potential of the perfusion fluid could then traverse the capacity of the glass capillary wall of the electrode and add a small increment to the recorded action potential.

Chandler and Hodgkin (1965) state that if a platinum wire is inserted into the electrode which is used to record membrane potential, its longitudinal resistance is greatly reduced. Using such a low resistance internal electrode to measure the potential, E_{AP}, at the peak of the spike overshoot, they found that E_{AP} was always less than E_{Na} and could be roughly approximated by:

$$E_{AP} = \frac{RT}{F} \ln \frac{a_{Na_o}}{a_{Na_i} + 0.2a_{K_i}} \tag{15-4}$$

with K-free ASW externally.

Variations in Other Cations as well as Sodium and Potassium.

Baker, Hodgkin and Shaw (1962b) internally perfused squid axons with isotonic solutions of Li, Na, K, Rb and Cs sulfate. It was found that the effects of these alkali metals on the amplitude of the action potential was different from their effects of the resting potential. Replacing internal isotonic K_2SO_4 with either Rb_2SO_4 or Cs_2SO_4 produced characteristic effects. The action potential duration first lengthened and then, as the resting potential declined, the action potential was abolished. An interesting finding with internal Cs_2SO_4 perfusion was that, before inexcitability occurred, the voltage at the peak of the action potential was increased by about 13 mv over normal, even though the resting potential was more depolarized than normal. Subsequently, Chandler and Meves (1965) claimed that these results could be explained using their voltage clamp measurements on axons internally perfused with RbCl and CsCl solutions (see page 296).

Tasaki, Singer and Watanabe (1966) have been able to record action potentials from axons internally perfused with 100 mM RbF or KF solutions (isotonicity maintained with glycerol) and externally perfused with a mixture of 300 mM hydrazinium chloride and 200 mM $CaCl_2$. Action potentials could also be recorded from RbF perfused axons when the external medium was a mixture of guanidinium chloride, tetramethylammonium chloride and $CaCl_2$.

All of the aforementioned results imply that the squid giant axon can produce action potentials in a variety of abnormal internal and external media. The following section describing voltage clamp results on perfused squid axons will attempt to provide some framework for a preliminary analysis of these action potential results.

Membrane Currents in Response to Voltage Clamping

Ions Contributing to the Initial Transient Current.
Chandler and Meves (1965) discovered that for squid axons internally perfused with solutions containing only potassium salts, for the initial transient current,

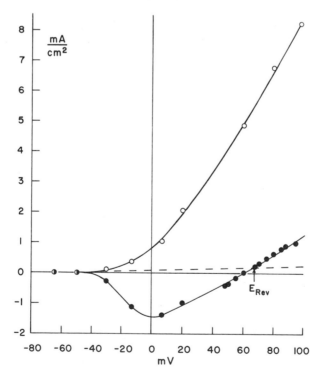

Fig. 15-4 *Current-voltage relations for peak initial transient current (filled circles) and steady-state delayed current (open circles) obtained from a squid giant axon perfused internally with 456 mM KF solution and externally with K-free ASW. Dashed line represents leakage current estimates; E_{rev} is the reversal potential for the initial transient current. Inward currents plotted negatively. See text.*

the reversal potential was lower than that predicted for a sodium equilibrium potential by the Nernst relation. They analyzed their results by assuming that, for the reversal potential, the Goldman-Hodgkin and Katz relation would hold (zero net current flow) and that the initial transient conductance was not specific to sodium ions but admitted some potassium ions. Applying the relation:

$$E_{rev} = \frac{RT}{F} \ln \frac{P_{Na}\, a_{Na_o} + P_K\, a_{K_o}}{P_{Na}\, a_{Na_i} + P_K\, a_{K_i}}, \qquad (15\text{-}5)$$

where E_{rev} is the potential at which the initial transient current reversed from inward to outward current flow, and the other terms have their usual significance, they were able to obtain consistent values for $P_{Na}:P_K$ of $1:0.08$ for their perfused axons.

Fig. 15-4 shows data obtained by Adelman and Senft from an axon perfused internally with 456 mM KF solution and externally with K-free ASW. Peak initial transient current and steady state delayed current values are plotted. From Eq. (15-5) and $E_{rev} = +67$, $P_{Na}:P_K$ is calculated as $1:0.082$, which is in complete agreement with Chandler and Meves' average value. Table 15-2 gives experimental values of E_{rev} and $P_K:P_{Na}$ ratios obtained from a number of axons internally perfused with a variety of KF and NaF concentrations (Adelman and Senft, 1966 and unpublished).

Table 15-2   Reversal potential values, E_{rev} for the initial transient current obtained upon voltage clamping squid giant axons internally and externally perfused with solutions containing various activities of sodium and potassium ions. Temperature = 5 °C

Axon	a_{K_o}	a_{K_i}	a_{Na_o}	a_{Na_i}	E_{rev}	P_K/P_{Na}
	mM	mM	mM	mM	mv	
65–82	0	175	240	0	+65	.086
65–87	0	175	240	0	+63	.095
65–91	0	175	240	0	+58	.117
65–92	0	175	240	0	+63	.095
65–66	5.5	175	234	0	+60	.106
65–68	5.5	175	234	0	+76	.054
65–93	5.5	175	234	0	+61	.101
65–94	5.5	175	234	0	+60	.106
66–59	0	169	240	0	+65	.089
65–95	0	150	240	19	+43	.132
65–94	5.5	150	234	19	+47	.087
65–96	5.5	150	234	19	+43	.126

In addition, Chandler and Meves (1965) voltage clamped axons internally perfused with Na, Li, Rb and Cs solutions. They found that the reversal potential was a function of the internal ionic species as well as the ionic concentration. From the Goldman relation, they calculated $P_{ion}:P_{Na}$ ratios and found the following series: $P_{Li}:P_{Na}:P_K:P_{Rb}:P_{Cs} = 1.1:1.0:0.08:0.025:0.016$. The implication in this finding is that the initial transient conductance is not tight or specific for sodium ions but admits other ions of Group IA as well.

Generally, one can examine the findings of Tasaki, Singer and Watanabe (1966) in this context as well. Presumably ions such as hydrazinium and ammonium act as charge carriers carrying current through the initial transient conductance. One should be able to produce an inward transient current with a slightly positive finite reversal potential by selecting a proper pair of cations and arranging their concentrations inside and outside in a considered manner. For example, if one were to perfuse the inside of an axon with 50 mM CsF and the outside with sea water containing 430 mM RbCl, substituted for NaCl, it should be possible to obtain a positive reversal potential value for the initial transient current.

One would want to find a pair of cations which would give all the same general characteristics as the pair found *in vivo* (Na^+ and K^+). The ideal pair would be an internal cation which has a resting permeability greater than the external cation and an active permeability less than the external cation. Of course, *in vivo* mechanisms involving membrane metabolic machinery (see chapter 3) allow for the distribution of Na^+ and K^+ and contribute to the maintenance of their concentration gradients.

Ions Contributing to Delayed Membrane Currents. Adelman and Fok (1964) internally perfused squid axons with solutions of various concentrations of K_2SO_4. The steady state values of the delayed outward current were shown to vary with the internal potassium concentration, approaching zero as c_{K_i} was reduced toward zero.

Chandler and Meves (1965) internally perfused squid axons with either 300 mM KCl or 300 mM RbCl solutions. They found that there was a marked reduction in the delayed outward current for depolarizing pulses with Rb^+ internally as compared with K^+ internally. Small negligible outward currents were seen with internal rubidium even with pulses to E_M values of $+88$ mv of long duration (106 msec). Chandler and Meves (1965) concluded that internal Rb^+ produced an inhibitory effect on the delayed conductance channel (see page 297 for further details).

With the exception of NH_4^+ (Binstock and Lecar, 1969), potassium ion seems to be the only ion capable of acting as a charge carrier for the delayed outward membrane current in the squid axon.

A Common Ion Carrying both Initial Transient and Delayed Membrane Currents. Ammonium ion has extremely interesting properties as a charge carrier contributing to membrane currents in the squid axon. Binstock and Lecar (1969) have demonstrated that NH_4^+ can partially substitute for Na^+ and K^+ in contributing to both the early transient and the delayed currents in response to step changes in the early and late membrane potential in internally perfused squid axons. The conductance of the early and late channels for NH_4^+ was found to be about 0.3 times that found for the normal charge carriers. External TTX was found to block the inward NH_4^+ current through the early transient conductance and internal TEA^+ was found to block the outward movement of NH_4^+ current through the delayed conductance. For perfusion with 430 mM NH_4^+-sea water externally and 500 mM KF internally, reversal potentials (E_{rev}) for the initial transient current were described by:

$$E_{rev} = \frac{RT}{F} \ln \frac{a_{Na_o} + y(a_{NH_{4o}}) + b(a_{K_o})}{a_{Na_i} + y(a_{NH_{4i}}) + b(a_{K_i})}, \tag{15-6}$$

where $y = P_{NH_4}/P_{Na} = 0.27$, $b = P_K/P_{Na} = 0.11$ and a_{NH_4} is the NH_4 ion activity. Other terms have been described previously. Notice that b is approximately equal to the value obtained by Chandler and Meves (1965) and Adelman and Senft (1966a).

The peak transient current vs. membrane potential relation found for axons perfused with 500 mM NH_4F internally and artificial sea water externally showed a negative slope region between $E_M = -40$ mv and $E_M = 0$ mv. However, the amplitude of the currents could not be predicted from Hodgkin and Huxley's independence principle. On the other hand, for axons perfused with 500 mM KF solution internally, substitution of NH_4^+ for Na^+ in the external sea water gave changes in the initial transient peak amplitude that could be predicted from the independence principle. Under these conditions, the time to peak (t_p) of the initial transient current for the onset of the clamping pulse was the same in 430 mM NH_4SW as in ASW. Since t_p is dependent mainly on τ_m and not strongly influenced by τ_h, one can conclude that τ_m is relatively unchanged upon changing the initial transient current charge carrier from Na^+ to NH_4^+.

Binstock and Lecar suggest that internal NH_4^+ may have an effect on the inactivation of the early transient current. They cite the effect of internal Cs^+ on the blockade of outward delayed potassium current, whereby outward transient currents are prolonged as outward delayed currents are diminished (Adelman and Senft, 1966b).

As external NH_4^+ mildly depolarized squid axons (in 430 mM NH_4SW, the average $E_{RP} = -46$ mv), action potentials were elicited by anodal break excitation. Similar action potentials could be predicted from the Hodgkin-Huxley equations. The values for the conductances and ion potentials for such computations were determined experimentally but the parameter time constants and normalized conductance functions were not changed from those given by Hodgkin and Huxley (1952b). The implications of such work is that the Hodgkin-Huxley parameters derived from membrane current characteristics may be used to describe membrane action potentials even in the presence of rather nonphysiological ions such as ammonium and even when the classical conductances are admitting a common ion.

INTERNAL PERFUSION WITH SALTS OF DIFFERENT ANIONS

The importance of anions and membrane function has been stressed by Tasaki, Singer and Takenaka (1965). These authors have examined the function of the membrane of the internally perfused squid giant axon during internal perfusion with a variety of salts or different anions. They differentiated "favorable" and "unfavorable" internal anions by the ability of the membrane to produce action potentials in the presence of the various anions. They were able to rank-order the anions according to their ability to support excitation (SCN $<$ 1 $<$ Br $<$ NO$_3$ $<$ Cl $<$ acetate $<$ SO$_4$ $<$ aspartate $<$ glutamate $<$ HPO$_4$ $<$ F), and were able to show that while the internal membrane surface was very sensitive to differences in anion species, the external surface was insensitive.

Adelman, Dyro and Senft (1966) examined the effects on membrane conductances of a sample "favorable" anion (F$^-$) and an "unfavorable" anion (Cl$^-$). Fig. 15-5 compares the effect on resting potential (E$_{RP}$) and on the voltage at the peak of the action potential (E$_{AP}$) of internal perfusion with a solution of 400 mM KF vs. 400 mM KCl. During the 15 min exposure to internal KCl, the resting axon membrane depolarized about 10 mv. After 10 min of KCl perfusion, the rate of E$_{AP}$ change accelerated and a marked rise in threshold took place. After 15 min of KCl perfusion the threshold rose to over 10 times that found previously for KF perfusion. These effects were reversible for perfusion times with KCl of less than 15 min. Irreversible loss of resting potential and excitability invariably occurred with KCl perfusion for 45 min or more. This result was in contrast to perfusion with KF solutions which could maintain excitability for as long as 4 hours.

Adelman, Dyro and Senft (1966) found that the leakage currents (I$_L$)

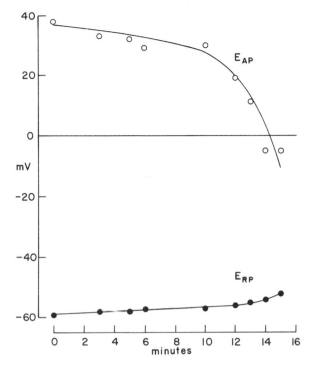

Fig. 15-5 *Effects on membrane potentials of a squid giant axon of changing internal perfusion solution from 400 mM KF to 400 mM KCl. E_{RP} — resting potential values, E_{AP} = voltage at the peak of the action potential. Time zero represents the time at which internal solution was changed from KF to KCl.*

were abnormally increased by internal Cl^- and decreased by internal F^- (Fig. 15-6). They also found that the initial transient conductance (normally representing a higher permeability to sodium than other ions) was influenced by these two anions such that Cl_i^- decreased g_{Na} even when $c_{Na_o} : c_{Na_i}$ was kept constant and the membrane was kept hyperpolarized for 50 msec before test voltage clamp pulses (Fig. 15-7). Internal F^- was shown to restore g_{Na} toward the normal values found in intact axons, and, in mixtures of fluoride and chloride salts, F^- was able to protect against the deleterious effects of Cl^- (Fig. 15-8). Adelman, Dyro and Senft (1966) concluded that there was an effect of internal anions on that part of the membrane structure relating to the ionic conductances.

It is interesting to note that $SO_4^=$ has somewhat the same properties as Cl^-. Diecke and Adelman in 1963 (unpublished) found that squid axons

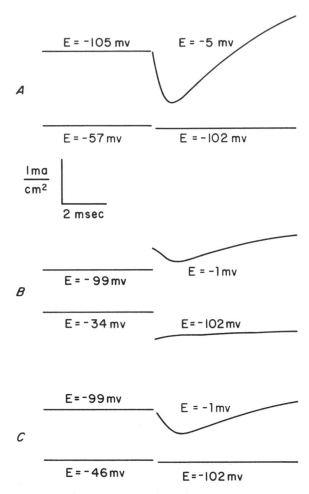

Fig. 15-6 *Comparison of voltage clamped membrane currents obtained during internal perfusion with potassium chloride with initial and final records with internal perfusion with KF. Three sets, A, B, and C, each of two records, are shown. Upper record of each set shows the membrane current flowing during a depolarizing step pulse in membrane potential following a hyperpolarizing prepulse of 50-msec duration; each lower record shows the membrane current flowing during a hyperpolarizing step pulse in membrane potential from the holding potential, which was set equal to the resting potential. Capacitive current transients are not shown as in E = clamped potential. A. Initial records of membrane current upon internal perfusion with potassium chloride solution. B. Records of membrane current obtained after 30-minute internal perfusion with potassium chloride solution. C. Records of membrane current obtained after 15-minute recovery following internal perfusion with potassium fluoride solution. Axon 65–84; temperature, 6°C. (From Adelman, Dyro and Senft, 1966, Copyright 1966 by the American Association for the Advancement of Science.)*

internally perfused with K_2SO_4 solutions containing radioactive K^{42} showed a gradually rising rate of resting K^{42} loss which roughly paralleled the rise in leakage current estimated by intermittent voltage clamping. Shaw (1966) found that squid axons internally perfused with K_2SO_4 solutions had net sodium influxes that were 2 to 3 times those recorded from intact axons.

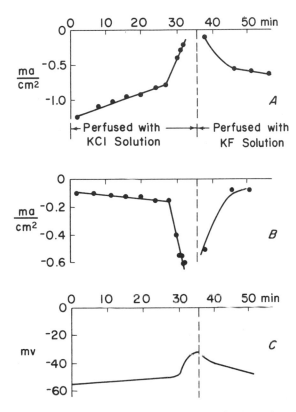

Fig. 15-7 *Alterations in selected membrane currents in the voltage clamp and in membrane resting potential, upon internal perfusion with potassium chloride solution, and their restitution by potassium fluoride. A. Maximum initial transient inward (sodium) current values (ma/cm²), during a brief voltage clamp period, plotted against the time elapsed in minutes for internal perfusion of axons with potassium chloride solution followed by potassium fluoride solution. Prepulses and depolarizing pulses, similar to those in Fig. 15-6; all values corrected for leakage currents. B. Step outward current (ma/cm²), upon a change in voltage clamped membrane potential, from a value equal to the resting potential to a hyperpolarized value of −102 mv plotted against the time elapsed for internal perfusion as in A. C. Resting potential values at the times between the points given in A and B. Axon 65–84. See text and Fig. 15-6. (From Adelman, Dyro and Senft, 1966. Copyright 1966 by the American Association for the Advancement of Science.)*

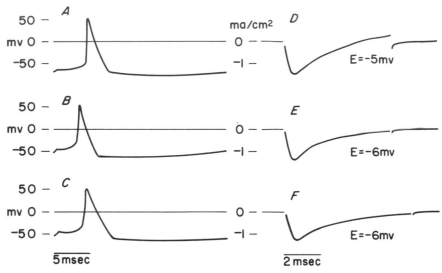

Fig. 15-8 *Demonstration of the protection against the deleterious effects of internal chloride afforded by a low concentration of fluoride in the chloride perfusion medium. Records A, B and C are membrane action potentials; D, E, and F are maximum sodium currents during a voltage clamp, depolarizing, step pulse in membrane potential, from the holding potential of −65 mv to the values shown on the records. Records A and D were obtained initially upon potassium fluoride perfusion; B and E after approximately 45 minute internal perfusion with a mixture of nine parts chloride and one part fluoride solution; C and F, 10 and 15 minutes, respectively, after reperfusion of the axon with fluoride solution. Axon 65–69; temperature, 4°C. (From Adelman, Dyro and Senft, 1966. Copyright 1966 by the American Association for the Advancement of Science.)*

INTERNAL PERFUSION WITH SOLUTIONS OF LOW IONIC STRENGTH

Resting Potential

In squid giant axons externally perfused with ASW solution and internally perfused with low K solutions prepared by diluting salts such as KCl with isotonic sugar solutions (dextrose or sucrose), the resting potential may be correlated with the internal K ion activity (a_{K_i}) (Baker, Hodgkin and Meves, 1964). The constant field equation [Eq. (15-1)] was used to calculate the $P_{Na}:P_K$ ratio in such experiments. $P_{Na}:P_K$ was shown to vary between 0.01 and 0.1, and the ratio tended to increase with the duration of internal perfusion. For variations in external potassium from $a_{K_o} = 0$ to $a_{K_o} = 13.6$ mM, correlations were made between recorded resting potential values (corrected for estimated junction and tip potentials) and Eq. (15-1). Values for the permeability coefficients used were $P_K:P_{Na}:P_{Cl} = 1:0.035:0.02$, for E_M values when $a_{K_i} < 30$ mM and $P_K:P_{Na}:P_{Cl} = 1:0.05:0.1$ when $a_{K_i} > 30$ mM.

Adelman, Dyro and Senft (1965*b*) perfused squid giant axons with low ionic strength solutions of phosphate salts. In these experiments, E_{RP} values were more negative internally with low ionic strengths than those obtained by Baker, Hodgkin and Meves (1964). Fig. 15-9 illustrates the relation between the resting potential, E_{RP}, and internal potassium ion activity, a_{K_i}, obtained from unpublished data of Adelman and Senft. Experimental E_{RP} values obtained from axons externally perfused with either ASW or K-free ASW are plotted. The curves through the experimental points were fit by the equation:

$$E_{RP} = \frac{RT}{F} \ln \frac{P_K a_{K_o} + P_{Na} a_{Na_o}}{P_K a_{K_i} + P_{Cl} a_{Cl_o}}, \qquad (15\text{-}7)$$

where $P_K = 1.0$, $P_{Na} = 0.05$ and $P_{Cl} = 0.1$. Notice that these permeability coefficients are the same values used by Baker, Hodgkin and Meves (1964) to fit curves for a_{K_i} values > 30 mM.

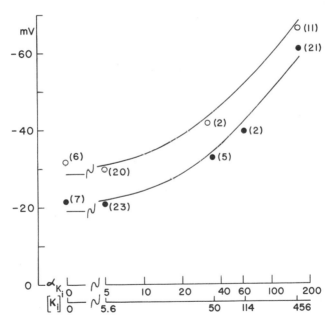

Fig. 15-9 *Relation between mean resting potential values (ordinate) and internal potassium ion activity (upper abscissa) or concentration (lower abscissa) in axons internally perfused with KF solutions of various ionic strengths. Open circles: external perfusion with K-free ASW, filled circles with 10 mM K-ASW. Numbers adjacent to the points represent the number of axons from which measurements were taken. Curves are plotted according to Eq. (15-7).*

The effect of a difference in internal anions on E_M at low internal ionic strengths can not be explained at this time. However, these experiments do make clear that the resting potential is not simply determined by a Nernst relation for potassium, but that other ions are permeable through the membrane in the resting state.

Action Potentials and Membrane Currents in Response to Voltage Clamping

While much of the early work on internally perfused axons was directed toward conventional analysis of membrane function in terms of an ionic basis for bioelectric potentials, other work was directed toward uncovering a new class of phenomena (Tasaki and Shimamura, 1962; Narahashi, 1963) which are now uniquely associated with the perfused axon. These new phenomena include the ability of axons perfused internally with low ionic strength solutions to produce either excited or spontaneous action potentials of large amplitude (relatively normal). The action potentials are produced despite low resting potential values which should result in extensive sodium inactivation, according to the Hodgkin and Huxley theory. A shift in the g_{Na} vs. E_M relationship of the sodium conductance along the voltage axis was found (Baker, Hodgkin and Meves, 1964; Chandler, Hodgkin and Meves, 1965; Moore, Narahashi and Ulbricht, 1964; and Narahashi, 1963). The displacement was roughly equal to the apparent depolarization brought about by perfusion with the low ionic strength solutions. Baker, Hodgkin and Meves (1964) suggested that the shift in the sodium conductance and the sodium inactivation curves is the result of a change in the double-layer potential at the inner membrane surface such that the measured potential from inside to outside reflected both the membrane potential drop and the double-layer contribution. Chandler, Hodgkin and Meves (1965) have calculated that the double-layer potential shift with change in ionic strength would predict the observed threshold shift upon assuming a fixed charge density at the inner boundary of -2.2 μcoul/cm^2. For further discussion of the role of fixed surface charges, see chapter 18.

Perfusing axons with low ionic strength solutions results in action potentials of abnormally long duration (Narahashi, 1963; Baker, Hodgkin and Meves, 1964; and Adelman, Dyro and Senft, 1965a,b). These action potentials exhibit a plateau after the spike peak, which can be accounted for by incomplete sodium inactivation. In this connection, Baker, Hodgkin and Meves (1964) showed that sodium inactivation was related to the internal potassium concentration. Adelman et al. (1965a,b) showed, in addition, that both the plateau of the action potential and the time constant of sodium

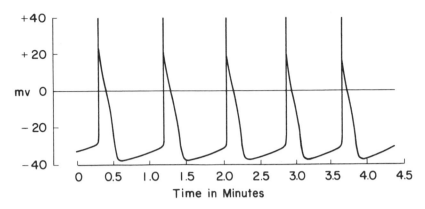

Fig. 15-10 *Spontaneous discharge recorded from an axon internally perfused with isosmotic dextrose having a* Tris Cl *concentration of 5* mM. *External solution was potassium-free ASW. (From Adelman, Dyro and Senft 1965a.)*

inactivation increased reversibly when the external potassium concentration was reduced.

Adelman, Dyro and Senft (1965a) have shown that squid axons internally perfused with low ionic strength solutions fire spontaneous trains of long duration action potentials similar in configuration to cardiac muscle fiber responses, whenever a_{K_i} and $a_{K_o} = 0$ (Fig. 15-10). At $5°C$, the frequency of these responses is from 0.5 to 2 per minute. The absolute refractory period is about equal to the plateau duration of approximately 15 sec. The amplitude and duration of additional responses externally stimulated during various phases of the activity indicated that the inactivation of the spike generating mechanism has a voltage dependent time constant of about 10 sec. Such a slow inactivating mechanism would account for both the response duration and the repetition frequency.

Adelman et al. carried out voltage clamp studies on axons where $a_{K_i} = 5.6$ mM and $a_{K_o} = 0$. These studies indicated that a long lasting sodium conductance was responsible for the spike plateau. When external sodium chloride was replaced with Tris Cl, sodium currents were reduced to zero. A formal sodium current separation was carried out to show that the long duration currents were carried by sodium ion.

The long duration responses elicited with low ionic strength solutions could be predicted from the time course of the sodium current-voltage relation (Adelman, Dyro and Senft, 1965a,b). The time constant of sodium inactivation, τ_h, was shown to be a function of the internal and external potassium ion concentrations. Adelman and Senft (1968) and Adelman and Palti (1969a,b) have suggested and presented evidence for a coupling between the potassium conductance and the sodium inactivation mechanism.

INHIBITORY EFFECTS OF SOME INTERNAL AND EXTERNAL CATIONS

It is generally well known and accepted (see chapter 20) that some agents and drugs are capable of blocking or inhibiting membrane permeability to various ions. That some cations are able to inhibit or interfere with membrane permeability to the alkali metal cations is probably not so well known or even expected. This section considers this latter problem and discusses some of the recent findings with an emphasis on membrane ionic conductances measured by means of the voltage clamp technique. The following inhibitory cations will be considered: Cs^+, Rb^+, Na^+, and quaternary ammonium ions such as tetraethylammonium ion (TEA^+).

Internal Cesium, Rubidium and Sodium

Membrane Potentials. Squid giant axons begin to show a decrease in resting potential when the internal perfusion fluid is changed from isotonic K_2SO_4 to isotonic Cs_2SO_4 (Baker, Hodgkin and Shaw, 1962b). Action potentials elicited during the decline of the resting potential with Cs_2SO_4 perfusion show an increase in duration (Baker et al., 1962b; Adelman and Senft, 1966b). Eventually, the axons perfused with isotonic Cs_2SO_4 become inexcitable when the resting potential reaches a steady depolarized level (ca. -20 mv).

Sjodin (1966) microinjected cesium sulfate solutions into squid axons achieving c_{Cs_i} ranging from 152 to 180 mM. If these axons were hyperpolarized to a steady membrane potential of from -60 to -70 mv, action potentials could be elicited by brief depolarizing currents superposed on the steady hyperpolarizing current. These action potentials were characterized by a rapid rate of rise to peak potentials of about $+50$ mv, followed by a decline to a slightly positive plateau potential ($+20$ to $+30$ mv) which was maintained for as much as 45 msec. The Cs_2SO_4 solution injected into the axon contained radioactive K^{42} and Cs^{134}. Resting effluxes of potassium were shown to be similar to those obtained in axons without Cs^+ internally. However, the potassium efflux obtained during the repetitive action potential period was only 7 to 20 per cent of that expected from axons without cesium ion internally. Sjodin (1966) concluded that the normal K efflux expected during the action potential duration was inhibited by internal cesium ions.

Membrane Currents in Response to Voltage Clamping. Chandler and Meves (1965) discovered, upon voltage clamping squid axons internally perfused with a cesium solution (150 mM KCl, 150 mM CsCl, and made isotonic by adding sucrose), that the steady state values of the delayed outward currents were 0.02 to 0.05 times the values obtained for internal perfusion with a 300 mM KCl solution. This reduction was greater than predicted for a simple halving of the internal K concentration. Assuming that K^+ is the charge carrier for the delayed current and that the independence principle applies, the following relation can be used to determine the predicted current:

$$I_K' = I_K \frac{[(a_{K_i'}/a_{K_i}) \exp\{(E_K - E_P)(F/RT)\}] - 1}{\exp[(E_K - E_P)(F/RT)] - 1} , \qquad (15\text{-}8)$$

where I_K' is the predicted potassium current for changing the internal activity from a_{K_i} to $a_{K_i'}$ (at constant ionic strength, concentrations may be used rather than activities) and for the original reversal potential, E_K, where E_P is the pulse potential, and I_K is the potassium current corresponding to original potassium activity.

Similar experiments with mixture of KCl and RbCl showed decreased delayed outward steady state currents that were also reduced below independence principle predictions. Experiments with internal KCl and NaCl mixtures revealed a small inhibitory effect of Na ions on K currents. Chandler and Meves (1965) concluded that delayed currents were inhibited by internal cations in the sequence $Na < Rb < Cs$.

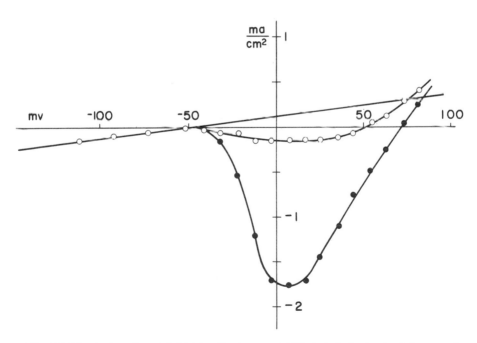

Fig. 15-11 *A plot of peak initial (sodium) current (filled circles), steady state current at 28 msec (open circles), and the leakage current (solid line above zero current axis at 0 mv) from an axon internally perfused with Cs-K solution. The holding potential, E_H, was −52 mv. Temperature, 4°C. (From Adelman and Senft, 1966b.)*

Adelman and Senft (1966*b*) internally perfused squid axons with mixtures of KF and Cs_2SO_4 and found similar effects on delayed currents to those observed by Chandler and Meves. Fig. 15-11 illustrates typical current-voltage relations obtained from a voltage clamped squid giant axon internally perfused with a solution in which $a_K = 53.5$ mM, $a_{Cs} = 86.8$ mM. Notice that while the values for the maximum initial transient currents appear normal, steady-state values of the delayed current are inward for membrane potential values between -50 and $+50$ mv. These current-voltage relations showed a negative slope over the potential range -50 mv to zero. Adelman and Senft substituted all the sodium in the external sea water with $Tris^+$ and obtained membrane currents in sodium-free sea water. These were subtracted from those found in normal sea water, thus separating the membrane currents into sodium and non-sodium components (after Hodgkin and Huxley, 1952*a*). The sodium currents, for any given membrane potential, for the first few msec were similar to those found without Cs^+ internally but then tended toward sustained plateaus. This indicated that sodium inactivation was incomplete when Cs^+ was present internally.

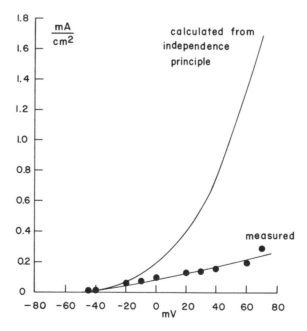

Fig. 15-12 *Comparison between steady-state values (filled circles) of separated delayed membrane currents and the values expected for a potassium current as calculated from the independence principle (upper curve). Experimental values obtained from a voltage clamped squid axon perfused internally with 86 mM KF, 140 mM Cs_2SO_4 solution and externally with Na-free Tris Cl sea water. See text.*

The nonsodium currents were outward over the range of membrane potentials from -50 to $+70$ mv. They were smaller in absolute amplitude than the inward sodium currents over this potential range. Thus, the sum of these currents gave the predominantly inward steady state currents seen in Fig. 15-11 (open circles). The persistent nature of the plateaus of sodium current accounted for the steady-state current-voltage relation developing a negative slope in the potential range where positive ohmic slopes are obtained normally. Adelman and Senft compared the steady-state values of potassium currents $[I_{ss} - (I_{Na} + I_L)]$ with those predicted by the independence principle for currents carried by potassium ions ($a_{K_i} = 53.5$ mM) and were able to confirm the finding of Chandler and Meves that I_K was less than predicted (Fig. 15-12). By plotting the ratio of predicted I_K to measured I_K, Adelman and Senft were able to show that the deviation from prediction became progressively worse as the membrane potential became more positive internally. Fig. 15-13 illustrates this relation. Adelman and Senft concluded that the effects of internal Cs^+ on potassium currents was dependent upon membrane potential.

Further evidence for this dependence was obtained by Adelman (1968). Squid giant axons internally perfused with 300 mM KF and 100 mM CsF were voltage clamped. Steady-state values of delayed currents increased with depolarizing pulses up to about $+30$ mv and then declined progressively with further depolarization. Fig. 15-14 is typical of membrane currents

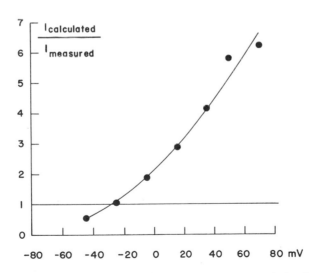

Fig. 15-13 *Ratio of calculated potassium current to measured steady-state current (from Fig.* 15-12*) plotted as a function of membrane potential. Axon internally perfused with Cs-K solution and externally with Na-free Tris Cl sea water. Horizontal line represents the independence principle prediction. Points indicate that the deviation between predicted and measured values becomes larger as membrane is depolarized. See text.*

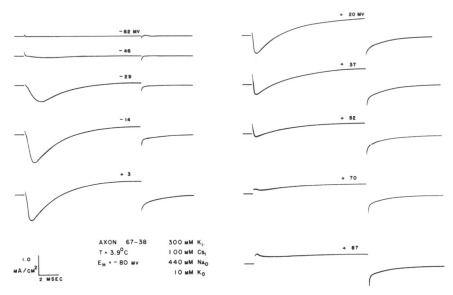

Fig. 15-14 *Typical membrane current records obtained upon voltage clamping squid giant axons internally perfused with* 300 mM KF, 100 mM CsF *solution and externally with* ASW. *Values of the membrane potential during pulses are given above each record. Pulses were made from a holding potential (E_H) of* -80 *mv. See text.*

obtained for 12 msec pulses of various potential amplitudes. Fig. 15-15 shows the peak and the steady-state values of these currents plotted against membrane potential. Notice that the steady state values rise and then fall with increasing depolarization. Fig. 15-15 was obtained from the same axon for internal perfusion with 400 mM KF solution. While the peak values of the initial transient currents in *A* and *B* are quite similar, there is a marked difference between the steady-state current values in *A* and *B*. It seems apparent that Cs^+ exerts its blocking effects on the delayed current as some function of the membrane potential. One should expect that a positive ion such as cesium should be driven into the membrane by increasing the electric field across the membrane such that the electrical gradient for Cs^+ is made larger. Then, it could be expected that cesium ion would not normally block potassium ion permeability until the membrane is depolarized. This would explain Sjodin's finding (1966) that the K^{42} effluxes in the presence of internal Cs are normal at the resting potential but are below prediction for periods of repetitive firing of action potentials. This implies that the inhibitory effect of Cs^+ on delayed membrane currents is dependent upon Cs ions being carried to blocking sites by current flow rather than on a simple chemical affinity of Cs^+ for such sites.

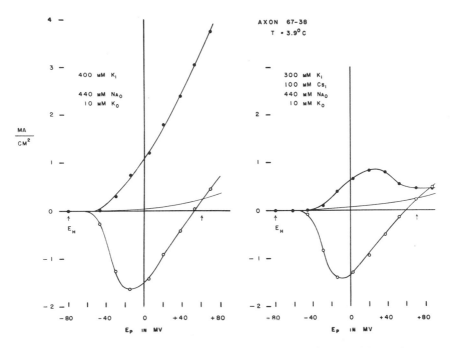

Fig. 15-15 *Current-voltage relations for peak values of the initial transient current (open circles) and steady-state values of the delayed current (filled circles) obtained upon voltage clamping a squid axon internally perfused with 400mM KF solution (A) and 300 mM KF and 100 mM CsF solution (B). E_H indicates the holding potential; E_P, the pulse potential; thin line, the leakage current estimates. See text.*

Membrane Currents as Influenced by either Internal or External Cs^+

One of the more interesting facts relating to Cs^+ blockade of the delayed current is that the effect is very much dependent on the direction of membrane current flow. Adelman and Senft (1968) and Adelman (1968) presented evidence that: 1) internal Cs^+ blocks only outward delayed steady-state currents but not inward, and 2) external Cs^+ blocks inward delayed currents but not outward. These results may be explained by assuming that the blocking ion is driven into the membrane from the side where its concentration is high but that it may be swept away from its blocking site by a current flow in the opposite direction. In other words, the delayed conductance channel can be blocked from either side by Cs^+ and may be unblocked by potassium ion flow from the opposite side.

The blockade thus has the electrical characteristics of a diode. Armstrong and Binstock (1965) proposed such a property to explain the blocking action of internal TEA$^+$. The diode proposed by Adelman (1968) would have a dynamic characteristic dependent on the potential gradient across the element (local field strength) and its rectification would be conferred according to the side on which the blocking ion (Cs$^+$) was present.

Analysis of the Prolonged Action Potential Seen with Internal Cs$^+$

The experiments of Adelman and Senft (1966b) provide information with which one can attempt to explain the action potential prolongation described above. It is apparent from voltage clamp measurements that there is a relatively persistent inward membrane current resulting primarily from the inhibition of the delayed (potassium) outward current and the incomplete inactivation of the initial transient (sodium) current.

We can say that the rising phase of such a prolonged action potential should result from a rise in the Na conductance in response to depolarization. As the sodium conductance rises, the membrane potential would begin to approach the sodium reversal potential. Unlike in the normal axon, this depolarization would not result in the slow increase in potassium conductance. However, at depolarized values of the membrane potential, the sodium conductance does decline to some extent. Thus, beyond the peak of the spike, the membrane potential tends to approximate some resultant of the sodium and leakage reversal potentials. This potential should remain relatively constant, as g_L is time invariant and g_{Na} at any potential tends toward a small constant value. The result should be a potential value less than but close to that at the peak of the spike. As the conductance to sodium ions declines further with slow inactivation, the membrane becomes more negative internally and eventually all conductances rapidly approach those found in the resting state. The membrane should then repolarize. In the most simplistic terms, 1) during the rising phase, and at the peak of the spike, $P_{Na} \gg P_L > P_K$, 2) during the fall from peak, $P_{Na} > P_L > P_K$, 3) during the plateau, $P_{Na} \cong (P_L + P_K)$ and 4) during repolarization, $P_K > (P_L + P_{Na})$. In somewhat more sophisticated terms, one may say that the prolonged membrane action potential seen in Cs perfused axons should result from the locus in time of stable potential points determined primarily by the persistence of the sodium conductance and the anomalous potassium conductance rectification.

If we consider that I_L is primarily carried by potassium ions, then the limits of membrane potential are close to E_{Na} during the spike peak and close to E_K during the undershoot following the termination of the plateau. This analysis is somewhat similar to that given by Armstrong and Binstock (1965) for long duration action potentials seen in axons injected internally with tetraethylammonium ions (TEA$^+$). The effects of internal TEA$^+$ on the behavior of squid axons are discussed in the following section.

Internal Quaternary Ammonium Ions

In 1957, Tasaki and Hagiwara injected tetraethylammonium ions (TEA^+) into a squid giant axon and found that recorded action potentials were prolonged and had plateaus similar to those described above for internal Cs^+ perfusion. Measurements of membrane resistance during the plateaus indicated that resistance during the plateau was the same as the resting resistance. However, the membrane resistance was greater than the resting resistance during the rapid return of the potential from the plateau to the resting level. One interesting aspect of these results was that the prolonged action potential with TEA injection produced no significant change in resting potential values. Tasaki and Hagiwara voltage clamped these TEA^+ injected axons and found that the delayed outward currents were greatly reduced as compared to those recorded from axons without TEA^+. From these results, one can say that internal TEA^+ produces a prolongation of the action potential which is related to a change in membrane conductance. Tasaki and Hagiwara (1957) concluded that the prolonged action potential recorded in squid giant axons injected with internal TEA^+ was similar to that recorded from vertebrate cardiac muscle.

Armstrong and Binstock (1965) made extensive voltage clamp measurements on axons injected with TEA^+. These measurements confirmed Tasaki and Hagiwara's finding that internal TEA^+ selectively blocked the delayed outward potassium membrane current. One of the most striking findings from Armstrong and Binstock's work was that at membrane potentials between E_{RP} and the initial transient current reversal potential, a small steady state membrane current remained which contained a prolonged inward sodium current component. Armstrong and Binstock showed (1965, Fig. 2) that the plateau durations of action potentials recorded from TEA^+ injected axons could be correlated with the degree to which the sodium conductance failed to completely inactivate. Longer duration action potentials were recorded from axons showing longer duration sodium current.

Armstrong and Binstock also studied TEA^+ injected axons externally perfused with an artificial sea water in which K^+ was substituted for Na^+. Under these conditions, the instantaneous current-voltage relations (presumed to be directly related to g_K) showed anomalous rectification. At membrane potentials more negative than E_K, in axons bathed at the same C_K, internal TEA^+ was shown to block outward currents but not inward currents. External TEA^+ was shown to have no effect on membrane currents.

Armstrong and Binstock concluded that TEA^+ acts only on the inner surface of the squid axon membrane, and that its action could be represented in the equivalent circuit proposed by Hodgkin and Huxley (1952b) by inserting a diode in series with g_K. The choice of this model followed from the fact that the anomalous rectification seen with internal TEA^+ was dependent on the direction of the current flow rather than on the membrane potential *per se*.

Armstrong (1969) has presented a model for the kinetics of the potassium conductance as influenced by internal quaternary ammonium ions. Using TEA^+ as the best known example, Armstrong states that its blocking effect has the following characteristics:

1. TEA^+ does not alter α_n and β_n, the Hodgkin and Huxley rate constants for the n process.
2. TEA^+ does not have an effect on g_K.
3. TEA^+ blocks only those channels that are in the n^4 state (i.e., are open).
4. All channels are freed of TEA^+ upon remaining at the resting potential for a sufficient time.
5. Channels are either fully open or closed.
6. TEA^+ blockade is a function of $c_{TEA i}^+$
7. TEA^+ may be cleared from the channel by driving potassium current inward.

With these characteristics in mind, Armstrong considers the following scheme:

$$u \underset{\beta}{\overset{4\alpha}{\rightleftarrows}} v \underset{2\beta}{\overset{3\alpha}{\rightleftarrows}} w \underset{3\beta}{\overset{2\alpha}{\rightleftarrows}} x \underset{4\beta}{\overset{\alpha}{\rightleftarrows}} y \underset{1}{\overset{k}{\rightleftarrows}} z, \tag{15-9}$$

where g_K is proportional to y and z is proportional to the number of TEA-blocked channels. Six linear first-order differential equations describe the scheme above. For example, for y:

$$\frac{dy}{dt} = -ky - 4\beta y + \alpha x + 1z. \tag{15-10}$$

Armstrong takes the condition that:

$$u + v + w + x + y + z = 1$$

in order to eliminate one differential equation. Thus, a single 5th order equation in y can be obtained. Without TEA^+, z can be eliminated to give a fourth order equation in y. Equivalence to the Hodgkin and Huxley formalism can be shown as follows:

$$y = n^4 = n_\infty - (n_\infty - n_o)e^{-t/\tau_n}. \tag{15-11}$$

Using axons bathed in 10^{-7} M TTX to block the initial transient current, Armstrong has shown that the effective dose of quaternary amines for blocking delayed membrane currents is a function of the length of one of the carbon chains around the central N in TEA analogs. Blocking effectiveness increases with chain length from one carbon (methyltriethylammonium = c_1) to five carbons (n-pentyltriethylammonium = c_5). Armstrong found that axons containing c_5 at 0.3 mM internal concentration had delayed currents which rose to a maximum and then declined to a plateau. He attempted to fit such curves with the kinetic scheme described by Eq. (15-9) and demonstrated that the experimental curves could be fit rather well.

Armstrong (1969) states that the development of C_5 inactivation of the delayed current has a sigmoidal time course for a given depolarizing pulse. In 1970 he compared the time course of the quaternary ammonium ion inactivation of the delayed conductance with that for inactivation of the initial transient conductance (g_{Na}) and found that both were delayed in their development. As the Hodgkin and Huxley (1952b) theory predicts an exponential development of inactivation of g_{Na} with a rapid onset, Armstrong (1970) considered this as evidence supporting his kinetic model for membrane conductance changes as opposed to the Hodgkin and Huxley formalism.

INTERNAL AND EXTERNAL PERFUSION WITH SOLUTIONS OF APPROXIMATELY EQUAL COMPOSITION AND CONCENTRATION

Isolated squid giant axons perfused internally with sodium fluoride solutions (300 to 430 mM) and externally with sea water exhibit large outward currents when pulsed to membrane potentials more positive than E_{Na} (Chandler and Meves, 1966; Adelman and Senft, 1966b). The axons were first hyperpolarized to overcome inactivation of the K-dependent early channel (Adelman and Palti, 1969b). These currents obtained at the test pulses exhibited an early peak followed by a decline to a sustained plateau. Pulsing the membrane to potentials more negative than E_{Na} resulted in inward currents having characteristics similar to the prolonged outward currents.

Fig. 15-16 illustrates typical records of membrane currents obtained upon voltage clamping an axon internally perfused with a solution containing 430 mM NaF and 10 mM KF and externally perfused with ASW. The membrane potential was held constant at -81 mv and pulsed twice a minute to various potential values.

Both the transient and the steady-state currents nulled at the same potential, which was about equal to the value calculated for a sodium equilibrium potential. Adelman and Senft (1966a) showed that the amplitudes of both the inward and outward initial transient and plateau currents were functions of the value of the holding potential as well as the pulse potential. These facts were taken as presumptive evidence that the plateau current was carried

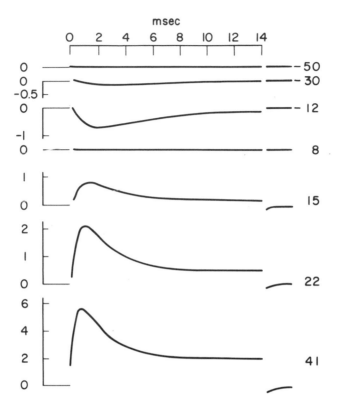

Fig. 15-16 *Voltage clamped membrane currents obtained during perfusion internally with a solution having a sodium activity of* 200 mM *and externally with ASW having a sodium activity of* 234 mM. *Current densities in* ma/cm² *are shown to the left of each record; negative values denote inward current flow and positive values outward current flow. Pulse potentials are given to the right of each record in* mv. *Potential values are taken internally with respect to an external reference electrode. Axon 65–94. Temperature,* 4° C. *(From Adelman and Senft, 1966a.)*

by the same conductance mechanism as the initial transient current. Further evidence for this contention was provided by experiments in which Tris⁺ was substituted for Na⁺ in the external ASW. Under these conditions, both the inward transient current and the inward plateau current were abolished but the outward currents were maintained. Similarly, upon comparing membrane currents obtained from axons internally perfused with Tris F, it was found that both the outward transient and plateau currents were abolished in Tris F but the inward currents remained.

Another curious property of axons internally and externally perfused with sodium solutions is that they can be made to exhibit hyperpolarizing

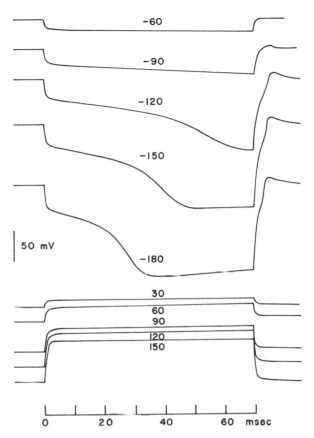

Fig. 15-17 *Typical membrane potentials in response to passing constant currents through the membrane of an axon perfused internally with* 430 mM NaF, 10 mM KF *solution and externally with* ASW. *Values of currents in* μa/cm² *are given above each record. Inward currents have negative signs and outward currents positive signs. Current duration indicated by time scale; onset of constant current at time zero. Initial segment of each record at* $E_{RP} = 0$ mv.

responses similar to those seen in intact axons externally perfused with 440 mM K sea water (Segal, 1958; Tasaki, 1959; and Moore, 1959) and in axons internally perfused with potassium solutions and externally with K sea water (Adelman, Dyro and Senft, 1965a). Fig. 15-17 illustrates a family of membrane potential records obtained by passing 69.5 msec duration hyperpolarizing and depolarizing currents through the axon membrane. Fig. 15-18 shows isochronal current voltage relations derived from the records shown in Fig. 15-17. A relation similar in configuration to the 69.5 msec

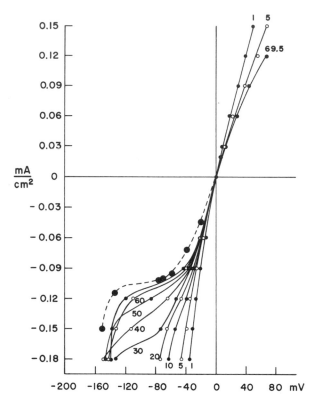

Fig. 15-18 *Isochronal current-voltage relations derived from records shown in Fig.* 15-17 *(small points). Values adjacent to each curve indicate the time after the onset of the constant current at which the potential was determined. Large points and dashed line indicate the steady-state values of membrane currents obtained upon voltage clamping the same axon.*

curve can be obtained from membrane currents obtained upon voltage clamping the membrane with 70 msec pulses from a holding potential set equal to the resting potential. The dashed line in Fig. 15-17 was obtained from the same axon a few minutes after the hyperpolarized records were obtained. This curve is a plot of the steady state current voltage relation for values of E_P more negative than $E_{HP} = E_{RP} = 0$.

The implication in this finding is that the appearance of a hyperpolarizing response is not specific to potassium ions in the squid giant axon. Rather, the response seems to be related to a general membrane-characteristic appearing when a depolarized membrane ($E_{RP} = 0$) is pulsed to negative potentials.

EXPERIMENTAL AND COMPUTATIONAL CONSIDERATIONS OF TRANSIENT IONIC CURRENTS SHOWING PLATEAUS

Early "Transient" Channel

As described in the previous section of this chapter, both Chandler and Meves (1966) and Adelman and Senft (1966a) have reported persistent or long duration membrane currents obtained from voltage clamped squid giant axons perfused internally with high ionic strength solutions containing predominantly NaF. These axons routinely had resting potentials close to zero and required previous membrane potential conditioning to hyperpolarized values before pulsing to various membrane potentials in order to elicit the long duration current.

Both Adelman and Senft (1966a) and Chandler and Meves (1966) had assumed that the persistence of the sodium current was due to an incomplete inactivation of sodium conductance through the early membrane channels. For this case, the time course of the long duration sodium current can be described by the kinetics of the early channel if one assumes that h_∞ does not go to zero at depolarized values of the membrane potential. The following procedures (Adelman and Senft, 1970) were designed to test this hypothesis: 1) To show that the prolonged sodium currents do not use the delayed conductance normally carrying potassium ions, axons were internally perfused with solutions containing sodium and cesium. The expectation was that internal Cs^+ should inhibit the delayed conductance, and if sodium ions were being carried through the delayed conductance, the plateau currents should be inhibited. Conversely, if no changes in plateau current occurred, sodium movement should not be through the delayed channel. 2) To show that the prolonged currents are carried through the early channel, the kinetics of these currents were predicted from the Hodgkin and Huxley hypothesis, without invoking any new mechanisms or parameters, but rather by making reasonable modifications in the early channel parameters consistent both with theory and with independent voltage clamp results.

Procedure (1) involved internally perfusing squid axons with 400 mM NaF, 130 mM CsF solutions. Membrane currents obtained upon voltage clamping these axons were similar to the plateau-type currents reported by Chandler and Meves (1966) and Adelman and Senft (1966a). Between pulses to various membrane potentials, the membrane was clamped at a holding potential (E_H) of $E_M = -90$ mv. The internal perfusion solution was then changed to 400 mM NaF, 130 mM Tris F solution. No significant changes in amplitude or time course of the membrane currents were observed. It was concluded that internal Cs^+ affected neither the initial transient nor the plateau phases of the membrane currents. Substituting 400 mM Tris, 130 mM CsF internally abolished the outward membrane currents and substituting Tris Cl for external NaCl abolished the inward membrane currents. It can be said that 130 mM CsF internally does not influence sodium ion movement through the squid axon membrane.

However, when 400 mM KF, 130 mM CsF was perfused internally, it was found that outward membrane currents were greatly reduced as compared to membrane currents obtained upon internal perfusion with 400 mM KF solution. This demonstrated that 130 mM Cs^+ internally does block potassium ion movement through the delayed channel. In fact, steady-state outward currents at comparable pulse potentials were much larger when 400 mM NaF, 130 mM CsF was the internal solution than when 400 mM KF, 130 mM CsF was the internal solution. If we can assume on the basis of these experiments that 130 mM CsF is an effective concentration for blocking outward current through the delayed channel, we can safely conclude that the plateau current results from the persistence of the early (sodium) conductance rather than that sodium ions are carried through the delayed conductance.

The previous experiment is not sufficient in itself to prove the hypothesis

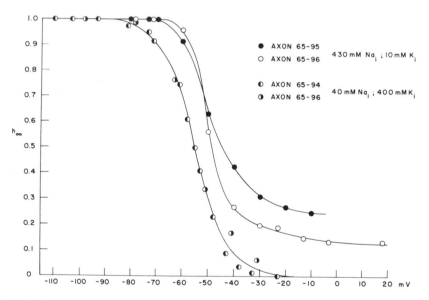

Fig. 15-19   *Values of the steady state level, h_∞, of parameter h plotted as a function of c_{Na_i} and c_{K_i} against membrane potential, in mv. h_∞ was determined as the ratio of the peak value of the outward initial transient current for a given prepulse potential to the maximum peak value of the outward initial transient current. Prepulse potentials were at least 3 seconds in duration (cf. Adelman and Palti, 1969b) and test pulses were always to +40 mv. Squid giant axons internally perfused with solutions indicated on figure. External solution was ASW.*

that the plateau currents are carried solely by the early channel, as there is the possibility that Cs^+ ions interact with K^+ ions, *per se*, and not with the delayed channel. Therefore, procedure (2) was carried out in order to demonstrate that the description of the kinetics of the early channel given by Hodgkin and Huxley (1952b) can be used to describe the locus of these persistent sodium currents. Procedure (2) was essentially analytical.

The analysis of the persistent sodium currents depended on the observation that curves of h_∞ vs. membrane potential (Hodgkin and Huxley, 1952b) obtained from axons internally perfused with 430 mM Na, 10 mM K solutions were shifted in the direction of depolarization as compared to h_∞ curves obtained when these axons were perfused with 400 mM K, 40 mM Na solution (Fig. 15-19). In addition, the h_∞ curves in Fig. 15-19 for high c_{Na}, low c_{K_i} perfusion showed that h_∞ did not go to zero for E_M values more depolarized than -20 mv. Rather, h_∞ tended toward small (0.1 to 0.3) but definite values even at $E_M = +20$ mv.

This finding was incorporated into the application of the Hodgkin and Huxley equations to persistent sodium currents in the following manner. The Hodgkin and Huxley formalism describes the sodium conductance as follows:

$$g_{Na} = m^3 h \, \bar{g}_{Na}, \tag{15-12}$$

$$\frac{dm}{dt} = \alpha_m (1 - m) - \beta_m \, m, \tag{15-13}$$

$$\frac{dh}{dt} = \alpha_h (1 - h) - \beta_h \, h. \tag{15-14}$$

For definition and descriptions of terms, see chapter 6. The solutions of Eq. (15-13) and (15-14) at boundary conditions $m = m_o$ and $h = h_o$ at $t = 0$ are:

$$m = m_\infty - (m_\infty - m_o) \, e^{-t/\tau_m}, \tag{15-15}$$

$$h = h_\infty - (h_\infty - h_o) e^{-t/\tau_h}. \tag{15-16}$$

Generally, in order to plot curves of g_{Na} as a function of time for E_M values more positive than -30 mv, m_o is neglected because m is much smaller in the resting state than during depolarization. Also, h_∞ is neglected because inactivation is very near complete when E_M values are more positive than -30 mv. Eq. (15-12), (15-15) and (15-16) may then be combined to give:

$$g_{Na} = g'_{Na} (1 - e^{-t/\tau_m})^3 e^{-t/\tau_h}, \tag{15-17}$$

where $g'_{Na} = \bar{g}_{Na} \, m^3 h_o$, the value the sodium conductance would obtain if h_∞ stayed at its initial level (h_o). However, as explained above, for the case of perfusion with high c_{Na_i} and low c_{K_i}, h_∞ does not go to zero with depolarizations beyond -30 mv. Thus, for these cases, Eq. (15-12), (15-15) and (15-16) must be combined to give:

$$g_{Na} = g'_{Na}(1 - e^{-t/\tau_m})^3[h_\infty - (h_\infty - h_o)e^{-t/\tau_h}]. \qquad (15\text{-}18)$$

Equation 15 can be restated as follows, since $I_{Na} = g_{Na}(E - E_{Na})$ where E is a given pulse potential:

$$I_{Na} = g'_{Na}(1 - e^{-t/\tau_m})^3[h_\infty - (h_\infty - h_o)\, e^{-t/\tau_h}](E - E_{Na}). \qquad (15\text{-}19)$$

Eq. (15-19) can be rearranged in a form more convenient for calculations:

$$I_{Na} = (1 - e^{-t/\tau_m})^3[K_1 e^{-t/\tau_h} + K_2(1 - e^{-t/\tau_h})](E - E_{Na}) \qquad (15\text{-}20)$$

where $K_1 = \bar{g}_{Na}m_\infty{}^3 h_o{}^2$ and $K_2 = \bar{g}_{Na}m_\infty{}^3 h_o h_\infty$. The value of K_2 is determined from the steady state level of I_{Na} during the pulse and the experimental value for I_{Na} at the peak of the current is used to solve for K_1.

It was possible to obtain calculated curves that superposed those obtained experimentally by substituting in Eq. (15-20) proper values for τ_m and τ_h. A computer program was written to facilitate curve-fitting and the output of the best fit curves was plotted by the computer. Fig. 15-20 shows experimental (left) and calculated (right) curves for a particular fit of Eq. (15-20) to membrane currents obtained for high $c_{Na\,i}$ perfusion.

Fig. 15-21 shows the values of τ_m and τ_h required to produce the calculated curves, plotted as a function of E_m. Also shown in Fig. 15-21 are curves taken from data of Hodgkin and Huxley (1952b). Notice that over the range of E_M considered, τ_m values are almost identical to the Hodgkin and Huxley values. τ_h values are somewhat larger, but are not in themselves sufficient to account for the existence of a sodium current plateau. The major factor accounting for the plateau is that h_∞ does not go to zero during depolarization.

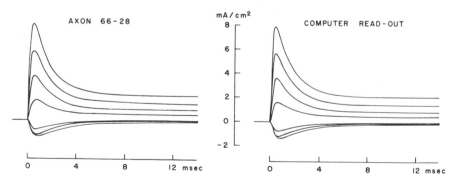

Fig. 15-20 Left: Membrane currents obtained in response to voltage clamping a squid giant axon perfused internally with 400 mM NaF and 130 mM CsF solutions and externally with ASW. Right: Computer read-out of membrane currents plotted according to Eq. (15-17). The holding potential, E_H, was -90 mv, and the pulse potentials, E_P, reading from top to bottom were: $+50$, $+40$, $+30$, $+20$, 0, -5 and -10 mv.

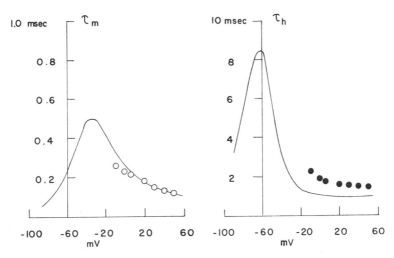

Fig. 15-21 *Values of τ_m (open circle) and τ_h (filled circles) used to fit curves shown in Fig. 15-20, plotted against membrane potential in mv. Continuous curves are plots of average values of τ_m and τ_h obtained from Hodgkin and Huxley (1952b). See text.*

Delayed "Steady-State" Channel

In the earlier section of this chapter on inhibitory effects of some internal and external cations, delayed membrane currents showing a transient phase followed by a plateau are described (Armstrong, 1969). These membrane currents were recorded from voltage clamped axons internally injected with various quaternary ammonium ions (TEA analogs). Armstrong proposed a kinetic scheme for deriving equations describing the locus in time of these "up-down plateau" delayed membrane currents. Previously, we have shown that "up-down plateau" sodium currents may be fitted with the Hodgkin-Huxley model if one simply assumes that the plateau results from incomplete inactivation (parameter h does not go to zero with sustained depolarizations lasting less than 100 msec). Armstrong (1969) suggested that the blocking action of quaternary ammonium ions had the form of a potassium inactivation.

Here, we wish to show that an equation similar to Eq. (15-20) may be used to fit the delayed currents. The equation was derived from the Hodgkin-Huxley equations for g_K by considering a potassium inactivation parameter, j, similar in form to parameter h, and that n_o is very small with respect to the value of n achieved during depolarization. For curve fitting purposes, this equation has the form:

$$I_K = (1 - e^{-t/\tau_n})^4 [K_3 e^{-t/\tau_j} + K_4 (1 - e^{-t/\tau_j})](E - E_K) \qquad (15\text{-}21)$$

where τ_j is the time constant of potassium inactivation, $K_3 = \bar{g}_K n^4 j_o{}^2$ and $K_4 = \bar{g}_K n^4 j_o j_\infty$. The value of K_4 may readily be obtained from the steady-state level of I_K during the plateau. The value of K_3 is determined from the peak value of the delayed current and incorporates the general effect of the blocking agent on resting potassium inactivation, j_o.

Fig. 15-22 shows the fit of Eq. (5-21) to the data of Armstrong (1969). This

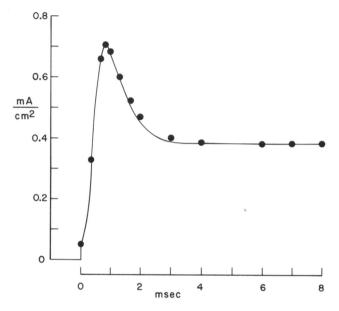

Fig. 15-22   Fit of Eq. (15-18) to data shown in Armstrong (1969, Fig. 8d, $E_m = +90$ mv). Line was traced from Armstrong's record of membrane current obtained from a squid axon having an internal C5 concentration of about 0.3 mM. Points were obtained from a solution of Eq. (15-18). See text.

fit is as good as that achieved by the Armstrong kinetic scheme. This is expected from the mathematical similarity between the two schemes. However Eq. (15-21) is symmetrical with Eq. (15-20) and thus allows the use of the full Hodgkin-Huxley equations for predictive purposes. The latter point is not trivial, as full programs exist for such predictions (see chapter 7) and there is general widespread familiarity with the use of this model system.

In addition, inactivation processes described by parameters h and j may be conceived of as having similar molecular mechanisms. Adelman and Palti (1969a,b) have shown that K_o^+ acts as an inhibitor of the initial transient channel and that this inhibition has profound effects on both h_∞ and τ_h. It seems quite apparent that quaternary ammonium ions may exert similar effects on the delayed membrane channel and this inhibition might be considered to be an inactivation having characteristic values for j_∞ and τ_j (j_o being that value of j_∞ for the initial E_M conditions).

Armstrong (1969, 1970) uses the delay in onset of inactivation to be the major support for his hypothesis. Thus, the most important conclusion to be derived from the exercise above is that the success of a given set of empirical equations in fitting membrane current curves is not to be taken as proof that the hypothetical scheme from which the equations are derived contains information as to the molecular machinery of the membrane channel (cf. chapter 17 for a description of the general problems of model making as related to molecular mechanisms).

GENERAL CONCLUSIONS

1.   Internally perfused squid giant axons have the same electrical properties as intact axons when the internal and external solutions approximate the ionic makeup found *in vivo*.

2.   With suitable modification, the ionic hypothesis can be used to describe the electrical properties of squid giant axons internally and externally perfused with a variety of nonphysiological solutions.

3.   The selectivity of the squid axon membrane to cations follows definite sequences specific for given conductance channels.

4.   Inactivation mechanisms appear to be dependent on the concentration and species of certain cations as well as on membrane potentials.

5.   The Hodgkin-Huxley formalism (1952b) can be modified to describe long duration action potentials, and initial transient and delayed plateau membrane currents.

Note added in proof: Recently, Chandler and Meves (1970a, b, c and d) have presented an analytical description of the finding that squid axons internally perfused with 300 mM NaF solutions exhibit a sodium conductance that is not fully inactivated by depolarizations having durations of tens of milliseconds. This analysis was based on voltage clamp experiments in which the steady-state level of the Hodgkin and Huxley parameter, h, was shown to have two components, h_1 and h_2. The values of h_2 were found to become larger with increasing depolarizations. From the kinetics of the h_1 and h_2 processes, Chandler and Meves (1970b) concluded that their results could be described upon assuming that the h process could be represented by the kinetic scheme

$$h_1 \underset{\alpha_{h_1}}{\overset{\beta_{h_1}}{\rightleftharpoons}} x \underset{\beta_{h_2}}{\overset{\alpha_{h_2}}{\rightleftharpoons}} h_2,$$

where the inactive state is represented by x, and the α's and β's are the appropriate forward and backward rate constants. Thus the sodium conductance, g_{Na}, is given by

$$g_{Na} = \bar{g}_{Na}m^3(h_1 + h_2).$$

An average value of 66 mmho/cm^2 was found for $\bar{g}_{Na}$, the maximal sodium conductance. The m process was found to behave as described by Hodgkin and Huxley (1952b), and the following new empirical relations for β_{h_1}, α_{h_2} and β_{h_2} at zero °C were given as follows:

$$\beta_{h_1} = 0.5/\{\exp[-(E + 32)/10] + D_1\exp(E/E_1)\},$$
$$\alpha_{h_2} = p \exp(E/E_2),$$
$$\beta_{h_2} = p \exp(E/E_2 - E/23.5) + pD_2,$$

for the condition that at 0 mv,

$$(\alpha_{h_2} + \beta_{h_2}) = p(D_2 + 2) = 0.55 \text{ msec}^{-1}.$$

Chandler and Meves found average experimental values of 3.6 for D_1, 240 mv for E_1, 0.08 msec^{-1} for p, and 70 mv for E_2.

In addition, Chandler and Meves (1970d) found a slowly inactivating process, which they labelled s, which inactivated with a rate constant of from 0.11 to 0.73 sec^{-1}. By assuming that this process could be incorporated into the relation for g_{Na} as follows:

$$g_{Na} = \bar{g}_{Na}m^3(h_1 + h_2)s,$$

they wrote equations which described the long duration membrane action potential recorded under these conditions. However, these equations did not accurately represent the repolarization phase of these action potentials. They suggested that this discrepancy might be accounted for by additional slow changes in membrane permeability not identified in their experiments.

ACKNOWLEDGMENT

The author wishes to express his gratitude to Drs. Joseph P. Senft and Frances Dyro for their collaboration in much of the original work discussed in this chapter. The technical assistance of Dr. Leon A. Cuervo, Miss Sandra Miller and Mrs. H. Nancy Laue is gratefully acknowledged. Support for original work was supplied by USPHS grant NB (S) 04601, USPHS fellowship 1 Flo NS 2204, and NSF grants GB 332, GB 3888 and GB 15588.

REFERENCES

ADELMAN, W. J., Jr. 1968. In the Squid Axon, Potassium Current Blockade by Internal Cesium is Dependent on Membrane Potential and Current Direction. *Biophys. Soc. Abstr., 12th Annu. Meet.* 133a.

ADELMAN, W. J., Jr., F. Dyro and J. P. SENFT. 1965a. Long Duration Responses Obtained from Internally Perfused Axons. *J. Gen. Physiol.* 48:1.

———. 1965*b*. Prolonged Sodium Currents from Voltage Clamped Internally Perfused Squid Axons. *J. Cell. Comp. Physiol. Suppl.* 66:55.

———. 1966. Internally Perfused Axons. Effects of Two Different Anions on Ionic Conductance. *Science.* 151:1392.

ADELMAN, W. J., Jr. and Y. B. FOK. 1964. Internally Perfused Squid Axons Studied Under Voltage Clamp Conditions. II. The Effects of Internal Potassium and Sodium on Membrane Electrical Characteristics. *J. Cell. Comp. Physiol.* 64:429.

ADELMAN, W. J., Jr. and D. L. GILBERT. 1964. Internally Perfused Squid Axons Studied Under Voltage Clamp Conditions. I. Method. *J. Cell. Comp. Physiol.* 64:423.

ADELMAN, W. J., Jr. and Y. PALTI. 1969*a*. The Influence of External Potassium on the inactivation of Sodium Currents in the Giant Axon of the Squid, *Loligo pealei. J. Gen. Physiol.* 53:685.

———. 1969*b*. The Effects of External Potassium and Long Duration Voltage Conditioning on the Amplitude of Sodium Currents in the Giant Axon of the Squid, *Loligo pealei. J. Gen. Physiol.* 54:589.

ADELMAN, W. J., Jr. and J. P. SENFT. 1966*a*. Effects of Internal Sodium on Ionic Conductance of Internally Perfused Axons. *Nature.* 212:614.

———. 1966*b*. Voltage Clamp Studies on the Effect of Internal Cesium Ions on Sodium and Potassium Currents in the Squid Giant Axon. *J. Gen. Physiol.* 50:279.

———. 1968. Dynamic Asymmetries in the Squid Axon Membrane. *J. Gen. Physiol.* 51:102.

———. 1970. Prolonged Na Currents in Perfused Squid Axons have Multiple Time Constants of Inactivation and are not Carried Through the K Conductance. *Biophys. Soc. Abstr. 14th Annu. Meet.* 184a.

ARMSTRONG, C. M. 1969. Inactivation of the Potassium Conductance and Related Phenomena Caused by Quaternary Ammonium Ion Injection in Squid Axons. *J. Gen. Physiol.* 54:553.

———. 1970. Comparison of g_K Inactivation Caused by Quaternary Ammonium Ion with g_{Na} Inactivation. *Biophys. Soc. Abstr., 14th Annu. Meet.* 185a.

ARMSTRONG, C. M., and L. BINSTOCK. 1965. Anomalous Rectification in the Squid Axon Injected with Tetraethylammonium Chloride. *J. Gen. Physiol.* 48:872.

BAKER, P. F., A. L. HODGKIN and H. MEVES. 1964. The Effect of Diluting the Internal Solution on the Electrical Properties of a Perfused Giant Axon. *J. Physiol. (London).* 170:541.

BAKER, P. F., A. L. HODGKIN and T. SHAW. 1961. Replacement of the Protoplasm of a Giant Nerve Fibre with Artificial Solutions. *Nature.* 190:885.

———. 1962*a*. Replacement of the Axoplasm of Giant Nerve Fibers with Artificial Solutions. *J. Physiol. (London).* 164:330.

———. 1962*b*. The Effects of Changes in Internal Ionic Concentrations on the Electrical Properties of Perfused Giant Axons. *J. Physiol. (London).* 164:355.

BINSTOCK, L., and H. LECAR. 1969. Ammonium Ion Currents in the Squid Giant Axon. *J. Gen. Physiol.* 53:342.

CALDWELL, P. C., and R. D. KEYNES. 1960. The Permeability of the Squid Giant Axon to Radioactive Potassium and Chloride Ions. *J. Physiol. (London).* 154:177.

CHANDLER, W. K., and A. L. HODGKIN. 1965. The Effect of Internal Sodium on the Action Potential in the Presence of Different Internal Ions. *J. Physiol. (London).* 181:594.

CHANDLER, W. K., A. L. HODGKIN and H. MEVES. 1965. The Effect of Charging the Internal Solution on Sodium Inactivation and Related Phenomena in Giant Axons. *J. Physiol. (London).* 180:821–836.

CHANDLER, W. K., and H. MEVES. 1965. Voltage Clamp Experiments on Internally Perfused Giant Axons. *J. Physiol. (London).* 180:788.

———. 1966. Incomplete Sodium Inactivation in Internally Perfused Giant Axons from *Loligo forbesi. J. Physiol. (London).* 186:121P.

———. 1970a. Sodium and Potassium Currents in Squid Axons Perfused with Fluoride Solutions. *J. Physiol. (London).* 211:623.

———. 1970b. Evidence for Two Types of Sodium Conductance in Axons Perfused with Sodium Fluoride Solution. *J. Physiol. (London).* 211:653.

———. 1970c. Rate Constants Associated with Changes in Sodium Conductance in Axons Perfused with Sodium Fluoride. *J. Physiol. (London).* 211:679.

———. 1970d. Slow Changes in Membrane Permeability and Long-Lasting Action Potentials in Axons Perfused with Fluoride Solutions. *J. Physiol. (London).* 211:707.

COLE, K. S. 1965. Introduction to "Newer Properties of Perfused Squid Axons." *J. Gen. Physiol.* 48: part 2.

CURTIS, H. J., and K. S. COLE. 1942. Membrane Resting and Action Potentials from the Squid Giant Axons. *J. Cell. Comp. Physiol.* 19:135.

FISHMAN, H. 1970. Leakage Current in the Squid Axon Membrane after Application of TTX and TEA. *Biophys. Soc. Abstr., 14th Annu. Meet.* 180a.

GOLDMAN, D. E. 1943. Potential, Impedance and Rectification in Membranes. *J. Gen. Physiol.* 27:37.

GOLDMAN, L., and L. BINSTOCK. 1969. Leak Current Rectification in *Myxicola* Axons. Constant Field and Constant Conductance Components. *J. Gen. Physiol.* 54:755.

HODGKIN, A. L. 1964. *The Conduction of the Nervous Impulse.* Charles C. Thomas, Springfield, Illinois.

HODGKIN, A. L., and A. F. HUXLEY. 1952a. Currents Carried by Sodium and Potassium Ions through the Membrane of the Giant Axon of *Loligo. J. Physiol. (London).* 116:449.

———. 1952b. A Quantitative Description of Membrane Current and its Application to Conduction and Excitation in Nerve. *J. Physiol. (London).* 117:500.

HODGKIN, A. L., and B. KATZ. 1949. The Effect of Sodium Ions on the Electrical Activity of the Giant Axon of the Squid. *J. Physiol. (London)*. 108:37.

MOORE, J. W. 1959. Excitation of the Squid Axon Membrane in Isosmotic Potassium Chloride. *Nature*. 183:265.

MOORE, J. W., T. NARAHASHI, R. POSTON and N. ARISPE. 1970. Leakage Currents in Squid Axon. *Biophys. Soc. Abstr., 14th Annu. Meet.* 180a.

MOORE, J. W., T. NARAHASHI and W. ULBRICHT. 1964. Sodium Conductance Shift in an Axon Internally Perfused with a Sucrose and Low-Potassium Solution. *J. Physiol. (London)*. 172:163.

NARAHASHI, T. 1963. Dependence of Resting and Action Potentials on Internal Potassium in Perfused Squid Giant Axons. *J. Physiol. (London)*. 169:91.

OIKAWA, T., C. SPYROPOULOS, I. TASAKI and T. TEORELL. 1961. Methods for Perfusing the Giant Axon of *Loligo pealei*. Acta Physiol. Scand. 52:195.

ROBINSON, R. A., and R. H. STOKES. 1959. *Electrolyte Solutions*. Butterworths Publications Ltd., London.

SEGAL, J. 1958. An Anodal Threshold Phenomenon in the Squid Giant Axon. *Nature*. 182:1370.

SHAW, T. I. 1966. Cation Movements in Perfused Giant Axons. *J. Physiol. (London)*. 182:209.

SJODIN, R. A. 1966. Long Duration Responses in Squid Axons Injected with ¹³⁴Cesium Sulfate Solutions. *J. Gen. Physiol.* 50:269.

TASAKI, I. 1959. Demonstration of 2 Stable States of the Nerve Membrane in Potassium-Rich Media. *J. Physiol. (London)*. 148:306.

TASAKI, I., and S. HAGIWARA. 1957. Demonstration of Two Stable Potential States in Squid Giant Axon under Tetraethylammonium Chloride. *J. Gen. Physiol.* 40:859.

TASAKI, I., and M. LUXORO. 1964. Intracellular Perfusion of Chilean Giant Squid Axons. *Science*. 145:1313.

TASAKI, I., and M. SHIMAMURA. 1962. Further Observations on Resting Action Potential of Intracellularly Perfused Squid Axon. *Proc. Nat. Acad. Sci. USA*. 48:1571.

TASAKI, I., I. SINGER and T. TAKENAKA. 1965. Effects of Internal and External Ionic Environment on Excitability of Squid Giant Axon. A Macromolecular Approach. *J. Gen. Physiol.* 48:1095.

TASAKI, I., I. SINGER and A. WATANABE. 1966. Observations of Excitation of Squid Giant Axons in Sodium-Free External Media. *Am. J. Physiol.* 211:746.

TASAKI, I. and T. TAKENAKA. 1963. Resting and Action Potential of Squid Giant Axons Intracellularly Perfused with Sodium-Rich Solutions. *Proc. Nat. Acad. Sci. USA*. 50:619.

TASAKI, I., A. WATANABE and T. TAKENAKA. 1962. Resting and Action Potential of Intracellularly Perfused Squid Giant Axon. *Proc. Nat. Acad. Sci. USA*. 48:1177.

THE EFFECT OF TEMPERATURE ON THE FUNCTION OF EXCITABLE MEMBRANES **16**

R. GUTTMAN

In this chapter, it is proposed to present first the presently known effects of temperature on the excitable membrane, with some comments on the experiments performed which suggested these effects, and then to outline what remains to be studied by investigators in this area.

The effect of temperature on various electrical parameters of the resting and active excitable membrane (i.e. upon the resting potential, spike height, rate of rise of spike, and Na and K conductances) will be treated. In addition, the relationship between temperature and the various *processes* that occur in excitable membranes (i.e. the excitation process, the time constant of excitation and accommodation) will be discussed. This discussion will concern itself mostly with the nerve axon membrane. Both Hoyt (1965) and Stein (personal communication) have emphasized the importance of the temperature characteristics of excitable membranes as guides to their structure and function.

Three experimental systems were used to elucidate what is known today about the effect of temperature on excitable membranes. These involved studying: 1) the propagated nerve impulse, 2) the space clamped axon and 3) the voltage clamped axon. Each of these systems has its own advantages.

One studies the propagating impulse because it most resembles the physiological condition and therefore is most directly related to the *in vivo* condition of the axon. Using this system, it has been possible to obtain information on the effect of temperature on impulse velocity, spike height and the maximum rate of rise of the spike.

In the space clamped condition, the impulse is made to stand still. This technique, then, does not resemble the physiological state. However, the space clamped condition provides an excellent test of the Hodgkin-Huxley equations. The mathematics is simpler than that describing propagated impulses. The space clamped condition has made possible a comparison of experimental data with calculations based on the Hodgkin-Huxley equations for: 1) accommodation, 2) frequency of oscillation and repetitive firing, and 3) threshold phenomena. Many factors affect threshold phenomena, and this system therefore provides a sensitive test of the Hodgkin-Huxley equations.

The voltage clamp provides a means of studying the basic property of excitable membranes associated with excitation, namely the N-shaped current voltage relation containing a negative resistance region. Voltage clamp data allow one to measure very subtle aspects of excitability. Data obtained using this method, therefore, provide one with the necessary information to formulate a model for nerve excitation. This model, derived from voltage clamped axons, may then be extended to predict the condition of the nerve in other modes.

COMPARISON OF EXPERIMENTAL DATA AND THEORETICAL PREDICTIONS

In comparing experimental data with computations based on the Hodgkin-Huxley equations, it should be remembered that the equations are merely a formulation based on the empirical information available at the time. Some have said that the temperature material is the weakest area in the equations. If this is so, it may be necessary to vary the parameters in the Hodgkin-Huxley equations until they give better agreement with experimental studies. On the whole, the equations have remained consistent with new information and have been a powerful tool to investigators in this area.

Voltage Clamped Axons

Based on results obtained from voltage clamp experiments, temperature coefficients of 1) the time constants, 2) the sodium and potassium conductances, 3) the resistances of membrane and axoplasm, and 4) the membrane capacity will be discussed.

First, the Q_{10} of the rate constants of the "sodium on", "sodium off", and "potassium on" processes (Table 16-1) will be considered. According to the Hodgkin-Huxley formulation (1952) for squid giant axons, the Q_{10} of the rate constants is 3. To date, the only careful experimental test of this

Table 16-1 Temperature dependence of processes determined in the voltage clamped axon

	Calculated (HHK)[1]	Modified
Rate constants of ionic conductances (Q_{10})	3,3,3	2,3,3 F & M[2]
Steady state conductances (Q_{10})	0	1.3–1.5 H[3], H & K[4]
		1.4 M[5]
Membrane capacity	0	1%°/C T&C[6], P&A[7]
Axoplasm resistance	0	2%/°C
E_{Na}, E_K	0	0.33%/°C

[1] Hodgkin, Huxley and Katz, 1952.
[2] Frankenhaeuser and Moore, 1963.
[3] Huxley, 1959.
[4] Hodgkin and Katz, 1949.
[5] Moore, 1958.
[6] Taylor and Chandler, 1962.
[7] Palti and Adelman, 1969.

value has been done by Frankenhaeuser and Moore (1963) who obtained a value of 2 for "Na on" and 3 for "Na off" and "K on" in *Xenopus* node. The time courses for the ionic currents are thus very dependent on temperature, increasing in speed by a factor of about 3 for a 10° C increase. This indicates the presence of a high energy process (almost 20 kcal/mole) as pointed out by Cole (1968).

For the temperature dependence of the sodium and potassium conductances themselves, Hodgkin, Huxley and Katz (1952) found that the apparent Q_{10}'s of the conductances were between 1.3 and 1.5. Moore's (1958) experimental figure comes in the middle of this range, 1.4.

With regard to membrane capacity, Taylor and Chandler (1962) have found the capacity increase with temperature to be about 1 per cent per degree. This has been confirmed by Palti and Adelman (1969). The axoplasm resistance is probably like that of any electrolyte, i.e. one would expect an increase of about 2 to 3 per cent per degree. From the Nernst equation, a value of about 1/3 per cent per degree can be assigned for the temperature dependence of the sodium and potassium potentials. Palti and Adelman (1969) have found a large and irreversible increase in both membrane resistance and membrane capacity at the high temperature of 40°C.

Since a decrease in temperature results in an extension of the duration of a propagating action potential, Shanes (1954) used a flame photometer to measure the expected increase of potassium loss with a decrease in temperature. At 6.1° C, he found a potassium loss of 11.4 pmole/cm²/impulse, which is much larger than the 3.7 pmole/cm²/impulse potassium loss at 24° C, giving a Q_{10} of $1/1.91$. Calculations based on the Hodgkin-Huxley equations by FitzHugh and Antosiewicz (1959) and by FitzHugh and Cole (1964) give a Q_{10} for K loss of $1/2.73$. This difference in temperature dependence suggests that while the Hodgkin-Huxley equations should be slightly refined in this area, the direct chemical measurement constitutes quite good support for the original equations.

Propagating Axons

Huxley (1959) calculated from theory a Q_{10} of 1.74 for the speed of propagation, θ, of an impulse in an axon. There are at present no experimental data from squid axons to substantiate or contradict this. Theoretical calculations suggest that velocity increases with temperature until about $33°$ C. Above this temperature, a reversible condition of heat block occurs. Hodgkin and Katz (1949) found something similar experimentally, i.e., propagation was abolished at about $38°$ C and restored when the fiber was brought below this temperature again.

Hodgkin and Katz (1949) were the first to study the effect of temperature on spike amplitude and found the amplitude to decrease with rising temperature. Huxley confirms this fairly well for the axon model.

That these agreements between theory and experimental results are not more precise is probably, according to Huxley, caused by the fact that no allowance was made in the Hodgkin-Huxley computations for an increase in sodium and potassium conductances with a rise in temperature. However, the fit is good enough to lead us to believe, even in the absence of voltage clamp data at higher temperatures, that the Q_{10}'s of the rate constants are still similar.

Space Clamped Axons

Although impulse velocity, spike height and the rate of rise of the spike are important physiologically, there are not much data on the effect of temperature on these aspects of nerve function. There are many more data available on threshold phenomena in space clamped axons. Here there has been an attempt at correlation of experimental results with calculated values in an effort to test the Hodgkin-Huxley formulation.

The Double Sucrose Gap. Guttman has studied threshold phenomena utilizing the double sucrose gap technique, developed by Julian, Moore and Goldman (1962) from an earlier version of Stämpfli (1954). Stämpfli's isosmotic sucrose gaps were in turn modified from earlier air gaps of Tasaki and Takeuchi (1941). All of these and the electronic methods of isolating nodes used by Dodge and Frankenhaeuser (1958) are modifications and variations of the space clamp originally described by Marmont in 1949.

The advantage of the double sucrose gap technique is that one can employ large, liquid, external electrodes. It is not necessary to use impaling electrodes or axial wires or perfusion and yet one is in effect getting measurements across the membrane. The drawbacks of the method are 1) that it results in hyperpolarization of the membrane (the resting potentials obtained are higher than normal and this must be taken into account), and 2) that prolonged use of glucose or sucrose ultimately alters the axon. Nevertheless, judicious use of the sucrose gap technique can give one good, reproducible results for about $1\frac{1}{2}$ hours.

In Fig. 16-1, we see a double sucrose gap arrangement. The ends of the axon, at *A* and *E*, are brought to zero potential by inactivating them with an artificial sea water containing twenty times the amount of KCl usually present in sea water. The axon in the central compartment, *C*, is the experimental node and here we have normal sea water or an experimental solution flowing.

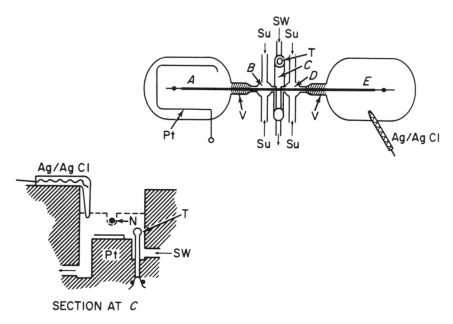

SECTION AT *C*

Fig. 16-1 *Mounting chamber used for studying temperature characteristics of excitation in space-clamped squid axons. Chamber is internally divided into five compartments,* A, B, C, D *and* E *by partitions provided with aligned clefts in which axon,* N, *rests.* Pt *are platinized platinum electrodes for application of current. The* Ag-AgCl *electrodes are used for potential measurement.* T, *thermistor. For further details, see text.* (*From Guttman, 1969.*)

In compartments *B* and *D* between the central, experimental portion of the axon and the killed ends, there is flowing isosmotic sucrose of very low conductivity, which has been sent through a deionizing resin. The sucrose isolates the experimental node from the killed ends of the fiber and prevents any current flow between them. Thus the impulse is made to stand still and we have what is called a space clamped condition. A thermistor is present in the central experimental compartment to record temperatures. (The needed accuracy of recording temperature for a Q_{10} of 3, which corresponds to about 12 per cent per degree, would require recording temperature to less than $0.5°$ *C* error, if one wishes τ, the time constant of excitation, to about 5 per cent accuracy.)

Platinum stimulating electrodes are inserted into compartments *A* and *C*. Recording electrodes of silver-silver chloride are present in *C* and *E*. The

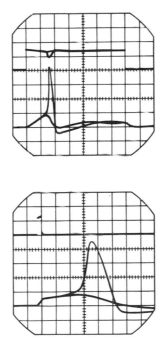

Fig. 16-2 *Typical records obtained when a rheobasic current of 7 msec (upper picture) and a 70 μsec short pulse (lower picture) are used to study threshold membrane voltage. In each, the upper trace represents the square wave pulse sent in, and the lower trace represents the records obtained when a just subthreshold pulse is followed by a threshold pulse and the two records are superimposed in a double exposure. The point where the trace of the local subthreshold response deviates from the action potential is used as a measure of threshold membrane voltage. Calibration per division in upper picture is 20 mv, 1 μa, and 1 msec in lower picture, 20 mv, 10 μa and 0.2 msec. (From Guttman, 1966.)*

stimulus picked up from the stimulating electrodes is recorded on the upper trace of a dual beam oscilloscope and the response picked up by the silver-silver chloride electrodes is recorded on the lower trace of the oscilloscope.

Temperature Effect on Threshold Phenomena. With such an arrangement it is possible to study the effect of temperature upon membrane current and membrane voltage threshold (Fig. 16-2). The stimulating current appears in the upper trace. In the upper picture there is a long rheobasic current stimulus and in the lower picture the stimulus is a short shock, a square wave of about 50 μsec duration. The lower trace in each case shows the response in millivolts. The just subthreshold response and the threshold spike are superimposed in a double exposure. Where the two first deviate we have a measure of the threshold membrane voltage (Guttman, 1962; 1966).

The effect of temperature upon the threshold depends upon the duration of the stimulating square wave pulse (Guttman, 1966). In Fig. 16-3, the duration of the stimulus, t, is plotted on the x axis and the strength of the stimulating current, I, causing a just threshold response, on the y axis. Instead of plotting

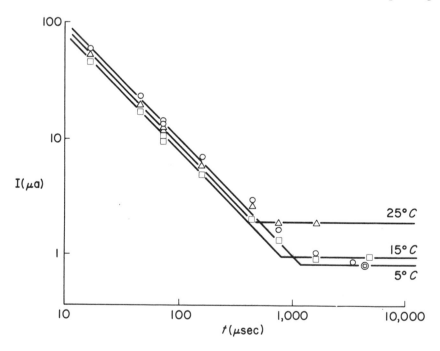

Fig. 16-3 *Temperature dependence of threshold membrane current when stimulating pulses of various durations are utilized. Intensity of the current in microamperes is plotted against duration in microseconds both on log scales. Runs were carried out at 5° C, 15° C, and 25° C on the same axon. (From Guttman, 1966).*

strength-duration curves on the customary linear plots, it is much more valuable to plot them on logarithmic scales. Then it can be seen that at long durations a constant current threshold is obtained. At short durations, the data fit 45° lines, approaching a constant quantity, Q, threshold, $It = Q$. Temperature affects the rheobasic threshold markedly but affects the constant quantity threshold only slightly. The Q_{10} for rheobasic threshold (2.35) agrees quite well with that of the axon of Cole and Marmont (1947) and the axon model computed by FitzHugh (1966).

The intersection of the Q and I_o lines gives the time constant of excitation: $\tau = Q/I_o$. In Fig. 16-4 we see that τ is very temperature dependent, decreasing markedly as the temperature is increased.

Fig. 16-5 shows theoretical computations by Cooley and Dodge (cf. Guttman, 1968a) for accommodation time. Accommodation time is defined as the ratio of rheobase to the minimum gradient of a linearly rising current that will just excite: $I_o/(dI/dt)$. $A = 1$ represents computations based on *Loligo forbesi* axons used by Hodgkin and Huxley (1952) at Plymouth, England and

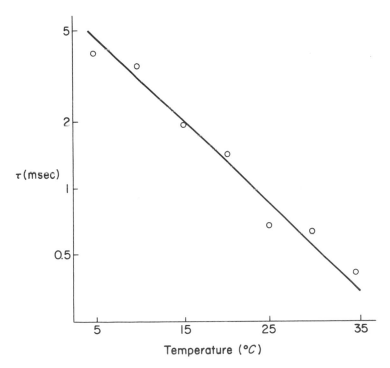

Fig. 16-4 *Temperature dependence of time constants of excitation in four axons. Average τ in milliseconds plotted logarithmically vs. temperature in degrees Centigrade. (From Guttman, 1966.)*

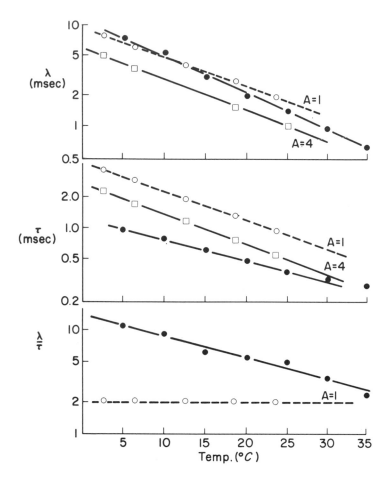

Fig. 16-5 *Comparison of experimental axon (heavy lines, solid circles, representing average values) with calculated axon (open circles for A = 1 condition; open squares for A = 4 condition) for λ, accommodation time; τ, excitation time, and λ/τ; all on logarithmic scale vs. temperature in C° on a linear scale. For discussion of conditions A = 1, A = 4, see text. (From Guttman, 1968a.)*

A = 4 represents computations for *Loligo pealei* axons obtained at Woods Hole. For *L. pealei* computations, the ionic conductances in the Hodgkin-Huxley equations were increased by a factor of 4 (FitzHugh, 1966). In the same figure we see the experimental results using *L. pealei* obtained for λ, accommodation time, which was measured using a linearly rising ramp for a stimulus.

The Q_{10} for accommodation and excitation are practically identical in the Hodgkin-Huxley model axon computed by Cooley and Dodge (Fig. 16-5).

For the experimental axon represented by the heavy continuous lines of the same figure, the Q_{10} of accommodation is 44 per cent higher than the Q_{10} of the excitation process. This discrepancy is brought out in the lowest graph of Fig. 16-5, where λ/τ is plotted against temperature. The Q_{10} of λ/τ computed for the model axon is 1.00 while that for the experimental axon is 1.62. Clearly, the Hodgkin-Huxley equations must be modified in this area.

Decreasing the sodium content of the sea water bathing the axon raises both the rheobasic and the constant quantity thresholds (Fig. 16-6) but the

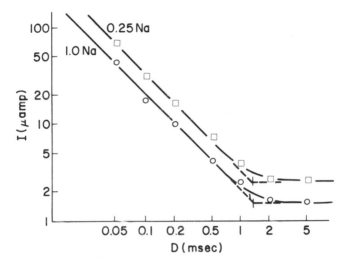

Fig. 16-6 *Strength-duration curves for axon in seawater (○), and in seawater in which 75 percent of the sodium has been replaced by choline (□). Duration of square wave pulses in milliseconds, D, is plotted against the current in microamperes, I, both on log scales. All points taken on the same axon at 5 °C. Vertical bars represent τ, as defined in the text, for the two different concentrations of sodium. (From Guttman, 1968b.)*

general shape of the strength-duration curve is not altered (Guttman, 1968b). The application of tetrodotoxin which blocks sodium conductance accomplishes the same thing and the blocking action of tetrodotoxin has a positive temperature coefficient (Fig. 16-7). On the other hand, Narahashi and Anderson (1967) were able to show that allethrin, a potent blocking agent which affects both sodium and potassium conductance, has a negative temperature coefficient.

In a paper published in 1966, Cooley and Dodge present calculations for the initiation of a propagated disturbance in the cable model. Strength duration curves for space clamped and propagated responses have the same form. As seen in Table 16-2, the time constants are almost the same. Because of the

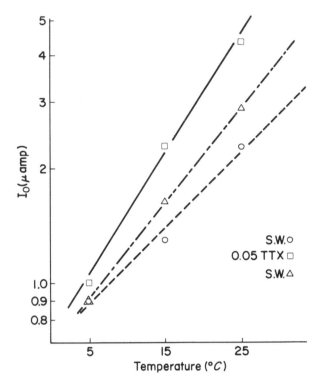

Fig. 16-7 *Temperature dependence of tetrodotoxin effect upon rheobase. Rheobasic current, I_o, in microamperes, is plotted against temperature in °C. First data were taken with the axon in seawater (○). Then the axon was treated with 0.05 μg/ml solution of tetrodotoxin (□). Then partial recovery in seawater (△) is shown. All data taken on the same axon. One of two similar experiments. (From Guttman, 1968b.)*

Table 16-2   Comparison of threshold parameters in cable model and in space clamp at two different temperatures

	$T(°C)$	$Q(ncoul)$	$I_o(\mu amp)$	$\tau(msec)$
Cable model	18.5	1.33	1.53	0.87
Space clamp	20.0	6.92	5.57	1.24
Cable model	6.3	1.71	.82	2.08
Space clamp	6.3	6.401	2.241	2.856

(From Fig. 5, Cooley, Dodge and Cohen, 1965)

difference in the actual experimental systems, the charge necessary for short shock threshold stimuli and rheobasic thresholds are expressed in different units. That is, the amount of charge needed to initiate a propagated response is about the same as the charge needed to achieve a threshold response in approximately one centimeter of space clamped axon.

Repetitive Responses. In discussing repetitive responses in the space clamped axon, one must first consider the classical work of Arvanitaki (1939) who found that subthreshold oscillation and repetitive spikes may be obtained if the membrane is bathed in low calcium solutions. In 1941, Cole and Baker found an inductive membrane reactance which permitted them to devise an analogous equivalent membrane circuit. Such a circuit is capable of oscillation.

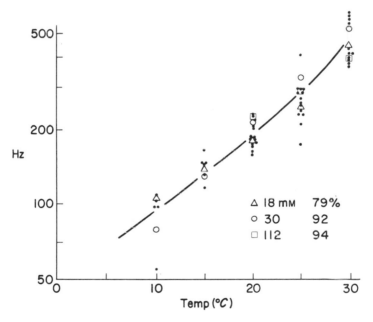

Fig. 16-8 *Effect of temperature upon frequency of subthreshold oscillation in space-clamped squid axons after near threshold current steps. Temperature in degrees centi-grade on a linear scale vs. frequency (Hz) plotted on a logarithmic scale. Sixteen threshold and three subthreshold temperature runs on eight axons were first plotted separately and then displaced vertically for best fit to the continuous line, which represents computed values. Dots indicate individual experimental points. Averages for axon in 18, in 30, or in 112 mM CaCl$_2$ are plotted by symbols as shown above. Percentages indicate amount of vertical displacement necessary for best fit to calculated eigenfrequencies in the case of each calcium concentration. (From Guttman, 1969.)*

In Fig. 16-8, the effect of temperature is seen upon the subthreshold oscillations following one threshold spike in axons bathed in low calcium sea water. The heavy continuous line represents computed results by FitzHugh and it is apparent that the experimental points fit rather well (Guttman, 1969).

The agreement between experiments and computations for repetitive spikes is also very good (Fig. 16-9). Here, period is plotted against temperature. The heavy continuous line represents computations by Cooley and Dodge based on the Hodgkin-Huxley equations. The experimental points fit quite well, except perhaps at the extreme temperatures (Guttman and Barnhill, 1970). [In a non space-clamped experiment, Coates, Altamirano and Grundfest (1954) found that the frequency of electrogenic discharges for knife fish depended upon temperature, with a Q_{10} of 1.5].

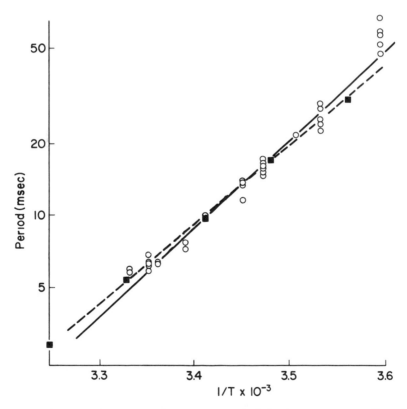

Fig. 16-9 *Period of repetitive firing vs. reciprocal of absolute temperature* (T) × 10⁻³. *Experimental data* (O———O) *and computed data* (■ - - - ■). *The ordinate pertains only to the computed axon, since experimental runs have been displayed vertically for best fit.* (*From Guttman and Barnhill,* 1970.)

In all the experimental results that have been presented near threshold and over a considerable temperature range, the data have not given any evidence of a critical temperature or of significant differences between two distinct ranges of temperature. In the Hodgkin-Huxley equations, also, the various parameters are rather gradual and smooth functions of temperature without any indication of a sharp transition in any temperature range. Thus, though it has been suggested by Tasaki (1968), Changeux, Thiery, Tung and Kittel (1966), and Adam (1968), that first order phase transitions in the membrane may be responsible for excitation, neither the Hodgkin-Huxley equations nor experimental observations give any support of such a process.

It is also interesting to note that recent work by Cole, Guttman and Bezanilla (1970) demonstrates that nerve membrane excitation is without threshold in the space clamped condition and that the all or none law does not hold for the nonpropagated impulse. This is true at all temperatures but is more apparent at high temperatures. It is in agreement with the BVP analogue of FitzHugh (1969) which shows a fractional response.

FUTURE EXPERIMENTAL CONSIDERATIONS

It has been suggested by R. Hoyt, D. Noble, Stein, FitzHugh, Cooley and Dodge that further experiments be performed using a voltage clamped axon to determine the effect of temperature on membrane conductance parameters.

Another example of an area that remains to be investigated is based on a finding by Spyropoulos reported in 1965 at a conference in Miami. Spyropoulos cooled or supercooled axons from room temperature down to 0 to $-5°$ C. At the critical temperature range of 0 to $-5°$ C, he found that the effects of temperature upon membrane resistance, membrane potential, rectification and the state of the membrane (resting or active), depend upon the concentrations of calcium and potassium in the external medium. That is, the effects of variations in temperature and in concentrations of potassium and divalent ions are interdependent. Thus, by a proper combination of a thermal source, a voltage source and of concentrations of calcium and potassium, any point of the I-V plane is an operating point. A response may be initiated at threshold by a sudden increase in potassium or a sudden decrease in calcium.

Guttman has found that sudden cooling causes depolarization in smooth muscle of *Mytilus edulis* bathed in sea water containing extra potassium. If the cooling is sudden enough, and if the depolarization is larger than a critical amount and rapid enough, a contraction of the muscle occurs (Guttman and Gross, 1956; Guttman and Ross, 1958). Application of an anodal current relaxes the contracted muscle (Guttman, Dowling and Ross, 1957).

Here then are a number of topics that should be looked into in the future: 1) temperature change as an effective stimulus, 2) the relation of temperature to concentration of potassium and divalent ions, and 3) the effect of temperature upon the voltage clamped axon.

The experiments described in this chapter on the whole support the Hodgkin-Huxley formulation quite well except in the realms of accommodation and oscillation phenomena. Theoreticians have long said that temperature is likely to be an important variable in any attempt to deduce a physical model for membrane phenomena.

ACKNOWLEDGMENTS

I have referred to my own work and that of others. Whenever my own work has been referred to, I wish to acknowledge the help of Robert Barnhill with instrumentation. I also had the cooperation of Richard FitzHugh and of James Cooley and Frederick Dodge, who did computations which I requested and who suggested to me experiments which should be done. Throughout, I had the good fortune of criticisms and suggestions by Kenneth S. Cole. I am most grateful to all these people.

REFERENCES

ADAM, G. 1968. Ionenstrom nach einem Depolarisierenden Sprung in Membran Potential. *Z. Naturforsch.* 23b:181.

ARVANITAKI, A. 1939. Recherches sur la réponse oscillatoire locale de l'axone géant isolé de "Sepia." *Arch. Intern. Physiol.* 49:209.

CHANGEUX, J. P., J. THIERY, Y. TUNG and C. KITTEL. 1966. On the Cooperativity of Biological Membranes. *Proc. Nat. Acad. Sci. USA.* 57:335.

COATES, C. W., M. ALTAMIRANO and H. GRUNDFEST. 1954. Activity in Electrogenic Organs of Knifefishes. *Science.* 120:845.

COLE, K. S. 1968. *Membranes, Ions, and Impulses.* Univ. of California Press, Berkeley.

COLE, K. S., and R. F. BAKER. 1941. Longitudinal Impedance of the Squid Giant Axon. *J. Gen. Physiol.* 24:771.

COLE, K. S., R. GUTTMAN and F. BEZANILLA. 1970. Nerve Membrane Excitation Without Threshold. *Proc. Nat. Acad. Sci. USA.* 65: 884.

COLE, K. S., and G. MARMONT. 1947. Unpublished data.

COOLEY, J. W., and F. A. DODGE, Jr. 1966. Digital Computer Solutions for Excitation and Propagation of the Nerve Impulse. *Biophys. J.* 6: 583.

COOLEY, J., F. DODGE and H. COHEN. 1965. Digital Computer Solutions for Excitable Membrane Models. *J. Cell. Comp. Physiol. (Suppl. 2)* 66:99.

DODGE, F. A., and B. FRANKENHAEUSER. 1958. Membrane Currents in Isolated Frog Nerve Fibre under Voltage Clamp Conditions. *J. Physiol.* 143:76.

FITZHUGH, R. 1966. Theoretical Effect of Temperature on Threshold in the Hodgkin-Huxley Model. *J. Gen. Physiol.* 49:989.

———. 1969. *In* H. Schwan [ed.] *Biological Engineering.* McGraw-Hill, New York, p.1.

FITZHUGH, R., and H. A. ANTOSIEWICZ. 1959. Automatic Computation of Nerve Excitation-Detailed Corrections and Additions. J. Soc. Indust. Appl. Math. 7:447.

FITZHUGH, R., and K. S. COLE. 1964. Theoretical Potassium Loss from Squid Axons as a Function of Temperature. *Biophys. J.* 4:257.

FRANKENHAEUSER, B., and L. E. MOORE. 1963. The Effect of Temperature on the Sodium and Potassium Permeability Changes in Myelinated Nerve Fibres of *Xenopus laevis. J. Physiol.* 169:431.

GUTTMAN, R. 1962. Effect of Temperature on the Potential and Current Thresholds of Axon Membrane. *J. Gen. Physiol.* 46:257.

———. 1966. Temperature Characteristics of Excitation in Space-Clamped Squid Axons. *J. Gen. Physiol.* 49:1007.

———. 1968a. Temperature Dependence of Accommodation and Excitation in Space Clamped Axons. *J. Gen. Physiol.* 51:759.

———. 1968b. Effect of Low Sodium, Tetrodotoxin and Temperature Variation upon Excitation. *J. Gen. Physiol.* 51:621.

———. 1969. Temperature Dependence of Oscillatory Behavior of Squid Axons. *Biophys. J.* 9:269.

GUTTMAN, R., and R. BARNHILL. 1970. Oscillation and Repetitive Firing in Squid Axons. *J. Gen. Physiol.* 55:104.

GUTTMAN, R., J. A. DOWLING and S. M. ROSS. 1957. Resting Potential and Contractile System Changes in Muscles Stimulated by Cold and Potassium. *J. Cell. Comp. Physiol.* 50:265.

GUTTMAN, R., and M. M. GROSS. 1956. Relationship between Electrical and Mechanical Changes in Muscle Caused by Cooling. *J. Cell. Comp. Physiol.* 48:421.

GUTTMAN, R., and S. M. ROSS. 1958. The Effect of Ions upon the Response of Smooth Muscle to Cooling. *J. Gen. Physiol.* 42:1.

HODGKIN, A. L., and A. F. HUXLEY. 1952. A Quantitative Description of Membrane Currents and its Application to Conduction and Excitation in Nerve. *J. Physiol. (London).* 117:500.

HODGKIN, A. L., A. F. HUXLEY and B. KATZ. 1952. Measurement of Current-Voltage Relations in the Membrane of the Giant Axon of *Loligo. J. Physiol. (London).* 116:424.

HODGKIN, A. L., and B. KATZ. 1949. The Effect of Sodium Ions on the Electrical Activity of the Giant Axon of the Squid. *J. Physiol. (London).* 109:240.

HOYT, R. 1965. Discussion Report. *J. Gen. Physiol.* 48:(5, Pt.q): 83.

HUXLEY, A. F. 1959. Ion Movements During Nerve Activity. *Ann. N.Y. Acad. Sci.* 81:221.

JULIAN, F. J., J. W. MOORE and D. E. GOLDMAN. 1962. Membrane Potentials of the Lobster Giant Axon Obtained by Use of the Sucrose-Gap Technique. *J. Gen. Physiol.* 45:1195.

MARMONT, G. 1949. Studies on the Axon Membrane. I. A New Method. *J. Cell. Comp. Physiol.* 34:351.

MOORE, J. W. 1958. Temperature and Drug Effects on Squid Axon Membrane Ion Conductances. *Federation Proc.* 17:113.

NARAHASHI, T., and N. C. ANDERSON. 1967. Mechanism of Excitation Block by the Insecticide, Allethrin, Applied Externally and Internally to Squid Giant Axons. *Toxicol. Appl. Pharmacol.* 10:529.

PALTI, Y., and W. J. ADELMAN, Jr. 1969. Measurement of Axonal Membrane Conductances and Capacity by Means of a Varying Potential Control Voltage Clamp. *J. Memb. Biol.* 1:431.

SHANES, A. M. 1954. Effect of Temperature on Potassium Liberation during Nerve Activity. *Am. J. Physiol.* 177:377.

SPYROPOULOS, C. S. 1965. The Role of Temperature, Potassium and Divalent Ions on the Current-Voltage Characteristics of Nerve Membrane. *J. Gen. Physiol.* 48:49.

STÄMPFLI, R. 1954. A New Method for Measuring Membrane Potentials with External Electrodes. *Experientia.* 10:508.

STEIN, R. 1969. Personal communication.

TASAKI, I. 1968. *Nerve Excitation.* Charles C. Thomas, Springfield, Ill.

TASAKI, I., and T. TAKEUCHI. 1941. Der am Ranvierschen Knoten entstehende Aktionsstrom und seine Bedeutung für die Erregungsleitung. *Arch. Ges. Physiol.* 244:696.

TAYLOR, R. E., and W. K. CHANDLER. 1962. Effect of Temperature on Squid Axon Membrane Capacity. *Biophys. Soc. Abstr.* TD1.

YAMASAKI, T., and T. NARAHASHI. 1953. *Botyu-Kagaku.* 19:39.

EXCITABILITY MODELS **17**

D. E. GOLDMAN

PHILOSOPHY AND PROCEDURES OF MODEL MAKING

This chapter deals with the problem of making models for excitable membranes. To begin with, we take a brief look at the meaning of the term model. The word derives from the Latin *modulus* or small measure, and current usage suggests the concept of a representation of something which cannot be directly observed. For such a representation to have scientific value it should not only provide a convenient way of visualizing a set of phenomena but must be based on realistic data and concepts and should permit calculation and predictions to be made.

The process of developing useful explanations, i.e., making meaningful models, obviously begins with intuitive ideas. It continues through the refinement of these ideas into analogies and hypotheses to generate a formal, verifiable theoretical structure. Intuitive notions which may arise in various ways and from different sources interact, clash and fuse into what we may refer to as conceptual models. This is the stage in which serious proposals develop as to how a system works, but without there being necessarily any immediate intent or ability to formalize them. Another procedure is the formulation of phenomenological analogues, which are often presented in mathematical form. Still another approach rests on the use of material models. These are also analogues but have the great advantage that they can

337

be made and then subjected to experimental treatment. On the other hand, one must then accept whatever properties are inherent in the material whether or not they are desirable or relevant. In the present context of membrane biophysics a reasonable goal would seem to be the development of a physico-chemical molecular model, based on physically comprehensible elements, logically combined, which relates structure and function in a quantitative way. Thus, one can try to explain in detail the phenomena observed and make useful predictions for further experimentation. There is, of course, no assurance that there is only one way of producing a given result; there is no guarantee of uniqueness. In fact, the only really complete model of any system may be said to be the original itself.

A very real question is why one should try to make models at all. To a considerable extent model making is a characteristic of the human mind and while some people are not interested in it, most people are, and for a variety of intellectual, practical and emotional reasons. A model, for example, may be a very satisfactory way of coordinating a complex mass of data. It may be valuable for predictive purposes. Beyond this many people seem to derive considerable esthetic satisfaction from the effort—and hopefully the results—of model making. It may also be a challenge and an outlet for competitive instincts.

A few comments about the processes of model making may also be made. We have spoken of the development of intuitive ideas and their subsequent refinement toward greater precision and accuracy. A major requirement is the intelligent use of whatever information may be available. In many areas serious problems arise as to the relevance, credibility and completeness of the information to which one may have access. Intelligent use implies the exercise of as much good judgment and as little prejudice as one can manage. One finds himself involved in a complex decision-making process with very incomplete information. However, after a model maker has made his choices he is still faced with the problem of analyzing the data in order to discover what underlying elements may be involved. In dealing with highly complex systems such as those usually found in biology, this process of analysis may be difficult and uncertain; it is nevertheless essential. Furthermore, the analysis must be carried to the point where the significant elements can be formulated clearly. In systems with strong interactions between many elements, the task almost invariably requires the use of approximations and this, again, requires considerable judgment and skill. It is often difficult to maintain a balance between oversimplification and extreme realistic detail. Finally, the elements must be put back together in analytical form and the consequences deduced as explicitly as possible. The consequences can then be compared with what is known and one may have to go through several cycles of modification and recalculation in order to reach acceptable results, both in terms of explanation and prediction.

Since we are dealing here with models for excitable membranes, we shall use them as examples of the modeling process. Since the field is extremely broad and complex, we shall confine the discussion primarily to axon membranes as a "target system". No one knows whether or not all excitable systems depend on the same mechanisms, nor indeed to what degree different axon membranes resemble each other. In order to make any progress, certain basic assumptions must be made. The selection process is difficult but whatever is selected must be explicitly and clearly understood.

SOME BASIC CHARACTERISTICS OF AXON MEMBRANES

In applying the procedural philosophy we have outlined to the specific problem of the axon membrane, we note first that there are a number of preliminary questions which must be raised and answered in order that one make any sense out of the situation. For example:

1. Are the mechanisms on which excitability depends localized in the axon and, if so, where?
2. Is there really an axon membrane or is there merely a phase boundary which has been modified slightly, if at all, and provides only the limit of the axon contents?
3. What are the energy sources for excitation? Should a distinction be made between direct and indirect energy supply? What thermodynamic and metabolic limitations must be applied to the problems of energetics?
4. How is electric current carried into and out of the axon? If the membrane is a finite region the question must be extended to include entry into the membrane, transfer through it and exit from the other side.
5. What particles constitute the actual carriers for the electric current?
6. What mechanisms control the current flow and its kinetics?
7. What are the significant relationships between excitability and general biological mechanisms such as growth, repair and protection?
8. To what extent are different membranes alike and to what extent unique? That is, how much general information derived from study of other membranes can be used in the analysis of axon membrane behavior?
9. Since we are dealing with transient phenomena related to action potentials, what time range is involved in what we may believe to be the significant aspects of excitability?

These questions do not now all have precise answers nor do they constitute all the significant questions to be asked. Nevertheless, we must establish some working basis and in accordance with our recognition of the importance of being explicit, we must produce some answers, or at least, estimates.

The axon appears to be a cylindrical body whose diameter is at least an order of magnitude less than its length, whose primary function is the generation and propagation of an action potential which is a transient change in the potential difference between the inside and the outside of the axon. The contents of the axon include water, dissolved salts, proteins and other organic substances as well as organelles. The environment has a more or less specified composition. There are many other properties of great importance. Not all of them, however, are necessarily related directly to the processes of excitability and propagation.

Electron micrographs of cross-sections of axons show a region roughly 100 Å thick whose appearance differs markedly both from the inside of the axon and the material outside it. The appearance of this region closely resembles what is seen in electron microscopy of the boundary regions of many other cells. This suggests very strongly that the membrane is a real material region and we shall accept this as a basis for further development. The membranes of certain other cells such as erythrocytes, liver cells and the myelin wrapping of many axons have been subjected to chemical analysis and have been shown to contain a variety of lipids, as well as protein-like material. However, no analyses have yet been made on reasonably uncontaminated axon membranes. Without going into detail, we may state that there is considerable indirect evidence consistent with the notion that the chemical composition of axon membranes is similar to what has been found in the other membranes studied. The lipids of the membrane presumably include phospholipids, neutral lipids and possibly glycolipids. Protein elements may include lipoproteins and glycoproteins. Other substances, such as polysaccharides and nucleic acids, may also be present.

As to the arrangement of these various molecular species, there is a strong suggestion that the lipids occur as a bilayer without any large breaks and that most of the protein elements occur along the surface of the lipid layer, possibly combined in various ways with at least some of the lipid elements. There may be a small amount of nonlipid material within the lipid bilayer. We may expect, on behavioral grounds, that the structure of the membrane is asymmetrical, i.e., that its two halves are not identical. The question as to how uniform the membrane is along its surface is completely open. Some experimental results can be interpreted to indicate the presence of small pores, holes or channels, or at least the occurrence of occasional small critical regions or sites in or on the membrane. Of the lipid elements, the phospholipids, being highly polar at one end, should be oriented with the polar regions at the surfaces and the hydrocarbon parts in the interior of the membrane. The extent to which there are mosaic, micellar or lattice-like arrangements is unknown. It does appear likely that the organization of the membrane can be modified by changes in membrane potential and by the immediate chemical environment. Optical anisotropy is known to be present and to vary with the membrane potential. Evidently, the extent to which the molecular structure of the axon membrane is understood is so limited that a wide range of speculation is possible.

When we turn to the behavioral aspects of axon excitability, the situation is entirely different. There is a tremendous amount of experimental data dealing with action potentials and current flows, and with their modifications by electrical and environmental changes. Much of the material has been discussed: early work by Katz (1939) and later work by Cole (1968). The cable equation provides a link between the propagation of the action potential and the current-voltage characteristics of the membrane. The fact that it has been possible to remove almost all of the axoplasm from certain large axons and to replace it by appropriate simple salt solutions without significantly interfering with normal function indicates very strongly that excitability is associated primarily with properties of the membrane and that the organic materials present in the axoplasm and environmental solutions are not directly concerned with excitability. This experimental work also suggests that no significant metabolic activity is directly involved in excitation. Further, metabolic inhibitors do not seem to interfere with excitation processes. It is true that these experiments have not been carried out on a wide variety of axons. Nevertheless, the observations are sufficiently striking and clear-cut to provide strong encouragement for generalization.

The voltage clamp technique provides a direct means of measuring the current-voltage behavior of a membrane. In view of the cable equation relationship, the study of excitability mechanisms is conceptually more simple, if experimentally more difficult, when based on voltage clamp data than when it is based on observations of action potentials.

There are a number of more specific statements, with which axonologists would generally agree on as being well established, and which seem unlikely to be overturned as a result of future experiments. Let us state them in the form of general propositions concerning the axon membrane. There are doubtless more, but as one looks further, the situation seems to become less and less clear and a definite cutoff point is difficult to find.

1. The axon, whose gross structure is considered to be known, is limited by a membrane which is real; i.e., the membrane contains distinct material, and is the seat of excitability phenomena.

2. The excitability process is associated with some molecular change in the structure of the membrane.

3. The membrane behaves electrochemically as a semipermeable lamina. It has an equivalent electrical capacitance which is relatively independent of applied voltage and changes in ion concentration. It has an equivalent parallel electrical resistance which is highly dependent on applied potential and on environmental ions and which responds in a characteristically time-dependent way to changes in applied potential.

4. The effects of ionic and electrical influences differ when applied to one side of the membrane versus the other. Thus, the membrane is asymmetrical.

5. Ionic concentrations differ on the two sides of the membrane. These concentration differences are the major internal energy source of the membrane potential. The ultimate source of energy, however, is metabolic activity.

6. Since the axon is a core conductor, the action potential can be derived with the use of a (nonlinear) cable equation from an appropriate set of current-time curves (voltage clamp curves) with membrane potential as the primary variable.

7. The total voltage clamped current consists of several components whose time course is regarded as known:

    a. A charging (I_C) current (attributable largely to the membrane capacitance).

    b. an early transient (I_E) current (mostly carried by sodium ions under normal conditions).

    c. a delayed (I_D) current (primarily carried by potassium ions under normal conditions).

    d. a small additional (I_L) current called leakage current with as yet undetermined current carriers.

    These components generally appear in the time range from a few microseconds to a few milliseconds. There is evidence for changes in some current elements which become significant at appreciably longer times. They have not yet been well characterized. We shall identify the current elements as I_C, I_E, I_D and I_L, as indicated above. These terms are intended to avoid implying any necessary relation to other phenomena.

8. The I_E and I_D currents generated in a voltage clamped system are carried almost entirely by univalent cations in accordance with their availability and the specificity of the relevant transfer mechanisms. The relative specificities are different for the two currents. Electrons, protons, polyvalent cations and large organic ions appear, in general, to contribute relatively little to the current.

9. Many small polyvalent cations, when placed in the external medium, appear to exert a controlling action on both the kinetics and voltage sensitivity of the I_E and I_D currents. Increases in concentration reduce the speeds of response and modify the voltage sensitivity of the conductance changes. In general, only very small concentrations of these ions are permissible internally without permanent damage to the membrane.

10. Certain organic substances have been shown to have predictable, reversible effects on axon conductance:
    a. Small to moderate molecular weight, nonionized fat soluble compounds reduce the amplitudes of the I_E and I_D components.
    b. Small to moderate molecular weight, ionized, fat-soluble, compounds have not only the action noted above but may also have an effect very similar to that of small polyvalent cations (calcium-like effect).
    c. Certain specific substances, such as tetrodotoxin, tetraethyl-ammonium ion, etc., appear to have highly selective effects on the I_E and I_D current elements.

11. Lytic enzymes, organic poisons, some heavy metals and many other substances may produce irreversible effects on the membrane, but few, if any, reliable generalizations have yet been made as to how they act.

THE ACTION OF ELECTRIC FIELDS ON MEMBRANE STRUCTURES

In searching for possible physical bases for membrane changes associated with excitability, we note first that the primary determinant of these changes is the electric field at and in the membrane. Magnetic and electromagnetic phenomena do not seem to play a significant role. While these phenomena may produce certain second order effects, we shall proceed on the basis that they are not particularly relevant.

All matter contains an intimate mixture of positive and negative charges so arranged that in many cases even an extremely small volume is electrically neutral. The application of an electric field to such a volume tends to move the negative charges in one direction and the positive charges in the other direction, thus producing an electric moment. This process is called polarization. The displacement of the charges is limited by the forces holding the various elements together and so, in a dielectric, there is no removal of charges from the region. If there are pre-existing dipoles, they will, of course, be stretched along the direction of the electric field or may be caused to rotate since the field produces a torque through its opposite action on the charges at the two ends of the dipole. To the extent that the medium is a conductor, the electric field drives any charged particles not permanently attached to the structure in an appropriate direction. In other words, an electric field produces polarization in a dielectric and current flow in a conductor. In addition, the application of electric fields to materials may produce significant mechanical stresses.

There are also a number of indirect effects of electric fields which may prove to be relevant. For example, the electrically generated stresses produce strains and, if the material has an appropriate structure, a piezoelectric effect may occur. This is an electromechanical coupling phenomenon in which the appearance of a charge distribution may produce a mechanical distortion proportional to the field strength. The converse is also true. There is an electrostriction effect proportional to the square of the electric field. There are electro-optical coupling phenomena, too, such as the Kerr effect in which strains produce or modify optical anistropy. There may be changes in light scattering if there is a change in the arrangement of scattering particles. Electrohydraulic coupling is also possible; namely, the production by the electric field of the flow of fluid through a system containing charges. Electrochemical coupling may appear as changes in chemical bond and activation energies with consequent modifications in equilibrium constants and reaction rates. Lastly, there must be thermal effects due to internal energy losses resulting from the motion of the charges whether the medium is a dielectric or a conductor.

More complex indirect effects of possible importance include cooperative phenomena in which the responses of neighboring individual particles to the electric field influence the interactions between these particles. Other effects may be of a feedback nature. For example, if the field drives a significant number of charged particles in or out of a given region, the local ionic strength may be changed significantly and this ionic strength change may have a marked action on the structure of certain large molecules, such as proteins, in the immediate area, whether or not the field has any direct action on the large particles themselves. At boundaries, we may also have significant accumulation or depletion of material in diffuse and adsorbed layers. Electric currents produced by fields not only produce significant changes in local concentrations but may move important amounts of material from one region to another.

The rapid survey given here of the effects of an electric field on materials, while useful for excluding a certain number of possible forms of behavior, still leaves many possibilities, any combination of which may be the significant factor in explaining excitability. Details are to be found in textbooks of electricity. At present, some of these phenomena appear much more likely than others, and those effects believed to be best justified by presently available evidence can be used for model construction.

MEMBRANE MODELS

We should now be ready to consider the selection of information about axon membranes which can be used for model construction. Again personal judgment necessarily plays a large role. We will try to restrict consideration here to data which are widely accepted and available in a relevant form. Where data are controversial, or not in a clearly specified form, they will be treated merely as possibilities. Hopefully, rapid progress in this field will be made in the near future.

On the basis of discussions given above, we will examine some of the choices to be made with respect to a structural model for the axon membrane. We may begin by considering the membrane as a simple phase boundary. However, excitability behavior is clearly so complex that such a concept hardly seems useful, even though a number of important properties are usually associated with phase boundaries. Next, we may treat the membrane as a thin lipid layer. This is essentially the early Overton-Meyer concept (Overton, 1902) which was based primarily on the solubility and penetration characteristics of small, uncharged molecules. The characteristics of such a membrane suggest that the flow of charged particles through it would be expected to follow the Nernst-Planck electrodiffusion regime. Let us call this model the Mark I; it seems to have many desirable features but has so far proved to be of little help in explaining excitability and is no longer an adequate representation of the generally known structural details of the axon membrane, as obtained from electron microscopy, x-ray diffraction, chemical analyses, etc. More recent modifications of the Mark I consider the lipid layer to be bimolecular in thickness and to have oriented neutral lipid and phospholipid elements. The original basis for this was the work of Fricke (1923) on the dielectric properties, and of Gorter and Grendel (1925) on the lipid analysis of erythrocyte membranes. It is thus easy to see that there must be distinct modifications of the simple diffusion concepts previously indicated since problems of molecular arrangement immediately arise and strongly suggest structural anisotropy.

A more modern concept, which we will call the Mark II, is based on the ideas of Danielli and Davson (1935) and more recently on material put together by Robertson (1960). For a very few membranes this has been backed up by x-ray diffraction studies. It has been labelled the "unit membrane" and is generally visualized as an oriented bimolecular layer, as described above, thinly coated with nonlipid material, mostly protein. Evidence in many forms has been presented by a number of people suggesting possible significant modifications of this model. For example, there may be pores, holes or channels through the membrane and these may either be aqueous in nature or may incorporate protein filaments. There may be specific transfer sites along the membrane; i.e., infrequently occurring molecular complexes with special properties related to ion transfer. Some of these sites may be electrically charged. If protein filaments occur within the body of the membrane, it is also possible that there may be fixed charges within the body of the membrane giving it certain characteristics of ion exchange resins. Such a modified Mark II membrane may also contain molecular elements which could move back and forth within it to function as carriers. Suggestions have also been made that there are micellar or mosaic substructures within the membrane possibly composed of protein material. Finally, the process of pinocytosis has been suggested as perhaps being relevant to ion transport. It is very difficult at present to find acceptable data capable of either proving or disproving the existence of any of these modifications. Nevertheless, the simple Mark II model does appear to have a number of inadequacies and some modification is clearly necessary. In any case, we do have reasonably good estimates of the membrane thickness and dielectric constant, as well as measurements of certain optical, mechanical and chemical properties and membrane conductance values, including their dependence on a number of controllable variables.

With respect to membrane behavior, we have access to voltage clamp data for normal axons and the effects of changes in both univalent and divalent cation concentrations. That the effects of these cations are different when the ions are placed on the two sides of the membrane strongly supports the idea that the membrane is asymmetrical. We have a substantial amount of data on the effects of organic substances, ranging from small uncharged molecules through enzymes, and we have also important information based on tracer flux measurements. Clearly, to proceed further, we must commit ourselves to some useful minimum concept of structure and must decide which of the many possible electric field effects play the major roles. There is a large body of detailed knowledge related more broadly to membrane transport phenomena on which we can draw, including the theory of electrolytes, surface electrochemistry, solid state theory (especially of semiconductors), molecular chemical theory (involving proteins and other polyelectrolytes), enzyme chemistry, etc. As a more recent addition, there has been some development of ideas about possible roles of conformational changes in molecular elements and complexes.

SOME EXAMPLES OF EXCITABILITY MODELS

Over a period of years, in fact ever since serious consideration was given to action potential propagation in nerve, many approaches have been considered and many suggestions have been offered to interpret the excitable behavior of the axon in rational terms. Since the development of the voltage clamp technique, attention has been given primarily to the characteristics of the axon membrane. An important early proposal was made by Bernstein (1902) based on the work of Nernst and others. He considered the membrane as the site of a diffusion potential due primarily to a potassium concentration gradient and proposed that an essential change associated with activity was a sharp increase in permeability and loss of ion specificity, thereby permitting the potential temporarily to approach zero. The basic ideas of his formulation are still sound, although the details and thus the consequences have been modified considerably. Later on, Hill (1910), Rashevsky (1931) and others developed phenomenological mathematical models of subthreshold electrical changes in the membrane and the appearance of an excitation threshold in terms of simple linear differential equations. Some attempts were made to interpret the equations in terms of experimentally measurable quantities. This approach is now mainly of historical interest, having been made obsolete by the development of the voltage clamp techniques and the consequent shift of attention to the current voltage relationships at the membrane.

Recently a very general mathematical approach to basic excitability characteristics has been used by FitzHugh (1961). His particular model is based on the phase plane treatment of the van der Pol equation for a relaxation oscillator. This approach abandons structural and electrophysiological realism and focuses instead on the topological relations essential to excitability. It has the great value of bringing out clearly certain fundamental requirements for excitable systems.

Karreman (1951) developed a treatment of the excitation process based on the concept of a lipid bilayer containing phospholipids which interacted with calcium to provide a current flow control mechanism. However, the difficulties encountered in attempting to calculate the action potential directly from a membrane structure proved to be too great for practical purposes. It was nevertheless one of the earlier essays toward a structural model.

The Hodgkin-Huxley (1952) formulation provided a fairly complete description and analysis of the voltage clamp currents of the axon membrane along with a demonstration that the action potential could be calculated from this formulation using the cable equation. The formulation was partly phenomenological and in a number of places detailed equations were given which were purely empirical. While the original material was based on the behavior of the squid axon under relatively normal physiological conditions, the system of equations has turned out to have a much wider application than might have been expected. Hodgkin and Huxley pointed out a number of mechanisms which might be involved but it was not then possible to eliminate any of them. In spite of the large dose of empiricism mixed with the physico-chemical analysis it remains from a practical viewpoint the most satisfactory model yet developed and has withstood a number of critical tests. It forms, in fact, a standard of comparison against which new models must be judged.

Mullins (1959) has offered an interpretation of certain aspects of the Hodgkin-Huxley system based on a distribution of pore sizes in the membrane with calcium as a surface blocking agent which affects the pore size. The sodium-potassium specificity problem is handled by relating the ion hydration radius to pore size. With this, Mullins has been able to match some aspects of the voltage clamp curves but the development has not been carried further. There appear to be serious problems related to the validity of the assumed relationships between hydration and pore penetration.

Hoyt (1963) has analyzed parts of the Hodgkin-Huxley system in an attempt to obtain further information relating to interactions which may exist between the flow mechanisms for sodium and potassium. Tille (1965) has offered a specific proposal for the control of potassium flux (I_D current) using a multiple-element gate along the lines suggested by Hodgkin and Huxley. Both of these approaches are primarily phenomenological although they are clearly directed toward problems of structure.

Agin and Schauf (1968) have proposed that the steady-state negative resistance characteristics of the axon membrane may be explained if it is assumed that diffusional ion transfer differs in mechanism from electrical transfer. The ion flow presumably involves both random diffusion and transfer through a series of unspecified sites. At the present time there is little evidence for such a mechanism. No attempt has been made to work out the transient current flow.

Lettvin, Pickard, McCulloch and Pitts (1964) have proposed a theory of passive ion fluxes by which sodium and potassium pass through separate channels in a lipid bilayer. An essential aspect of the behavior relates to the steric characteristics of the hydrated ions. Calcium is considered to shunt back and forth through the membrane providing a control mechanism. Although the proposal is qualitatively fairly explicit, no attempt appears to have been made to develop it quantitatively.

The fact that the membrane has many properties similar to those of semiconductors suggests the use of transistor mechanisms as an analogue. Such an analogue has been proposed by Wei (1969). As in the case of all analogues, it is extremely difficult to extend it beyond a certain point and until the calculations have been carried to the point of correlation with a variety of experimental data, it will not be possible to make any kind of serious evaluation.

A number of material models have also been used, the earliest being the iron wire model of R. S. Lillie (1936). It has excitable properties and has therefore been of great interest even though it is obviously unsuitable for biological structural purposes. Membranes made of various plastics ranging from collodion to ion exchange materials have also been used, along with layers of lipid materials. They do not exhibit excitability and their chief value to the biologist has been to provide simple means of studying steady-state transport phenomena. Plastic material membranes have been used by Teorell (1957) in systems in which electrical, osmotic and diffusion processes are combined. Oscillatory phenomena appear which are of great interest, but in terms of the current state of the art seem too simple to be very help-ful in studies of excitation.

Monolayers and lipid bilayers have also provided elementary models. Blank (1965) has suggested a possible close analogy between ion penetration through membranes and penetration through monolayers. So far this sug-gestion has not led to useful developments. Mueller and Rudin (1963) and others have shown that the addition of appropriate nonlipid materials to lipid bilayers may result in the appearance of excitability. The fact that these membranes do consist of lipid bilayers with protein or other nonlipid material has opened very exciting possibilities although at the present time it is impossible to predict to what extent the phenomena observed will prove to be relevant to excitability in biological systems.

Shashoua (1969) has recently prepared membranes of nonlipid materials capable of producing electrical activity under appropriate environmental conditions. Both the modified lipid bilayers and these nonlipid membranes exhibit steady-state current-voltage curves which have the negative resistance characteristics essential to electrical activity. Shashoua has referred to a possible molecular mechanism for the excitability of his membranes based on a concept put forward by Katchalsky and Oster (1969) that current flow may induce ionic strength changes in polyelectrolyte regions, thereby producing conformational changes. The usefulness of this notion clearly depends on whether or not there are in fact protein or polyelectrolyte filaments of an appropriate type within the body of the membrane. No evidence on this subject, pro or con, is now available.

Changeux, Thiery, Tung and Kittel (1967) have recently discussed certain general aspects of the cooperative interaction between micellar units, possibly consisting of lipoprotein, arranged in a two-dimensional lattice. The treat-ment is general and largely phenomenological. One can show the possibility of "flip-flop" transformations between conformational states. No specific application of the model has been made to excitable processes, though it provides an example of the use of phase transitions to explain action potential phenomena. Related to this is the idea suggested by Tasaki and Hagiwara (1957) that in view of the macromolecular structure of the membrane there may be two stable conformational states interconvertible through the action of an appropriate electrical field. This has not been reduced to quantitative form.

Wobschall (1968) has invoked the idea of electrets consisting of phospholipid dipole domains. This is another example of an approach based on a study of cooperative interactions. He has expressed his ideas in analytical form and has carried out some computer calculations. Again, it is difficult to evaluate this kind of approach until it has been carried to the point of being able to explain a considerably greater variety of experimental observations than at present.

The use of cooperative phenomena as a basis for excitability has also been developed by Adam (1968) based on a lattice of cation exchanger units which bind or release calcium competitively with sodium, thereby controlling the current flow. This also leads to a "flip-flop" system and limited attempts have been made to calculate some aspects of the voltage clamped current curves.

Among other ideas there should be mentioned that of Ling (1962) that the membrane is essentially a phase boundary and that the properties relevant to excitation reside primarily in the structure of the protoplasm. Nachmansohn (1968) considers that acetylcholine is the essential trigger for excitation. Kavanau (1965) considers the membrane to contain an arrangement of phospholipid or lipoprotein micelles which change their shape under the influence of electric fields. None of these provide any clear means of explaining function in terms of structure and physical principles; for the time being they remain exhibits of undeveloped approaches.

It will be evident from the above sketch that several classes of concepts have been invoked as possible ways in which axon excitability could be explained. There have been others as well. At the present time there appears to be only one model (Goldman, 1964) in which an attempt is made to develop a complete quantitative formulation. It also includes a way of handling a variety of modifications of the ionic environment. Certain aspects of this model will be described in the next section as an illustration of the methodology of model making.

A DETAILED PHYSICO-CHEMICAL MODEL FOR EXCITABILITY

In an attempt to specify the membrane characteristics essential to excitable behavior, we may start with the statement that the membrane is indeed a real thin material layer containing mainly lipid and protein elements. We would like to know how ions get into and out of the membrane, pass through its bulk, and by what method the flow of ions is controlled.

We shall adopt the view that the "unit membrane", with appropriate modifications, is to form the basis for our developemnt. The following seems to be the simplest structure we dare postulate which is not inconsistent with facts generally accepted at the present time. The bulk of the membrane probably consists of the nonpolar parts of lipid molecules, appropriately oriented. There may or may not be small pores present and there may or may not be nonlipid filaments or more complex structures within it. The simplest position to take is that such special elements are not necessary for excitability, or at least that their presence has relatively little effect on the complexities of the situation. Thus we postulate that the bulk of the membrane is an uncharged, low dielectric constant region through which particles may pass, although with considerably greater difficulty than through an aqueous medium. Such a view implies that those properties necessary to the existence of excitability reside in the surface layers of the membrane which presumably contain the protein material and the polar heads of the phospholipids. Because of the suggestive evidence that ions enter and leave the membrane only at relatively infrequent sites, the surface layer will be regarded as a relatively impermeable barrier to ion flow, except at these sites which then act as ports of entry or exit for the significant ions. The freedom with which ions pass through is under the control of the membrane potential. Since, for different ions, the ion flow differs considerably in time characteristics and voltage sensitivity and since special substances appear to be able to block these flows in specific ways, it appears likely that the precise loci of passage for say, sodium and potassium are different. Nevertheless, the rarity of situations in which the time characteristics and voltage behavior of these two ion flows can be separately controlled, suggests that the actual ports for sodium and potassium are close together, and that there may well be a single control mechanism operating on both. If there are separate control mechanisms, they are closely coupled.

In seeking a convenient way for the electric field to act on such a site region, we note that there are many phospholipid and protein elements available and that a striking characteristic of such molecules is that they may have large electric moments. Accordingly, we specify that the ion flow control mechanism incorporates an effective dipole whose rotation under the influence of the electric field opens or closes the appropriate ports. Although not necessary, it is convenient to treat the elements as equivalent phospholipid dipoles. We observe further that the ion specificity of the site complexes appears to be modified by changes in the electric field passing, with increasing depolarization, from a closed state through a sodium specific state to a potassium specific state, although this specificity is not absolute. Suspecting also that the membrane is asymmetrical, we have the possibility that control systems may occur on both sides of the membrane, not necessarily behaving in the same way, and even possibly a cooperative action of the two is required to produce the observed current sequences. Nevertheless, in anticipation of the extreme complexity of the physical formulation to which we are being driven, we retreat arbitrarily, at least until otherwise forced, to the pretense that the control mechanism is on one side. Because of the strong influence of external calcium on the control system, we put the mechanism on the outside of the membrane. The dipolar configurations generally appear to have their positive charge associated with a nitrogen atom and their negative charge associated with an oxygen which is either part of a phosphate or carboxyl group. We require that the dipole have at least three configurational states; i.e., one closed to ion flow, one sodium specific and one potassium specific. In the closed state the oxygen may combine with the calcium, lock the gate, and convert the structure into a form which is not a dipole. Depending on the binding constant of the calcium to the oxygen, some fraction of the complexes will remain dipoles and it is only these which can be rotated by a change in the electric field strength. Clearly, if left to their own devices, the dipole complexes will reach a steady-state distribution in which interaction primarily with potassium can occur but with an intermediate state in which sodium, primarily, can enter. During repolarization the dipoles revert to their former closed configuration, but need not do so by passing through the sodium configuration. There are several ways in which this can occur. Inspection of a molecular model of a phospholipid, for example, shows that there is more than one region of the molecule in which folding or unfolding can occur easily. Thus, the complex can pass from one configuration to another in different ways depending on local molecular conditions as well as on the electric field itself. The gate complexes thus have ion specificity, voltage sensitivity and ion exchange properties. They are embedded in the polar region of the membrane adjacent to the environmental solutions and their dielectric properties are affected by the polarity and dielectric properties of their immediate neighbors with whom they may interact cooperatively.

There is evidence that the membrane surface contains fixed charges and it is also obvious that there must be diffuse double layers in the immediate vicinity. Because of the relatively high ion concentration in biological media, these double layers are necessarily very thin, being usually no more than the thickness of the diameter of a partly hydrated small ion. Since the double layer potential can be roughly estimated from the membrane potential and the ionic properties of the medium, a convenient procedure is to ignore the double layers as such but to calculate an appropriate correction for the membrane potential.

Let us now try to develop from this model a system of equations to describe its behavior. First, as to the bulk of the membrane, the logical step appears to be the application of the Nernst-Planck equations for sodium and potassium. Let us use them in the simplest possible form. In view of the behavior of the I_L current and the presence of a high concentration of chloride ions, it is convenient to write a third equation for chloride as representing nonspecific anionic elements. The diffusion (or mobility) coefficients in the equations may be directly those of ions in the lipid or may reflect anisotropy of flow and the fact that entry occurs only at limited regions. Since the membrane is a condenser as well as a conductor, Poisson's equation is necessary. Since we are concerned with dynamic behavior, we include the equations of continuity which relate ion accumulation in the membrane to nonuniformity of current in the direction of flow. This gives us seven differential equations with seven variables: three current elements, three concentrations and the potential, as follows:

$$I_{Na} = QD_{Na} \left(\frac{\partial c_{Na}}{\partial x} + c_{Na} \frac{\partial \theta}{\partial x} \right)$$

$$I_K = QD_K \left(\frac{\partial c_K}{\partial x} + c_K \frac{\partial \theta}{\partial x} \right)$$

$$I_{Cl} = QD_{Cl} \left(-\frac{\partial c_{Cl}}{\partial x} + c_{Cl} \frac{\partial \theta}{\partial x} \right)$$

$$\frac{\partial^2 \theta}{\partial x^2} = \frac{-4\pi Q^2}{\varepsilon kT} (c_{Na} + c_K - c_{Cl})$$

$$\frac{\partial I_{Na}}{\partial x} = Q \frac{\partial c_{Na}}{\partial t}$$

$$\frac{\partial I_K}{\partial x} = Q \frac{\partial c_K}{\partial t}$$

$$\frac{\partial I_{Cl}}{\partial x} = -Q \frac{\partial c_{Cl}}{\partial t}$$

where c_{Na}, c_K and c_{Cl} are the concentrations of sodium, potassium and chloride respectively, at a point x, in the membrane at a time, t. I_{Na}, I_K and I_{Cl} are the corresponding currents and θ is QV/kT where V is the potential. D_{Na}, D_K and D_{Cl} are the corresponding diffusion coefficients in the membrane, Q is the electronic charge, k, the Boltzmann constant, T, the absolute temperature and ε, the dielectric constant. We treat the flow across the barriers and the behavior of the control system as time dependent boundary conditions for the system.

To develop these boundary conditions we consider the problem of ion transfer across the surface layer. The analysis requires two steps. First, a consideration of the process whereby ions are transferred across the barriers and, second, an investigation of the way in which the barrier configurations are redistributed and modified by the action of the electric field. The energy barrier at the surface layer prevents the passage of ions which do not have sufficient kinetic energy. The barrier energy may be separated into a chemical and an electrical term. The chemical term refers to the local steric and molecular structure. The electrical term is supplied by the field and raises or lowers the height of the barrier. The net flow of ions across the barrier can then be calculated as the difference between two terms each involving the concentrations at a side of the barrier and the appropriate Boltzmann factor which assures that the particles have enough energy. A proportionality constant represents the "barrier permeability". It should be pointed out that when a site in the potassium configuration is occupied by a potassium ion, potassium flow occurs through collision transfer. We note, incidentally, that because of the difference between the aqueous and the lipid media, we must take into account the existence of a distribution coefficient and it appears likely that the distribution coefficient between lipid and water will be quite small. To get the total current we multiply the flow per site by the number of sites in the proper configuration per unit of membrane.

$$I_{Na} = Q\lambda_{Na}v_{Na}(c_{Na_i}e^{(1-\beta)\theta_1} - \xi_{Na}c_{Na_o}e^{-\beta\theta_1})$$

$$I_K = Q\lambda_K v_K(c_{K_i}e^{(1-\beta)\theta_1} - \xi_K c_{K_o}e^{-\beta\theta_1})$$

$$I_{Cl} = -Q\lambda_{Cl}(c_{Cl_i}e^{-(1-\beta)\theta_1} - \xi_{Cl}c_{Cl_o}e^{\beta\theta_1})$$

Here, the λ's are appropriate barrier permeabilities, v_{Na} and v_K are the surface densities of sites in the sodium and potassium configurations, θ_1 is the (normalized) potential just inside the barrier, β is the fraction of θ_1 found at the top of the barrier, the ξ values are the distribution coefficients. The subscripts i refer to the inner edge of the barrier and the subscripts o refer to the external medium at the outer edge of the barrier. At the axoplasmic side of the barrier we can specify the concentrations as related to the values in the axoplasm directly or with a Donnan correction if we wish to assume that the inner barrier is relatively unimportant. A more complex alternative would be to use a formulation analogous to that applied to the outer barrier.

To obtain the number of sites in the various configurations at any time, we work out the kinetics of the transformations between the states, treating the states as though they were chemical entities differing in energy and between which there are activation energy barriers. Since we have assumed that there are three basic configurations of the dipoles, we find ourselves with a triangular reaction network which contains six rate constants. Since the barrier heights depend in part on the potential across them, we may invoke the theory of absolute reaction rates and separate out an electric term. This term will then appear exponentially in the expressions for the rate constants with opposite signs in the two opposite directions. If we refer to the diagram,

$$\text{ICa} \underset{k_m}{\overset{k_m'}{\rightleftharpoons}} \text{Ca} + \text{I} \quad \begin{array}{c} \overset{k_1}{\underset{k_2}{\rightleftarrows}} \text{II} \\ \overset{k_3}{\underset{k_5}{\swarrow}} \Big\Vert k_4 \\ \overset{k_5}{\underset{k_6}{\swarrow}} \\ \text{III} + \text{K} \underset{k_K'}{\overset{k_K}{\rightleftharpoons}} \text{IIIK} \end{array}$$

it is clear that the rate constants connecting configurations II and III need not depend on the potential, at least for our purposes, and so we require only the other four to be voltage dependent. In view of the ion exchange interactions which may occur between the different configurations and the relevant ions, we should add six more rate constants for this purpose. Fortunately, there is substantial reason to believe that sodium does not interact with the complex so that we may omit the corresponding rate constants although this may not be possible if other ions are substituted for sodium. Further, because the basic network is cyclic, only five of the six rate constants are independent; this results from the principle of detailing balancing. We can now write the appropriate rate equations for the configurational changes and thus determine the distribution of sites between the different states.

If the fractions of the total number of sites in a form corresponding to each of the five states is represented by $\omega_{Ca}, \omega_I, \omega_{II}, \omega_{III}, \omega_K$ we can write

$$\dot{\omega}_{Ca} = -k_m'\omega_{Ca} + k_m c_{Ca}\omega_I$$

$$\dot{\omega}_I = k_m'\omega_{Ca} - (k_m c_{Ca} + k_1 + k_6)\omega_I + k_2\omega_{II} + k_5\omega_{III}$$

$$\dot{\omega}_{II} = k_1\omega_I \qquad\qquad -(k_2 + k_3)\omega_{II} + k_4\omega_{III}$$

$$\dot{\omega}_{III} = k_6\omega_I \qquad\qquad + k_3\omega_{II} \qquad -(k_4 + k_5 + k_K\overline{K})\omega_{III} + k_K'\omega_K$$

$$\dot{\omega}_K = \qquad\qquad\qquad\qquad\qquad\qquad k_K\overline{K}\omega_{III} - k_K'\omega_K$$

The quantity, $\overline{K}$, is the average of the concentrations of potassium ion immediately available at the two sides of the barrier. c_{Ca} is the calcium concentration. Of course, the sum of all the ω's is unity. k_1, k_6 are exponential in θ_1 and k_2, k_5 are exponential in $-\theta_1$.

In addition, we note that we have fixed ion concentrations and specified potentials on the two sides of the membrane. Referring back to our original set of equations we note that one more boundary condition is needed. This we obtain by assuming that the membrane itself is electrically neutral and thus that the field strength at the two boundaries must be the same. We should observe also that the external barrier is considered to be very much thinner than the membrane, and that chloride transfer occurs at the external boundary primarily as a generalized leak. At the internal boundary we may have a Donnan correction and here no specific ports are now considered. A small amount of interstitial leakage of sodium and potassium may of course occur at the external boundary as well. While there is significant evidence that there may be specific sites or ports on the axoplasmic side of the membrane it does not at the present state of development seem advisable to treat a separate control system there.

Finally our initial conditions can be specified by noting that, in any steady state we have a time independent condition and that at zero time we produce an instantaneous change in one quantity, namely, the membrane potential. We then follow the system through its transient to another steady state. We can, however, use any time variation of potential we please, such as AC, a ramp, a rectangular wave, etc.

We now have a formal system of equations in as simple a form as seems possible, based on an explicitly defined mechanism. In order to make use of it we need to solve the equations and compare the results with appropriate experimental data. The system contains nonlinear partial differential equations and requires a large computer. The system also contains a large number of numerical parameters, a few of which can be estimated independently either from theory or from experimental results. Others, however, can only be obtained at the present time by curve fitting. In principle, the determination of twenty-one parameters (which is what the system has in its simplest form) requires the use of twenty-one independent sets of experimental data. The methods used for their determination require matching of functional forms of curves and of numerical values. It is obviously essential that all numerical values of parameters so obtained be physically acceptable in relation to the approximations used. Assuming that these procedures have been carried out, it then becomes possible to verify the usefulness of the system, first by comparing it with still other independent experiments which have been carried out and then by predicting the results of experiments which have not yet been carried out.

It is not the purpose of this discussion to evaluate the formulation described here. It is presented primarily to demonstrate a procedure, part of which consists of breaking up the problem into workable parts, formulating them separately and then recombining them. This procedure not only forces one to specify exactly what one is talking about but also points out and permits modifications in specifics which may later prove to be necessary. Further details may be found elsewhere (Goldman, 1964, 1965, 1969). If one compares this model with some of the ideas referred to in the broader discussion given earlier, it will be evident that there may be several ways of describing model elements. It is indeed possible that the equations developed, or equations much like them, could be derived from other bases. Uniqueness, as has already been mentioned, is scarcely attainable. Here it is only possible to point out that there are a number of alternatives and that they should be investigated in detail.

REFERENCES

ADAM, G. 1968. Ionenstrom nach einem depolarisierenden Sprung im Membranpotential. *Z. Naturforsch.* 236:181–197.

AGIN, D., and C. SCHAUF. 1968. Concerning Negative Conductance in the Squid Axon. *Proc. Nat. Acad. Sci. USA.* 59:1201–1208.

BERNSTEIN, J. 1902. Untersuchungen zur Thermodynamik der bioelektrischen Ströme. *Pflügers Arch. Ges. Physiol.* 92:521–562.

BLANK, M. 1965. A Physical Interpretation of the Ionic Fluxes in Excitable Membranes. *J. Colloid Sci.* 20:933–949.

CHANGEUX, J. P., J. THIERY, Y. TUNG and C. KITTEL. 1967. On the Cooperativity of Biological Membranes. *Proc. Natl. Acad. Sci. USA.* 57:335–341.

COLE, K. S. 1968. *Membranes, Ions, and Impulses.* Univ. of Calif. Press, Berkeley.

DANIELLI, J. F., and H. DAVSON. 1935. A Contribution to the Theory of Permeability of Thin Films. *J. Cell. Comp. Physiol.* 5:495–508.

FITZHUGH, R. 1961. Impulses and Physiological States in Theoretical Models of Nerve Membrane. *Biophys. J.* 1:445–466.

FRICKE, H. 1923. The Electrical Capacity of Cell Suspensions. *Phys. Rev.* 21:708–709.

GOLDMAN, D. E. 1964. A Molecular Structural Basis for the Excitation Properties of Axons. *Biophys. J.* 4:167–188.

———. 1965. Gate Control of Ion Flux in Axons. *J. Gen. Physiol.* 48:75–77.

———. 1969. Physico-Chemical Models of Excitable Membranes, pp. 259–279. *In* D. C. Tosteson [ed.] *The Molecular Basis of Membrane Function.* Prentice-Hall, New Jersey.

GORTER, E., and F. GRENDEL. 1925. Biomolecular Layers of Lipoids on Chromocytes of Blood. *J. Exp. Med.* 41:439–443.

HILL, A. V. 1910. A New Mathematical Treatment of Changes of Ionic Concentration in Muscle and Nerve under the Action of Electric Currents, With a Theory as to Their Mode of Excitation. *J. Physiol.* 40:190–324.

HODGKIN, A. L., and A. F. HUXLEY. 1952. A Quantitative Description of Membrane Current and Its Application to Conduction and Excitation in Nerve. *J. Physiol. (London).* 117:500–544.

HOYT, R. 1963. The Squid Giant Axon. Mathematical Models. *Biophys. J.* 3:399–431.

KARREMAN, G. 1951. Contributions to the Mathematical Biology of Excitation with Particular Emphasis on Change in Membrane Permeability and on Threshold Phenomena. *Bull. Math. Biophys.* 13:189–243.

KATCHALSKY, A., and G. OSTER. 1969. Chemico-Diffusional Coupling in Membranes, pp. 1–44. *In* D. C. Tosteson [ed.] *The Molecular Basis of Membrane Function.* Prentice-Hall, New Jersey.

KATZ, B., 1939. *Electrical Excitability in Nerve.* Oxford University Press, London.

KAVANAU, J. L. 1965. *Structure and Function in Biological Membranes* (Ch. 2, 5, 12). Holden-Day, New York.

LETTVIN, J. Y., W. F. PICKARD, W. S. McCULLOCH and W. PITTS. 1964. A Theory of Passive Ion Flux through Axon Membranes. *Nature.* 202:1338–1339.

LILLIE, R. S. 1936. The Passive Iron Wire Model of Protoplasmic and Nervous Transmission and Its Physiological Analogs. *Biol. Rev.* 11:181–209.

LING, G. 1962. *A Physical Theory of the Living State.* Blaisdell, New York.

MUELLER, P., and D. O. RUDIN. 1963. Induced Excitability in Reconstituted Cell Membrane Structure. *J. Theoret. Biol.* 4:268–280.

MULLINS, L. J. 1959. An Analysis of Conductance Changes in Squid Axon. *J. Gen. Physiol.* 42:1013–1035.

NACHMANSOHN, D. 1968. Proteins in Bioelectricity: The Control of Ion Movements Across Excitable Membranes. *Proc. Nat. Acad. Sci. USA.* 61:1034–1041.

OVERTON, E. 1902. Beiträge zur allgemeinen Muskel-und nervenphysiologie. *Pflügers Arch. Ges. Physiol.* 92:115–280.

RASHEVSKY, N. 1931. On the Theory of Nervous Conduction. *J. Gen. Physiol.* 14:517–528.

ROBERTSON, J. D. 1960. The Molecular Structure and Contact Relationship of Cell Membranes. *Prog. Biophys.* 10:343–418.

SHASHOUA, V. E. 1969. Electrically Active Protein and Polynucleic Acid Membranes, pp. 147–159. In D. C. Tosteson [ed.] *The Molecular Basis of Membrane Function.* Prentice-Hall, New Jersey.

SINGER, I., and I. TASAKI. 1968. Nerve Excitability and Membrane Macromolecules. Chapter 8. *In* D. Chapman [ed.] *Biological Membranes.* Academic Press, New York.

TASAKI, I., and S. HAGIWARA. 1957. Demonstration of Two Stable Potential States in the Squid Giant Axon under Tetraethylammonium Chloride. *J. Gen. Physiol.* 40:859–885.

TEORELL, T. 1957. On Oscillatory Transport of Fluid across Membranes. *Acta Soc. Med. Upsalien.* 62:60–66.

TILLE, J. 1965. A New Interpretation of the Dynamic Changes of the Potassium Conductance in the Squid Giant Axon. *Biophys. J.* 5:163–171.

WEI, L. Y. 1969. Molecular Mechanisms of Nerve Excitation and Conduction. *Bull. Math. Biophys.* 31:39–58.

WOBSCHALL, D. 1968. An Electret Model of the Nerve Membrane. *J. Theoret. Biol.* 21:439–448.

FIXED SURFACE CHARGES

18

D. L. GILBERT

The study of charges on the surface of biological membranes in recent years has become increasingly important in our understanding of how membrane surfaces influence various aspects of biological behavior. Thus, muscle contraction might be greatly influenced by a potential charged surface in the myosin filaments (Elliott, 1968). It has also been suggested that the surface of the muscle fiber contains fixed charges (Hagiwara and Takahashi, 1967; Lorković, 1967). The importance of fixed charges on subcellular particles has also been emphasized (McLaren, 1960). This chapter will attempt to show how surface charges might be responsible for some electrobiological phenomena of the membrane of nerve cells.

If a membrane has negative charges on its surface, then it would be expected that positive cations will be attracted to the surface and therefore the concentration of the cations at the surface of the membrane will be greater than in the bulk solution. Similarly, negative anions will have a lower concentration in the vicinity of the membrane than in the bulk solution. For a positively charged membrane, the roles of the cation and anion are reversed. The potential at the surface with respect to the potential in the bulk solution will be negative for a negatively charged membrane and will increase as a function of distance until it equals the potential in the bulk solution.

Attempts to measure this potential, in which there is a relative movement between the membrane surface and the bulk solution are called electrokinetic measurements. The calculated potential from these measurements is called the zeta potential. Since it occurs at the shearing plane, which occurs some distance from the membrane surface, the absolute magnitude of the zeta potential may not be as large as the absolute magnitude of the surface potential. A derivation of the zeta potential occurs at the end of this chapter.

359

THE GOUY-CHAPMAN THEORY

The Gouy-Chapman theory or the Diffuse Double Layer Theory is a quantitative treatment of the relationship between the potential and surface charge. It is assumed that ions are point charges and can reach the surface within any distance no matter how small the distance is. It is also assumed that the discrete charges on the membrane will have a negligible effect upon the assumption that the potential at the surface of the membrane is continuous in the neighborhood of measurement (Cole, 1969). The theory accounts only for the potential up to the plane of closest approach of the center of the ions. This plane has been designated as the outer Helmholtz plane. From the membrane to this plane is the region of the compact double layer, which is also called the Helmholtz double layer or the inner double layer. The Stern theory takes into account the compact double layer, but for simplicity, we will not discuss it. Several authors have discussed the limitations of the simplified treatment which we will present (Davies and Rideal, 1961; Delahay, 1965; Grahme, 1947; Haydon, 1964; Osipow, 1962; Overbeek, 1952a; Overbeek and Lyklema, 1959).

In this derivation, only the dimension perpendicular to the charged surface will be discussed. The relationship between the ion concentration at a perpendicular distance x from the membrane and in the bulk solution (where x approaches infinity) is given by (Gilbert, 1960):

$$\frac{c_i}{c_{x_i}} = r_x^{z_i}, \tag{18-1}$$

where i = ion species

$\quad\quad x$ = perpendicular distance from the membrane

$\quad\quad c_i$ = ion concentration in bulk solution of species i

$\quad\quad c_{x_i}$ = ion concentration at distance x from the membrane

$\quad\quad z_i$ = valence of ion

$\quad\quad r_x$ = Donnan ratio

The Donnan ratio is defined by:

$$r_x = e^{HV_x}, \tag{18-2}$$

where V_x = electrical potential at distance x minus electrical potential in bulk solution.

$$H = \frac{F}{RT}, \tag{18-3}$$

where $F =$ Faraday $= 96,500$ coul/mole

$R =$ gas constant $= 8.314$ joule/$°K \cdot$ mole

T $=$ temperature in $°K$.

Actually activities should be used instead of concentrations. If the total concentration is less than 10 mM then the Gouy-Chapman theory is valid using concentrations (Haydon, 1964). When the temperature is $10°$ C, then H $= 0.041$ mv^{-1}. The Boltzman factor is equal to the Donnan ratio raised to the power of z_i. We will need later to consider the boundary conditions of x.

When x equals zero, then $r_x = r$, $c_{x_i} = c_{M_i}$, $V_x = V$, and $dV_x/dx = dV/dx$. When x approaches infinity, then $r_x = 1$, $c_{x_i} = c_i$, $V_x = 0$ and $dV_x/dx = 0$.

The net volume charge density in solution is:

$$\rho = \sum_{i=1}^{n} z_i F c_{x_i}, \tag{18-4}$$

where $\rho =$ net charge density in solution.

n $=$ number of species in solution.

The relationship between potential and charge density is given by the Poisson equation, providing the dielectric constant is independent of the field strength (Buckingham, 1956):

$$\frac{d^2V_x}{dx^2} = -\frac{4\pi\rho}{\varepsilon D_o}, \tag{18-5}$$

where ε is the dielectric constant of water. The value of ε is 84.11 at $10°$ C. D_o is the permittivity of free space [(4π) $(8.85 \times 10^{-12}$ coul^2/newt $\cdot$ m^2)].

Derivation of Gauss' Law

Integration of Eq. (18-5) using the boundary conditions of x is:

$$\frac{dV}{dx} = \frac{4\pi}{\varepsilon D_o} \int_0^\infty \rho dx. \tag{18-6}$$

The integration assumes that the dielectric constant is constant even close to the membrane where it probably decreases (Grahme, 1950). However, this assumption will not greatly influence the potential, V, unless the surface charge is large (Bolt, 1955).

The total charge density in units of electronic charge/area is

$$\sigma_s = -\frac{N}{F} \int_0^\infty \rho dx = -\bar{\sigma}. \tag{18-7}$$

where σ_s = total diffuse double layer charge in the solution in units of electronic charge/area

$\bar{\sigma}$ = membrane surface charge density in units of electronic charge/area

N = Avogadro's number = 6.02×10^{23} electron/mole

Eq. (18-7) means that the charge on the membrane is neutralized by the charges in solution. Thus, there is no net charge in the system. If some of the charge is neutralized by charges within the membrane or on the other surface of the membrane, then correction factors have to be introduced into the theory (Chandler, Hodgkin and Meves, 1965; Haydon, 1961a). The conversion factor for coulombs to electronic charge is $-N/F$. Rearrangement of Eq. (18-7) is

$$\int_0^\infty \rho dx = \frac{F\bar{\sigma}}{N}. \tag{18-8}$$

Substitution of Eq. (18-8) into Eq. (18-6) leads to the Gauss equation:

$$\frac{dV}{dx} = \frac{4\pi F\bar{\sigma}}{\varepsilon D_o N}. \tag{18-9}$$

Relationship Between the Surface Charge and Surface Electrical Potential

In order to show this relationship, it is convenient to first point out the mathematical relations:

$$\frac{d\left[\frac{dV_x}{dx}\right]^2}{dx} = 2 \frac{dV_x}{dx} \cdot \frac{d\left[\frac{dV_x}{dx}\right]}{dx} = 2\left(\frac{dV_x}{dx}\right)\left(\frac{d^2V_x}{dx^2}\right). \tag{18-10}$$

Combining Eq. (18-1), (18-4), and (18-5) results in

$$\frac{d^2V_x}{dx^2} = -\frac{4\pi}{\varepsilon D_o} \sum_{i=1}^{n} z_i Fc_i r_x^{-z_i}. \tag{18-11}$$

Using Eq. (18-10) and (18-11) results in:

$$2dV_x\left(\frac{d^2V_x}{dx^2}\right) = d\left[\frac{dV_x}{dx}\right]^2 = -\frac{8\pi}{\varepsilon D_o} \sum_{i=1}^{n} z_i Fc_i r_x^{-z_i} dV_x. \tag{18-12}$$

Integration of Eq. (18-12) using the boundary conditions of x and Eq. (18-1), (18-2), and (18-3) produces:

$$\left[\frac{dV}{dx}\right]^2 = \frac{8\pi RT}{\varepsilon D_o} \sum_{i=1}^{n} [c_i(r^{-z_i} - 1)]. \tag{18-13}$$

In the integration, it is assumed that ε is constant, which, as mentioned, will not appreciably alter the result.

Squaring Eq. (18-9) and combining it with Eq. (18-3) and (18-13) yields Eq. (18-14), the Gouy-Chapman equation:

$$\bar{\sigma}^2 = \frac{1}{G^2} \sum_{i=1}^{n} [c_i(r^{-z_i} - 1)], \tag{18-14}$$

where

$$G = \frac{1}{N} \left[\frac{2\pi FH}{\varepsilon D_o} \right]^{1/2}. \tag{18-15}$$

When the surface charge is in units of electronic charge per square Angstrom and the concentration is in moles/liter, then at $10°\ C$, the value of G is 270 (Å^2/electronic charge)(mole/liter)$^{1/2}$. Eq. (18-14) shows how the bulk ionic concentration and the Donnan ratio, which is a function of the surface electrical potential, are related to the surface charge. If the absolute magnitude of the potential at the surface is greater than 100 mv, then the Gouy-Chapman theory has to be modified (Haydon, 1964).

Relationship of V_x to V

If the value of $z_i HV_x$ is small, then

$$\exp[-z_i HV_x] = 1 - z_i HV_x. \tag{18-16}$$

Combining Eq. (18-1), (18-2), (18-4), (18-5), and (18-16) together gives

$$\frac{d^2V_x}{dx^2} = -\frac{4\pi}{\varepsilon D_o} \sum_{i=1}^{n} z_i Fc_i(1 - z_i HV_x). \tag{18-17}$$

Electroneutrality in the solution requires that

$$\sum_{i=1}^{n} c_i z_i = 0. \tag{18-18}$$

Thus, Eq. (18-17) simplifies with the aid of Eq. (18-18) to:

$$\frac{d^2V_x}{dx^2} = \frac{4\pi FH}{\varepsilon D_o} V_x \sum_{i=1}^{n} c_i z_i^2. \tag{18-19}$$

Eq. (18-19) can be simplified by introducing another definition:

$$K^2 = \frac{4\pi FH}{\varepsilon D_o} \sum_{i=1}^{n} c_i z_i^2 = \frac{8\pi FH}{\varepsilon D_o} \mu, \tag{18-20}$$

where K = Debye reciprocal length

$$\mu = 0.5 \sum_{i=1}^{n} c_i z_i^2 = \text{ionic strength.} \tag{18-21}$$

Substitution of Eq. (18-20) into Eq. (18-19) is

$$\frac{d^2 V_x}{dx^2} = K^2 V_x. \tag{18-22}$$

Integration of Eq. (18-22) using the boundary conditions gives

$$V_x = V \, e^{-Kx}. \tag{18-23}$$

Thus, increasing the ionic strength increases the Debye reciprocal length and makes the electrical potential approach zero at small distances from the membrane.

The ion atmosphere radius is the reciprocal of the Debye reciprocal length. At $10°$ C the ion atmosphere radius equals 3.067 divided by the square root of the ionic strength where the ion atmosphere radius is in Å and μ is in moles/liter. For an ionic strength of 0.5 M, the ion atmosphere radius is 4.34 Å at $10°$ C.

REMOVAL OF SURFACE CHARGE BY A TITRATING ION

Gilbert and Ehrenstein (1969) have assumed that the external surface of the squid axon contains negative fixed charged sites which are neutralized by the addition of divalent cations, such as calcium and magnesium. This section is taken from their expanded model (Gilbert and Ehrenstein, 1970).

Addition of the neutralizing cation produces a shift in the surface electrical potential which should also produce the same electrical potential shift of the electrical potential dependent conductance parameters. This condition is expressed in Eq. (18-24):

$$V = V_{1/2} - B, \tag{18-24}$$

where $V_{1/2}$ is a measured electrical potential at a given value of an electrical potential dependent conductance parameter, and B is an arbitrary electrical potential, which is constant and dependent upon how $V_{1/2}$ is measured.

An alternate way of expressing the surface charge is

$$\bar{\sigma} = \frac{1}{A_e} = \frac{1}{d^2}, \tag{18-25}$$

where A_e is the average area/electronic charge and d is the average spacing between electronic charges. It is assumed that the sites are neutralized by the simple relationship:

$$M_M + S \rightleftarrows MS, \qquad (18\text{-}26)$$

where S = free negatively charged site concentration
M_M = neutralizing cation concentration at the membrane
MS = neutralized site concentration.

The equilibrium constant, K, which is the reciprocal of the dissociation constant, for reaction (18-26) is:

$$K = \frac{(MS)}{(M_M)(S)} . \qquad (18\text{-}27)$$

If the titrating cation is the hydrogen ion, then the pK is equal to the negative logarithm of the dissociation constant. Hence, the pK is equal to the logarithm of K, when K is in units of M^{-1}.

The relation of the cation concentration at the membrane surface to the cation concentration in the bulk solution is:

$$M_M = Mr^{-z}, \qquad (18\text{-}28)$$

where M is the neutralizing cation concentration in the bulk solution and z is the valence of M. The total concentration of sites is

$$S_t = S + MS, \qquad (18\text{-}29)$$

where t refers to the state when all the sites are in the charged form and S_t is the total concentration of sites. When the value of MS is zero, then S_t equals S. Since the surface charge is due only to the charged sites, it follows that:

$$\frac{\bar{\sigma}}{\bar{\sigma}_t} = \frac{S}{S_t}, \qquad (18\text{-}30)$$

where $\bar{\sigma}_t$ is the maximum surface charge.
Combining Eq. (18-27) and (18-29) gives

$$\bar{\sigma} = \frac{\bar{\sigma}_t}{KM_M + 1} . \qquad (18\text{-}31)$$

The relationship between the cation concentration in the bulk solution and the measured electrical potential, $V_{1/2}$, is obtained by combining Eq. (18-3), (18-14), (18-24), (18-25), (18-28) and (18-31):

$$d_t^2(KM\exp[zH(B-V_{1/2})] + 1) = \frac{G}{[\sum_{i=1}^{n} c_i(\exp[z_iH(B-V_{1/2})] - 1)]^{1/2}} \qquad (18\text{-}32)$$

where d_t is the minimum spacing between electronic charges. Experimentally, there are three unknowns, B, K, and dt in Eq. (18-32). At least three different values of the neutralizing cation concentration and the corresponding values of $V_{1/2}$ must be obtained in order to solve Eq. (18-32).

Influence of the Three Unknown Parameters in Eq. (18-32)

Eq. (18-32) is best understood by plotting $V_{1/2}$ against the logarithm of the titrating bulk cation concentration. The parameter B is merely a reference electrical potential and only shifts the potential axis up or down on this graph.

Fig. 18-1 illustrates that changing the value of the equilibrium constant,

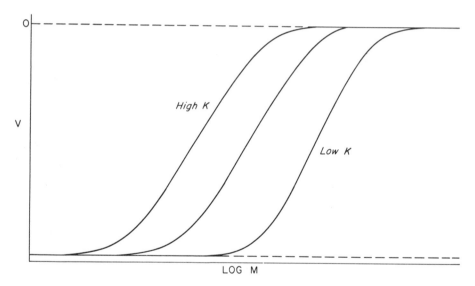

Fig. 18-1 *Effect of equilibrium constant.*

K, shifts the curve to the right or left. When the equilibrium constant is large, the sites become saturated at low values of the titrating cation concentration and the curve is shifted to the left. If the sites become saturated at high values of the titrating cation concentration (i.e., the equilibrium constant is small), then the effect of the large titrating cation concentrations will increase the c_i terms on the right hand side of Eq. (18-32) and the slope will be increased. This increase in the slope will be minimal, however, unless the titrating cation concentrations are very large in relation to the other ions in the solution.

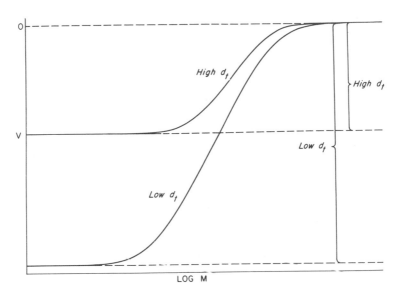

Fig. 18-2. *Effect of charge separation.*

Fig. 18-2 shows that the minimum spacing between electronic charges, d_t, determines the maximum electrical potential shift and the slope. As stated above, the slope is also slightly influenced by the equilibrium constant, when the equilibrium constant is small.

The dependence of the maximum electrical potential shift on the average fixed charge separation is illustrated in Fig. 18-3 for the case when practically all the solution is composed of uni-univalent salt solution of concentration c in the bulk phase. In this situation for a uni-univalent ionic solution, Eq. (18-32) simplifies to:

$$d_t^2(K M \exp[zH(B - V_{1/2})] + 1) = \frac{G}{2c^{1/2}\sinh\left[\dfrac{H(B - V_{1/2})}{2}\right]}, \qquad (18\text{-}33)$$

where c = bulk cation concentration = bulk anion concentration.

The maximum value of $V_{1/2}$ occurs when the titrating cation concentration is high enough to neutralize the fixed negative charges; then $V_{1/2}$ equals B. The minimum value of $V_{1/2}$ occurs when the titrating bulk cation concentration is zero. Then from Eq. (18-33) it can be seen that the maximum electrical potential shift is:

$$V_{max} = \frac{2}{H}\sinh^{-1}\frac{G}{2c^{1/2}d_t^2}. \qquad (18\text{-}34)$$

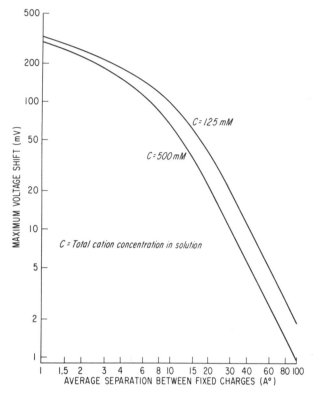

Fig. 18-3 *Dependence of maximum voltage shift on average fixed charge separation.*

V_{max} is the maximum electrical potential shift. Notice that in Fig. 18-3 and also in Eq. (18-34) that an increase in c decreases the maximum electrical potential shift. Chandler, Hodgkin and Meves (1965) determined the internal surface charge of the squid axon by changing c and measuring the voltage shift. It should be mentioned that changing the surface charge on one side of the membrane will have little effect on the other side (Chandler, Hodgkin and Meves, 1965).

Substitution of Eq. (18-24) into Eq. (18-33) and subsequent differentiation yields:

$$\frac{dV}{d(\ln M)} = \frac{1}{zH - \dfrac{H \coth\left(\dfrac{H}{2}V\right)}{2\left[1 + \dfrac{2d_t{}^2c^{1/2}\sinh\left(\dfrac{H}{2}V\right)}{G}\right]}} \qquad (18\text{-}35)$$

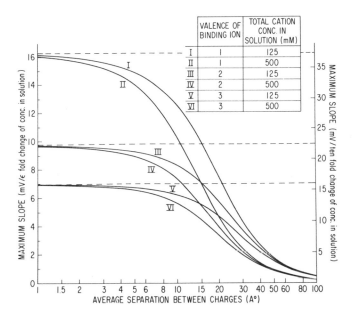

VALENCE OF BINDING ION	TOTAL CATION CONC. IN SOLUTION (mM)	
I	1	125
II	1	500
III	2	125
IV	2	500
V	3	125
VI	3	500

Fig. 18-4 *Dependence of maximum rate of change of voltage shift on average fixed charge separation. (From Gilbert and Ehrenstein, 1970.)*

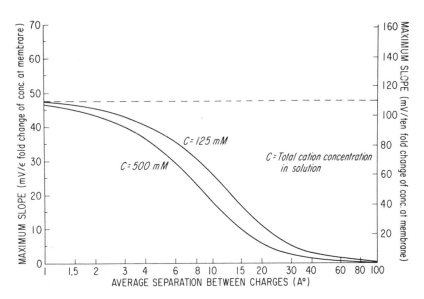

Fig. 18-5 *Dependence of maximum rate of change of voltage shift with change of membrane concentration on average fixed charge separation.*

Notice that Eq. (18-35) gives the slope of the graphs in Fig. 18-1 and 18-2. Notice also from these figures that the slope is relatively constant over a considerable range and equals the maximum value of the slope. Fig. 18-4 illustrates the dependence of maximum rate of change of electrical potential shift on the average fixed charge separation for an uni-univalent salt solution of concentration equal to c. Observe that in Eq. (18-35), the slope is not a function of the equilibrium constant. Hence, if only the maximum slope is known, it is possible to determine the average charge separation. Notice that the maximum slope increases with a decrease in the ionic concentration and also with a decrease in the valence of the titrating cation (Gilbert and Ehrenstein, 1970). If the titrating cation concentration at the membrane surface is used instead of the bulk titrating ion concentration, then the slope of the resulting graph (Fig. 18-5) is increased and is independent upon the valence of the titrating cation.

Application of the Model

A shift in the potassium conductance curves of squid axons immersed in isosmotic solutions of potassium chloride containing various amounts of divalent cation was observed by Gilbert and Ehrenstein (1969). The data

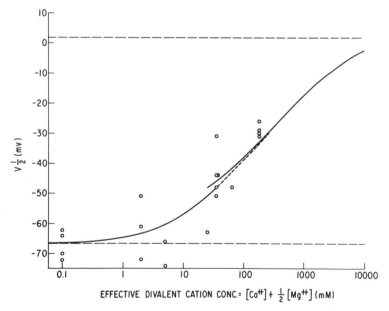

Fig. 18-6 *Theoretical fit of experimental voltage shifts for various divalent cation concentrations. The lower solid curve was obtained for an ionic concentration corresponding to the solutions containing a low divalent cation concentration. The upper solid curve was obtained for an ionic concentration corresponding to the solutions containing a high divalent cation concentration. (From Gilbert and Ehrenstein, 1969.)*

points were fitted by this model as shown in Fig. 18-6. The average charge separation was found to be 11 Å and the equilibrium constant was equal to 0.0001 mM^{-1}.

Previously, Frankenhaeuser and Hodgkin (1957) also observed a voltage shift of the sodium conductance due to calcium ions, which according to the model is consistent with an average charge separation of 11 Å.

Gilbert and Ehrenstein (1970) also analyzed the data of Hille (1968) on the voltage shift of the sodium conductance parameter, m_∞, as influenced by pH in the frog node. These results are shown in Fig. 18-7. The analysis provided

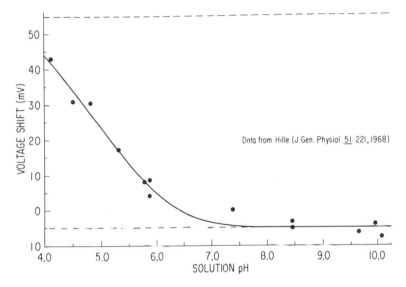

Fig. 18-7 *Effect of pH on voltage shift of m_∞. (From Gilbert and Ehrenstein, 1970.)*

an average charge separation of 15 Å and an equilibrium constant of 39.8 mM^{-1}, which corresponds to a pK of 4.6.

For the internal membrane surface of the squid nerve, Chandler, Hodgkin and Meves (1965) found that by changing the internal ionic strength, the average charge separation was 27 Å.

These results are consistent with an internal surface electrical potential of -13 mv and an external surface electrical potential of -46 mv for the squid axon. Thus, as illustrated in Fig. 18-8, a measured resting potential of -65 mv indicates that the controlling electrical potential across the membrane of the squid giant axon is only -32 mv. Ions might affect the inner membrane surface charge (point *b*, Fig. 18-8) or the outer membrane surface charge (point *c*) without affecting the potential from point *a* to point *b*. Hence, ions might regulate the controlling potential for ionic conductance without necessarily influencing the resting potential.

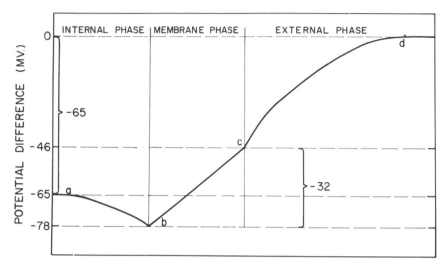

Fig. 18-8 *Potential difference across squid axon membrane. The distance is not drawn to scale.*

ZETA OR ELECTROKINETIC POTENTIALS

The electrokinetic potential, as measured by producing some motion of the membrane surface with respect to the diffuse double layer in the solution, occurs at the plane of shear which is some distance away from the surface. As a first approximation, we will assume that the plane of shear is at the surface, and then the zeta or electrokinetic potential will be the same as the surface potential.

The zeta potential can be measured by applying an electric field and observing the motion of a particle with a charged surface (electrophoresis) or by observing the motion of the fluid past a stationary charged surface (electroosmosis). Alternately, a potential can be measured by applying some nonelectrical force to a system in which either the particle with a charged surface can move (sedimentation potential, or migration potential, or Dorn potential) or the fluid can move past a stationary charged surface (streaming potential). In the next section, we will derive the fundamental equation which gives the relation of the velocity in either electrophoresis or electroosmosis to the zeta potential. The treatment will be a simple one, and for more details, limitations and discussion of a more sophisticated approach, other works can be consulted (Davies and Rideal, 1961; Jirgensons and Straumanis, 1962; MacInnes, 1961; Osipow, 1962; Overbeek, 1952b; Overbeek and Wiersema, 1967).

If one knows the electrical conductance in the diffuse double layer, i.e., surface conductance, and the conductance in the bulk solution, then it is possible to show the relation of the zeta potential to either the sedimentation potential or the streaming potential (Davies and Rideal, 1961).

Derivation of the Helmholtz-Smoluchowski Equation for Electrophoresis

In electrophoresis, the frictional and electrical forces act upon the diffuse double layer. The frictional force is:

$$f_F = \eta A \frac{dv_x}{dx} ,$$ (18-36)

where f_F = frictional force in the diffuse double layer

η = coefficient of viscosity

A = area, which is perpendicular to the direction of x

v_x = velocity which is perpendicular to the surface and at a distance x from the membrane. It is the velocity of the liquid with respect to the surface.

Differentiation of Eq. (18-36) gives:

$$\frac{df_F}{dx} = \eta A \frac{d^2 v_x}{dx^2} .$$ (18-37)

Eq. (18-38) next considers the electrical force in the diffuse double layer

$$\frac{df_E}{dx} = EA\rho,$$ (18-38)

where f_E = electrical force in the diffuse double layer
E = applied electric field in a direction opposite to v_x for a positively charged surface

The sum of the electrical plus the frictional forces is equal to zero and hence, Eq. (18-37) plus Eq. (18-38) can be set equal to zero resulting in, after rearrangement,

$$E\rho = -\eta \frac{d^2 v_x}{dx^2} .$$ (18-39)

Substitution of Eq. (18-5) into Eq. (18-39 yields

$$\frac{d^2 V_x}{dx^2} = \frac{4\pi\eta}{\varepsilon D_o E} \frac{d^2 v_x}{dx^2}.$$ (18-40)

Eq. (18-40) is next integrated between the values of $x = x$ and $x \to \infty$. When $x \to \infty$, $dV_x/dx = 0$ and $dv_x/dx = 0$, so it follows that

$$\frac{dV_x}{dx} = \frac{4\pi\eta}{\varepsilon D_o E} \frac{dv_x}{dx}. \tag{18-41}$$

A further integration is next performed using the boundary conditions of $x = 0$ and $x \to \infty$. When $x = 0$, then $V_x = V$ and $v_x = 0$. When $x \to \infty$, then $V_x = 0$ and $v_x = v$. Thus, the integration gives

$$V = -\frac{4\pi\eta v}{\varepsilon D_o E}. \tag{18-42}$$

Actually, the value of x when v_x equals zero is defined as the shearing plane and the potential at the shearing plane is the zeta potential. It is assumed here that at the plane of shear the value of x is approximately equal to zero. It is also assumed in the integrations of Eq. (18-41) and (18-42) that the ratio of the viscosity to the dielectric constant is the same in the diffuse double layer as it is in the bulk solution. If the surface charge is not greater than 0.00625 electronic charges per square angstrom, then this assumption is valid (Haydon, 1960).

The velocity of the particle with the charged surface moves in the opposite direction to the liquid element we are considering, so

$$v = -v, \tag{18-43}$$

where v is the velocity of the particle. Substitution of Eq. (18-43) into Eq. (18-42) yields the Helmholtz-Smoluchowski equation

$$V = \frac{4\pi\eta}{\varepsilon D_o} \frac{v}{E} \tag{18-44}$$

At $10°$ C, the coefficient of viscosity is 0.0013077 kg/m-sec, and hence, the zeta potential in volts equals 1.757×10^6 times the velocity in m/sec divided by the electric field in volts/m. At $25°$ C, the factor 1.757×10^6 is changed to 1.287×10^6 since the coefficient of viscosity is 0.008937 kg/m-sec and the dielectric constant is 78.49. The term, electrophoretic mobility, is often used and is defined as the velocity of the particle divided by the applied electric field.

Eq. (18-44) can also be used for the phenomenon of electroosmosis providing that v is now defined as the liquid velocity in a direction opposite to that of the applied electric field.

It is possible to modify Eq. (18-44) with some additional assumptions to make it applicable for giving the dependence of cither the streaming potential or the migration potential upon the zeta potential where a nonelectrical force produces the velocity.

If the bathing solution is composed almost entirely by a uni-univalent salt solution, then the zeta potential will equal the maximum electrical potential shift in Eq. (18-34) and Fig. 18-3.

Zeta Potential Measurements of Cells

The zeta potential of a number of biological surfaces has been measured by electrophoresis (Abramson, Moyer and Gorin, 1964; Brinton and Lauffer, 1959); Vassar et al., 1967). Elul (1967) has measured many cells, such as muscle fibers and neurones in tissue culture and observed that the surface charge density is about equal to the charge density on the red cell, which corresponds to about one electronic charge per 2000 square Angstroms. Under the appropriate conditions, the absolute magnitude of the zeta potential is increased by decreasing the ionic strength (see Fig. 18-3) or by increasing the pH in the red cell (Heard and Seaman, 1960). Thus, hydrogen ions appear to neutralize the charged groups as in our model. Titration of the charged groups on the membrane has actually caused reversal of the zeta potential on some surfaces of cells (Bangham, Pethica and Seaman, 1958; Dan, 1947). The effect of enzymes on the zeta potential has been used to determine more about the nature of the surface charge on red cells (Haydon and Seaman, 1967; Seaman, Jackson and Uhlenbruck, 1967).

Segal (1968) observed that the zeta potential of the external surface of the squid axon was approximately −1 mv. Examination of Fig. 18-3, assuming a concentration of about 500 mM, indicates that this value would correspond to about one electronic charge per 8000 to 9000 square Angstroms. He made similar measurements on the external surface of the lobster axon and found that the surface charge density was greater than for the squid axon, corresponding to about one electronic charge per 3500 to 4000 square Angstroms. However, Segal's measurements most probably reflect the effect of the connective tissue and Schwann cells.

In any event, it might be that the external charge density near the conduction sites of sodium and potassium on the squid axon membrane is greater than on other parts of the membrane surface. Haydon (1961b) has postulated that perhaps for the surface of the micro-organism there might be only a small number of ionizing groups on a large area of nonionogenic material.

SUMMARY

A surface charge model has been presented to explain some of the observed voltage shifts of some of the electrical potential dependent conductance parameters of the nerve fiber. According to this model, the surface charge density can be neutralized by some cations.

The surface charge density for the average surface of the red cell membrane is much smaller than the surface charge density surrounding the sodium and potassium sites of conduction. Perhaps, the axon also has a low surface charge density except near these conduction sites. At these sites, it appears that the electric potential difference across the membrane is much smaller than the electrical potential difference measured between the bulk phases of the external and internal environments. The electric potential difference across the membrane at the sites is the controlling potential for the ionic conductances and can be regulated by ions without necessarily affecting the measured membrane potential.

ACKNOWLEDGMENT

Acknowledgment is given to Dr. Gerald M. Ehrenstein with whom the author has had helpful discussions upon the content of this chapter.

REFERENCES

ABRAMSON, H. A., L. S. MOYER and M. H. GORIN. 1964. *Electrophoresis of Proteins and the Chemistry of Cell Surfaces*. Hafner Publ. Co., New York.

BANGHAM, A. D., B. A. PETHICA and G. V. F. SEAMAN. 1958. The Charged Groups at the Interface of Some Blood Cells. *Biochem. J.* 69:12.

BOLT, G. H. 1955. Analysis of the Validity of the Gouy-Chapman Theory of the Electric Double Layer. *J. Colloid Sci.* 10:206.

BRINTON, C. C., Jr., and M. A. LAUFFER. 1959. The Electrophoresis of Viruses, Bacteria, and Cells, and the Microscope Method of Electrophoresis. *In* C. C. Brinton, Jr. and M. A. Lauffer [ed.] *Electrophoresis. Theory, Methods, and Applications*. Academic Press, Inc., New York, p. 427.

BUCKINGHAM, A. D. 1956. Theory of the Dielectric Constant at High Field Strengths. *J. Chem. Phys.* 25:428.

CHANDLER, W. K., A. L. HODGKIN and H. MEVES. 1965. The Effect of Changing the Internal Solution on Sodium Inactivation and Related Phenomena in Giant Axons. *J. Physiol.* 180:821.

COLE, K. S. 1969. Zeta Potential and Discrete vs. Uniform Surface Charges. *Biophys. J.* 9:465.

DAN, K. 1947. Electrokinetic Studies of Marine Ova. VI. The Effect of Salts on the Zeta Potential of the Eggs of Strongylocentrotus Pulcherrimus. *Biol. Bull.* 93:267.

DAVIES, J. T., and E. K. RIDEAL. 1961. *Interfacial Phenomena*. Academic Press, New York.

DELAHAY, P. 1965. *Double Layer and Electrode Kinetics*. Interscience Publ., John Wiley & Sons, Inc., New York.

ELLIOTT, G. F. 1968. Force-Balances and Stability in Hexagonally-Packed Poly-electrolyte Systems. *J. Theor. Biol.* 21:71.

ELUL, R. 1967. Fixed Charge in the Cell Membrane. *J. Physiol.* 189:351.

FRANKENHAEUSER, B., and A. L. HODGKIN. 1957. The Action of Calcium on the Electrical Properties of Squid Axons. *J. Physiol. (London).* 137:218.

GILBERT, D. L. 1960. Relationship between Ion Distribution and Membrane Potential during a Steady State. *Bull. Math. Biophys.* 22:323.

GILBERT, D. L., and G. EHRENSTEIN. 1969. Effect of Divalent Cations on Potassium Conductance of Squid Axons: Determination of Surface Charge. *Biophys. J.* 9:447.

_____. 1970. Use of a Fixed Charge Model to Determine the pK of the Negative Sites on the External Membrane Surface. *J. Gen. Physiol.* 55:822.

GRAHME, D. S. 1947. The Electrical Double Layer and the Theory of Electro-capillarity. *Chem. Rev.* 41:441.

GRAHME, D. 1950. Effects of Dielectric Saturation upon the Diffuse Double Layer and the Free Energy of Hydration of Ions. *J. Chem. Physics.* 18:903.

HAGIWARA, S., and K. TAKAHASHI. 1967. Surface Density of Calcium Ions and Calcium Spikes in the Barnacle Muscle Fiber Membrane. *J. Gen. Physiol.* 50:583.

HAYDON, D. A. 1960. A Study of the Relation between Electrokinetic Potential and Surface Charge Density. *Proc. Roy. Soc. A.* 258:319.

_____. 1961a. The Surface Charge of Cells and Some Other Small Particles as Indicated by Electrophoresis. I. The Zeta Potential-Surface Charge Relation-ships. *Biochim. Biophys. Acta.* 50:450.

_____. 1961b. The Surface Charge of Cells and Some Other Small Particles as Indicated by Electrophoresis. II. The Interpretation of Electrophoretic Charge. *Biochim. Biophys. Acta.* 50:457.

_____. 1964. The Electrical Double Layer and Electrokinetic Phenomena. *Recent Prog. Surface Sci.* 1:94.

HAYDON, D. A., and G. V. F. SEAMAN. 1967. Electrokinetic Studies on the Ultra-structure of the Human Erythrocyte. I. Electrophoresis at High Ionic Strengths —The Cell as a Polyanion. *Arch. Biochem. Biophys.* 122:126.

HEARD, D. H., and G. V. F. SEAMAN. 1960. The Influence of pH and Ionic Strength on the Electrokinetic Stability of the Human Erythrocyte Membrane. *J. Gen. Physiol.* 43:635.

HILLE, B. 1968. Charges and Potentials at the Nerve Surface. Divalent Ions and pH. *J. Gen. Physiol.* 51:221.

JIRGENSONS, B., and M. E. STRAUMANIS. 1962. *A Short Textbook of Colloid Chem-istry*. The Macmillan Co., New York. 2nd edition.

LORKOVIĆ, H. 1967. The Influence of Ionic Strength on Potassium Contractures and Calcium Movements in Frog Muscle. *J. Gen. Physiol.* 50:883.

MACINNES, D. A. 1961. *The Principles of Electrochemistry*. Dover Publ., Inc., New York.

McLaren, A. D. 1960. Enzyme Action in Structurally Restricted Systems. *Enzymologia* 21:356.

Osipow, L. I. 1962. *Surface Chemistry. Theory and Industrial Applications.* Reinhold Publ. Corp., New York.

Overbeek, J. T. G. 1952a. Electrochemistry of the Double Layer, p. 115. *In* H. R. Kruyt [ed.] *Colloid Chemistry.* Volume I. *Irreversible Systems.* Elsevier Publ. Co., New York.

———. 1952b. Electrokinetic Phenomena, p. 194. *In* H. R. Kruyt [ed.] *Colloid Chemistry.* Volume I. *Irreversible Systems.* Elsevier Publ. Co., New York.

Overbeek, J. T. G., and J. Lyklema. 1959. Electric Potentials in Colloidal Systems. p. 1. *In* M. Bier [ed.] *Electrophoresis: Theory, Methods, and Applications.* Academic Press Inc., New York.

Overbeek, J. T. G., and P. H. Wiersema. 1967. The Interpretation of Electrophoretic Mobilities, p. 1. *In* M. Bier [ed.] *Electrophoresis: Theory, Methods, and Applications.* Volume II. Academic Press, New York.

Seaman, G. V. F., L. J. Jackson and G. Uhlenbruck. 1967. Action of α-Amylase Preparations and some Proteases on the Surface of Mammalian Erythrocytes. *Arch. Biochem. Biophys.* 122: 605.

Segal, J. R. 1968. Surface Charge of Giant Axons of Squid and Lobster. *Biophys. J.* 8: 470.

Vassar, P. S., G. V. F. Seaman, W. L. Dunn and L. Kanke. 1967. Electrokinetic Properties of Nuclear Surfaces. A Comparison of Nuclei from Normal and Regenerating Rat Liver. *Biochim. Biophys. Acta.* 135: 218.

ION EXCHANGE PROPERTIES AND EXCITABILITY OF THE SQUID GIANT AXON

19

L. D. CARNAY and I. TASAKI

The excitable membrane of a nerve cell is considered to be composed of macromolecular complexes of proteins and phospholipids (Hodgkin, 1964; Lenard and Singer, 1966; Rouser et al., 1968). In the physiological pH range the amphoteric nature of such macromolecules results in a large excess of negative charge, probably within the external membrane layer (Tasaki, Singer and Takenaka, 1965). This excess of negative charge thus can confer upon the excitable membrane the property of a cation exchanger: the ability to exchange equivalent amounts of cations with the electrolyte media contiguous with the membrane surfaces. The theory describing ion exchange processes in membranes has greatly evolved since 1935, when Teorell first proposed a quantitative theory for charged membranes (Teorell, 1935; Helfferich, 1962). A familiarity with the basic principles of ion exchange membranes is essential to the understanding of how such factors as ion mobilities, selectivity coefficients and fixed charge densities can affect the movement of ions across a biological membrane with fixed charge.

It is reasonable to assume that the factors which influence the interaction of a charged membrane with its ionic environment can also play an important role in the mechanism responsible for the abrupt changes in membrane resistance and potential associated with the process of nerve excitation. Therefore, we shall first present a brief description of the electrochemistry of charged membranes. Then, we shall discuss some electrical and optical data obtained from giant axons of squid which are consistent with the view

379

that the process of excitation involves a rapid, reversible cation exchange process (Tasaki, 1968; Singer and Tasaki, 1968; Lerman, Watanabe and Tasaki, 1969).

PROPERTIES OF IONIC MEMBRANES

Transmembrane Fluxes

When two aqueous solutions containing an uncharged substance, species i, at different concentrations are brought into contact with each other across a liquid junction, the direction and magnitude of the flux of species i across the liquid junction is governed by the chemical potential of the species. The chemical potential of species i is defined as the rate of change of Gibbs free energy of a system with respect to the number of moles n_i when the temperature, pressure, and number of moles of other components are held constant (Daniels and Alberty, 1961). For a nonideal solution the chemical potential, μ_i, is described by the equation:

$$\mu_i = \mu_i^0 + RT \ln a_i, \qquad (19\text{-}1)$$

where μ_i^0 is the reference value of the chemical potential of i, a_i is the activity of species i, R is the gas constant and T is the absolute temperature. Substances spontaneously diffuse from regions of high chemical potential to regions of low chemical potential, and thus, the driving force for the transfer of species i across the liquid junction is the gradient of the value of μ_i within the junction. The quantity of species i diffusing per unit time through a unit area of the liquid junction is denoted by j_i, which is related to the chemical potential by the equation:

$$j_i = -u_i c_i \,\text{grad}\, \mu_i = -u_i c_i RT \frac{d \ln a_i}{dx}. \qquad (19\text{-}2)$$

The proportionality coefficient u_i is the mobility of species i, c_i is the concentration of i at point x in the solution, and x represents the coordinate in the direction of the flux.

If the uncharged species i in the above example is replaced by a salt composed of the univalent ions, cation k and anion a, the tendency of each ionic species to spontaneously diffuse across the liquid junction is no longer determined only by the chemical potential gradient of the individual species. The two fluxes, j_k and j_a, are not independent of each other. When no external electric field is applied, i.e., when there is no net current across the junction, the two fluxes are bound by the condition:

$$j_k - j_a = 0. \qquad (19\text{-}3)$$

In other words, the flux of the cations must equal the flux of the anions across the junction even when the mobilities of the two ion species are very different. The ion of greater mobility tends to move faster across the junction, but, as a minute separation between k and a is produced, the electrostatic forces of attraction between oppositely charged ions prevent any further separation. The resultant electric field slows down the faster ion and speeds up the slower ion so that no net transfer of charge occurs as the salt diffuses. It is important to note that the actual separation of charge across the liquid junction is in the order of molecular dimensions and involves only an unmeasurably small amount of ions. Any deviation from electroneutrality at the junction occurs within 10^{-9} sec; after this period, electroneutrality conditions become applicable (Hafeman, 1965).

The influence of an electric field on species k and a in each of the two solutions is described by the addition to the chemical potential of a term which depends on the electric potential of the solution. Thus, the electrochemical potential of a charged species, $\tilde{\mu}_k$, in a solution is given by the equation:

$$\tilde{\mu}_k = \mu_k^0 + RT \ln a_k + zF\phi, \qquad (19\text{-}4)$$

where z is the valence of species k, F is the Faraday constant and ϕ is the electric potential. The driving force determining the flux of each charged species across the junction is the gradient of the electrochemical potential, and the flux equation describing the diffusion of species k is given by the Nernst-Planck equation:

$$j_k = - u_k c_k \frac{d\tilde{\mu}_k}{dx}$$

or $\qquad\qquad\qquad\qquad\qquad\qquad\qquad\qquad\qquad\qquad\qquad\qquad (19\text{-}5a)$

$$j_k = - u_k c_k \left[RT \frac{d \ln a_k}{dx} + z_k F \frac{d\phi}{dx} \right].$$

The flux of the anion species a across the liquid junction is also given by the Nernst-Planck equation:

$$j_a = - u_a c_a \left[RT \frac{d \ln a_a}{dx} + z_a F \frac{d\phi}{dx} \right]. \qquad (19\text{-}5b)$$

When a membrane containing macromolecules with a surplus of fixed negative charge is introduced between two dilute electrolyte solutions of the type $k_1 a$ and $k_2 a$, the effect of fixed charge on the flux of ions across the membrane has to be considered. If the concentration of the fixed charge within the membrane is significantly greater than anion concentrations in the solutions contiguous with the membrane, the anions are mostly excluded from the membrane. There is, however, a continuous interdiffusion of the cation species across the membrane. The net negative charge of the membrane must be balanced by the counter ions (cations k_1 and k_2) from the solutions. If the concentration of fixed charge is represented by $\bar{x}$, and the concentration of the two cations in the membrane by $\bar{c}_{k_1}$ and $\bar{c}_{k_2}$, then the condition of electroneutrality is given by:

$$\bar{c}_{k_1} + \bar{c}_{k_2} = \bar{x}, \tag{19-6}$$

where $\bar{c}_{k_1}$ and $\bar{c}_{k_2}$ vary with the coordinate normal to the surface of the membrane. One of the implications of this relationship is that the total concentration of the cations in the membrane remains essentially unaffected by changes in the concentration of these cations in the solutions contiguous with the membrane surfaces. Thus, the concentration of the cations in the system is discontinuous at the boundaries between the charged membrane and the solutions. (The significance of this discontinuity on the equations describing membrane potential will be discussed later.) Also, the ratio of $\bar{c}_{k_1}$ to $\bar{c}_{k_2}$ in the membrane is usually different from the ratio of these ions in the bulk solutions, and is determined by the selectivity of the negative sites in the membrane for oppositely charged (counter) ions. In such a model this property of membrane selectivity is mainly due to the electrostatic (coulombic and other) interactions between the negative sites and the cations. Therefore, such factors as size (ionic radius and hydration number) and valence of the cations influence the preference of the membrane for a particular cation (Helfferich, 1962). For example, a divalent cation is preferred over a univalent cation, and a univalent cation with a smaller outer shell of water molecules (i.e., K-ions) is usually preferred over a cation which has a smaller ionic radius but a larger hydration number (i.e., Na-ions).

The selectivity of the membrane with two univalent cations, k_1 and k_2, may be determined in the following manner. If the surface layer of the fixed charge membrane is in equilibrium with the contiguous bulk solution, the electrochemical potentials of ions k_1 and k_2 in the membrane at the surface are equal to the electrochemical potentials of k_1 and k_2 in the solution. Therefore, the relationship between the concentrations $\bar{c}_{k_1}$ and $\bar{c}_{k_2}$ in each surface layer of the membrane, and the activities of these cations of the same valence in the contiguous solution, may be represented by

$$\frac{\bar{c}_{k_2}(0)}{\bar{c}_{k_1}(0)} = K_{k_1}^{k_2}\frac{a_{k_2}{}'}{a_{k_1}{}'} \; ; \; \frac{\bar{c}_{k_2}(\delta)}{\bar{c}_{k_1}(\delta)} = K_{k_1}^{k_2}\frac{a_{k_2}{}''}{a_{k_1}{}''}, \tag{19-7}$$

where the two surfaces are denoted by $x = 0$ and $x = \delta$. $K_{k_1}^{k_2}$, the proportionality coefficient, describes the selectivity of the membrane for ion k_2 over ion k_1, and is greater than 1 if k_2 is preferred by the membrane.

The fluxes of ions k_1 and k_2 *within* the membrane can be described by the Nernst-Planck relation:

$$j_{k_1} = - \bar{u}_{k_1} \bar{c}_{k_1} \left[RT \frac{d \ln \bar{a}_{k_1}}{dx} + F \frac{d\phi}{dx} \right];$$

(19-8)

$$j_{k_2} = - \bar{u}_{k_2} \bar{c}_{k_2} \left[RT \frac{d \ln \bar{a}_{k_2}}{dx} + F \frac{d\phi}{dx} \right],$$

where $\bar{u}_{k_1}$ and $\bar{u}_{k_2}$ are the mobilities, $\bar{c}_{k_1}$ and $\bar{c}_{k_2}$ are the concentrations of k_1 and k_2 at point x in the membrane and $\bar{a}_{k_1}$ and $\bar{a}_{k_2}$ are ion activities in the membrane. As in the example of the diffusion of a cation and anion across a liquid junction, it is important to recognize that the fluxes of the cations within a negatively charged membrane are bound by the conditions of no net transfer of charge, $j_{k_1} + j_{k_2} = 0$, as well as the condition of electroneutrality, $\bar{c}_{k_1} + \bar{c}_{k_2} = \bar{x}$. These conditions mean that the fluxes of the cations j_{k_1} and j_{k_2} *are not* due to individual ions moving *independently* of each other across the fixed charge membrane. Assuming the mobilities $\bar{u}_{k_1}$ and $\bar{u}_{k_2}$ and selectivity $K_{k_1}^{k_2}$ are constant throughout the membrane, the Nernst-Planck equations may be solved to obtain the interdiffusion flux j of the cations k_1 and k_2 within the membrane, $[j = j_{k_1} = -j_{k_2}]$

$$j = \frac{RT\bar{x}\bar{u}_{k_1}\bar{u}_{k_2}}{(\bar{u}_{k_1} - \bar{u}_{k_2})\delta} \ln \frac{\left(a_{k_1}{}'' + K_{k_1}^{k_2} \frac{\bar{u}_{k_2}}{\bar{u}_{k_1}} a_{k_2}{}'' \right) \left(a_{k_1}{}' + K_{k_1}^{k_2} a_{k_2}{}' \right)}{\left(a_{k_1}{}' + K_{k_1}^{k_2} \frac{\bar{u}_{k_2}}{\bar{u}_{k_1}} a_{k_2}{}' \right) \left(a_{k_1}{}'' + K_{k_1}^{k_2} a_{k_2}{}'' \right)}$$

(19-9)

where $a_{k_1}{}'$, $a_{k_1}{}''$, $a_{k_2}{}'$ and $a_{k_2}{}''$ are the activities of the individual cations in the two bulk solutions.

This equation is greatly simplified in the special case where the activity of k_1 is equal to 0 in the solution on one side of the membrane and the activity of k_2 is equal to 0 in the solution on the other side, $[a_{k_1}{}' \rightarrow 0, a_{k_2}{}'' \rightarrow 0]$. In such a case, the equation for the interdiffusion flux reduces to:

$$j = \frac{RT\bar{x}\bar{u}_{k_1}\bar{u}_{k_2}}{(\bar{u}_{k_1} - \bar{u}_{k_2})\delta} \ln \frac{\bar{u}_{k_1}}{\bar{u}_{k_2}}.$$

(19-10)

It is apparent from this equation that the flux of cations across the membrane is not simply proportional to the concentration gradient of a particular cation in the solutions contiguous with the membrane surfaces. But rather, the rate of movement of cations across the membrane is determined by the mobilities and concentrations of the cations within the fixed charge membrane. Thus, the concept of membrane permeability based on the Nernst equilibrium potential of a particular cation has no practical value in a system containing a fixed charge membrane. The inadequacy of equations for membrane permeability and potentials that do not consider discontinuities of ion activities at the boundary between bulk solution and membrane phase will become more evident in subsequent sections of this article.

The membrane resistance in the special case cited above may be expressed (Tasaki, 1968) by the equation:

$$R_M = \frac{(\bar{u}_{k_1} - \bar{u}_{k_2})\delta}{F^2 \bar{x} \bar{u}_{k_1} \bar{u}_{k_2} \ln (\bar{u}_{k_1}/\bar{u}_{k_2})}. \tag{19-11}$$

Thus, the resistance of a charged membrane is determined by the ion mobilities and concentrations within the membrane. A comparison of this equation with the preceding equation for the interdiffusion flux reveals that at a given temperature the resistance-flux product is equal to a universal constant:

$$R_M j = \frac{RT}{F^2}. \tag{19-12}$$

This relationship is valid irrespective of the mobilities of the two interdiffusing cations. The implications of this relationship on membrane conductance will be discussed later.

Membrane Potentials

In a cation exchange membrane separating a solution of univalent salt $k_1 a$ from another solution containing univalent salt $k_2 a$, a diffusion potential enforcing the equivalence of the cation fluxes is developed within the membrane. But this intramembrane diffusion potential is only one of the components of the potential difference measured when a pair of recording electrodes are placed across the membrane. The components of the membrane potential are easily understood by introducing the concept of phase-boundary potential. It was previously mentioned that under the conditions of complete anion exclusion, the concentration of cations in the membrane must equal the fixed charge of the membrane:

$$\bar{c}_{k_1} + \bar{c}_{k_2} = \bar{x}. \tag{19-6}$$

Therefore, the cation concentration is much larger and the concentration of the mobile anion is far smaller in the membrane than in the bulk solutions on either side of the membrane. Any tendency of the cation at the outer membrane surface to diffuse into the solution creates an electric potential difference across the interface due to separation of an immeasurably small amount of cation from the membrane and further diffusion is prevented.

This electric potential difference which exists across each of the two surfaces of the membrane contiguous with the bulk solutions, is the Donnan or phase boundary potential. The membrane potential must then be considered to be the sum of the two phase boundary potentials (Donnan potentials) and the intramembrane diffusion potential (Teorell, 1953).

$$E_M = E_i + E'_d + E''_d. \tag{19-13}$$

Although the Donnan phase boundary potentials, E'_d and E''_d and the intramembrane diffusion potential E_i are not accessible to separate and independent measurements, the dependence of the membrane potential on these components may be more readily appreciated from the following example.

In Fig. 19-1, a cation exchange membrane with a fixed negative charge

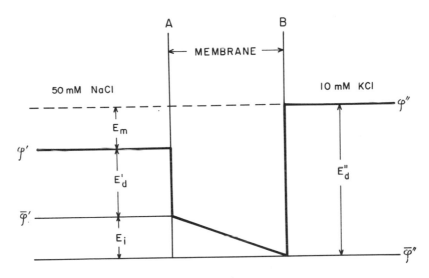

Fig. 19-1 *Schematic electric potential profile in a system containing a cation exchange membrane with a fixed charge concentration of* 300 meq/liter, *separating a* 50 mM NaCl *solution from a* 10 mM KCl *sucrose solution. The membrane potential* E_M *consists of the intramembrane diffusion potential,* $\overline{\varphi}'' - \overline{\varphi}'$ *and the Donnan potentials* $(\overline{\varphi}' - \varphi'')$ *and* $(\varphi'' - \overline{\varphi}'')$ *at the phase boundaries. Note that the interdiffusion potential and the membrane potential have opposite signs.*

concentration of 300 meq/liter, separates a 50 mM NaCl solution and a 10 mM KCl solution. Large sugar molecules are added to the KCl solution to maintain isotonicity, thereby preventing movement of water across the membrane. If it is assumed that the selectivity coefficient of the membrane is equal to unity, that the fixed charge density, $\bar{x}$, in the membrane is constant, and that the mobility ratio, $\bar{u}_K/\bar{u}_{Na}$, of the cations in the membrane is equal to the value in free solution, 1.46, the interdiffusion potential is given by:

$$E_i = \bar{\varphi}'' - \bar{\varphi}' = -\frac{RT}{F} \ln \frac{\bar{u}_K}{\bar{u}_{Na}} \approx -6 \text{ mv}, \qquad (19\text{-}14)$$

where $\bar{\varphi}''$ and $\bar{\varphi}'$ represent the potentials of the cation exchange membrane in the vicinity of each surface. The diffusion potential is negative on the side of the membrane facing the solution containing the ion of greater mobility, the K ion. Assuming complete co-ion exclusion, the electroneutrality condition requires that $\bar{c}_K + \bar{c}_{Na} = 300$ meq. But the distribution of the two cations within the membrane is not uniform, for the concentration of Na ions at the surface of the membrane contiguous with the NaCl solution is 300 mM and that at the other surface is zero. Similarly, the concentration of K ions varies between zero and 300 mM within the membrane. Assuming that the activity coefficients of these ions are unity, the potential differences $(\bar{\varphi}' - \varphi')$ and $(\varphi'' - \bar{\varphi}'')$ at the phase boundaries may be given by the equations:

$$E_d' = \bar{\varphi}' - \varphi' = -\frac{RT}{F} \ln \frac{300}{50} \approx -45 \text{ mv} \qquad (19\text{-}15a)$$

$$E_d'' = \varphi'' - \bar{\varphi}'' = -\frac{RT}{F} \ln \frac{10}{300} \approx +90 \text{ mv}, \qquad (19\text{-}15b)$$

where φ' and φ'' represent the potentials of the solutions contiguous with the membrane surfaces. Therefore, $E_M = -6 + (90 - 45) \approx 39$ mv. Note that the membrane potential is significantly larger than the intramembrane diffusion potential and is even opposite in sign. It is thus invalid to use the value of the measured membrane potential to calculate the intensity of the electric field within the membrane.

The equation for membrane potential across a membrane of fixed negative charge separating mixtures of salts of two univalent cations has been derived by Wyllie (1954)

$$E_M = -\frac{RT}{F} \ln \frac{a_{k_1}'' + Q a_{k_2}''}{a_{k_1}' + Q a_{k_2}'}, \qquad (19\text{-}16)$$

where Q equals the membrane selectivity mobility ratio product, $K_{k_1}^{k_2} \bar{u}_{k_2}/\bar{u}_{k_1}$ and a_{k_1}', a_{k_1}'', a_{k_2}' and a_{k_2}'' are the activities of the individual cations in the two bulk solutions.

In the special case where $a_{k_1}' = a_{k_2}'' = 0$ this equation can be reduced to

$$E_M = \frac{RT}{F} \ln \frac{Q a_{k_2}'}{a_{k_1}''} \qquad (19\text{-}17)$$

When cation k_1 is *divalent* and cation k_2 is univalent, integration of the Nernst-Planck flux equations for a uniform membrane together with the phase boundary potentials yields (Tasaki, 1968):

$$\Delta\varphi = \frac{RT}{2F} \ln \frac{a'_{k_2}}{(a''_{k_1})^2} + \frac{RT}{2F} \ln (2K\bar{x}) + \frac{RT}{F} \frac{\bar{u}_{k_1} - \bar{u}_{k_2}}{\bar{u}_{k_1} - 2\bar{u}_{k_2}} \ln \frac{2\bar{u}_{k_2}}{\bar{u}_{k_1}} \quad (19\text{-}18)$$

where K, $\bar{x}$ and $\bar{u}_{k_2}/\bar{u}_{k_1}$ are assumed to be constant throughout the membrane. The experimental results for membrane potential and cation fluxes obtained with inanimate systems agree with this theoretical equation (Helfferich and Ocker, 1957). This equation for a bi-ionic potential, implies that at a given concentration c_{k_1} on one side and c_{k_2} on the other side of the membrane, a change in membrane potential may be explained in terms of a variation in one or more of the following quantities: membrane selectivity coefficient K, membrane fixed charge $\bar{x}$, and mobility ratio of cations within the membrane $\bar{u}_{k_1}/\bar{u}_{k_2}$. (Sudden alterations of these quantities in the excitable membrane can account for the abrupt change in membrane potential associated with nerve excitation.) The mechanisms responsible for a reversible abrupt change in K, $\bar{x}$ or $\bar{u}_{k_1}/\bar{u}_{k_2}$ may be explained by the introduction of the concepts of cooperative ion exchange and phase transition.

Cooperative Ion Exchange Process

Phase transition has been well demonstrated in rigid ion exchanges. For example, the lattice constant of a divalent cation-rich zeolite is different from that of a univalent cation-rich zeolite (Olson and Sherry, 1968). As the mole fraction of the divalent cation in a solution bathing the univalent rich zeolite is *continuously* increased, an *abrupt* change in the lattice constant occurs. When the divalent rich zeolite is converted back to the univalent rich form, the reverse transition takes place at a different mole fraction of univalent cation in the solution. The occurrence of such an abrupt transition and hysteresis suggests that the presence of a particular cation at a negative site in the membrane energetically favors the accumulation of similar cations at other negative sites in the immediate vicinity; namely, the exchange of univalent and divalent cations is cooperative (Changeaux, Thiery, Tung and Kittel, 1967).

Phase transition in physiological fixed charge membranes in which the negative sites are occupied by divalent cations may be explained by a similar cooperative ion exchange process. The theoretical ion exchange isotherm (Tasaki, 1963) for such a process is illustrated in Fig. 19-2. A *continuous* increase in the equivalent fraction of univalent cations in the bulk solution in which the membrane is immersed can produce a *discontinuous* change in the concentration of the univalent cation in the membrane. This discontinuity

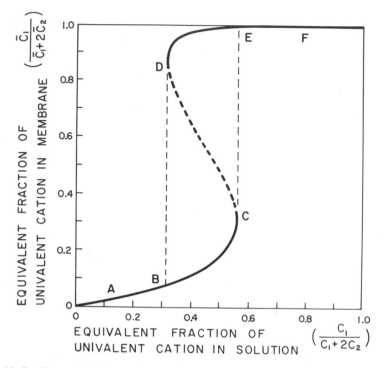

Fig. 19-2 *Theoretical ion-exchange isotherm calculated for a cation-exchanger membrane immersed in a salt solution containing univalent and divalent cations. In the calculation it was assumed that occupancy of two neighboring charge sites in the membrane by two cations of different valencies is energetically unfavorable. c_1 and c_2 represent the concentrations in the solution of uni- and divalent cations respectively. $\bar{c}_1$ and $\bar{c}_2$ represent the concentrations within the membrane. When the equivalent fraction of the univalent cation in the solution is increased continuously from zero to one the corresponding fraction in the membrane increases along the course, 0, A, B, C, E, and F. When the equivalent fraction in the solution is decreased from one to zero, the corresponding fraction in the membrane changes along F, E, D, B, A, 0. (Adapted from Tasaki, 1963.)*

is represented by C--E in the figure. Any further increase in the external univalent concentration increases the intramembrane univalent concentration continuously. The selectivity coefficient of this membrane (K) is given by an equation similar to Eq. (19-7) assuming the activity of the ions in the solution phase to be unity:

$$\frac{(\bar{c}_1)^2}{\bar{c}_2} = K \frac{(c_1)^2}{c_2} \tag{19-19}$$

where $\bar{c}_1$ and $\bar{c}_2$ are the uni- and divalent concentrations within the membrane and c_1 and c_2 are the uni- and divalent concentrations in the bulk solution.

As $(\bar{c}_1)/(\bar{c}_1 + 2\bar{c}_2)$ increases along the line ABC, K varies continuously. When the univalent concentration abruptly increases along C--E, there is a sudden increase in the membrane selectivity coefficient K. From Eq. (19-18) it is apparent that any sudden change in K or $\bar{x}$ can cause a sudden change in membrane potential. As c_1 in the bulk solution is decreased, $\bar{c}_1$ decreases along EDB indicating the presence of hysteresis.

With this brief presentation of ion exchange theory, the experimental results obtained with the squid giant axon can be more readily appreciated and are less likely to be misinterpreted.

ELECTROPHYSIOLOGICAL PROPERTIES OF INTERNALLY PERFUSED AXONS

Ion Flux Measurements

The development of the technique of intracellular perfusion of the squid axon has made it possible to remove the protoplasm from the axon almost completely without affecting the ability of the axon to produce action potentials (Oikawa, Spyropoulos, Tasaki and Teorell, 1961). It is therefore possible to study the excitation process with well-defined solutions on both sides of the membrane. Since under such conditions it is possible to have no ion species common to both the internal and external media, radiotracer determinations of ion fluxes across the squid axon membrane can be made without ambiguity. [If a nonradioactive ion is present on both sides of the membrane, a radioisotope of the same species does not trace the (net) flux of that species across the membrane.]

Ion fluxes across the unstimulated axon membrane are measured using an isotonic internal medium of 0.5 M KF and glycerol (pH 7.3) and an isotonic external medium of 0.4 M NaCl and 0.08 M CaCl$_2$ (pH 8.0) (Tasaki, 1963; Tasaki, Singer and Watanabe, 1967). The appropriate isotope is then added to the internal perfusion fluid for efflux studies and to the external medium for influx determinations. The measured efflux of K-ion is 150 to 200 pmole cm^{-2} sec^{-1} and the influx of Na-ions is of the same order of magnitude. The influx of Ca-ion is less than 1/20 that of the Na-ion and the anion fluxes are less than 1/10 the univalent cation fluxes. As discussed in the preceding section, when there is no net current across the membrane, the fluxes must satisfy the equation

$$j_{Na} + j_K + 2j_{Ca} - j_F - j_{Cl} = 0 \qquad (19\text{-}20)$$

Since the univalent cation fluxes are much greater than the other fluxes, this equation can be reduced to:

$$j_{Na} + j_K \approx 0 \qquad \text{or} \qquad j_{Na} \approx -j_K \qquad (19\text{-}21)$$

Determination of ion fluxes during the action potential is limited by the poor time resolution of the radioisotope technique due to slow diffusion of ions across connective tissue and Schwann cells outside the membrane. In spite of this difficulty, the extra cation flux (flux above the resting level) per impulse can be obtained unambiguously by dividing the extra flux collected during a given period by the number of impulses elicited during the period. The extra efflux of K ions and influx of Na ions associated with the production of an action potential in a squid axon is in the order of 10–17 pmole cm^{-2}/impulse. The extra Ca ion influx is less than 0.05 pmole cm^{-2}/impulse and the fluxes of anions are not affected during stimulation (Tasaki, Watanabe and Lerman, 1967). Thus, the flux across the membrane during excitation may be represented by the relationship:

$$j_{Na}' \approx -j_K' \qquad (19\text{-}22)$$

Since the impedance loss during an action potential has a triangular time course, and since, as noted, the $R_m j$ product at a given temperature is a constant, the interdiffusion flux j rises and falls along a similar time course. Thus, the peak value of the cation fluxes is twice the average flux increase. This peak value of $(2.0\text{-}3.4) \times 10^{-8}$ mole cm^{-2} sec^{-1} is 150 to 200 times the resting level of the interdiffusion fluxes.

The amount of electricity carried across the excited area of the membrane by the local current associated with the propagation of a nerve impulse can be calculated from the product of the apparent membrane capacity (1 μf/cm^2) times the change in the intracellular potential during excitation (Tasaki, 1968). This product, 10^{-7} coul cm^{-2} divided by the Faraday constant is the extra cation flux associated with the local current. This value is about 1 pmole cm^{-2}/impulse and is 1/10 to 1/17 the value of either the K efflux or Na influx during the action potential. Therefore it is quite reasonable to assume that $j_{Na}' \approx -j_K'$ during excitation. Thus, although both fluxes are greatly increased during excitation, net transfer of charge across the membrane is negligibly small as compared with Fj_{Na} or Fj_K.

Membrane Resistance and Membrane Conductance

By using the voltage clamp method (Hodgkin, Huxley and Katz, 1952) and A.C. impedance method (Cole and Curtis, 1939) for determining the resistance of the membrane of unperfused axons, it has been shown that the membrane resistance falls from a resting value of 1500 ohm cm^2 to about 7–17 ohm cm^2 at the peak of excitation: a decrease in resistance of 100–200 times. Thus the ratio of the membrane resistance in the resting state to that at the peak of excitation is equal to the reciprocal of the ratio of the interdiffusion flux in the resting state to that at the peak of excitation. This relationship is expressed:

$$R_M j = R_{M_a} j_a. \tag{19-23}$$

As seen in Eq. (19-12), the theoretically expected value of the $R_M j$ product is equal to RT/F^2 (2.62×10^{-7} mole sec^{-1} ohm at $20°C$). For the unstimulated axon membrane, the experimental estimate of the $R_M j$ product is $2-4 \times 10^{-7}$ mole sec^{-1} ohm at $18-22°C$. This close agreement between theoretical and experimental $R_M j$ values is strong evidence in support of the fixed charge character of the major diffusion barrier in the axon membrane.

Since membrane conductance is the reciprocal of membrane resistance, the resistance-interdiffusion flux product may be written as:

$$g = \frac{F^2 j}{RT}. \tag{19-24}$$

If we assume that the membrane is ideally permselective and uniform, the value of g in the bi-ionic case [see Eq. (19-10)] is given by:

$$g = \frac{F^2 \bar{x} \bar{u}_{k_1} \bar{u}_{k_2}}{(\bar{u}_{k_1} - \bar{u}_{k_2})\delta} \ln \frac{\bar{u}_{k_1}}{\bar{u}_{k_2}}, \tag{19-25}$$

where $\bar{u}_{k_1}$ and $\bar{u}_{k_2}$ are the intramembrane mobilities of the two univalent cations, $\bar{x}$ is the density of fixed negative charge available for interdiffusing univalent cations and δ is the thickness of the membrane. It is seen in this equation that the conductance, g, of a fixed charge membrane is determined by the products of the ion mobilities and concentrations of cations in the membrane. (Although not explicitly given in the equation for g, $\bar{c}_{k_1}$ and $\bar{c}_{k_2}$ are dependent on $\bar{x}$ as well as on the selectivity coefficient K.) There is no potential term in the above equation. Therefore, a sudden increase in membrane conductance must be a reflection of either an abrupt increase in the density of negative sites available for univalent cations (an increase in $\bar{x}$) and/or a sudden decrease in the compactness of the membrane (an increase in $\bar{u}_{k_1}$ and $\bar{u}_{k_2}$). (The concept of "voltage-dependent conductances" is not applicable to this system.)

Ion Requirement for Production of Action Potentials

According to the "Na-theory", the initiation of an action potential in a normal squid axon *in vivo* is explained on the basis of an increase of the membrane permeability specifically to the Na ion (Hodgkin and Huxley, 1952). Although the concept of selective membrane permeability to univalent cations has been generalized to explain the process of action potential production in other invertebrate and vertebrate excitable membranes, limits to such a generalization exist. There are many excitable tissues and cells [i.e. crustacean muscle fibers (Fatt and Katz, 1953), insect muscle fibers (Werman, McCann and Grundfest, 1961), frog dorsal root ganglia (Koketsu, Cerf and Nishi, 1959), and *Nitella* internodes (Osterhout and Hill, 1933)], which maintain their ability to produce action potentials in media containing only salts of divalent cations. The complexity of intact cells in their natural state prevents us from analyzing the process of excitation from a physicochemical point of view. However, with the use of internally perfused axons, it is possible to establish the ionic prerequisites for maintaining excitability of the axon membrane.

In Fig. 19-3 are the records from an experiment demonstrating all-or-none

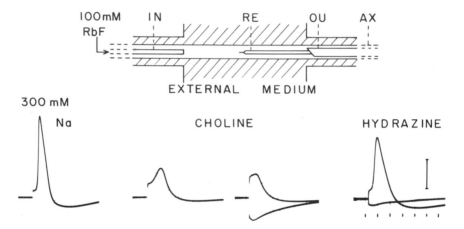

Fig. 19-3 Top: Experimental arrangement (not to scale) used for demonstration of excitability of squid giant axon immersed in Na-free medium and internally perfused with 100 mM RbF. AX, giant axon; RE, recording electrode (wire enclosed in glass capillary); OU, outlet cannula, IN, inlet cannula. Bottom: An example of the action potential records obtained with the above arrangement. The composition of the external fluid medium was 300 mM NaCl and 200 mM CaCl₂ (left record), 300 mM choline chloride and 200 mM CaCl₂ (center records) and 300 mM hydrazine chloride and 200 mM CaCl₂ (right record). The latter two media were Na-free. The bar represents 50 mv and the time markers are 1 msec apart. The perfusion zone was 12 mm in length, and the axon diameter was 620μ. (From Tasaki, Singer and Watanabe, 1965.)

action potentials by an internally perfused squid axon in external media free of Na ion (Tasaki, Singer and Watanabe, 1965). The internal perfusion fluid was a 100 mM RbF solution. For the record on the left, the external medium was 300 mM NaCl and 200 mM CaCl₂. The action potentials under these conditions were similar to those of the unperfused axon. When the external medium was replaced by a solution containing 300 mM choline chloride and 200 mM CaCl₂, the action potential was suppressed (center records). Prompt restoration of the action potentials was obtained when hydrazinium was substituted for choline. Action potentials were obtained when NH_4 or guanidinium ions were substituted for the external univalent cations. Therefore, production of the action potential in the perfused squid axon is not dependent on specific univalent cation. While performing these experiments, it was established that excitation in the absence of external divalent cations was not possible. It was reasonable then to examine if squid axon membranes were excitable in external media containing only the salt of a divalent cation.

Using the intracellular perfusion technique, it was found by Watanabe, Tasaki and Lerman (1967) and later confirmed by Conti (personal communication) that squid giant axons immersed in solutions containing only divalent cation (Ca or Sr) salts were capable of producing all-or-none action

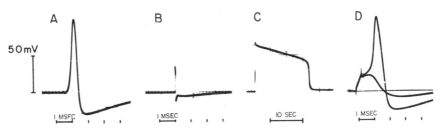

Fig. 19-4 *Oscillograph records demonstrating development of an all or none action potential in an axon perfused internally with sodium phosphate solution and externally with CaCl₂ solution. Records A and B were taken before initiation of internal perfusion; the external medium contained 300 mM NaCl and 100 mM CaCl₂ in A and only 100 mM CaCl₂ in B. The time markers are 1 msec apart. Record C was taken approximately 12 minutes after the onset of internal perfusion with 10 mM sodium phosphate; the external medium was 100 mM CaCl₂. The time marker represents 10 seconds. Record D was obtained from the same axon after switching the internal perfusion fluid to 400 mM KF and the external medium to the solution used in A; sub and suprathreshold responses are superposed. Time markers are 1 msec apart. The stimulus duration was 0.01 msec for A and B, 100 msec for C, and about 0.3 msec for D. Temperature, 21° C. (From Watanabe et al., 1967.)*

potentials when internally perfused with a dilute salt solution containing one of the following univalent cations: Na, Cs, Li, choline, guanidinium, or tetramethylammonium. The all-or-none electrical responses evoked under these conditions in which only two critical cations are involved in the process of excitation are term "bi-ionic potentials". In Fig. 19-4, records from an experiment demonstrating bi-ionic action potentials are presented.

Record *A* was obtained from the unperfused axon immersed in a solution containing 300 mM NaCl and 100 mM CaCl₂. Note that the action potential is large in amplitude and short in duration. When the external medium was switched to a 100 mM $CaCl_2$ solution the action potential of the yet unperfused axon was suppressed within one minute (Record *B*). Next, intracellular perfusion with a dilute Na phosphate solution was begun. Under these conditions, with the continuous flow of internal and external media, the axon regained and maintained the ability to produce all-or-none action potentials (Record *C*). Such action potentials could be evoked for more than 60 min. The amplitude and duration of the bi-ionic action potential will be discussed later. At present we are interested in the essential requirements for excitability and not in the factors which influence the magnitude or time course of the action potential. Record *D* was obtained after the internal perfusion solution was changed to 400 mM KF and the external medium to the original 300 mM NaCl–100 mM $CaCl_2$ solution. The action potential under these conditions is similar to the action potential recorded prior to the initiation of internal perfusion. This indicates that the bi-ionic conditions produced no permanent alteration of the axon membrane.

Characteristics of the Resting Membrane

Having simplified and described the conditions essential for the production of an all-or-none response in the squid axon membrane, the ionic composition within the substance of the axon membrane in the resting state can be analyzed. According to the charged membrane theory, the negative sites in the axon membrane have a greater affinity for divalent cations than for univalent cations. Thus, one would expect that the negative sites at the major diffusion barrier in the unstimulated axon membrane are predominantly occupied by divalent cations from the external medium. That the squid axon membrane is calcium rich is suggested by demonstration of the dependence of the resting membrane potential on the divalent concentration in the external medium.

As discussed earlier, the potential across a fixed charge membrane consists of two phase boundary potentials and an intramembrane diffusion potential. In such a system, the excess fixed negative charge of the macromolecules in the resting membrane must be balanced by the mobile cations from the external solution. Thus, under bi-ionic conditions, the divalent cation concentration within the membrane is expected to remain constant and, within a wide range, independent of the external divalent cation concentration. Therefore a dilution of the external medium would result in a change in the phase boundary potential at the external surface of the membrane. Consequently, the resting membrane potential is expected to change. In Fig. 19-5 are the records of such an experiment (Tasaki, Watanabe and Lerman, 1967). An axon is internally perfused with a 100 mM KF solution and immersed in a rapidly circulating 200 mM $CaCl_2$ solution. When the external

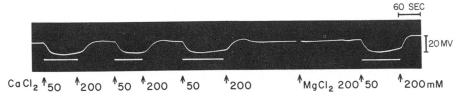

60 SEC

20 MV

CaCl₂ ↑50 ↑200 ↑50 ↑200 ↑50 ↑200 ↑MgCl₂ 200↑50 ↑200mM

Fig. 19-5 *The effect of reduction of the external divalent cation concentration on membrane potential in an axon internally perfused with 100 mM KF solution (phosphate buffer). Downward deflections represent increased negativity within the axon. The lower beam indicates the onset and duration of changes of divalent cation concentration in the external medium. The horizontal and vertical bars represent 60 seconds and 20 mv, respectively. External media (Tris buffer) were prepared by mixing a 400 mM solution of either MgCl₂ or CaCl₂ with 12 per cent (vol) glycerol. The brief (approximately 10 seconds) gap in the recording marks the change in external solution from 200 mM CaCl₂ to 200 mM MgCl₂. Note that the replacement of 200 mM CaCl₂ with 200 mM MgCl₂ solution did not produce a significant change in the resting potential. Temperature, 20° C. (Modified from Tasaki, Watanabe and Lerman, 1967.)*

medium was changed to a rapidly flowing 50 mM CaCl₂ solution, an increase in the intracellular negativity was recorded. As seen in the records, this change is reversible. The four-fold decrease in the divalent cation concentration resulted in potential changes between 12–18 mv. Similar results were obtained with dilution of an external medium containing MgCl₂ as the sole electrolyte (see right side of records).

The change in the Donnan potential at the phase boundary between the external medium and the outer surface of the membrane for two different concentrations of divalent cations is given by:

$$\Delta E_d' = \frac{RT}{zF} \ln \frac{c_1 \gamma_1}{c_2 \gamma_2}, \qquad (19\text{-}26)$$

where c_1 and c_2 represent the concentrations of divalent cations in the external medium, γ_1 and γ_2 the activity coefficients of the two salt solutions and z represents the valence of the cations. By substituting 200 mM and 50 mM for c_1 and c_2 and by approximating the activity coefficients for the CaCl₂ solutions, the potential change predicted by the above equation is 14.9 mv at 21° C. The close agreement of this value with the experimentally determined range of 12–18 mv indicates that the high density of fixed negative charge is located near the outer surface of the membrane. (Dilution of the internal perfusion fluid produces much smaller changes in the resting potential than is predicted by the Donnan potential equation. This finding may be attributed to the presence of a membrane layer of low fixed charge density separating the internal surface from the high fixed charge density area near the outer membrane surface.)

The results obtained from dilution of the external media under bi-ionic conditions strongly support the view that the negatively charged sites in the critical membrane layer are predominantly calcium rich in the resting state. If indeed the resting membrane behaves as a cation exchanger with fixed negative sites occupied mainly by divalent cations derived from the external medium, a process that involves reversible displacement of the intra-membrane divalent cations may account for the drastic changes in membrane properties which occur during the excitation process. An analysis of the phenomenon of abrupt depolarization will allow us to propose a mechanism by which such a reversible cation exchange process can occur.

The Osterhout Phenomenon

While investigating the effect of K ion on the internode of *Nitella*, Osterhout and Hill (1938) observed that when the KCl concentration of the medium was increased by small amounts, the potential difference across the membrane of a single internodal cell changed suddenly at a critical K ion concentration.

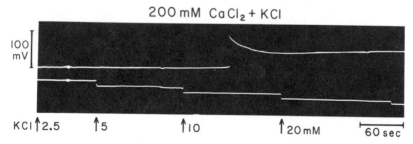

Fig. 19-6 *Abrupt depolarization induced by external application of KCl (upper oscillograph trace). The concentration of KCl in the rapidly flowing external medium was doubled at the time marked by the lower oscillograph trace. The internal perfusion fluid was 25 mM CsF. No electric stimuli were delivered while continuous recording of the membrane potential was made. The potential jump produced by 10 mM KCl was 80 mv initially, and was followed by a gradual fall of approximately 40 mv Temperature, 16° C. (From Tasaki, Takenaka and Yamaghshi, 1968.)*

Similar examples of abrupt depolarization have been demonstrated with nodes of Ranvier treated with traces of nickel or cobalt (Tasaki, 1959). Based on these observations, the effect of the addition of univalent cations to the external media of internally perfused axons was analysed.

Fig. 19-6 shows the records obtained from experiments made on a squid axon internally perfused with a dilute Cs salt solution and immersed in a $CaCl_2$ solution. The preparation was capable of producing all-or-none "bi-ionic" action potentials, but in the following observations no electric current was applied to the membrane. The external medium was switched to another 200 mM $CaCl_2$ solution containing small amounts of KCl. As the K ion concentration was increased in two-fold steps, there was very little change in the membrane potential until the KCl concentration reached a critical level (in this experiment 10 mM). At this concentration of K ions, a large abrupt rise (80 mv) in the intracellular potential was recorded (Tasaki, Takenaka and Yamagishi, 1968). After this potential jump, a further increase in KCl concentration failed to produce another sudden change in membrane potential and the axon membrane was no longer capable of developing an action potential in response to outward (depolarizing) current.

Similar results can be obtained with other common univalent cations. However, the ability of these cations to abruptly depolarize the axon membrane was weaker than that of K ion. The cation sequence arranged in accordance with their ability to depolarize the membrane is: $K > Rb > NH_4 > Cs > Na, Li$.

Measurements of membrane impedance during abrupt depolarization indicate that the membrane resistance suddenly falls (conductance suddenly increases) at the moment the abrupt change in membrane potential occurs (Tasaki, Takenaka and Yamagishi, 1968). As we have discussed earlier, conductance of a fixed charge membrane is determined by the products of the concentrations and ion mobilities of cations in the membrane, which in turn are dependent on the membrane selectivity coefficient, the fixed charge density and the compactness of the membrane. The abrupt increase in conductance may be a reflection of an abrupt change in these quantities when the mole fraction of K ion (or other univalent cations) reaches a critical level in the outer layer of the membrane. The abruptness and *discontinuity* of these changes as the external K ion concentration is gradually and *continuously* increased may be explained by cooperative cation exchange. (See earlier discussion of ion exchange isotherm.) With the addition of K ions to the external medium, the fraction of negative sites in the outer layer of the membrane occupied by K ions increases. Occupation of negative sites by univalent cations is now favored over occupation by divalent cations. At a critical univalent-to-divalent cation concentration ratio in the membrane, the univalent concentration in the membrane suddenly increases, in a manner similar to *C--E* in Fig. 19-2.

The cooperative cation exchange may be regarded as an indication of a (first order) phase transition of the macromolecules in a layer with a high charge density in the membrane. Such a phase transition can account for the abrupt change in membrane fixed charge density, membrane selectivity coefficient, membrane compactness and water content. The existence of a phase transition is supported by the finding that when the K ion concentration in the external medium is lowered in small steps after the membrane is abruptly depolarized, the membrane potential and conductance return to their original level at a K ion concentration far below the critical K ion concentration. This example of hysteresis is similar to the curve in Fig. 19-2 and suggests the existence of a metastable state between the resting, divalent cation-rich state and the depolarized, univalent cation-rich state.

The critical univalent-to-divalent cation ratio at which abrupt depolarization occurs is also dependent on the divalent concentration of the external medium. Under the same conditions as described for the experiment in Fig. 19-6, a decrease in the Ca ion concentration in the external medium resulted in a decrease in the concentration of K ions required for the production of abrupt depolarization. An increase in the Ca ion concentration in the external medium during the period following abrupt depolarization resulted in a return of the membrane potential to the original level.

The well-known K-Ca antagonism (Baylis, 1924) can be explained by the competition of both Ca ions and K ions for the same negative charged sites in the axon membrane. As the divalent cation concentration in the external medium is increased, occupation of negative sites in the membrane by divalent cations is favored over the univalent cations. Phase transition of the membrane macromolecules to the univalent cation-rich state may thus be opposed or reversed.

Electric Stimulation

It is evident from the properties of the axon membrane during abrupt depolarization (sudden increase in conductance, loss of excitability in response to outward current, etc.) that the process of abrupt depolarization is related to the process of action potential production evoked by electric stimulation under bi-ionic conditions. The abruptly depolarized membrane in Fig. 19-6 may be considered to correspond to the plateau stage of the action potential in Fig. 19-4.

This similarity may be explained in terms of the effect of current applied to the resting membrane of an axon perfused with a dilute univalent salt and immersed in the salt solution of a divalent cation. An outward directed (stimulating) current drives the intracellular univalent cation into and across the resting membrane. The extracellular anions are driven toward the membrane but are effectively excluded from the membrane because of the fixed charge density of the macromolecules at the outer layer of the membrane. Therefore, the ionic distribution at the outer membrane surface produced by an electric current is similar to that produced by the addition of the univalent salt to the external medium: the univalent cations within the membrane outer surface are increased and the divalent cations at the negatively charged sites are decreased. At a critical intramembrane concentration ratio of univalent-to-divalent cations, the membrane macromolecules undergo a phase transition from the resting divalent cation-rich state to the univalent cation-rich excited or depolarized state. The phase transition results in drastic changes in membrane selectivities, fixed charge densities and mobilities which are responsible for the discontinuous changes in the measured electrical properties of the axon membrane.

The obvious difference between abrupt depolarization and the action potential evoked by electric stimulation is that in the former the membrane is not repolarized. We have noted that the excited membrane may be returned to its resting state as the intramembrane divalent cation concentration is increased or the univalent cation concentration is decreased. In the cation exchange model, during an action potential evoked by outward current, the intracellular univalent cation accumulates only in the vicinity of the outer membrane surface and is rapidly reduced as the membrane resistance returns toward the resting level during the plateau, restoring the membrane to the original resting state. In the case of abrupt depolarization, the univalent cation distribution in the external medium is uniform and is thus not significantly reduced at the outer membrane surface by the increased interdiffusion associated with the excited state.

Membrane Subunits and Nonuniform Excitation

The spread of the excitation process along the axon membrane is due to the difference in the properties between those parts of the membrane which are in the excited state and those in the resting state. In the fixed charge membrane spontaneous fluctuation in the univalent-divalent cation population within the membrane macromolecules, reinforced by the cooperative nature of the ion exchange process, tends to lead to the formation of a small number of "active patches" or "excited subunits" in the predominantly resting membrane (Tasaki and Singer, 1966). Since the electric conductance through the excited subunits is high and a difference in membrane potential exists between the excited subunits and the resting area, local (or eddy) currents are generated, which flow inwardly across the few excited subunits and outwardly through the remaining subunits. The membrane as a whole remains in the resting state because the randomly distributed subunits are repolarized by the high intensity of inward current. (Passage of inward current increases the divalent cation concentration in the membrane.)

Upon application of a brief outward current across the membrane the number of negative sites occupied by the intracellular univalent cation increases, thereby increasing the number of excited subunits. That only some, and not all, of the subunits are excited at a given current intensity may be due in part to the various morphological irregularities of the membrane which tend to make the effect of the stimulating current nonuniform. Using a voltage clamp and measuring the difference in membrane potential and conductance between the resting and active state of the membrane, it is possible to determine that the fraction of the membrane in the active state at the critical membrane potential (threshold) is approximately 1 per cent of the total membrane (Tasaki, 1968).

A diagrammatic example of this cooperative process is seen in Fig. 19-7.

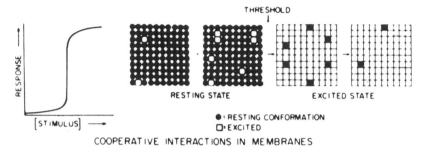

Fig. 19-7 Diagram showing cooperative interactions in membrane structure. (*From Lehninger,* 1968, *Proc. Nat. Acad. Sci. USA,* 60 : 1069.)

The term "stimulus" in the figure may represent either the univalent cation fraction or the intensity of outward current. The "response" can be membrane conductance or membrane potential. The white squares in the diagrams represent membrane subunits in the active state.

Resting Potential under Multi-ionic Conditions

Using the intracellular perfusion technique, the number of variables affecting the experimental measurements has been decreased. The membrane phenomena were analyzed in terms consistent with the theoretical treatment of fixed charge membranes. The factors influencing the resting and active membrane under natural multi-ionic conditions should now be more readily understood.

Much significance has been attributed to the absolute value of the resting potential. However, measurements of resting potentials are fraught with ambiguities arising from liquid junction potentials, single ion activities and suspension effects (Tasaki and Singer, 1968). The reason for the emphasis on the measured resting potential (-50 to -60 mv) is its close agreement with the potassium equilibrium potential described by the Nernst equation at $20°\ C$:

$$\frac{RT}{F} \ln \frac{c_{K_i}}{c_{K_e}} \approx -90 \text{ mv}, \tag{19-27}$$

where c_{K_i} is the concentration of K ion in squid axoplasm, 400 mM, and c_{K_e} is the K ion concentration in sea water, 10 mM. [Ion activities rather than total concentrations are usually substituted in Eq. (19-27) resulting in a smaller equilibrium potential. However, it is well known that the activity coefficients of single ions in a mixed salt solution are inaccessible to thermodynamic measurements without ambiguity (Robinson and Stokes, 1959).] If, indeed, the Nernst equation [rather than an equation similar to Eq. (19-18)] describes the resting potential of an excitable axon membrane, it is expected that every two-fold change in either the internal or external K ion concentration would result in an 18 mv change in membrane potential. The following experiments on perfused axons permit us to test this assumption.

A squid giant axon was perfused with a 400 mM KF solution (Tasaki, 1968). The concentration ratio of KCl-NaCl in the external medium was varied while the sum of the two salts was maintained constant at 400 mM. In addition to these salts, the external medium contained 150 mM MgCl$_2$ and 50 mM CaCl$_2$. As the external K ion concentration was increased from 0 to 30 mM the membrane potential was hardly affected. At about 50 mM there was a rise in the intracellular potential and the ability of the axon to develop action potentials was lost. With a further increase in K ion the membrane potential was found to vary with the external K ion concentration logarithmically. Thus, The Nernst equation describes the membrane potential only when the membrane is in the depolarized, K ion rich state.

It is also possible to vary the internal K ion concentration by altering the internal K-Na ratio while maintaining the total univalent concentration at 400 mM (Tasaki and Takenaka, 1963). The external medium in such an experiment contained 300 mM NaCl, 45 mM MgSO$_4$ and 22 mM CaCl$_2$. As seen in Fig. 19-8, the resting potential is practically unaffected by changes in the internal K ion concentration. In this experiment, the ability of the axon to develop all-or-none action potentials was maintained throughout the experiment. These results are inconsistent with the view that the axon membrane in the resting state is specifically permeable to K ions and that the resting membrane potential is determined by the potassium equilibrium potential (Hodgkin, 1964). In terms of ion exchange theory, the membrane potential departs from the Nernst equation because the selectivity of the fixed charge membrane in the resting state for K ions is not overwhelmingly large as compared to other ions.

Amplitude of the Action Potential under Multi-ionic Conditions

Although the presence of external univalent cations are not essential to the process of excitation, the amplitude of the action potential does depend on the concentration and degree of hydrophilicity of the univalent species in

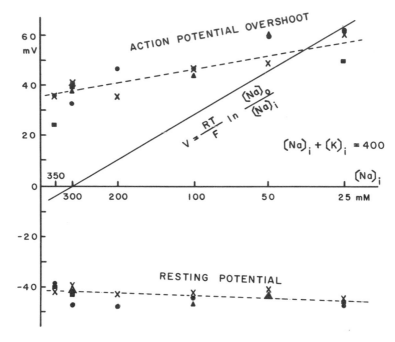

Fig. 19-8 *Resting potential and overshoot of action potential of intracellularly perfused squid axons plotted as a function of Na concentration in the perfusing fluid, c_{Na_i}. The sum of the internal Na and K concentrations was held at a constant level of 400 meq/liters throughout. The perfusing fluid contained 470 mM glycerol besides Na and K salts in the glutamate form. The external medium contained 300 mM NaCl, 45 mM MgSO$_4$, and 22 mM CaCl$_2$. A 50 per cent increase in the Mg and Ca concentrations outside did not alter the results significantly. (From Tasaki and Takenaka, 1963.)*

the external medium. Reduction of the external Na ion concentration by replacing the Na ion with a polyatomic univalent cation surrounded by hydrophobic side-groups (e.g., choline) decreases the action potential amplitude (Hodgkin, 1951) in accordance with the equation:

$$E = \frac{RT}{F} \ln c_{Na_e} + \text{constant.} \tag{19-28}$$

The decrease in amplitude does not necessarily imply that the abrupt changes in membrane properties associated with action potential are brought about by the movement of the Na ion; rather, the variation in potential indicates that the phase boundary potential at the outer surface of the *excited membrane* is determined predominantly by the univalent cation concentration in the external medium. We have noted that the membrane negative sites are more selective for univalent cations in the excited state. Thus, the Na ion concentration at the outer membrane layer contiguous with the external solution is increased at the peak of excitation. Since the concentration of Na ion in the active membrane is determined primarily by the fixed charge density, the intramembrane Na ion concentration is independent of the external Na ion concentration over a given range. Therefore, the outer phase boundary potential should vary with the Na ion concentration in the external medium according to Eq. (19-28). This relationship is valid only in the range where the Na ion concentration is larger than the divalent cation concentration of the medium. (Note also that under the experimental conditions described in Fig. 19-8, the amplitude of the action potential is seen to decrease as the Na ion concentration is increased internally. However, the observed concentration dependence of the action potential amplitude is far smaller than the value of 58 mv for a ten-fold change expected from the sodium theory.)

As the choline concentration in the external medium is increased, the divalent cation-rich resting membrane tends to accumulate these hydrophobic polyatomic cations. [Macromolecules with a high Ca ion content are known to lose water molecules bound by hydration (Ikegami and Imai, 1962) which leads to increased hydrophobic interactions.] Rapid transition of the membrane to the active state is hindered by the presence of additional hydrophobic forces in the membrane, although the membrane subunits eventually undergo transition to the excited state if a strong outward current is maintained. The action potentials that are obtained upon the addition of a small amount of choline are due to transition of resting subunits still occupied by Na and Ca ions (Tasaki, 1968).

The Voltage Clamp

Thus far, we have analyzed the mechanism of excitation without using the voltage clamp. It is considered by many investigators that the analysis of the inward and outward currents across an excitable membrane, observed when the potential is clamped at a specified level, provides a means of defining the excitation process. This view is correct to the extent that one is aware of the limitations of the technique. Now we are ready to discuss the relationship between these membrane currents and the states of the membrane subunits.

A voltage clamp device consists of a feedback amplifier connected to a current passing electrode which is inserted into the interior of a large excitable cell such as the squid axon, and a pair of recording electrodes which measure the potential across the membrane. It is widely assumed that, with such a device, the forces acting upon individual ion species can be changed in controlled steps, and that the abrupt change of potential associated with the process of excitation can be prevented by current automatically supplied by the feedback amplifier (Cole, 1949). In voltage clamping a charged membrane, however, it is not justifiable to assume that each of the components of the membrane potential remains at a constant level each time the clamping pulse is altered. Under voltage clamp conditions the membrane potential consists of the following four components.

1. Donnan potential at the phase boundary of the outer membrane surface.
2. Intramembrane diffusion potential.
3. Intramembrane IR drop due to current passed across the membrane by the feedback amplifier.
4. Donnan potential at the phase boundary of the inner membrane surface.

The clamping device tends to keep the sum of the phase boundary potentials, the diffusion potential and the intramembrane IR drop at a constant level, but there is no means of maintaining each of the components unaltered during clamping. As we have discussed in the previous section, a strong outwardly directed membrane current at the onset of a depolarizing pulse is expected to change the phase boundary potential at the outer membrane surface. Under voltage clamping, this change in the phase boundary potential is automatically compensated by a change in the intramembrane IR drop. In a membrane with fixed charges, the phase boundary potentials play a dominant role in determining the membrane potential (see section on Properties of Ionic Membranes). The clamping device can directly alter only the intramembrane potential by superposing an IR-drop on the diffusion potential.

The term "membrane emf" is used to describe the clamped level of the membrane potential minus the superposed IR-drop. The time course of the membrane emf during clamping can be determined by superposing a weak sinusoidal voltage on the rectangular clamping voltage (Tasaki and Spyropoulos, 1958). The advantage of this method is that the entire time course of the membrane emf, $E(t)$, can be determined from a single oscillograph record (see Fig. 19-9) using the derived relationship $E(t) = V - A(I_s/I_a)$, where V is the potential of the clamping pulse, A is the amplitude of the superposed sinusoidal voltage, I_a is the amplitude of the sinusoidal component of the membrane current and I_s is the slow component of the membrane current. [The factor A/I_a represents the membrane resistance; therefore, $A(I_s/I_a)$ is the intramembrane IR-drop.] The intensity and the time course of the slow component, I_s, is unaffected by the superposition of I_a on the clamping pulse.

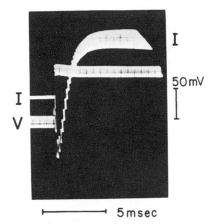

Fig. 19-9 *Voltage clamp with a high frequency sinusoidal wave superposed on rectangular voltage pulses. Upper trace displays the time course of the membrane current; lower trace, the clamping voltage. The frequency of A.C. was 8 kHz. The vertical bar represents 50 mv for the voltage trace and 2 ma/cm$_2$ for the current trace. Temperature 5° C. (From Tasaki and Spyropoulos, 1958.)*

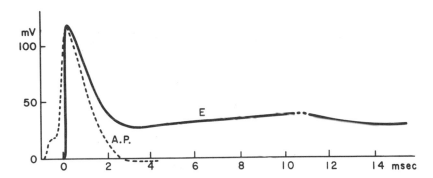

Fig. 19-10 *The result of the determination of E(t) by the A.C. method. The trace showing the time course of the membrane emf under voltage clamp conditions was obtained by calculating the value of V — A(I_s/I_a), where V is the potential of the clamping pulse and A(I_s/I_a) is the intramembrane IR-drop. The broken line shows the time course of the action potential recorded from the same giant axon. (From Tasaki and Spyropoulos, 1958.)*

It is seen in Fig. 19-10 that the initial portion of the time course E(t), determined by the A.C. method, is similar to the normal action potential recorded under unclamped conditions. The latter portion of the curve which deviates from the action potential will be discussed later. This direct demonstration of the change in membrane emf while the clamping pulse remains constant requires one to reconsider the interpretation of the inward and outward currents observed during voltage clamping.

Fig. 19-11A is a typical current-voltage (I-V) curve obtained from plotting the peak inward currents against clamping potential. The straight portion of the curve, $I = g_r(V - E_{RP})$ represents the ohmic character of the nerve in the resting state. When the membrane potential is made more negative than the resting level $(V < E_{RP})$ by the passage of inward current, additional divalent cations from the external medium are driven into the already divalent cation-rich membrane. Since the axon membrane remains in the resting state in this case, the observed change in membrane potential is directly proportional to the current applied. When the membrane potential is made less negative by the passage of weak outward current across the membrane, the outward current remains roughly proportional to the potential until the clamping pulse is raised to 5 mv above the resting potential. With a clamping voltage pulse above 15 mv, an inward current across the membrane is observed which increases as the clamping level is raised until potential level b in 19-11A is reached. At higher clamping pulses the intensity

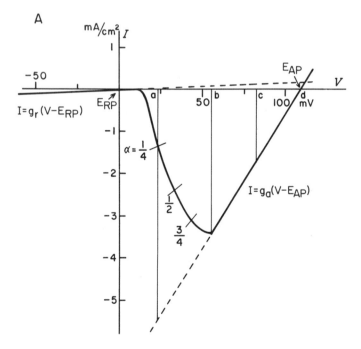

Fig. 19-11A   *Analysis of* $I - V$ *relationship in terms of nonuniformity of axon membrane and changes in membrane emf under voltage clamp. Current voltage diagram illustrating the ohmic behavior of the resting state,* $I = g_r(V - E_{RP})$, *and of the active,* $I = g_a(V - E_{AP})$. *In the range of depolarizations between 20 and 45 mv, the fraction of membrane subunits in the active state* α *is less than unity.* E_{RP} *represents the potential level of the resting membrane and* E_{AP} *the peak value of the action potential.* g_r *and* g_a *are the conductances in the resting and active states, respectively, and* V *is the clamped voltage.*

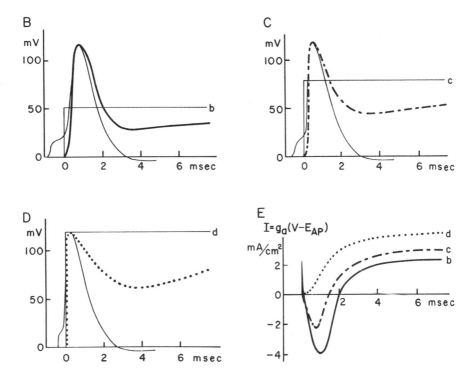

Fig. 19-11 B–E: B, C and D. *Time course of membrane emf determined by the A.C. method at clamping levels corresponding to b, c, and d in A. The action potential recorded from the unclamped axon is given in the three records. E. Early inward and delayed outward currents observed at clamping levels b, c and d. The inward and outward currents are supplied by the feedback amplifier to create an intramembrane IR-drop which is equal to the differences between the membrane emf and clamping potential levels. Potential level d corresponds to E_{AP}; there is no initial inward current in this case. Also note that the divergence of the falling phase of the membrane emf from the action potential increases as the intensity of the delayed outward current is increased (see text).*

of the inward current decreases, and the I-V relationship can be described by the equation $I = g_a(V - E_{AP})$, which represents the ohmic behavior of the membrane in the uniformly excited state.

When the membrane potential is clamped at the peak level of the action potential, namely, when $V = E_{AP}$, the peak inward current is zero; the membrane emf is equal to the voltage V in this case (see 19-11D and the dotted line in 19-11E). As the clamping pulse is decreased from *d* to *c* and then to *b* in Diagram *A*, the difference between the clamping level and the peak of the action potential is increased and consequently, the intensity of inward current increases (see Diagrams *C* and *B*). The I-V relationship is linear between *b* and *d*, indicating that both E_{AP} and g_a are constant (independent of V) in this range; hence, $E_{AP} = V - (I/g_a)$. It is obvious, therefore, that the membrane emf, E_{AP} is equal to the clamping level, V, less the IR-drop, I/g_a.

We now turn to the temporal changes in $E(t)$. As seen in 19-11B, C, and D, when the membrane emf shifts toward the resting level, $E(t)$ falls below the clamping pulse level, V. The feedback device must now provide an outward current across the membrane to create an IR-drop of the reversed sign to maintain the constant potential difference between the recording electrodes. According to the cation exchange membrane model, this maintained outward current raises the potassium concentration at the negative sites and the membrane remains steadily depolarized. This explains why the time course $E(t)$ diverges from the normal action potential. It is expected that, as the intensity of the outward current is increased, the divergence of $E(t)$ becomes larger. (Note Fig. 19-11B, C and D.)

The intermediate portion of the I-V curve which has a negative slope can be explained in terms of the increasing number of membrane subunits which become excited as the clamping voltage is raised above the resting potential. If α represents the fraction of excited subunits and $(1 - \alpha)$ the fraction of resting membrane subunits in a given area of membrane, the total membrane current should be given by the sum of the current through the membrane area in the excited state and the current through the resting portion of the membrane:

$$I \approx (1 - \alpha)g_r(V - E_{RP}) + \alpha g_a(V - E_{AP}) \qquad (19\text{-}29)$$

where V is the clamped voltage and g_r, E_{RP} and g_a, E_{AP} are conductances and membrane potentials in the resting and active states respectively, and are thus treated as constants. It is seen from this equation that in the intermediate range of the I-V curve corresponding to the mixed state of the membrane, *the membrane current is a function of the clamping level, V, and the active fraction α.* As V is increased from about 15 to 60 mv, α increases from approximately zero to unity. For a given value of α, the membrane conductance $\delta I(V, \alpha)/\delta V$ is always positive. Since a weak high frequency sinusoidal component superposed on the rectangular clamping pulse is considered not to alter the active fraction, α, the conductance $\delta I(V, \alpha)/\delta V$ can be determined directly by the A.C. method (Tasaki and Spyropoulos, 1958).

In Fig. 19-11A, the I-V relationship for a fixed value of α (i.e., $\alpha = 1/4$, 1/2 and 3/4) is shown by short lines with positive slopes. When the membrane is clamped at a voltage pulse corresponding to level a, the intensity of inward current supplied by the feedback amplifier is determined by the magnitude $|V - E_{AP}|$ and by the particular value of α at level a. If at this clamping level, it were physiologically possible to increase α from 1/4 to 1, the intensity of inward current would be about 4 times the observed values, represented by the intersection of the vertical line at a, with the extrapolated linear segment of the I-V plot corresponding to the uniformly excited state of the membrane. The reason that α does not increase from $1/4$ to 1 at this clamping level is explained in terms of the effect of inward current on the ratio of univalent to divalent cations within the membrane. As the difference $|V - E_{AP}|$ increases, the inward current tends to increase. However, a strong inward current brings about changes in the normal pattern of the interdiffusion fluxes [see Eq. (19-20)]. (This change becomes appreciable when $V - E_{AP}$ exceeds RT/F, which is about 25 mv at room temperature.) A strong inward current tends to increase the Ca ion concentration in the outer layer of the membrane. Therefore, the subunits which are in the excited state are forced to undergo transition to the resting state when the density of the inward current exceeds a certain limit.

The concept of active fractions derives directly from our basic postulate that the axon membrane consists of discrete subunits which can be in either of two states, excited or resting (see Fig. 19-7). In this theory of nerve excitation, the quantity α is not a mere adjustable parameter to fit the observed I-V curve to Eq. (19-29). It is found from Eq. (19-29) that

$$\frac{\partial I}{\partial V} \approx (1 - \alpha)g_r + \alpha g_a. \tag{19-30}$$

Because g_r is $1/100$ to $1/200$ the value of g_a, Eq. (19-30) may be written

$$\frac{\partial I}{\partial V} = \alpha g_a. \tag{19-30a}$$

Since g_a is known, the active fraction α is a quantity which is accessible to direct determination by the A.C. method. In fact, it has been shown (Tasaki and Spyropoulos, 1958) that the intermediate portion of the observed I-V relationship can be reconstructed from the values of α introduced in Eq. (19-29).

The above interpretation of the I-V curve is consistent with the notion that there is a strong interdiffusion flux across the axon membrane in the excited state (see section on ion flux measurements). In *unclamped* axons internally perfused with a 400 mM K salt solution and immersed in a medium containing 400 mM NaCl and 100 mM CaCl$_2$, the fluxes of Ca and anions are small; therefore,

$$j_K + j_{Na} \approx 0, \tag{19-21}$$

since the unclamped membrane carries no current. The peak value of this interdiffusion flux is estimated to be about 3×10^{-8} equivalent cm^{-2} sec^{-1}; hence the product Fj_{Na} is roughly 3 ma/cm^2. (The interdiffusion flux is estimated from the results of isotope measurements which require long repetitive stimulation of the axon; it is reasonable to assume that the fluxes associated with an isolated single stimulus are accompanied by much stronger interdiffusion fluxes.)

A peak inward current observed *under voltage clamp* is given by:

$$I \approx F(j_{Na} + j_K). \tag{19-31}$$

By comparing this equation with Eq. (19-21), it is apparent that an inward current of the order of a few ma/cm^2 is carried by *both Na and K ions*. In other words, the net inward current observed during clamping represents both a small decrease in K efflux and a small increase in Na influx.

It must be re-emphasized that the early inward and delayed outward currents observed during clamping are not present during the normal action potential under unclamped conditions. The local currents which are responsible for the propagation of the nerve impulse are *outward* through the resting regions of the membrane and *inward* through the active regions. As previously noted, the ion fluxes associated with these currents are very small and account for only 1/10 to 1/17 of the interdiffusion flux in the excited state.

It is clear from the analysis of the I-V relationship that any maneuver which abolishes or delays the abrupt transition of the membrane macromolecules from the resting to the excited state will eliminate the early inward current. Thus, when choline is added to the external medium in sufficient quantity, the rate of transition of the subunits from the resting to the excited state is diminished (see previous section) and the early inward current observed during clamping can be abolished. The puffer fish toxin, tetrodotoxin, has a similar effect.

OPTICAL STUDIES

Studies of the secondary and tertiary structures of polypeptides and other macromolecules and analyses of the forces which determine the stability or liability of the conformation of these macromolecules have during the past two decades provided new insights into our understanding of many biological processes. Therefore, it should come as no surprise that the process of excitation can also be described in terms of conformational changes of labile membrane macromolecules. Many of the techniques used by the biochemist for the study of molecular conformation involve the interaction of electromagnetic waves with macromolecules. Recently, high time resolution optical and electrical equipment have been developed which make it possible to determine changes in turbidity, birefringence, and fluorescence in nerves during an action potential (Cohen, Keynes and Hille, 1968; Tasaki, Watanabe, Sandlin and Carnay, 1968; Berestovskii, Lunevskii, Razhin, Musienko and Liberman, 1969).

Changes in Light Scattering and Birefringence during Nerve Activity

In our studies of light scattering and birefringence, crab leg nerves, lobster leg nerves, squid fin nerves and squid giant axons were used. The nerves were mounted in a chamber with two pairs of electrodes; one pair was used to stimulate the nerve and the other pair to record the action potential (see Fig. 19-12). A 3–6 mm portion of the nerve preparation was illuminated with quasi-monochromatic light obtained by using interference filters and either a Xenon-Mercury lamp or a D.C. operated quartz iodine lamp. The wave lengths of the monochromatic light used was varied between 220 and 850 nm. Changes in light intensity were detected with a photomultiplier placed above the nerve at either 0 or 90 degrees to the incident light. With all the wave lengths examined there was a transient increase in the intensity of scattered light at 90° during the action potential. These changes in light scattering are expected if either the thickness or refractive index of the cylindrical wall of the individual nerve fibers increase during impulse propagation (Born and Wolf, 1959).

In the birefringence studies, a polarizer was positioned below the nerve chamber and an analyzer in the crossed polar position was placed between the nerve and the photomultiplier. When the nerve was placed at an angle of 45° to the plane of polarization of the monochromatic light, there was a decrease in the intensity of transmitted light through the analyzer during the action potential. The optical signal may be interpreted as indicating a decrease in birefringence of the nerve during the action potential. The birefringence of the nerve fibers is considered to be derived mainly from longitudinally oriented structures smaller than the wave length of the incident polarized light but larger than the size of individual molecules. Thus, the decrease in birefringence may result from a transient decrease in the refractive index between the filamentous structures and the surrounding environment.

Changes in Fluorescence during Nerve Activity

Studies of the changes in the extrinsic fluorescence of nerve during impulse propagation offer a more direct means of investigating the molecular events occurring in the membrane during the transition from the resting to the excited state (Tasaki, Carnay, Sandlin and Watanabe, 1969; Tasaki, Barry and Carnay, 1969; Tasaki, Carnay and Watanabe, 1969). Nerve trunks were stained with fluorescent dyes by immersing the nerves in sea water containing the dye for a period of 10–15 min. The concentration of the dyes employed was usually between 0.05 and 0.3 mg/ml. Covalent bonding of a dye with the membrane could be achieved by immersing a nerve in sea water containing fluorescein isothiocyanate for a period of 1 to 3 hours. When internal application of a dye was required, squid giant axons were used. The dye solution was introduced into the interior of the axon by the intracellular perfusion technique and was left inside during the subsequent studies. The fluorescence of the nerve was examined at either 90° or 135° to the direction of the incident light. The nerve chamber was similar to that used in the light scattering experiments. Light from the quartz iodine or Xenon-Mercury source was converted into quasi-monochromatic light corresponding in wave length to the absorption maximum of the particular dye employed. A secondary filter was placed between the nerve and the photomultiplier to interrupt the scattered incident light and pass emitted light.

Three examples of the optical signals obtained by the procedure mentioned above are presented in Fig. 19-12. The results of the fluorescence studies are summarized in Table 19-1. The names and chemical structures of the compounds are given as well as the wave lengths of excited quasi-monochromatic light used and filter half band widths. Signs and relative magnitudes of the observed optical signals, which are between 5×10^{-5} and 10^{-5} times the light intensity observed before stimulation, are given for three different nerve preparations. The dyes were applied intracellularly in the case of the squid giant axons.

It is seen in Fig. 19-12 that nerves stained externally with either pyronin B or rhodamine B give rise to a transient decrease in fluorescence intensity during the action potential. It is noted from the results in Table 19-1, that intracellularly applied pyronin B or 8-anilinonaphthalene sulfonate (ANS) gives rise to signals opposite to those obtained from crab and squid fin nerves externally stained with pyronin B or ANS.

The molecular basis of emission of light by the dye molecules and the mechanism of quenching in the nerve membrane are complex and may be different for each fluorochrome, but some interpretation of the data can be made. It is well known from studies of relatively pure proteins stained with fluorescent dyes that a change in the molecular conformation of the protein drastically changes the intensity of fluorescence of the bound dye molecules (Edelhoch and Steiner, 1964). The magnitudes of the observed optical signals obtained from nerve fibers do not necessarily indicate that the extent of structural change in the nerve membrane is very small. Only a small fraction of the cellular elements in the nerve, representing less than 0.1 per cent of the total volume of the nerve, is involved in the excitation process. Therefore, the structural change in the axolemma and the layers in its immediate vicinity may indeed be quite large.

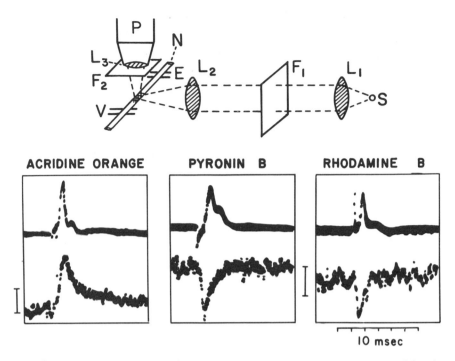

Fig. 19-12 *(Top) Schematic diagram of the experimental arrangement used for the detection of fluorescence changes associated with nerve conduction. The letter S represents the light source; L_1, L_2 and L_3, lenses; F_1 and F_2, optical filters; N, nerve; E and V, stimulating and recording electrodes, respectively and P, photomultiplier tube. (Bottom) Action potentials (top) and fluorescence changes (bottom) obtained from crab nerves stained with acridine orange, pyronin B and rhodamine B. A CAT computer was used to record both action potentials and optical signals. The amplitude of the action potentials was 2 mv. An upward deflection of the lower trace indicates an increase in the intensity of fluorescence. The vertical bars represent a change of 5×10^{-5} times the fluorescence intensity before stimulation. The time scale applies to all three records. Temperature 22° C.*

There was no direct relationship between the sign of the optical signal and the electric charge of the dye molecules or the type of binding. For example, acridine orange molecules are positively charged and are assumed to bind at negatively charged sites of the membrane macromolecules. The binding of negatively charged ANS is hydrophobic in nature. Yet, positive optical signals are obtained when nerves are externally stained with these two different fluorochromes.

Conformational changes in various macromolecules have been detected using the dye ANS which fluoresces more efficiently when it is bound to the nonpolar sites of macromolecules (Weber and Laurence, 1954). In the perfused squid axon it is reasonable to assume that the ANS molecules are bound in the hydrophobic layer of the membrane. Thus, the decrease in

Table 19-1   Compounds used to demonstrate fluorescence changes in nerve. Symbols used: ANS, 8-anilinonaphthalene-1-sulfonate; DNS, 1-dimethylaminonaphthalene-5-sulfonyl chloride; FIT, fluorescein isothiocyanate; LSD, lysergic acid diethylamide; TNS, 2-*p*-toluidinylnaphthalene-sulfonate. Half band widths of filters used are given in parentheses. Strong, medium and weak optical signals are shown by difference in number of + (increase) and − (decrease) symbols; the absence of a measurable signal is shown by 0.

Compound	Excitation wave length in nm	Giant axon	Fin nerve	Crab nerve
Acridine Orange	465(20)	+++	+++	+++
Acridine Yellow	450(20)	0	+	++
ANS	365(10)	− − −	++	+++
Auramine O	450(5)	0	0	++
DNS	365(10)	0	0	+

Compound	Excitation wave length in nm	Giant axon	Fin nerve	Crab nerve
FIT	500(20)	+(ext)	++	++
LSD	365(10)	0	+	++
Pyronin B	550(5)	++	-- --	--- --
Pyronin Y	550(5)	0	--	--
Rhodamine B	550(5)	--	---	---

Compound	Excitation wave length in nm	Giant axon	Fin nerve	Crab nerve
Rhodamine G	500(20)	0		++
Rose Bengal	550(5)	0		++
TNS	365(20)	–		–

fluorescence following stimulation may be explained by the large increase in the water content which may occur when the membrane is in the excited state. The difference between internally and externally applied ANS may indicate that, when externally applied, the negatively charged ANS molecule is excluded from the hydrophobic layer of the membrane.

Fluorescence Polarization

Direct information about the viscosity of the medium surrounding the fluorescent molecules in the resting and excited states of the membrane can be obtained by measurements of fluorescence polarization. The principles upon which such determinations are based may be briefly summarized. When the incident light is polarized, it excites predominantly those dye molecules whose direction of absorption oscillator is nearly parallel to the electric vector of polarized light. The emitted light will remain polarized if the dye molecules do not tumble rapidly relative to the lifetime of the excited state of the molecules (10^{-8} sec). If the viscosity of the medium is low, the molecules are free to tumble due to their rotational Brownian motion and they become randomly oriented before appreciable fluorescence is emitted. Thus, the polarization of the emitted fluorescent light is diminished as the viscosity of the medium surrounding the dye molecules is decreased.

The degree of polarization of fluorescent light from dye molecules in the nerve can be determined by modifying the experimental setup described earlier for detection of fluorescence signals. A polarizer is placed between the light source and the nerve so that the electric vector of the exciting light is parallel to the long axis of the nerve. An analyzer is positioned between the secondary filter and the photomultiplier (see Fig. 19-13). In the resting state,

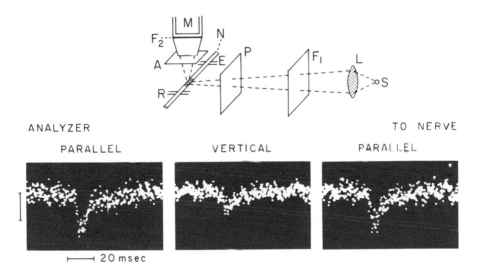

Fig. 19-13 (*Top*) *Schematic diagram showing the experimental setup used to measure changes in fluorescence polarization associated with production of action potential in a crab nerve stained with pyronin B. The letter S represents the light source L, lens; F_1 primary filter; F_2, secondary filter; M, multiplier phototube; P, polarizer; A, analyzer; N, nerve; E, stimulating electrodes; V, recording electrodes. (Bottom) Optical signals obtained with the principal axis of the analyzer (E-vector of fluorescent light) parallel to the long axis of the nerve (Records 1 and 3) and perpendicular to the nerve (Record 2). The quasi-monochromatic light used for excitation of dye molecules was polarized (with its E-vector) in the direction parallel to the nerve. A CAT computer was used for recording. For records 1 and 3, the vertical bar represents 2×10^{-4} times the background intensity. Record 2 was taken under the same experimental conditions except that the analyzer was rotated by 90°. Temperature, 19° C. (Adapted from Tasaki, Carnay and Watanabe, 1969.)*

the degree of polarization of fluorescent light, P_r, is given by the equation:

$$P_r = \frac{I_r' - I_r''}{I_r' + I_r''},\qquad(19\text{-}32)$$

where I_r' and I_r'' are the intensities of fluorescent light polarized longitudinally and vertically to the nerve.

The degree of polarization from randomly oriented immobile fluorochromes, with the absorption and emission oscillators parallel to each other, is known to be 0.5. A value close to this theoretical upper limit was obtained from fluorescein isothiocyanate coupled covalently with the nerve membrane in the resting state.

The value of P_r for pyronin B molecules in nerves at rest was 0.16. The degree of polarization during the excitation process is given by the equation $P_a = (I_a' - I_a'')/(I_a' + I_a'')$ where I_a' and I_a'' are the intensities of light observed during the action potential with the analyzer parallel and vertical to the nerve respectively. By expressing the fluorescence change as the ratio of the change in fluorescence during the action potential to the background intensity, the following relationship for the difference in polarization between the active and resting state may be obtained (Tasaki, Carnay and Watanabe, 1969).

$$P_a - P_r = \frac{1}{2}\left[\frac{I_a' - I_r'}{I_r'} - \frac{I_a'' - I_r''}{I_r''}\right](1 - P_r^2).\qquad(19\text{-}33)$$

The change in the degree of polarization of pyronin B molecules in crab nerve during the action potential was determined in the following manner. The stained nerve was irradiated with quasi-monochromatic light polarized in the longitudinal direction of the nerve. Initially, changes in the intensity of emitted polarized light were measured during the action potential with the electric vector of the emitted light parallel to the nerve (see Fig. 19-13, record 1). The analyzer was then rotated through 90° and a measurement was made of the fluorescence change during impulse conduction (record 2). A third measurement was made with the analyzer in the original position (record 3). It was found that the signal obtained with the analyzer parallel to the nerve is almost twice as large as that obtained with the analyzer vertical to the nerve. By substituting the appropriate values in Eq. (19-33), it can be shown that the degree of polarization of the dye molecules decreases during the action potential. This finding strongly suggests that the increase in the rotational Brownian movement of the dye molecules during the action potential is due to a decrease in the viscosity in the micro-environment of the dye molecules in the nerve membrane. This change in viscosity is consistent with the electrophysiological data indicating that there is an increase in intramembrane ion mobilities during impulse conduction.

SUMMARY

The development of the intracellular perfusion technique has resulted in a better understanding of the ionic requirements of excitation and has made possible an approach to the elucidation of the molecular mechanism of the excitation process.

A theory of nerve excitation is presented which is based on the presence of negatively charged membrane sites and the ability of the membrane to function as a cation exchanger. It is therefore postulated that ion fluxes, resistance and potentials are determined by the fixed charge density, cation selectivity and ion mobilities *within* the membrane.

In the resting state, the negative sites are predominantly occupied by divalent cations which form a hydrophobic complex with the membrane macromolecules. The excitation process is triggered by the electrophoresis of internal univalent cations into the membrane by the passage of outward current across the membrane. At a critical univalent-divalent cation concentration ratio in the membrane, a cooperative exchange occurs whereby the major portion of the negative sites become occupied by univalent cations. When the hydrophobic complex is reversibly disrupted in this manner, the fixed charge density, ion selectivity and ion mobilities within the membrane abruptly change. Consequently, there is a large increase in conductance and univalent interdiffusion during the physiological excited state. The change in membrane potential is due mainly to the variation in the phase boundary potentials and partly to the change in the intramembrane diffusion potential. The cooperative cation exchange model is thus very different from the view that the membrane is "permeable" specifically to a particular ionic species during the rising phase of the action potential.

The increased interdiffusion during excitation alters the ionic milieu in and near the membrane changing the membrane properties as divalent cations reaccumulate at the membrane negative sites. This exchange of divalent for univalent cations results in the return of the molecular conformation to the resting state.

Recent observations on the changes in the optical properties of nerves during the action potential are consistent with the view that there are changes in macromolecular conformation during nerve excitation.

REFERENCES

BAYLIS, W. M. 1924. *Principles of General Physiology.* Longmans, Green, London.
BERESTOVSKII, G. N., V. Z. LUNEVSKII, V. C. RAZHIN, V. S. MUSIENKO and E. A. LIBERMAN. 1969. *Pushchino Viniti* (USSR).
BORN, M. and E. WOLF. 1959. *Principles of Optics.* Pergamon Press, New York.

CHANGEUX, J. P., J. THIERY, Y. TUNG and C. KITTEL. 1967. On the Cooperativity of Biological Membranes. *Proc. Nat. Acad. Sci. USA.* 57:335–41.

COHEN, L., R. D. KEYNES and E. HILLE. 1968. Light Scattering and Birefringence during Nerve Activity. *Nature (London).* 218:433–41.

COLE, K. S. 1949. Dynamic Electrical Characteristics of the Squid Axon Membrane. *Arch. Sci. Physiol. (Paris).* 3:253–58.

COLE, K. S., and H. J. CURTIS. 1939. Electric Impedance of the Squid Giant Axon during Activity. *J. Gen. Physiol.* 22:649–70.

DANIELS, F., and R. ALBERTY. 1961. *Physical Chemistry.* John Wiley and Sons, New York.

EDELHOCH, H., and R. F. STEINER. 1964. *Electronic Aspects of Biochemistry.* Academic Press, New York.

FATT, P., and B. KATZ. 1953. The Electrical Properties of Crustacean Muscle Fibers. *J. Physiol. (London).* 120:171–204.

HAFEMAN, D. R. 1965. Charge Separation in Liquid Junctions. *J. Phys. Chem.* 69:4226–38.

HELFFERICH, F. 1962. *Ion Exchange.* McGraw-Hill, New York.

HELFFERICH, F., and H. D. OCKER. 1957. Ionenaustauschermembranen in Bi-ionischen Systemen. *Z. Phys. Chem.* (Neue Folge). 10:213–35.

HODGKIN, A. L. 1951. The Ionic Basis of Electrical Activity in Nerve and Muscle. *Biol. Rev.* 26:339–409.

————. 1964. *The Conduction of the Nervous Impulse.* Liverpool University Press, Liverpool.

HODGKIN, A. L., and A. F. HUXLEY. 1952. A Quantitative Description of Membrane Current and its Application to Conduction and Excitation in Nerve. *J. Physiol. (London).* 117:500–14.

HODGKIN, A. L., A. F. HUXLEY and B. KATZ. 1952. Measurement of Current-Voltage Relations in Membrane of the Giant Axon of *Loligo. J. Physiol. (London).* 116:424–48.

IKEGAMI, A., and N. IMAI. 1962. Precipitation of Polyelectrolytes by Salts. *J. Polymer. Sci.* 56:133–52.

KOKETSU, K., J. A. CERF and S. NISHI. 1959. Effects of Quaternary Ammonium Ions on Electrical Activity of Spinal Ganglion Cells in Frogs. *J. Neurophysiol.* 22:177–94.

LENARD, J., and S. J. SINGER. 1966. Protein Conformation in Cell Membrane Preparations as Studied by Optical Rotatory Dispersion and Circular Dichroism. *Proc. Nat. Acad. Sci. USA.* 56:1828–35.

LERMAN, L., A. WATANABE and I. TASAKI. 1969. Intracellular Perfusion of Squid Giant Axons. *In* S. Ehrenpreis [ed.] *Neurosciences Research,* Vol. 2. Academic Press, New York.

OIKAWA, T., C. S. SPYROPOULOS, I. TASAKI and T. TEORELL. 1961. Methods for Perfusing the Giant Axon of *Loligo pealei. Acta Physiol. Scand.* 52:195–196.

OLSON, D. H., and H. S. SHERRY. 1968. An X-ray Study of Strontium-Sodium Ion Exchange in Linde X: An Example of a Two Phase Xeolite System. *J. Phys. Chem.* 72:4095–4104.

OSTERHOUT, W. J. V., and S. E. HILL. 1933. Anesthesia Produced by Distilled Water. *J. Gen. Physiol.* 17:87–98.

_____. 1938. Calculations of Bioelectric Potentials. *J. Gen. Physiol.* 22:139–46.

ROBINSON, R. A., and R. H. STOKES. 1959. *Electrolyte Solutions.* 2nd ed. Buttersworth, London.

ROUSER, G., G. NELSON, S. FLEISCHER and G. SIMON. 1968. Composition of Membranes, Organelles and Organs. *In* D. Chapman [ed.] *Biological Membranes.* Academic Press, London.

SINGER, I., and I. TASAKI. 1968. Nerve Excitability and Membrane Macromolecules. *In* D. Chapman [ed.] *Biological Membranes.* Academic Press, London.

TASAKI, I. 1959. Demonstration of Two Stable States of the Nerve Membrane in Potassium-Rich Media. *J. Physiol. (London).* 148:306–31.

_____. 1963. Permeability of Squid Giant Axon Membrane to Various Ions. *J. Gen. Physiol.* 46:755–72.

_____. 1968. *Nerve Excitation: A Macromolecular Approach.* C. Thomas, Springfield. Ill.

TASAKI, I., W. BARRY and L. CARNAY. 1969. Optical and Electrophysiological Evidence for Conformational Changes in Membrane Macromolecules during Nerve Excitation. *In The Proceedings of the Coral Gables Conference on the Physical Principles of Biological Membranes.* Gordon and Breach, New York.

TASAKI, I., L. CARNAY, R. SANDLIN and A. WATANABE. 1969. Fluorescence Changes during Conduction in Nerves Stained with Acridine Orange. *Science.* 163: 683–85.

TASAKI, I., L. CARNAY and A. WATANABE. 1969. Transient Changes in Extrinsic Fluorescence of Nerve Produced by Electric Stimulation. *Proc. Nat. Acad. Sci. USA.* In press.

TASAKI, I., and I. SINGER. 1966. Membrane Macromolecules and Nerve Excitability: A Physico-Chemical Interpretation of Excitation in Squid Giant Axons. *Ann. N.Y. Acad. Sci.* 137:792–806.

_____. 1968. Some Problems Involved in Electrical Measurements of Biological Systems. *Ann. N.Y. Acad. Sci.* 148:36–53.

TASAKI, I., I. SINGER and T. TAKENAKA. 1965. Effects of Internal and External Ionic Environment on Excitability of Squid Giant Axon. *J. Gen. Physiol.* 48:1095–1123.

TASAKI, I., I. SINGER and A. WATANABE. 1965. Excitation of Internally Perfused Squid Giant Axons in Sodium-Free Media. *Proc. Nat. Acad. Sci. USA.* 54:763–69.

_____. 1967. Cation Interdiffusion in Squid Giant Axon. *J. Gen. Physiol.* 50: 989–1007.

TASAKI, I., and C. S. SPYROPOULOS. 1958. Membrane Conductance and Current-Voltage Relation in the Squid Axon under Voltage Clamp. *Am. J. Physiol.* 193:318–27.

TASAKI, I., and T. TAKENAKA. 1963. Resting and Action Potential of Squid Giant Axons Intracellularly Perfused with Sodium-Rich Solutions. *Proc. Nat. Acad. Sci. USA.* 50:619–26.

TASAKI, I., T. TAKENAKA and S. YAMAGISHI. 1968. Abrupt Depolarization and Bi-ionic Action Potentials in Internally Perfused Giant Axons. *Am. J. Physiol.* 215:152–59.

TASAKI, I., A. WATANABE and L. LERMAN. 1967. A Study of the Role of Divalent Cations in Excitation of Squid Giant Axons. *Am. J. Physiol.* 213:1465–74.

TASAKI, I., A. WATANABE, R. SANDLIN and L. CARNAY. 1968. Changes in Fluorescence Turbidity and Birefringence Associated with Nerve Excitation. *Proc. Nat. Acad. Sci. USA.* 61:883–88.

TEORELL, T. 1935. An Attempt to Formulate a Quantitative Theory of Membrane Permeability. *Proc. Soc. Exp. Biol. Med.* 33:282–85.

———. 1953. Transport Processes and Electrical Phenomena in Ionic Membranes. *Progr. Biophys.* 3:305–69.

WATANABE, A., I. TASAKI and L. LERMAN. 1967. Bi-ionic Action Potentials in Squid Giant Axons Internally Perfused with Sodium Salts. *Proc. Nat. Acad. Sci. USA.* 58:2246–52.

WEBER, G., and D. J. R. LAURENCE. 1951. Fluorescent Indicators of Adsorption in Aqueous Solution and on the Solid Phase. *Biochem. J.* 56:xxxl.

WERMAN, R., F. V. MCCANN and H. GRUNDFEST. 1961. Graded and All or None Electrogenesis in Arthropod Muscle. *J. Gen. Physiol.* 44:979–95.

WYLLIE, M. R. J. 1954. Ion Exchange Membrane. I. Equation for Multi-ionic Potential. *J. Phys. Chem.* 58:67–73.

NEUROPHYSIOLOGICAL BASIS FOR DRUG ACTION: IONIC MECHANISM, SITE OF ACTION AND ACTIVE FORM IN NERVE FIBERS

20

T. NARAHASHI

There are a number of aspects of nerve function which should be studied to elucidate the mechanism of action of chemicals on nerve fibers. First of all, one might ask a question as to *how* a chemical exerts any particular effect on the nerve fiber. For example, if the chemical blocks the nerve conduction, the mechanism may lie in a decrease in the resting membrane potential, or if the resting potential remains unchanged, the blockage may be due to an interaction of the chemical with certain membrane ionic conductances. Even after the conductance mechanism has been elucidated, questions still remain as to how this effect on the membrane conductance is brought about. Some physico-chemical interactions of the chemical with the membrane macromolecule components are no doubt involved in the conductance blockage. On the other hand, if the resting potential is decreased by the chemical, the mechanism could be due to some metabolic disturbances in the nerve fiber which may eventually lead to changes in membrane permeability and in ionic concentration gradients across the membrane.

In the second place, one might ask a question as to the *site* of action of the chemical in the nerve fiber. It could be inside the axon, or in the nerve membrane. In the latter case, the site of action may be on the external surface, internal surface or inside the membrane.

The third question one might ask is which *form* is responsible for the action in case the chemical exists in more than one form. For example, if the drug is dissociated into charged and uncharged forms, only one of these forms may be active or both forms may be active.

These three major questions about the mode of action of chemicals on the excitability of the nerve fibers may be answered, at least to a limited extent, by appropriate experimentation. The present article describes the results of experiments performed with a variety of neuroactive agents. The scope here is strictly limited to the mechanism of drug action in terms of membrane ionic conductances, the site of action and the active form. No attempt will be made to explore the mechanisms beyond this scope such as those at the physico-chemical and molecular levels.

IONIC MECHANISM OF DRUG ACTION

Tetrodotoxin

Tetrodotoxin (TTX) is the active ingredient of the poison from the puffer fish. It is contained in the ovary and liver, and also in the skin and intestines in some species. The existence of the poison has been known since the early Egyptian era and was recorded even in the first Chinese pharmacopea, The Herbal (2838–2698 B.C.) (Kao, 1966). Especially in Japan, the poison has drawn much attention because the puffer fish is served as one of the most delicious fish. Even nowadays, some 150 cases of accidental death caused by the puffer fish poison occur every year. For this reason, the toxin and its mode of action have been extensively studied in Japan since the early part of this century. The history and study of TTX until 1965 were reviewed in detail by Kao (1966).

Tahara extracted the poison for the first time in 1910, and since then a number of pharmacological studies have been performed. However, it was not until 1960 that the mechanism of action at the cellular level was unveiled. Using the sartorius muscle fibers of the frog, TTX was found to block the action potential without affecting the resting potential and delayed rectification (Narahashi, Deguchi, Urakawa and Ohkubo, 1960). It was suggested that TTX selectively blocks the mechanism whereby the membrane conductance to sodium is increased upon depolarization. Voltage clamp experiments with the giant axon of the lobster have indeed demonstrated that TTX blocks the peak transient (sodium) current without any effect on the late steady-state (potassium) current (Narahashi, Moore and Scott, 1964). This was later confirmed by other investigators with squid giant axons (Nakamura, Nakajima and Grundfest, 1965a), with lobster giant axons (Takata, Moore, Kao and Fuhrman, 1966), and with frog nodes of Ranvier (Hille, 1968).

The chemical structure of TTX was studied by two Japanese and two American groups, and their results were reported at the International

Conference on Chemistry of Natural Products held in Kyoto, Japan, in 1964. The chemical structure is shown in Fig. 20-1. It contains a guanidinium group and is unique in having a hemilactal link between C_5 and C_{10}. It exists in two cation forms and a zwitterion form, and the pK_a value is estimated as 8.76 (Goto, Kishi, Takahashi and Hirata, 1965), or 8.84 (Tsuda, Ikuma, Kawamura, Tachikawa, Sakai, Tamura and Amakasu, 1964). Another unique property is that TTX is not soluble in most organic solvents.

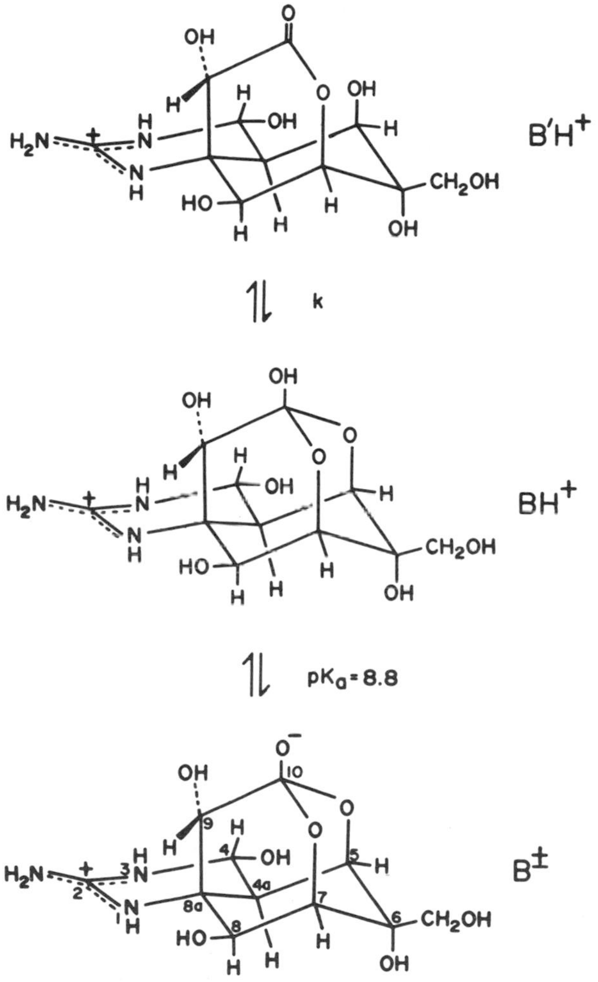

Fig. 20-1 Chemical structures and dissociation of tetrodotoxin molecule. (Modified from Narahashi, Moore and Frazier, 1969 © 1969, The Williams & Wilkins Co., Baltimore, Md. 21202, U.S.A.)

This hindered the chemical study somewhat, but gave an excellent opportunity to locate its site of action in the nerve membrane (see page 439).

As described before, TTX selectively blocks the peak transient conductance change at very low concentrations (in the order of 10 nM). Several questions may be raised regarding the characteristics of this blocking action:

1. Although early experiments showed that TTX blocks the inward transient sodium currents, it was not clear whether it also blocks the outward transient sodium current (Narahashi et al., 1964). There was some indication that the outward transient current is less affected by TTX than the inward transient current (Nakamura et al., 1965a). In other words, the question is whether TTX has any affinity for the *direction* of transient sodium current.

2. Although TTX blocks the transient sodium current, it is not known whether TTX has an affinity for *sodium ion* per se or TTX has an affinity for the *channel* through which sodium ions flow under normal conditions.

3. Similar questions may be raised with regard to the late potassium current, although TTX has been demonstrated to have no effect on the late steady-state outward potassium current. It is not clear whether TTX has no affinity for *potassium ion* per se or TTX has no affinity for the *channel* through which potassium ions flow under normal conditions.

Experiments have been designed to answer these questions (Moore, Blaustein, Anderson and Narahashi, 1967). The first experiment simply involved the observation of the peak transient currents associated with step depolarizations beyond the sodium equilibrium potential. The outward as well as inward transient sodium currents are inhibited by application of TTX (Fig. 20-2 and 20-12). The result was confirmed by Rojas and Atwater (1967). Larger outward peak transient currents can be observed when cesium is substituted for sodium in the external medium. Since cesium can only be poorly transported through the peak transient or late steady-state channels (Pickard, Lettvin, Moore, Takata, Pooler and Bernstein, 1964; Chandler and Meves, 1965), sodium current flows in an outward direction at depolarized levels while potassium current remains unchanged. It has clearly been shown that the outward sodium current in the isotonic cesium solution is effectively suppressed by TTX, whereas the outward potassium current is unaffected.

Secondly, lithium was replaced for sodium in the bathing medium. It is well known that nerve fibers are capable of producing the normal-sized action potential in an isotonic lithium solution (Overton, 1902; Hodgkin and Katz, 1949; Narahashi, 1963). Transient lithium currents were observed under voltage clamp conditions, and these currents were also blocked by application of TTX (Fig. 20-2) (Hille, 1968). The peak transient channel allows potassium, rubidium and cesium ions to pass through to a lesser extent than sodium ions (Chandler and Meves, 1965) and the potassium current through

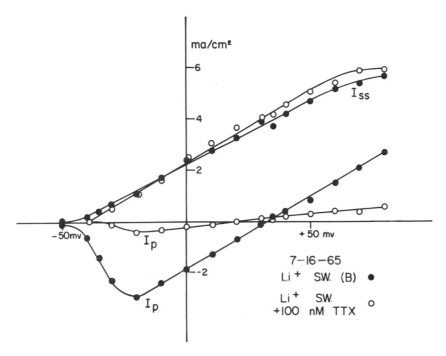

Fig. 20-2 *Current-voltage relationships before and during application of* 100 *nM tetrodotoxin* (TTX) *in a squid giant axon bathed in a sea water* (S.W.) *in which lithium is substituted for sodium.* I_p *and* I_{ss} *refer to peak transient and late steady-state currents, respectively.* (*From Moore, Blaustein, Anderson and Narahashi,* 1967.)

the peak transient channel is also blocked by TTX (Rojas and Atwater, 1967). Guanidine and hydrazine can flow through the peak transient channels and these currents are also inhibited by TTX (Tasaki and Singer, 1966). These observations have led to the conclusion that it is not sodium ion per se but the peak transient channel for which TTX has a special affinity.

Thirdly, the steady-state current was allowed to flow in an inward direction by increasing the concentration of potassium outside. The inward steady-state potassium current is also insensitive to TTX. Thus it can be said that TTX has no affinity for the late steady-state potassium current regardless of the direction of current flow. The inward rubidium current observed in the isotonic rubidium solution was not affected by TTX either. Thus, TTX has no affinity for the late steady-state channel regardless of the kind of ions flowing through it.

It is therefore concluded that TTX is able to block the peak transient channel for whatever ion may carry the current and whichever direction the current may flow. The effect is exerted at a very low TTX concentration

(10 nM) which is almost 10^{-5} times the effective concentration of procaine necessary to block nerve conduction.

Experiments to date indicate that TTX has no effect on the kinetics of the peak transient current and on the steady-state sodium inactivation (Takata et al., 1966).

Saxitoxin

Saxitoxin (STX) is the active ingredient of the poison isolated from the Alaska butter clam, *Saxidomas giganteus.* There is evidence that STX originally derives from the dinoflagellate, *Gonyaulax catanella* (Kao, 1966; Schantz, Lynch, Vayvada, Matsumoto and Rapoport, 1966). The molecular formula is $C_{10}H_{17}N_7O_4 \cdot 2HCl$, and is different from that of TTX $(C_{11}H_{17}N_3O_8)$.

The mechanism of action of STX on the nerve fiber is essentially the same as that of TTX; the peak transient conductance increase is selectively blocked by STX at very low concentrations in lobster giant axons (Narahashi, Haas and Therrien, 1967), in squid giant axons (Narahashi and Frazier, unpublished), in electroplaques (Nakamura, Nakajima and Grundfest, 1965b) and in nodes of Ranvier (Hille, 1968).

There are, however, some minor differences between STX and TTX in their mode of action on the nerve fibers. First of all, the recovery of the action potential after washing the nerve preparation is faster with STX treatment than with TTX treatment (Narahashi, Hass and Therrien, 1967; Kao, 1966). Secondly, the nerves from the puffer fish, *Spheroides maculatus*, and from the newt, *Taricha torosa*, are highly resistant to TTX, whereas they are sensitive to STX (Kao and Fuhrman, 1967).

Local Anesthetics

Procaine and cocaine have been found to inhibit both peak transient and late steady-state currents of squid axons under voltage clamp conditions (Taylor, 1959; Shanes, Freygang, Grundfest and Amatniek, 1959). With procaine, the time for the transient sodium current to reach its peak is prolonged, whereas the steady-state sodium inactivation is unaffected (Taylor, 1959).

Voltage clamp analyses have since been extended to other local anesthetics (Blaustein, 1968b; Hille, 1968), and also to those applied internally (Narahashi, Anderson and Moore, 1967; Narahashi, Moore and Poston, 1969). The effects of several local anesthetics and other blocking agents on a variety of membrane conductance parameters of squid and lobster axons are summarized in a paper by Narahashi, Moore and Poston (1969).

All of the local anesthetics used in the above studies have been found to inhibit both peak transient and late steady-state components of membrane conductance changes when applied from either side of the nerve membrane. These local anesthetics include procaine, dibucaine, tropine *p*-tolyl acetate hydrochloride (tertiary tropine), and tropine *p*-tolyl acetate methiodide (quaternary tropine). An example of the current-voltage relationship before and during internal application of tertiary tropine on the squid axon is illustrated in Fig. 20-3. The relative sensitivity of the peak transient and steady-state conductances to the blocking action varies slightly from one anesthetic to another, but the difference is not remarkable.

Most of the local anesthetics tested slow the time for the transient current to reach its peak. In Fig. 20-4, the time to peak current is plotted as a function of membrane potential before, during and after application of procaine internally. Dibucaine is an exception in that the kinetics are not affected.

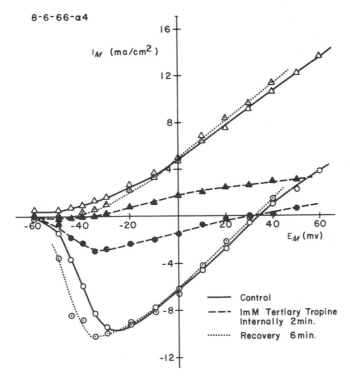

Fig. 20-3 *Current-voltage relationships before and during internal application of 1 mM tropine p-tolyl acetate hydrochloride, and after washing with normal internal medium in a squid giant axon. Circles refer to peak transient currents, and triangles refer to late steady-state currents.*

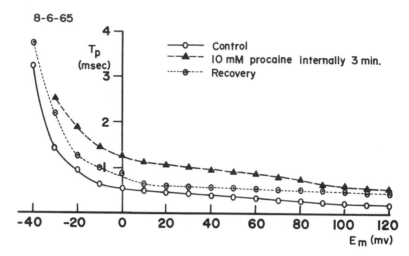

Fig. 20-4   *The time for transient current to reach its peak (T_p) plotted as a function of membrane potential in a squid giant axon before and during application of 10 mM procaine internally, and after washing with normal internal medium. (From Narahashi, Anderson and Moore, 1967.)*

One of the remarkable features of local anesthetic action on the membrane conductance is the appearance of a hump during the late current (Fig. 20-5). This effect was observed with dibucaine, tertiary tropine and quaternary tropine. When the current magnitude is plotted against membrane potential at the hump and at the steady-state, a linear current-voltage relationship is obtained only at the hump. This suggests that the appearance of a hump represents the occurrence of a voltage dependent inactivation process during the late potassium current. A more remarkable hump was observed with squid axons when pentyltriethylammonium iodide was internally applied (Armstrong, 1969).

The observation that the above mentioned local anesthetics are effective from either side of the nerve membrane is to be expected because tertiary amine local anesthetics can freely penetrate the membrane in the uncharged molecular form (see page 445). Permanently charged quaternary tropine is also able to block the conductances from either side of the membrane. This is probably related to its large binding affinity for phospholipids (Blaustein, 1967). Friess, Durant and Martin (1966) interpret the action of the quaternary tropine in blocking action potentials in nodes of Ranvier as being due to the ability of the tropine carbocyclic cage to insulate the permanently charged nitrogen moiety to the extent that the positive charge is no longer effective in being repelled by the membrane fixed charges.

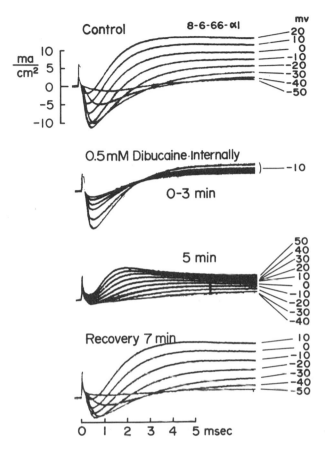

Fig. 20-5 *Families of membrane currents associated with step depolarizations in a squid giant axon before and during application of* 0.5 mM *dibucaine internally, and after washing with normal internal medium. The second set from the top shows changes in membrane current associated with a constant potential step during the course of blockage.* (*From Narahashi, Moore, and Poston,* 1969.)

In a later section, an attempt is made to determine the site of action and active form of local anesthetics using two tertiary lidocaine derivatives, 6211 and 6603 (Fig. 20-15), two quaternary lidocaine derivatives, QX-314 and QX-572 (Fig. 20-19), and hemicholinium-3 (HC-3) (Fig. 20-19). The tertiary anesthetics are able to block the action potential of the squid axon from either side of the membrane, whereas the quaternary anesthetics and HC-3 block the action potential much more strongly from inside than from outside the membrane. When the action potential is blocked by any of these chemicals, both peak transient and late steady-state conductances are inhibited (Narahashi, Frazier and Moore, 1968; Frazier, Narahashi and Moore, 1968, 1969). Therefore, it has been demonstrated that in squid and lobster giant axons, both components of membrane conductances are inhibited by all of the local anesthetics so far studied.

Barbiturates

Although barbiturates are not used as local anesthetics, they can exhibit blocking action when applied to isolated nerve preparations. Pentobarbital and thiopental have so far been studied for their ionic mechanism of action by means of voltage clamp techniques with squid and lobster giant axons (Blaustein, 1968a; Narahashi, Moore and Poston, 1969). These barbiturates inhibit both peak transient and late steady-state conductance increases and increase the time necessary for the transient current to reach its peak. Unlike some of the local anesthetics mentioned in the previous section, the barbiturates do not cause a hump during the course of the late current.

It should be noted that, although barbiturates differ from tertiary amine local anesthetics in their being dissociated into anions instead of cations, they can exert similar blocking effects on the membrane conductances. This does not, however, imply that barbiturates inhibit the membrane conductances by the same mechanism as local anesthetics.

DDT

In 1949, Shanes observed a prolongation of externally recorded action potentials when the crab nerve was exposed to DDT. The effect was later confirmed with cockroach nerves by means of external electrodes and internal microelectrodes (Yamasaki and Ishii, 1952b; Yamasaki and Narahashi, 1957a, b). From observations of delayed rectification and measurements of membrane conductance during the prolonged action potential, it was suggested that the late steady-state potassium conductance, or the sodium inactivation, or both, are inhibited by DDT, thereby causing the prolongation of the action potential (Narahashi and Yamasaki, 1960a, b). An increase in the negative after-potential is at least in part responsible for repetitive discharges which are observable in many types of nerves poisoned with DDT.

Voltage clamp experiments with lobster giant axons have indeed supported this hypothesis (Narahashi and Haas, 1967, 1968). Similar observations of DDT action were made on the nodes of Ranvier of the frog (Hille, 1968). Families of membrane currents associated with various magnitudes of step depolarizations before and during application of DDT are illustrated in Fig. 20-6. After application of DDT at a concentration of 5×10^{-4} M, the late steady-state currents which are normally outward at depolarized levels (Fig. 20-6, top) flow in an inward directon at certain depolarized levels (Fig. 20-6, second from top). These steady-state inward currents cannot be due to potassium ions because the resting potential remains unchanged, suggesting that the potassium concentration gradient across the nerve membrane is maintained normally. This problem can best be studied by application of TTX to the DDT-poisoned axon. After application of TTX at a concentration of 3×10^{-7}M, the peak transient current is blocked while the steady-state

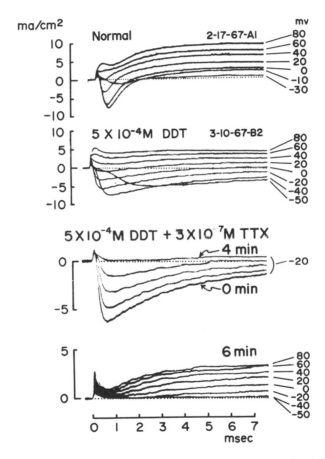

Fig. 20-6 *Families of membrane currents associated with step depolarizations in a normal lobster giant axon and in another lobster axon treated with DDT and with DDT and tetrodotoxin (TTX). The third set of records shows changes in membrane current associated with a constant potential step during the course of DDT + TTX action. The dotted line in each set refers to the zero base line. (Modified from Narahashi and Haas, 1967, Copyright 1967 by the American Association for the Advancement of Science.)*

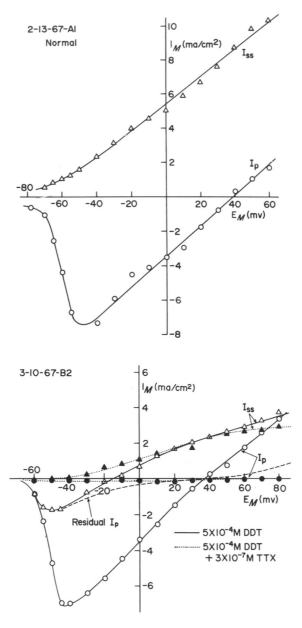

Fig. 20-7 *Current-voltage relationships for peak transient* (I_p) *and steady-state* (I_{ss}) *currents in a normal lobster giant axon* (A) *and in another axon* (B) *treated with* 5×10^{-4} M DDT *or with* 5×10^{-4} M DDT *and* 3×10^{-7} M *tetrodotoxin* (TTX). *The broken line shows the residual component of the peak transient current and was obtained by subtracting* I_{ss} *in DDT plus TTX from* I_{ss} *in DDT.* (*From Narahashi and Haas, 1967, Copyright 1967 by the American Association for the Advancement of Science.*)

inward current is converted into a small outward current (Fig. 20-6, third from top). Therefore, the difference between the current in DDT (zero min record in DDT + TTX) and the current in DDT plus TTX (4 min record in DDT + TTX) represents the sodium current which is turned off very slowly. The bottom record of Fig. 20-6 depicts a family of membrane currents in DDT- and TTX-poisoned axon, indicating that the steady-state potassium currents are suppressed by DDT.

The current-voltage relationships for the peak transient and the steady-state currents are shown in Fig. 20-7. The peak transient current is essentially unaffected by DDT, whereas the steady-state currents flow inwardly at the membrane potentials ranging between -60 mv and -15 mv. The potassium component in the steady-state current observed after application of TTX is seen to be suppressed. The difference between the steady-state current observed in DDT and that observed in DDT plus TTX gives the residual current. It should be pointed out that the residual current reverses its sign at the same potential as the reversal potential for the peak transient current. Thus it is clear that the residual current is carried by sodium ions.

DDT also slows the turn-on processes of the peak transient and steady-state currents, but the effect is much less than that on the sodium inactivation.

These effects of DDT on the sodium inactivation and on the potassium conductance increase are responsible for the prolongation of the action potential.

Allethrin

The insecticide allethrin is a derivative of pyrethrin I, which is one of the active principles of the natural insecticide, pyrethrum. It is the allethrolone ester of chrysanthemum-monocarboxylic acid. Pyrethrins have long been known to produce repetitive discharges in nerve and to eventually block nerve conduction (Hayashi, 1939; Lowenstein, 1942; Yamasaki and Ishii, 1952a), and these effects were considered to be directly responsible for the insecticidal action of pyrethrins. In 1962, the first study of the cellular mechanism of action of allethrin was carried out using intracellular microelectrodes on cockroach giant axons (Narahashi, 1962a, b). Three major effects of allethrin were unveiled: 1) increase in negative after-potential, 2) repetitive after-discharge, and 3) conduction block. Voltage clamp experiments with squid giant axons have clearly demonstrated that both peak transient and late steady-state conductance increases are inhibited by allethrin (Narahashi and Anderson, 1967).

Allethrin is said to be highly lipid-soluble, yet it exerts an effect additional to the blockage of both conductance increases when applied inside the perfused squid axon. The records of membrane currents associated with step depolarizations before and during internal application of allethrin are illustrated in Fig. 20-8. In the allethrin-treated axon, the peak transient current is followed, at certain membrane potentials, by a steady-state *inward* current, which is similar to that observed in the DDT-treated axon (previous section).

The current-voltage relation is shown in Fig. 20-9. The steady-state current in the allethrin-poisoned axon flows in an inward direction in the potential range of -45 mv to -5 mv (filled triangles in Fig. 20-9). Since the potassium concentration gradient across the nerve membrane is maintained constant by continuous perfusion in both external and internal phases, the inwardly flowing steady-state currents cannot be ascribed to potassium ions. Hence, the inward steady-state current is most probably carried by sodium ions. On this assumption, the potassium component of the steady-state current in the allethrin-poisoned axon is obtained as follows:

Since little or no steady-state outward (potassium) current flows in the membrane potential range of -50 mv to -25 mv in the control axon before treatment with allethrin (open triangles in Fig. 20-9), it is assumed that the steady-state inward currents in this potential range after treatment with allethrin are carried by sodium ions (filled triangles in Fig. 20-9). Thus the potassium current flows outwardly by depolarizations beyond -25 mv. The peak transient current reverses its sign at $+40$ mv, and therefore the steady-state current at that potential does not contain a sodium component. Hence, the current-voltage relationship for steady-state potassium current can be approximated by drawing a straight line connecting the zero current at -25 mv and the apparent steady-state current at $+40$ mv (dotted line in Fig. 20-9). Comparing this line to the open triangles, it then follows that the conductance for the steady-state potassium current is suppressed by application of allethrin.

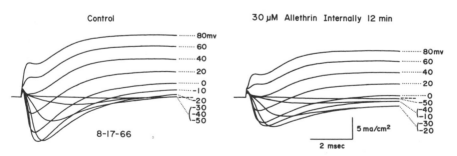

Fig. 20-8 *Families of membrane currents associated with step depolarizations in a squid giant axon before and during application of 30 μM allethrin internally. The broken line on the right of each set of records refers to the zero base line. (From Narahashi and Anderson, 1967, © 1967, The Academic Press, Inc., New York.)*

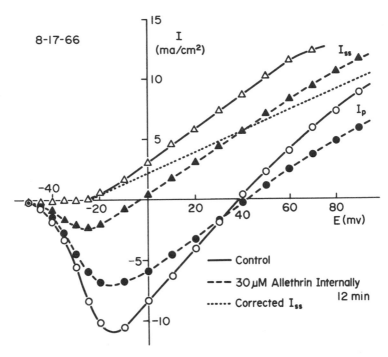

Fig. 20-9 *Current-voltage relationships in a squid giant axon before and during application of 30 μM allethrin internally.* I_p *and* I_{ss} *refer to peak transient and steady-state currents, respectively. The dotted line represents the potassium component in the steady-state current in allethrin obtained as described in the text. (From Narahashi and Anderson, 1967, © 1967, The Academic Press, Inc., New York.)*

In recent experiments with crayfish giant axons (Murayama, Abbott and Narahashi, unpublished), TTX was applied to the axon treated with allethrin externally to dissociate the membrane current into sodium and potassium components. The results clearly show that the residual current is indeed carried by sodium ions in support of the notion described above.

The kinetics of the peak transient current are also affected by allethrin. The time for the current to reach its peak is prolonged slightly by either external or internal application of allethrin to squid giant axons.

In summary, allethrin can exert at least four actions on the membrane conductances:

1. It inhibits the increase in peak transient sodium conductance.
2. It inhibits the increase in late steady-state potassium conductance.
3. It slows the time course of the sodium inactivation.
4. It slows the time course of the transient sodium conductance increase.

Condylactis Toxin

The Bermuda anemone, *Condylactis gigantea*, contains a potent neuroactive toxin. The *Condylactis* toxin (CTX) has a molecular weight of about 10,000–15,000, and is able to prolong action potentials from lobster giant axons and crayfish stretch receptors (Shapiro, 1968; Shapiro and Lilleheil, 1969). It is suggested that slowing of the sodium inactivation is responsible for the prolongation of the action potential in the CTX-poisoned nerve (Shapiro and Lilleheil, 1969). Voltage clamp experiments with crayfish giant axons have demonstrated that this actually is the case (Narahashi, Moore and Shapiro, 1969).

An example of membrane current records before and after exposure to CTX is shown in Fig. 20-10. In CTX, the peak transient current starts

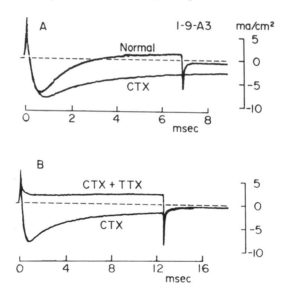

Fig. 20-10 *Membrane currents associated with a constant amplitude of step depolariz-ation in a crayfish giant axon.* A: *Before and during application of Condylactis toxin* (CTX) *at a concentration of* 0.2 mg/ml. B: *During application of* CTX, *and during application of* CTX *and* 3×10^{-7} M *tetrodotoxin* (TTX). (*From Narahashi, Moore and Shapiro, 1969, Copyright 1969 by the American Association for the Advancement of Science.*)

flowing normally but attains a higher level, and is followed by an *inward* steady-state current. To dissociate the inward steady-state current into sodium and potassium components, TTX was applied to the CTX-poisoned axon. The peak transient current is now blocked and the inward steady-state current is converted into an outward current (Fig. 20-10). The sodium component of the membrane current can be obtained by subtraction of the current in CTX plus TTX from that in CTX. It is clearly shown that the sodium inactivation is greatly slowed down. The amplitude of the potassium current, as given by the membrane current in CTX plus TTX (minus leakage current), is not appreciably affected by CTX poisoning, However, in some other experiments, the potassium current is slightly suppressed by CTX.

In summary, the major effect of CTX is to slow the sodium inactivation. This enables the transient sodium current to reach a value greater than the control, and prolongs the action potential. It is of great interest to note that in the CTX-poisoned axon the membrane current is shut off as quickly as in the normal axon upon termination of depolarization. This is in sharp contrast with the DDT-poisoned nodes of Ranvier in which the sodium current is shut off very slowly upon repolarization (Hille, 1968).

SITE OF ACTION AND ACTIVE FORM

Tetrodotoxin and Saxitoxin

Site of action.   Tetrodotoxin and saxitoxin are not soluble in most organic solvents (Tsuda et al., 1964). It is therefore reasonable to assume that TTX and STX are not freely permeable through the lipid phase of the nerve membrane. This gives us an excellent opportunity to study the site of action of TTX and STX in the nerve membrane.

In spite of their highly potent blocking action when applied to the external surface of the nerve membrane, TTX and STX are inert when applied to the inside of the squid axon (Narahashi, Anderson and Moore, 1966, 1967; Narahashi and Frazier, unpublished). An example of internal perfusion with 10^{-6}M STX is shown in Fig. 20-11, in which the resting potential and the amplitude and maximum rate of rise of the action potential undergo no appreciable change during 30 minutes of perfusion with STX. Internal applications of TTX or STX at a concentration of 10^{-6}M have no effect on either the peak transient current or late steady-state current during 30 minutes of observation. Internally applied TTX at a higher concentration of 10^{-5}M again does not cause any change in the action potential (17 minutes of observation). On the other hand, externally applied TTX or STX can block the action potential at a concentration of 3×10^{-8}M in a few minutes.

Based on these observations, it is concluded that the site of action of TTX and STX is located somewhere on or near the external surface of the nerve membrane. One might assume a gate mechanism, located on the external membrane surface for the peak transient conductance. This gate would open upon depolarizing stimulation, thereby allowing sodium ions to flow according to the concentration gradient.

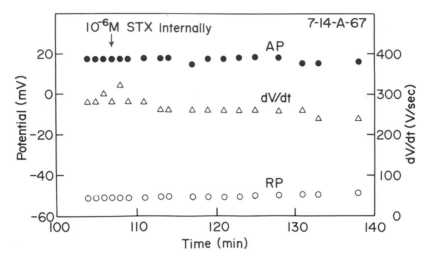

Fig. 20-11 *The effect of internal perfusion of* 10^{-6} M *saxitoxin (STX) on the resting potential (RP), the action potential (AP), and the maximum rate of rise of the action potential (dV/dt) of a squid giant axon. The abscissa represents the time after the beginning of internal perfusion. This axon had been used for experiments with other reversible drugs before the STX experiment.*

Active form. As described before (page 425), TTX exists in two cation forms and a zwitterion form (Fig. 20-1). Because the pK_a value for dissociation into the cations is estimated to be 8.8, it is possible to affect the ratio of the total cation concentration to the zwitterion concentration by changing the pH of the medium. The ratio between TTX forms is calculated by the equation

$$\log \frac{c_{BH^+} + c_{B'H^+}}{c_{B\pm}} = pK_a - pH, \qquad (20\text{-}1)$$

where c_{BH^+} and $c_{B'H^+}$ are the concentrations of the two cation forms, and $c_{B\pm}$ is the concentration of the zwitterion form. It would therefore be possible to determine the active form by observing the effect of pH change on the blocking action of TTX (Narahashi, Moore and Frazier, 1969).

In order to eliminate the possible effect of titration of the membrane macromolecule components with alteration of the pH, experiments were performed on squid giant axons in a narrow range of pH change. No significant change is observed in the peak transient and late steady-state currents when the pH of the external medium is altered between 7 and 9 (Fig. 20-12). However, the blocking action of 3×10^{-8}M TTX is stronger at pH 7 than at 9 (Fig. 20-12). This result is compatible with the concept that the cation forms are responsible for the blockage.

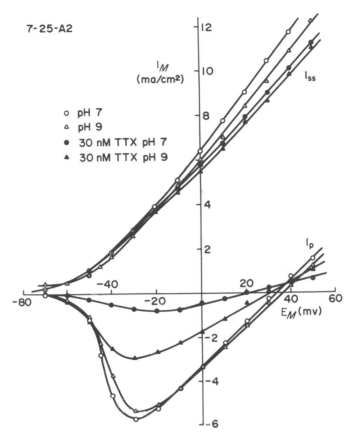

Fig. 20-12 *Current-voltage relationships for peak transient* (I_p) *and steady-state* (I_{ss}) *currents in a squid giant axon at* pH 7 *and* pH 9 *with and without* 30 nM *tetrodotoxin* (TTX). *(From Narahashi, Moore and Frazier, 1969, © 1969, The Williams & Wilkins Co., Baltimore, Md. 21202, U.S.A.)*

To explore this problem in a more quantitative manner, the percentage of inhibition of the peak transient current obtained at pH 7 and pH 9 is plotted as a function of the logarithmic concentration of TTX cations (triangles in Fig. 20-13). The data with lobster giant axons at a constant pH (Takata et al., 1966) are recalculated and plotted for comparison (filled circles and squares in Fig. 20-13). It is clearly shown that both dose-response curves agree fairly well and support the notion that the cation forms are active. This conclusion agrees with that obtained by Camougis, Takman and Tasse (1967), by Hille (1968) and by Ogura and Mori (1968) with different nerve preparations. It is not possible, however, to determine which cation form is active without

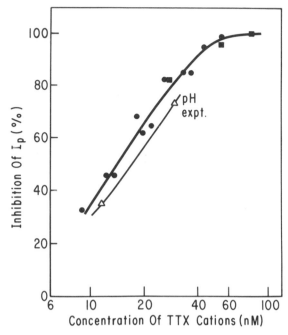

Fig. 20-13 *Per cent inhibition of peak transient current as a function of the concentration of tetrodotoxin* (TTX) *cations. Filled symbols represent the recalculated data from Takata, Moore, Kao and Fuhrman* (1966) (*Modified from Narahashi, Moore and Frazier, 1969,* © *1969, The Williams & Wilkins Co., Baltimore, Md. 21202, U.S.A.*)

having accurate information as to the equilibrium constant (K) between the two cation forms.

The experiments described above demonstrate that TTX in its cation forms blocks the peak transient conductance increase from outside the nerve membrane. It can then be predicted that pH changes in the internal medium should not affect the blocking potency of TTX applied outside. This has been demonstrated to be the case as is shown in Fig. 20-14, where the maximum rate of rise of the action potential is plotted against time. Tetrodotoxin is applied externally in a concentration of 1×10^{-8}M at a constant pH of 8 while changing the internal pH between 7 and 8. There is no effect of internal pH change on the blockage caused by TTX, and the maximum rate of rise of the action potential declines smoothly after application of TTX and eventually attains a steady-state level despite changes in the internal pH between 7 and 8.

Local Anesthetics

History.   Most local anesthetics which are commonly in use are basic tertiary amines, so that they are partially dissociated into cations, depending

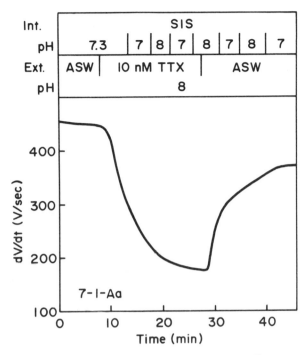

Fig. 20-14 *Effect of* 10 nM *tetrodotoxin* (TTX) *applied externally at a constant external pH and varying internal pH values in an internally perfused squid giant axon. The maximum rate of rise of the action potential (dV/dt) is plotted in the ordinate as a measure of excitability against time in the abscissa.*

on the pH of the medium and the pK_a of the compound. The ratio of the charged cation form to the uncharged molecular form is given by the Henderson-Hasselbach equation

$$\log \frac{c_{BH^+}}{c_B} = pK_a - pH, \qquad (20\text{-}2)$$

where c_{BH^+} and c_B represent the concentrations of charged and uncharged forms, respectively. Because most local anesthetics have a pK_a value somewhere between 7 and 9, they exist in two forms in the physiological pH range. Thus a question arises as to which form is responsible for the nerve blocking action.

This problem has been a subject for investigators since early this century. Since the ratio of the charged to uncharged form can be changed by altering the pH, the blocking potency should be affected by pH changes if only one of the forms is active. A number of pH experiments were performed with a variety of nerve and muscle preparations. Some investigators reported that local anesthetics block the excitability more strongly at higher pH values than at lower pH values, whereas others obtained the opposite results. Several different interpretations have been proposed for these seemingly controversial observations.

Experimental results in the literature will not be reviewed in detail here, since they are discussed elsewhere (Ritchie and Greengard, 1966; Ariëns, Simonis and van Rossum, 1964; Narahashi and Frazier, 1971). In short, earlier studies show that the blocking potency of various local anesthetics is augmented by raising the pH (e.g. Trevan and Boock, 1927; Skou, 1954a). These and other related experimental results are taken as indicating that uncharged molecules are mainly responsible for the nerve blockage (Skou, 1954b, c, d, e, f, 1961).

However, a model for a cationic active form was first proposed by Krahl, Keltch and Clowes (1940). Cell division of fertilized eggs and movement of larvae in Arbacia punctulata were used as indices of local anesthetic potency. They concluded that local anesthetics penetrate the membrane in the uncharged molecular form and exert their blocking action in the charged form from inside the cell.

Since 1960, the evidence in support of the concept that the charged form is the active one in nerves started accumulating. Observations along this line include those by Bartel, Dettbarn, Higman and Rosenberg (1960), Dettbarn (1962), Ritchie and Greengard (1961) and Ritchie, Ritchie and Greengard (1965a, b). Ritchie and Greengard put forward the hypothesis in which local anesthetics acting on nerves penetrate the diffusion barrier in the uncharged form and exert their blocking action in the charged form. Because the nerve sheath is an effective diffusion barrier, the effect of pH on the local anesthetic potency greatly depends on whether the sheathed or desheathed nerve preparation is used as material.

However, another situation has developed recently. Bianchi and Strobel (1968) and Ritchie and Ritchie (1968) suggest that both forms of certain local anesthetics could be active in blocking the action potential.

An important aspect of the present problem has drawn rather little attention, i.e. the site of action in the nerve fiber. Since the nerve membrane has fixed charges, it constitutes a strong diffusion barrier for penetration of charged local anesthetics. Therefore, if the site of action of local anesthetics were located on the inner surface of the nerve membrane, the anesthetic molecules would have to penetrate two diffusion barriers, the nerve sheath and the nerve membrane, before exerting their blocking action. The most straightforward approach to this problem would be to compare the blocking potency when a local anesthetic is applied to the external or internal phase at different controlled pH values. The internal perfusion techniques developed for squid giant axons by Baker, Hodgkin and Shaw (1961) and by Oikawa, Spyropoulos, Tasaki and Teorell (1961) allow such experiments to be performed. The present section describes the results of such internal perfusion experiments with squid giant axons (Narahashi and Frazier, 1968; Narahashi and Yamada, 1969; Narahashi, Yamada and Frazier, 1969; Narahashi, Frazier and Yamada, 1970; Frazier, Narahashi and Yamada, 1970).

Theory and Approach. Two assumptions are made in the present experimental analyses:

1. Only the uncharged form is freely permeable to the nerve membrane. This is documented by a number of experiments (Krahl et al., 1940; Ritchie et al., 1965a, b; Eldefrawi and O'Brien, 1967; O'Brien, 1967; Bianchi and Bolton, 1967; Rothenberg, Sprinson and Nachmansohn, 1948; Schanker, Nafpliotis and Johnson, 1961; Whitcomb, Friess and Moore, 1958).

2. The charged and uncharged forms of local anesthetics and the hydrogen ions are distributed uniformly in external and internal phases of the axon. This is in fact not true because of the effect of negative fixed charges on the nerve membrane surface, but it has been shown that this factor does not affect the conclusion derived from the analyses based on the uniform distribution (Narahashi and Frazier, 1971).

Based on these assumptions, at least three forms are possibly active, i.e. the uncharged form (present internally or externally), the charged form present externally and the charged form present internally. To determine which form is active, three experimental and analytical approaches have been attempted.

1. The relationship between the blocking potency and pH is calculated for the three assumptions and compared with observations. The calculations should fit the observations if the assumption on the active form is correct.

2. Dose-response curves are plotted using the concentration of each of the three possible active forms. If the assumption on the active form is correct, the measurements should fall into a single dose-response curve.

3. Permanently charged quaternary forms of local anesthetics are tested for their blocking potency when applied outside or inside of the nerve membrane. They are expected to exert blocking effect only from the membrane surface where the site of action is located.

Experimental Results and Analyses. Two tertiary derivatives of lidocaine, 6211 and 6603, were first used in experiments. Their chemical structures are given in Fig. 20-15. 6211 has an unusually low pK_a value of 6.3, so that it exists mostly in the uncharged form in the physiological pH range. On the other hand, 6603 exists mostly in the charged form because of its unusually high pK_a of 9.8. Therefore, the blocking potency of these two tertiary derivatives will be affected differently by pH changes.

6211
(pK$_a$ = 6.3)

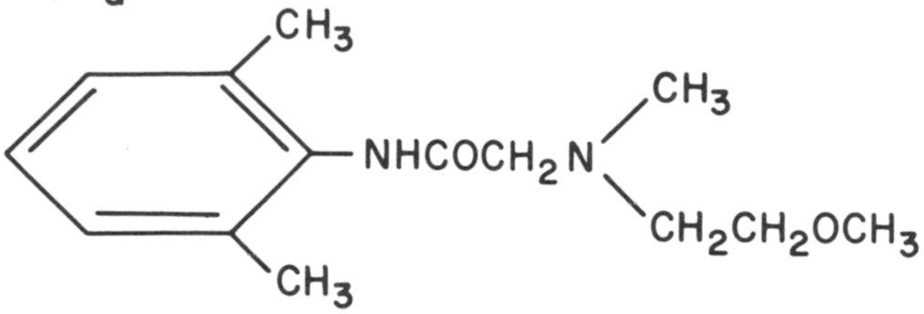

6603
(pK$_a$ = 9.8)

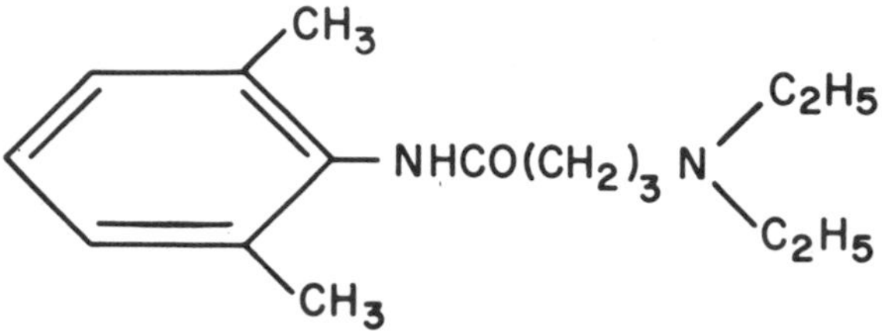

Fig. 20-15 *Chemical structures of tertiary derivatives of lidocaine. (From Narahashi, Frazier and Yamada, 1970, © 1970, The Williams & Wilkins Co., Baltimore, Md. 21202, U.S.A.)*

Fig. 20-16 illustrates the effect of pH on the blocking potency. The ordinate represents the maximum rate of rise of the action potential as the percentage of the control value, and the abscissa represents the pH value in the phase where the anesthetics are applied while maintaining the pH in the opposite phase at a constant value. The maximum rate of rise of the action potential is measured instead of the amplitude of the action potential, because the former is proportional to the inward ionic (sodium) current at that moment (Hodgkin and Huxley, 1952), and also because the former is generally more sensitive than the latter to environmental changes. In separate experiments without anesthetics, it has been shown that a pH change between 7 and 9 in the external phase or between 7 and 8 in the internal phase has no significant effect on the maximum rate of rise of the action potential. Thus, the observations shown in Fig. 20-16 clearly indicate the following:

1. When applied externally, the blocking potency of 6211 is insignificantly affected by pH, whereas when applied internally, it is augmented by lowering the pH.

2. When applied externally, the blocking potency of 6603 is augmented by raising the pH, whereas, when applied internally, it is insignificantly pH dependent.

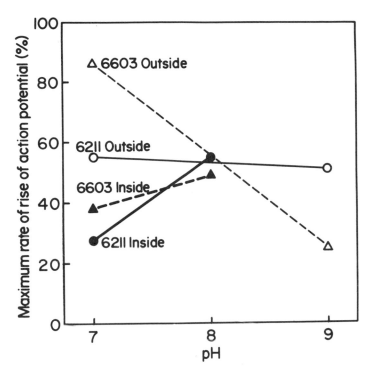

Fig. 20-16 *The effect of* pH *changes on the blocking potency of tertiary lidocaine derivatives 6211 and 6603 applied internally or externally to squid giant axons. The ordinate represents the mean value (3 to 9 measurements) for the maximum rate of rise of the action potential as the percentage of the control without anesthetics. The abscissa represents the* pH *value in the phase where the anesthetics were applied while maintaining the* pH *in the opposite phase constant. The concentrations are 1* mM *and 10* mM *(data combined together) for 6211 and 3* mM *for 6603. The differences between the* pH *values are significant for 6211 inside (P < 0.001) and for 6603 outside (P < 0.01), but insignificant for 6211 outside (P > 0.02) and for 6603 inside (P > 0.02) (From Narahashi, Yamada and Frazier, 1969.)*

In order to analyze these observations in terms of the potency-pH relationship, calculations were made of the effective dose at various pH values for both external and internal applications of 6211 of 6603. The calculated values are shown in Fig. 20-17, in which the ordinate represents the logarithm of the effective dose 50 (ED50) which is the amount of anesthetic necessary to block the maximum rate of rise of the action potential 50 per cent, and the abscissa represents the pH values of the phase where the anesthetics were applied. Three sets of calculated curves are shown for the three assumed active forms. The concentrations are normalized so that a value of one (log of the concentrations = 0) is assigned to the effective concentration in the phase where the receptor is located.

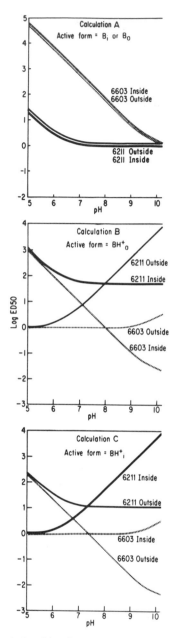

Fig. 20-17 *Calculated relationships between the logarithm of the effective dose 50 (ED50) to block the maximum rate of rise of the action potential 50 per cent and the pH of the phase where the anesthetics are applied while maintaining the pH in the opposite phase constant. Log ED50-pH curves are shown for internal and external applications of 6211 or 6603. In calculation A, internally or externally present uncharged form, B_i or B_o, is assumed as the active form. In B, externally present charged form, BH_o^+, is assumed as active. In C, internally present charged form, BH_i^+ is assumed as active. See text for further explanation. (From Narahashi, Frazier and Yamada, 1970, © 1970, The Williams & Wilkins Co., Baltimore, Md. 21202, U.S.A.)*

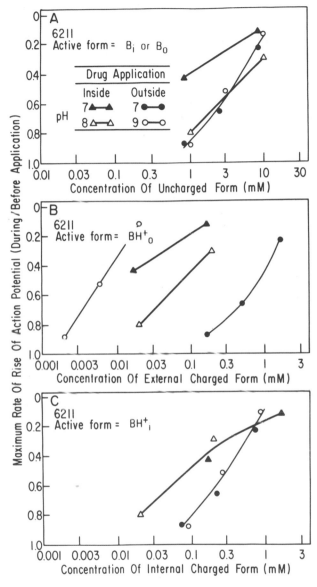

Fig. 20-18 *Dose-response curves for internally and externally applied 6211 in perfused squid giant axons. The ordinate represents the maximum rate of rise of the action potential relative to that before application of the anesthetic. The abscissae represent the concentrations of the assumed active form which is internally or externally present uncharged form, B_i or B_o, in A, externally present charged form, BH_o^+, in B, and internally present charged form, BH_i^+, in C. (From Narahashi, Frazier and Yamada, 1970 © 1970, The Williams & Wilkins Co., Baltimore, Md. 21202, U.S.A.)*

Calculation A in Fig. 20-17 represents the case where the uncharged form is assumed active. However, this set of curves does not fit the set of observations shown in Fig. 20-16. Calculation B represents the case where the externally present charged form is assumed as active, and again this does not fit the observations. Calculation C represents the case where the internally present charged form is assumed as active, and this does fit the set of observations given in Fig. 20-16. Thus, the observations can best be accounted for if the internally present charged form is assumed as being responsible for the nerve blocking action.

In order to analyze the observations in terms of the dose-response curve, the data for 6211 are replotted in Fig. 20-18, in which the maximum rate of rise of the action potential, relative to that before application of the anesthetic, is plotted as a function of the concentration of the form assumed to be active. If the assumption on the active form is correct, the measurements for external and internal applications should fall on a respective single dose-response curve. If the uncharged form is assumed active, the measurements for external application fall on a single dose-response curve, whereas for internal application, the measurements fall on two curves. Therefore, the uncharged form cannot be active. If the externally present charged form is assumed active, the measurements for both external and internal applications do not fall on single curves. Thus, the charged form in the external phase cannot be active either. If, however, the internal charged form is active, the measurements for external and internal applications do fall on single curves. Thus,

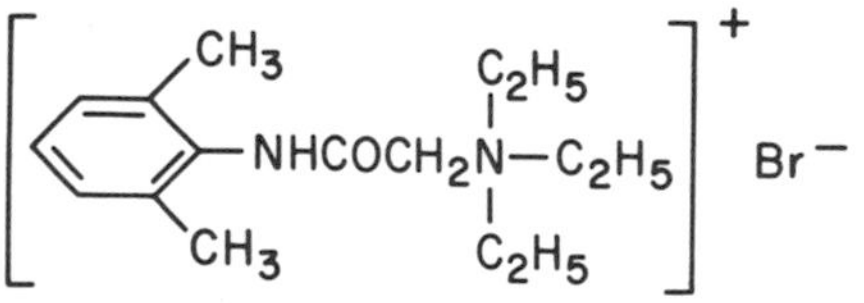

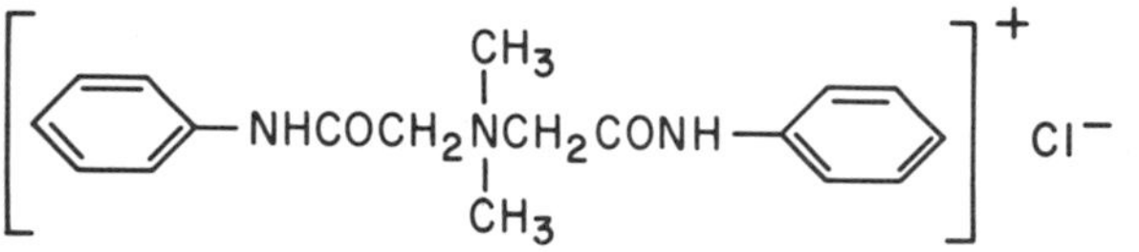

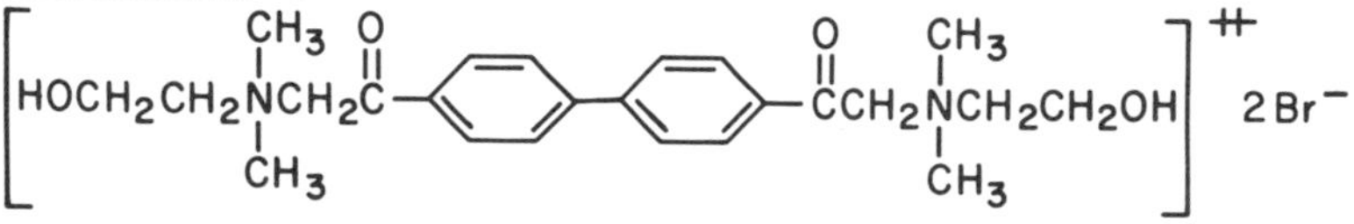

Fig. 20-19 *Chemical structures of quaternary derivatives of lidocaine and hemicholinium-3 (HC-3). (From Frazier, Narahashi and Yamada, 1970, © 1970, The Williams & Wilkins Co., Baltimore, Md. 21202, U.S.A.)*

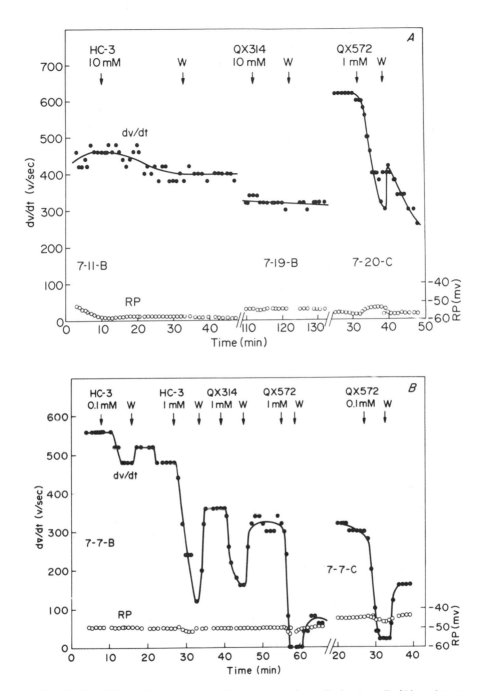

Fig. 20-20 *Effects of quaternary amine compounds applied externally* (A) *or internally* (B) *in squid giant axons. The maximum rate of rise of the action potential* (dV/dt) *and the resting potential* (RP) *are plotted against time. W refers to washing with normal media without compounds.*

the analysis based on the dose-response curves again supports the concept that the internally present charged form of local anesthetics is responsible for the nerve blockage.

The hypothesis presented here would predict that the blocking potency of local anesthetics applied externally would be affected by internal pH change, whereas the potency of anesthetics applied internally would be independent of external pH change, because it is the internally present anesthetic cation that exerts the blocking action. This has been proven to be the case for both 6211 and 6603.

Three quaternary amine compounds were tested for their blocking potency from either side of the nerve membrane. Their chemical structures are shown in Fig. 20-19. Two of them, QX-314 and QX-572, are derivatives of lidocaine, and the other is hemicholinium-3 (HC-3), a potent neuromuscular blocking agent. HC-3 has been shown to inhibit the synthesis of acetylcholine at the nerve terminals by prevention of uptake of choline (MacIntosh, Birks and Sastry, 1956; MacIntosh, 1961; Schueler, 1960).

All of these quaternary compounds are much more effective inside rather than outside the squid axon in blocking the action potential. Examples of external and internal applications are illustrated in Fig. 20-20A and B, respectively. Although HC-3 and QX-314 have no appreciable effect on the maximum rate of rise of the action potential when applied externally at a concentration of 10 mM, they exert profound blocking action when applied internally at an even lower concentration of 1 mM. Externally applied QX-572 partially blocks at 1 mM, whereas when applied internally it exhibits a much stronger blocking action even at a concentration of 0.1 mM. The results strongly support the present hypothesis that the charged form of local anesthetics blocks the action potential from inside the nerve membrane.

The present hypothesis would predict that the blocking potency of internally applied quaternary compounds should be independent of pH changes in either external or internal medium. This has been demonstrated to be the case.

Discussion. The factors which might affect the present analysis have been discussed in detail elsewhere (Narahashi, Frazier and Yamada, 1970; Narahashi and Frazier, 1971). It has also been shown that most of the data in the literature concerning the effect of pH on the blocking potency of local anesthetics can be accounted for by the present model if the internal pH changes to various extents upon changing the external pH (Narahashi, Frazier and Yamada, 1970; Narahashi and Frazier, 1971).

Only a general idea, helpful in interpreting existing data in the literature in view of the present model, will be given here. Fig. 20-21 depicts the log ED50-pH curves for internal and external applications of a tertiary amine local anesthetic having a pK_a of 8.7. The "Inside" curve represents the

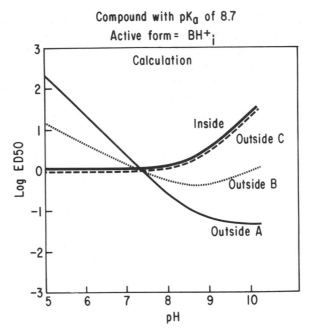

Compound with pK$_a$ of 8.7

Active form = BH^+_i

Fig. 20-21 *Calculated relationships between the logarithm of the effective dose 50 (ED50) to block the maximum rate of rise of the action potential 50 per cent and the pH of the phase where the compound with pK$_a$ of 8.7 is applied. Internally present charged form, BH_i^+, is assumed as the active form. Inside: Inside application of the drug while maintaining the outside pH constant. Outside A: Outside application of the drug while maintaining the inside pH constant. Outside B: Outside application while the inside pH follows the outside pH by 50 per cent. For example, if the outside pH is altered by one unit, the inside pH changes by 0.5. Outside C: Outside application while the inside pH freely follows the outside pH.*

case where the anesthetic is applied inside at various internal pH levels while maintaining the external pH constant. The curve labelled "Outside A" represents the case where the anesthetic is applied outside at various external pH values while maintaining the internal pH constant. The curve labelled "Outside C" also represents the external application of the anesthetic at various external pH values, but the internal pH freely follows the external pH change. Curve "Outside B" represents the external anesthetic application at various external pH values, but the internal pH only partially follows the external pH change. Thus depending on the degree of internal pH change when the external pH is altered, the blocking potency of an externally applied anesthetic could increase by increasing the external pH up to the

pK_a value (curve "Outside A"), or it could be independent of the external pH up to the pK_a and then decrease beyond it (curve "Outside C"). It is also possible to obtain a maximum potency at a certain pH value (curve "Outside B"). The degree of internal pH change may depend on a variety of factors such as fiber diameter, diffusion barriers, etc. The change will be greater in a small fiber than in a large fiber because the surface to volume ratio is greater in the former than in the latter.

The log ED50-pH curves presented in Fig. 20-21 can easily be modified to adapt to other tertiary amine local anesthetics having different pK_a values. Curve "Outside A" may be shifted along the 45° line so that the inflection occurs at the pK_a. Curve "Outside C" may also be shifted along the abscissa so that the inflection occurs at the pK_a. Curve "Outside B" may be plotted so that the ratio of vertical distance $(A - B)/(A - C)$ at any pH level coincides with the ratio of the internal vs. external change.

The experimental data in the literature obtained with nerve preparations having strong diffusion barriers such as the nerve sheath show a potency-pH relation similar to "Outside A", whereas those with small nerve preparations without the nerve sheath show a relation similar to "Outside C" or "Outside B".

Barbiturates

Barbiturate anesthetics dissociate into anions. The ratio of the uncharged form to the dissociated anionic form is expressed by the equation

$$\log \frac{c_A}{c_{A^-}} = pK_a - pH, \tag{20-3}$$

where c_A and c_{A^-} represent the concentrations of the uncharged and charged forms, respectively.

The question as to the active form of barbiturates has drawn much less attention than that of basic tertiary amine local anesthetics. Based on the observations on the cleavage of sea urchin eggs and on the movement of sea urchin larvae, Clowes, Keltch and Krahl (1940) proposed a model in which barbiturates act as the uncharged molecular form. On the contrary, in observation of action potentials in lobster giant axons, Blaustein (1968a) concluded that the anionic form of barbiturate anesthetics is responsible for nerve blockage.

In order to solve this problem, it was necessary to perform more straightforward pH experiments with internally perfused squid giant axons. The results of such experiments have been reported (Narahashi, Frazier, Deguchi, Cleaves and Ernau, 1970, 1971).

When pentobarbital is applied either externally or internally at varying pH values while maintaining the pH in the opposite phase constant, the blocking potency is higher at lower pH's than at higher pH's for either external or internal applications. When the barbiturate is applied at a constant pH and the pH in the opposite phase is altered, the blocking potency is independent of the pH change. These observations are compatible with the notion that pentobarbital penetrates the nerve membrane in the uncharged form and exerts the blocking action also in the uncharged form. At the same pH level, the potency of internal application is much higher (by a factor of 6 to 10) than of external application. This suggests that the site of action is located on the internal membrane surface.

CONCLUSIONS

The results and conclusions described in the present paper are summarized in Table 20-1. TTX exerts its highly specific blocking effect on the peak transient conductance increase ($g_{p\uparrow}$) only from outside the nerve membrane when in its cation forms. STX has a very similar action, but its active form remains to be studied. Basic tertiary amine local anesthetics block both peak transient and late steady-state conductance increases ($g_{ss\uparrow}$) from inside the nerve membrane in the cationic form. On the contrary, barbiturate anesthetics

Table 20-1   Summary of the mode of action of various neuroactive agents

	Ionic mechanism			Site of Action	Active form
Agents	$g_{p\uparrow}$	$g_{p\downarrow}$	$g_{ss\uparrow}$		
Tetrodotoxin	—	0	0	Outside	Cations
Saxitoxin	—	0	0	Outside	?
Local anesthetics	—	0	—	Inside	Cation
Barbiturates	—	0	—	Inside	Uncharged
DDT	0	—	—	?	....
Allethrin	—	—	—	?	....
Condylactis Toxin	0	—	—	?	?

$g_{p\uparrow}$: Mechanism whereby peak transient conductance is increased upon depolarization.
$g_{p\downarrow}$: Mechanism whereby peak transient conductance is decreased upon sustained depolarization.
$g_{ss\uparrow}$: Mechanism whereby late steady-state conductance is increased upon depolarization.
   0 : No effect
   — : Inhibition

block $g_{p\uparrow}$ and $g_{ss\uparrow}$ in the uncharged molecular form, possibly from inside the nerve membrane. DDT does not block $g_{p\uparrow}$, whereas it slows the inactivation of g_p ($g_{p\downarrow}$) and suppresses $g_{ss\uparrow}$, thereby prolonging the action potential. Allethrin exhibits these two actions and in addition blocks $g_{p\uparrow}$; it prolongs and eventually blocks the action potential. The sites of action of DDT and allethrin remain to be determined; difficulties in this determination are predicted because of their high lipid-solubility. The question as to the active form does not apply to these insecticides. *Condylactis* toxin slows $g_{p\downarrow}$ and slightly suppresses $g_{ss\uparrow}$ without affecting $g_{p\uparrow}$, thereby prolonging the action potential. Its site of action and active form remain to be determined.

It is of great interest that ionic channels could be blocked in a variety of manners. Here the term "channel" simply refers to a conceptual pathway through which ions flow according to the electrochemical gradient. It by no means reflects any anatomical feature. The peak transient channel can be blocked from the outside by TTX cations, or from the inside by local anesthetic cations, and by barbiturate or allethrin molecules. The late steady-state channel can be blocked from the inside by local anesthetic cations, as well as by barbiturate, DDT, or allethrin molecules. This provides strong evidence that the nerve membrane is asymmetrical.

It is not possible to generalize the physico-chemical mechanism of channel or conductance blockage by a variety of chemicals, because the peak transient and steady-state channels could be blocked from different sides of the membrane, and by entirely different chemical species, i.e. by cations or by uncharged molecules. The blockage of ionic conductances which can be observed under voltage clamp conditions is one of the final products brought about by a variety of physico-chemical interactions between blocking agents and membrane macromolecular components.

ACKNOWLEDGMENTS

This study was supported by grants from the National Institutes of Health (NS03437 and NS06855) and from the Grass Foundation, and by contract with the National Institute of Environmental Health Sciences (PH43-68-73). The experiments with squid giant axons were performed at the Marine Biological Laboratory, Woods Hole, Massachusetts.
I thank Mrs. C. A. Munday for her secretarial assistance.

REFERENCES

ARIËNS, E. J., A. M. SIMONIS and J. M. VAN ROSSUM. 1964. Drug-Receptor Interaction: Interaction of One or More Drugs with Different Receptor Systems, pp. 287–393. *In* E. J. Ariëns [ed.] *Molecular Pharmacology*, Vol. 1. Academic Press, New York and London.

ARMSTRONG, C. M. 1969. Inactivation of the Potassium Conductance and Related Phenomena caused by Quaternary Ammonium Ion Injection in Squid Axons. *J. Gen. Physiol.* 54: 553–75.

BAKER, P. F., A. L. HODGKIN and T. I. SHAW. 1961. Replacement of the Protoplasm of a Giant Nerve Fibre with Artificial Solutions. *Nature.* 190: 885–87.

BARTELS, E., W. D. DETTBARN, H. HIGMAN and P. ROSENBERG. 1960. Acetylcholine Receptor Protein and Nerve Activity. II. Cationic Group in Local Anesthetics and Electrical Response. *Biochem. Biophys. Res. Commun.* 2: 316–19.

BIANCHI, C. P., and T. C. BOLTON. 1967. Action of Local Anesthetics on Coupling Systems in Muscle. *J. Pharmacol. Exp. Therap.* 157: 388–405.

BIANCHI, C. P., and G. E. STROBEL. 1968. Modes of Action of Local Anesthetics in Nerve and Muscle in Relation to their Uptake and Distribution. *Trans. N. Y. Acad. Sci., Ser. II.* 30: 1082–92.

BLAUSTEIN, M. P. 1967. Phospholipids as Ion Exchangers: Implications for a Possible Role in Biological Membrane Excitability and Anesthesia. *Biochim. Biophys. Acta.* 135: 653–68.

––––––. 1968a. Barbiturates Block Sodium and Potassium Conductance Increases in Voltage-Clamped Lobster Axons. *J. Gen. Physiol.* 51: 293–307.

––––––. 1968b. Action of Certain Tropine Esters on Voltage-Clamped Lobster Axon. *J. Gen. Physiol.* 51: 309–19.

CAMOUGIS, G., B. H. TAKMAN and J. R. P. TASSE. 1967. Potency Difference between Zwitterion Form and the Cation Forms of Tetrodotoxin. *Science.* 156: 1625–27.

CHANDLER, W. K., and H. MEVES. 1965. Voltage Clamp Experiments on Internally Perfused Giant Axons. *J. Physiol. (London).* 180: 788–820.

CLOWES, G. H. A., A. K. KELTCH and M. E. KRAHL. 1940. Extracellular and Intracellular Hydrogen Ion Concentration in Relation to Anesthetic Effects of Barbituric Acid Derivatives. *J. Pharmacol. Exp. Therap.* 68: 312–29.

DETTBARN, W. D. 1962. The Active Form of Local Anesthetics. *Biochim. Biophys. Acta.* 57: 73–76.

ELDEFRAWI, M. E., and R. D. O'BRIEN. 1967. Permeability of the Abdominal Nerve Cord of the American Cockroach, *Periplaneta americana* L., to Quaternary Ammonium Salts. *J. Exp. Biol.* 46: 1–12.

FRAZIER, D. T., T. NARAHASHI and J. W. MOORE. 1968. The Mode of Action of Quaternary Compounds on the Excitability of Squid Axons. *Proc. Intern. Union Physiol. Sci.*, Vol. 7, XXIV Intern. Congr., p. 143.

––––––. 1969. Hemicholinium-3: Non-Cholinergic Effects on Squid Axons. *Science.* 163: 820–21.

FRAZIER, D. T., T. NARAHASHI and M. YAMADA. 1970. The Site of Action and Active Form of Local Anesthetics. II. Experiments with Quaternary Compounds. *J. Pharmacol. Exp. Therap.* 171: 45–51.

FRIESS, S. L., R. C. DURANT and H. L. MARTIN. 1966. Potency Convergence and Crossover in Tertiary vs. Quaternary Tropine Amino Esters. *Toxicol. Appl. Pharmacol.* 9: 240–56.

GOTO, T., Y. KISHI, S. TAKAHASHI and Y. HIRATA. 1965. Tetrodotoxin. XI. *Tetrahedron.* 21:2059–88.

HAYASHI, I. 1939. The Action of Pyrethrin I on Insect Nerve. *Shokubutsu Oyobi Dobotsu.* 7:2001–8.

HILLE, B. 1968. Pharmacological Modifications of the Sodium Channels of Frog Nerve. *J. Gen. Physiol.* 51:199–219.

HODGKIN, A. L., and A. F. HUXLEY. 1952. A Quantitative Description of Membrane Current and its Application to Conduction and Excitation in Nerve. *J. Physiol. (London).* 117:500–44.

HODGKIN, A. L., and B. KATZ. 1949. The Effect of Sodium Ions on the Electrical Activity of the Giant Axon of the Squid. *J. Physiol. (London).* 108:37–77.

KAO, C. Y. 1966. Tetrodotoxin, Saxitoxin and their Significance in the Study of Excitation Phenomena. *Pharmacol. Rev.* 18:997–1049.

KAO, C. Y., and F. A. FUHRMAN. 1967. Differentiation of the Actions of Tetrodotoxin and Saxitoxin. *Toxicon.* 5:25–34.

KRAHL, M. E., A. K. KELTCH, and G. H. A. CLOWES. 1940. The Role of Changes in Extracellular and Intracellular Hydrogen Ion Concentration in the Action of Local Anesthetic Bases. *J. Pharmacol. Exp. Therap.* 68:330–50.

LOWENSTEIN, O. 1942. A Method of Physiological Assay of Pyrethrum Extract. *Nature.* 150:760–62.

MACINTOSH, F. C. 1961. Effect of HC-3 on Acetylcholine Turnover. *Federation Proc.* 20:562–68.

MACINTOSH, F. C., R. I. BIRKS and P. B. SASTRY. 1956. Pharmacological Inhibition of Acetylcholine Synthesis. *Nature.* 178:1181.

MOORE, J. W., M. P. BLAUSTEIN, N. C. ANDERSON and T. NARAHASHI. 1967. Basis of Tetrodotoxin's Selectivity in Blockage of Squid Axons. *J. Gen. Physiol.* 50:1401–11.

MURAYAMA, K., N. J. ABBOTT and T. NARAHASHI. Unpublished.

NAKAMURA, Y., S. NAKAJIMA and H. GRUNDFEST. 1965a. The Action of Tetrodotoxin on Electrogenic Components of Squid Giant Axons. *J. Gen. Physiol.* 48:985–96.

———. 1965b. Analysis of Spike Electrogenesis and Depolarizing K Inactivation in Electroplaques of *Electrophorus electricus,* L. *J. Gen. Physiol.* 49:312–49.

NARAHASHI, T. 1962a. Effect of the Insecticide Allethrin on Membrane Potentials of Cockroach Giant Axons. *J. Cell. Comp. Physiol.* 59:61–65.

———. 1962b. Nature of the Negative After-Potential Increased by the Insecticide Allethrin in Cockroach Giant Axons. *J. Cell. Comp. Physiol.* 59:67–76.

———. 1963. The Properties of Insect Axons, pp. 175–256. *In* J. W. L. Beament, J. E. Treherne and V. B. Wigglesworth [ed.] *Advances in Insect Physiology,* Vol. 1. Academic Press, London and New York.

NARAHASHI, T., and N. C. ANDERSON. 1967. Mechanism of Excitation Block by the Insecticide Allethrin Applied Externally and Internally to Squid Giant Axons. *Toxicol. Appl. Pharmacol.* 10:529–47.

NARAHASHI, T., N. C. ANDERSON and J. W. MOORE. 1966. Tetrodotoxin Does Not Block Excitation from Inside the Nerve Membrane. *Science.* 153:765–67.

——. 1967. Comparison of Tetrodotoxin and Procaine in Internally Perfused Squid Giant Axons. *J. Gen. Physiol.* 50:1413–28.

NARAHASHI, T., T. DEGUCHI, N. URAKAWA and Y. OHKUBO. 1960. Stabilization and Rectification of Muscle Fiber Membrane by Tetrodotoxin. *Am. J. Physiol.* 198:934–38.

NARAHASHI, T., and D. T. FRAZIER. 1968. Site of Action and Active Form of Local Anesthetics in Nerve Fibers. *Federation Proc.* 27:408.

——. 1971. Site of Action and Active Form of Local Anesthetics. *In* S. Ehrenpreis and O. C. Solnitzky [ed.] *Neurosciences Research,* Vol. 4. Academic Press, New York and London. In press.

NARAHASHI, T., D. T. FRAZIER, T. DEGUCHI, C. A. CLEAVES and M. C. ERNAU. 1970. Active Form of Pentobarbital in Nerve Fibers. *Federation Proc.* 29:483 Abs.

NARAHASHI, T., D. T. FRAZIER, T. DEGUCHI, C. A. CLEAVES and M. C. ERNAU. 1971. The Active Form of Pentobarbital in Squid Giant Axons. *J. Pharmacol. Exp. Therap.* 177:25–33.

NARAHASHI, T., D. T. FRAZIER and J. W. MOORE. 1968. A Model for the Action of Local Anesthetics. *Proc. Intern. Union Physiol. Sci.,* Vol. 7, XXIV Intern. Congr., p. 313.

NARAHASHI, T., D. T. FRAZIER and M. YAMADA. 1970. The Site of Action and Active Form of Local Anesthetics. I. Theory and pH Experiments with Tertiary Compounds. *J. Pharmacol. Exp. Therap.* 171:32–44.

NARAHASHI, T., and H. G. HAAS. 1967. DDT: Interaction with Nerve Membrane Conductance Changes. *Science.* 157:1438–40.

——. 1968. Interaction of DDT with the Components of Lobster Nerve Membrane Conductance. *J. Gen. Physiol.* 51:177–98.

NARAHASHI, T., H. G. HAAS and E. F. THERRIEN. 1967. Saxitoxin and Tetrodotoxin: Comparison of Nerve Blocking Mechanism. *Science.* 157:1441–42.

NARAHASHI, T., J. W. MOORE and D. T. FRAZIER. 1969. Dependence of Tetrodotoxin Blockage of Nerve Membrane Conductance on External pH. *J. Pharmacol. Exp. Therap.* 169:224–28.

NARAHASHI, T., J. W. MOORE and R. N. POSTON. 1969. Anesthetic Blocking of Nerve Membrane Conductances by Internal and External Applications *J. Neurobiol.* 1:3–22.

NARAHASHI, T., J. W. MOORE and W. R. SCOTT. 1964. Tetrodotoxin Blockage of Sodium Conductance Increase in Lobster Giant Axons. *J. Gen. Physiol.* 47:965–74.

NARAHASHI, T., J. W. MOORE and B. I. SHAPIRO. 1969. *Condylactis* Toxin: Interaction with Nerve Membrane Ionic Conductances. *Science.* 163:680–81.

NARAHASHI, T., and M. YAMADA. 1969. Role of External and Internal pH in Nerve Blockage Caused by Local Anesthetics. *Federation Proc.* 28:476.

NARAHASHI, T., M. YAMADA and D. T. FRAZIER. 1969. Cationic Forms of Local Anaesthetics Block Action Potentials from Inside the Nerve Membrane. *Nature.* 223:748–49.

NARAHASHI, T., and T. YAMASAKI. 1960a. Mechanism of Increase in Negative After-Potential by Dicophanum (DDT) in the Giant Axons of the Cockroach. *J. Physiol. (London).* 152:122–40.

———. 1960b. Behaviors of Membrane Potential in the Cockroach Giant Axons Poisoned by DDT. *J. Cell. Comp. Physiol.* 55:131–42.

O'BRIEN, R. D. 1967. Barrier Systems in Insect Ganglia and their Implication for Toxicology. *Federation Proc.* 26:1056–61.

OGURA, Y., and Y. MORI. 1968. Mechanism of Local Anesthetic Action of Crystalline Tetrodotoxin and its Derivatives. *Europ. J. Pharmacol.* 3:58–67.

OIKAWA, T., C. S. SPYROPOULOS, I. TASAKI and T. TEORELL. 1961. Methods for Perfusing the Giant Axon of *Loligo pealei. Acta Physiol. Scand.* 52:195–96.

OVERTON, E. 1902. Beiträge zur allgemeinen Muskel und Nervenphysiologie. *Arch. Ges. Physiol.* 92:346–86.

PICKARD, W. F., J. Y. LETTVIN, J. W. MOORE, M. TAKATA, J. POOLER and T. BERNSTEIN. 1964. Caesium Ions Do Not Pass the Membrane of the Giant Axon. *Proc. Nat. Acad. Sci. USA.* 52:1177–83.

RITCHIE, J. M., and P. GREENGARD. 1961. On the Active Structure of Local Anesthetics. *J. Pharmacol. Exp. Therap.* 133:241–45.

———. 1966. On the Mode of Action of Local Anesthetics. *Ann. Rev. Pharmacol.* 6:405–30.

RITCHIE, J. M., B. RITCHIE and P. GREENGARD. 1965a. The Active Structure of Local Anesthetics. *J. Pharmacol. Exp. Therap.* 150:152–59.

———. 1965b. The Effect of the Nerve Sheath on the Action of Local Anesthetics. *J. Pharmacol. Exp. Therap.* 150:160–64.

RITCHIE, J. M., and B. R. RITCHIE. 1968. Local Anesthetics: Effect of pH on Activity. *Science.* 162:1394–95.

ROJAS, E., and I. ATWATER. 1967. Effect of Tetrodotoxin on the Early Outward Currents in Perfused Giant Axons. *Proc. Nat. Acad. Sci. USA.* 57:1350–55.

ROTHENBERG, M. A., D. B. SPRINSON and D. NACHMANSOHN. 1948. Site of Action of Acetylcholine. *J. Neurophysiol.* 11:111–16.

SCHANKER, L. S., P. A. NAFPLIOTIS and J. M. JOHNSON. 1961. Passage of Organic Bases into Human Red Cells. *J. Pharmacol. Exp. Therap.* 133:325–31.

SCHANTZ, E. J., J. M. LYNCH, G. VAYVADA, K. MATSUMOTO and H. RAPOPORT. 1966. The Purification and Characterization of the Poison Produced by *Gonyaulax catenella* in Axenic Culture. *Biochemistry.* 5:1191–95.

SCHUELER, F. W. 1960. The Mechanism of Action of the Hemicholiniums. *Intern. Rev. Neurobiol.* 2:77–97.

SHANES, A. M. 1949. Electrical Phenomena in Nerve. II. Crab Nerve. *J. Gen. Physiol.* 33:75–102.

SHANES, A. M., W. H. FREYGANG, H. GRUNDFEST and E. AMATNIEK. 1959. Anesthetic and Calcium Action in the Voltage Clamped Squid Giant Axon. *J. Gen. Physiol.* 42:793–802.

SHAPIRO, B. I. 1968. Purification of a Toxin from Tentacles of the Anemone *Condylactis gigantea*. *Toxicon.* 5:253–59.

SHAPIRO, B. I., and G. LILLEHEIL. 1969. The Action of Anemone Toxin on Crustacean Neurons. *Comp. Biochem. Physiol.* 28:1225–41.

SKOU, J. C. 1954a. Local Anesthetics. I. The Blocking Potencies of Some Local Anaesthetics and of Butyl Alcohol Determined on Peripheral Nerves. *Acta Pharmacol. Toxicol.* 10:281–91.

————. 1954b. Local Anaesthetics. II. The Toxic Potencies of Some Local Anaesthetics and of Butyl Alcohol, Determined on Peripheral Nerves. *Acta Pharmacol. Toxicol.* 10:292–96.

————. 1954c. Local Anaesthetics. III. Distribution of Local Anaesthetics between the Solid Phase/Aqueous Phase of Peripheral Nerves. *Acta Pharmacol. Toxicol.* 10:297–304.

————. 1954d. Local Anaesthetics. IV. Surface and Interfacial Activities of Some Local Anaesthetics. *Acta Pharmacol. Toxicol.* 10:305–16.

————. 1954e. Local Anaesthetics. V. The Action of Local Anaesthetics on Monomolecular Layers of Stearic Acid. *Acta Pharmacol. Toxicol.* 10:317–24.

————. 1954f. Local Anaesthetics. VI. Relation between Blocking Potency and Penetration of a Monomolecular Layer of Lipoids from Nerves. *Acta Pharmacol. Toxicol.* 10:325–37.

————. 1961. The Effect of Drugs on Cell Membranes with Special Reference to Local Anaesthetics. *J. Pharm. Pharmacol.* 13:204–17.

TAHARA, Y. 1910. Über das Tetrodongift. *Biochem. Z.* 10:255–75.

TAKATA, M., J. W. MOORE, C. Y. KAO and F. A. FUHRMAN. 1966. Blockage of Sodium Conductance Increase in Lobster Giant Axon by Tarichatoxin (Tetrodotoxin). *J. Gen. Physiol.* 49:977–88.

TASAKI, I., and I. SINGER. 1966. Membrane Macromolecules and Nerve Excitability: A Physico-Chemical Interpretation of Excitation in Squid Giant Axons. *Ann. N. Y. Acad. Sci.* 137:792–806.

TAYLOR, R. E. 1959. Effect of Procaine on Electrical Properties of Squid Axon Membrane. *Am. J. Physiol.* 196:1071–78.

TREVAN, J. W., and E. BOOCK. 1927. The Relation of Hydrogen Ion Concentration to the Action of the Local Anaesthetics. *Brit. J. Exp. Pathol.* 8:307–15.

TSUDA, K., S. IKUMA, M. KAWAMURA, R. TACHIKAWA, K. SAKAI, C. TAMURA and O. AMAKASU. 1964. Tetrodotoxin. VII. On the Structures of Tetrodotoxin and its Derivatives. *Chem. Pharm. Bull. (Tokyo).* 12:1357–74.

WHITCOMB, E. R., S. L. FRIESS and J. W. MOORE. 1958. Action of Certain Anticholinesterases on the Spike Potential of the Desheathed Sciatic Nerve of the Bullfrog. *J. Cell. Comp. Physiol.* 52:275–99.

YAMASAKI, T., and T. ISHII (Former name of T. Narahashi). 1952*a*. Studies on the Mechanism of Action of Insecticides. IV. The Effects of Insecticides on the Nerve Conduction of Insect. *Oyo Kontyu (J. Nippon Soc. Appl. Entomol.)* 7:157–64.

————. 1952*b*. Studies on the Mechanism of Action of Insecticides. V. The Effects of DDT on the Synaptic Transmission in the Cockroach. *Oyo Kontyu (J. Nippon Soc. Appl. Entomol.)* 8:111–18.

YAMASAKI, T., and T. NARAHASHI. 1957*a*. Increase in the Negative After-Potential of Insect Nerve by DDT. Studies on the Mechanism of Action of Insecticides. XIII. *Botyu-Kagaku.* 22:296–304.

————. 1957*b*. Intracellular Microelectrode Recordings of Resting and Action Potentials from the Insect Axon and the Effects of DDT on the Action Potential. Studies on the Mechanism of Action of Insecticides. XIV. *Botyu-Kagaku.* 22:305–13.

EXCITABILITY IN LIPID BILAYER MEMBRANES

21

G. EHRENSTEIN

Lipid bilayer membranes have been used as models for a wide variety of biological membrane properties. Two of these properties, selectivity and excitability, are intimately related to the functioning of an axon. By selectivity, I refer to permeability differences between cations, and by excitability, the variation of conductance with membrane potential. This description of excitability may need some explanation. In neurophysiology, excitability refers to the ability to generate action potentials. But for an understanding of the physical mechanism of excitability, it is desirable to concentrate attention on the fundamental properties underlying the generation of action potentials. A threshold phenomenon is essentially a nonlinear one (FitzHugh, 1969) and the *sina qua non* of an action potential is a nonlinear current-voltage curve.

The excitability of axons (Hodgkin and Huxley, 1952) and their selectivity (Chandler and Meves, 1965) have been well described phenomenologically, and investigators are now primarily concerned with elucidating the underlying mechanisms—on a molecular level if possible. In this presentation, I will try to describe the progress made so far in using lipid bilayer membranes to elucidate the property of excitability.

METHODS

The technique for making lipid bilayer membranes was developed about ten years ago by Mueller and Rudin. Their elegantly simple apparatus is shown in Fig. 21-1 (Mueller, Rudin, Tien and Wescott, 1964). Membranes are

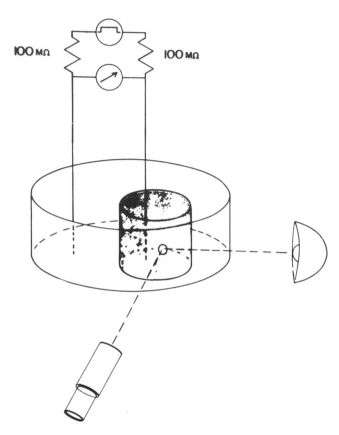

Fig. 21-1 *Membrane Chamber. A* 5 ml *polyethylene cup rests inside a glass petri dish* (60 × 20 mm). *Saline solution level is above the hole across which the membrane solution is brushed on and observed at* 10× *magnification under reflected light. Conventional pulsing and recording circuit is shown.* (*From Mueller et al.,* 1964, © 1964 *The Academic Press Inc., New York.*)

formed across a 1 mm diameter hole in a 5 ml polyethylene cup, which rests inside a glass petri dish. The saline solution level is above the hole, across which the lipid solution is applied with a small sable brush. Detailed descriptions of their methods, and of alternative schemes for bilayer formation have recently been published (Bangham, 1968; Rothfield and Finkelstein, 1968; Tien and Diana, 1968; Mueller and Rudin, 1969). These reviews also describe the kinds of lipids that may be employed.

PROPERTIES OF LIPID BILAYERS

A useful analog for an axon membrane is a system which is similar in most of its important properties, but yet has some differences which may be exploited. A comparison of properties of lipid bilayers and axon membranes is summarized in Table 21-1.

Table 21-1   A comparison of properties of axon membranes and lipid bilayers

Properties	Axon membranes	Lipid bilayers
Electron microscopy	Railroad tracks	Railroad tracks
Electrical capacitance	1 $\mu f/cm^2$	0.3–1.4 $\mu f/cm^2$
Thickness	75Å	75Å
Water permeability	0.14–1.1 $\times$ 10^{-3}	1 $\times$ 10^{-3}cm/sec
Electrical resistance	10^3 ohm cm^2	10^8 ohm cm^2

Electron microscopy of lipid bilayers (Henn, Decker, Greenawalt and Thompson, 1967) indicates the same type of railroad track appearance and the same thickness—about 75 Å—that have been observed with cell membranes (Robertson, 1960). Capacitance measurements on lipid bilayers range from 0.3–1.4 $\mu f/cm^2$ (Mueller and Rudin, 1969), in good agreement with the axon membrane capacitance of 1 $\mu f/cm^2$ (Cole, 1968). This value of capacitance is also consistent with the thickness determined by electron microscopy. Another way to determine the thickness of a lipid bilayer is to measure the light reflectance, and hence the difference in optical path length between light rays reflected from the two surfaces of the bilayer. Using this method, Huang and Thompson (1965, 1966) obtained a value of 72 Å for the membrane thickness, in good agreement with the value determined by electron microscopy.

Water permeability has also been measured for both axon membranes and lipid bilayers, but in both cases there has been some ambiguity. Permeabilities determined from experiments using osmotic pressure differences have been consistently larger than permeabilities measured in tritiated water experiments. For the lipid bilayer case, this situation has been resolved by Cass and Finkelstein (1967), who showed that the difference was caused by an unstirred layer near the membrane. This was done by deliberately varying the thickness of the unstirred layer using different membrane supports, and observing differences in apparent permeability. Cass and Finkelstein obtained an unambiguous water permeability of 1 $\times$ 10^{-3} cm/sec. It is not yet known whether the difference between the two types of permeability measurements in natural membranes has the same explanation. In any event, the range of permeability values for axon membranes is $0.14–1.1 \times 10^{-3}$ cm/sec (Villegas and Villegas, 1960), differing from the lipid bilayer value by less than an order of magnitude.

The evidence presented above indicates that lipid bilayers and axon membranes are quite similar in structure and thickness. But there is a very large difference in electrical resistance. A typical resting axon has an electrical resistance of about 1000 ohm cm^2, whereas the electrical resistance of a lipid bilayer is about 5 orders of magnitude larger. If an active axon membrane is considered, the difference between lipid bilayer and axon membrane is even larger.

ADVANTAGES OF LIPID BILAYERS

In attempting to learn about one system by studying an analogous system, there is the serious problem of the quality of the analogy. Since the analogy between axons and lipid bilayers is not perfect, any results from the latter must be regarded as suggestions of possible axon mechanisms, requiring verification. Why then should one bother to study the model system at all?

One of the advantages of the lipid bilayer is the lack of extraneous tissue. Indeed, the reflectance method of thickness measurement and the determination of the effects of unstirred layers, both of which were mentioned above, have been performed on lipid bilayers. These measurements have not yet been extended to cell membranes, largely because of the added complication of extraneous layers of tissue.

Perhaps the most useful difference between axon membranes and lipid bilayers is the low ionic permeability of the lipid bilayer. This provides a much smaller background current for the study of permeability properties of various additives. In particular, this allows the use of very small amounts of an additive, and makes possible the study of statistical properties of small numbers of channels.

Because of the low permeability of ions through lipids, it has often been suggested that ions must travel through lipid membranes by means of carriers or channels. Many materials whose presence in a lipid bilayer enhances ionic permeability have been found. Mueller and Rudin (1969) have summarized these materials, which they call translocators. A number of translocators, notably cyclic antibiotics such as valinomycin, impart selectivity properties. Only a very few produce excitability. Three excitability-inducing translocators—alamethicin, monazomycin, and EIM—will now be considered.

ALAMETHICIN

Alamethicin, a cyclic polypeptide with a known amino acid content (Meyer and Reusser, 1967), imparts to a lipid bilayer a large potential-dependent cation conductance. Mueller and Rudin (1968a) described the electrical properties of alamethicin, as well as its more complex behavior in the presence of protamine, where action potentials can be evoked. For simplicity, this discussion will be restricted to alamethicin without protamine.

Conductance-voltage curves for two concentrations of alamethicin are shown in Fig. 21-2. A semilog scale is used because the conductance ranges over 4 orders of magnitude. The positive potentials shown in the figure correspond to current flow from the side of the membrane containing alamethicin to the side not containing alamethicin. The conductances for negative potentials are approximately the same as for positive potentials, but a somewhat larger magnitude of negative potential is required for a given conductance. Conductance is also a very steep function of alamethicin concentration and of salt concentration, depending on approximately the sixth power of each. Furthermore, the rate constant for the change of conductance following a change of potential depends on approximately the sixth power of alamethicin concentration.

In the same paper that described these properties, Mueller and Rudin proposed a model to explain them. According to this model, a monomer of a single cation and a single alamethicin molecule is formed in the aqueous phase. A positive potential then drives the charged monomers into the hydrocarbon region, where they may aggregate into oligomers of six cations and six alamethicin molecules. Only the oligomers transfer cations to the other side of the membrane, perhaps because they form a sort of bridge across the hydrocarbon region, which may subsequently break up and deposit cations on the other side. It is not yet clear whether alamethicin acts as a channel or a carrier. In any event, limiting ion transfer to the oligomers explains the steep concentration dependences of conductance and rate constant. In this model, the potential dependence of the conductance arises from the requirement that oligomers form only in the hydrocarbon region, since the fraction of monomers in this region depends upon the potential.

Mueller and Rudin were able to fit their model to the experimental conductance-voltage data reasonably well, as shown in Fig. 21-2, where the curves are calculated and the points are experimental.

Further experimental support for this model has recently been provided. Chapman, Cherry, Finer, Hauser, Phillips and Shipley (1969), using monolayers, demonstrated a high lipid solubility for alamethicin, and Pressman (1968) showed that alamethicin extracts radioactively labelled alkali metal cations from water into butanol-toluene.

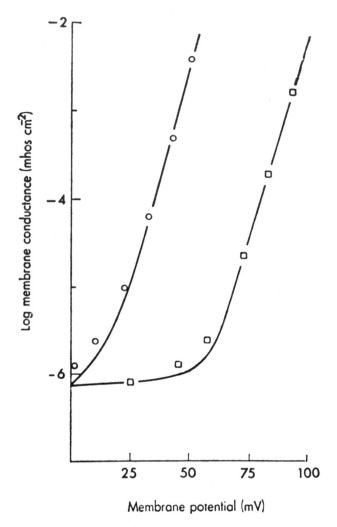

Fig. 21-2 *The relation between membrane conductance and applied potential at two different concentrations of alamethicin in the aqueous phase (circles* 5×10^{-6} *g/ml, squares* 10^{-6} *g/ml). Ionic concentrations in both aqueous compartments,* 0.1 M *sodium chloride. (From Mueller and Rudin, 1968a.)*

MONAZOMYCIN

Monazomycin is an antibiotic of unknown structure with formula $C_{62}H_{119}O_{20}N$ (Akasaki, Karasawa, Watanabe, Yonehara and Umezawa, 1963). Little has been reported about its electrical properties, but it is included in this discussion because of its unusual conductance-voltage relation, recently described qualitatively by Mueller and Rudin (1969). For positive potentials, the monazomycin conductance-voltage curve is quite similar to that for alamethicin, but for negative potentials, the conductance with monazomycin is constant, equal to the zero-potential conductance. Also, monazomycin is a water soluble molecule, in contrast with alamethicin, which is much more soluble in organic solvents than in water.

EIM

A bacterial extract called "excitability inducing material" or EIM (Mueller and Rudin, 1963) has a conductance-voltage curve that is opposite to that of alamethicin. As shown in Fig. 21-3, the conductance is maximum at zero membrane potential, and decreases as potential increases (Ehrenstein, Lecar and Nossal, 1970). The conductances for negative potentials are not shown in the figure, but are qualitatively the same as for positive potentials. Interesting electrical responses, including action potentials can be evoked when protamine is added to EIM (Mueller and Rudin, 1967). As in the discussion of alamethicin however, the discussion of EIM will be limited to properties found in the absence of protamine.

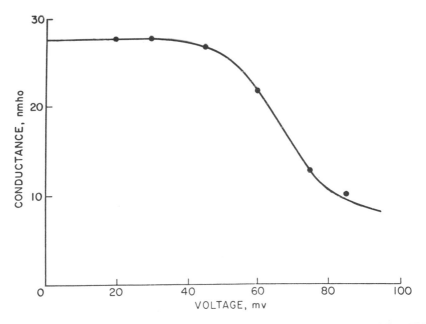

Fig. 21-3 *Conductance-voltage curve for a lipid bilayer membrane containing* EIM. *(From Ehrenstein et al.,* 1970.)

There are conflicting results for the dependence of conductance on EIM concentration (Bean, Shepherd, Chan and Eichner, 1969), but there is other evidence that EIM acts by forming channels through the membrane. This evidence will now be considered.

The enhanced conductance caused by EIM can be reduced by adding proteolytic enzymes to either side of the membrane, even when EIM has been added only on one side (Mueller and Rudin, 1968b; Bean, Chan and Eichner, 1968).

Bean et al. (1969), using low concentrations of EIM, detected discrete steps of conductance during the process of EIM adsorption, as shown in Fig. 21-4. These "formation bumps" average 4×10^{-10} mhos, and correspond to a step change in current of about 10^8 ions/sec. If the steps represent additions of carriers to the membrane, each carrier would be required to transport an unreasonably large number of ions (Stein, 1968). If the steps represent additions of channels, however, a reasonable channel diameter of about 20 Å is required under the assumption that ionic mobilities in the channel are similar to those in water.

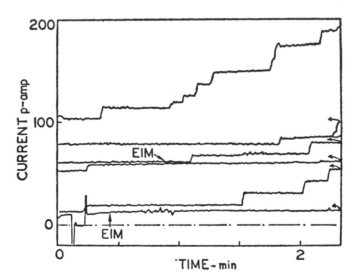

Fig. 21-4 *Long-lived conductance steps during membrane interaction with EIM. Membrane composition, brain phospholipid, 1.5%; tocopherol, 15%. Polarizing potential, −15 mv. Temperature, 36°C. Chart speed, 3 inches/min. EIM, to about 20 × 10⁻⁹ g/ml, was added to the reference compartment at the two points indicated. The recording was made on an X-Y recorder, necessitating the overlapping, sequential scans. This particular batch of brain lipids consistently gave remarkably uniform and distinct step reactions similar to those shown here. (From Bean et al., 1969.)*

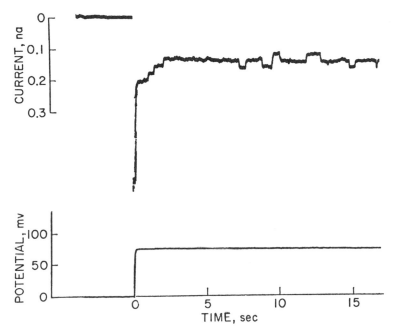

Fig. 21-5 *Current and potential records showing discrete conductance jumps for a lipid bilayer with* 4 EIM *channels. (From Ehrenstein et al.,* 1970.)

More recently, Ehrenstein et al. (1970) confirmed the channel hypothesis by observing "transition bumps" in a lipid bilayer whose EIM composition was no longer changing. Transition bumps for a membrane with four channels are shown in Fig. 21-5. Studies of the heights and time distributions of these transition bumps as a function of membrane potential provide more detailed information about the nature of EIM channels. The bump heights were found to be directly proportional to membrane potential, showing that individual EIM channels are ohmic. This suggests that, as potential increases, the conductance decreases because of the decreased likelihood that a channel is in the open state. This notion was tested by measuring the relative time a channel is open for several membrane potentials. In Fig. 21-6 the relative open time as a function of voltage is compared to an EIM conductance-voltage curve, whose scale has been appropriately adjusted. The relatively good fit confirms that the voltage-dependent conductance of EIM arises from the opening and closing of individual ohmic channels. Fig. 21-6 is taken from the paper of Ehrenstein et al. (1970), but includes additional experimental points. That paper also presents statistical evidence that the channels have only two states—"open" and "closed"—and that the channels are independent.

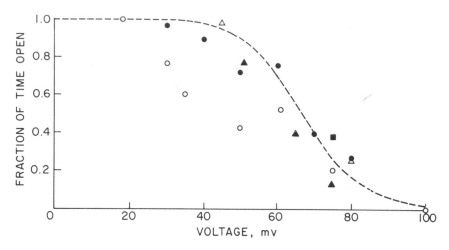

Fig. 21-6 *Voltage dependence of the fraction of time an EIM channel is open. Each symbol represents a different membrane. (Modified from Ehrenstein et al., 1970.)*

The latter result is expected simply because of the small number of channels in the membrane.

In summary, EIM acts in a lipid bilayer by providing independent channels through which cations can pass. The channels can be either open or closed. The probability that a channel is open decreases as the membrane potential is increased from zero.

COMPARISON WITH AXON EXCITABILITY

We shall now compare the excitability properties of lipid bilayers containing the translocators discussed above with the excitability properties of axons. Fig. 21-7 shows the steady-state conductance-voltage curve for a squid axon. Comparison among Fig. 21-2, 21-3, and 21-7 indicates that the steady-state conductance-voltage curves for EIM and for a squid axon are qualitatively similar. In both cases the conductance is large near zero membrane potential, and decreases as the absolute value of the membrane potential increases. The apparent difference in sign of the membrane potential is of little significance, especially because the sign convention is quite arbitrary. It is not yet clear whether the side of the lipid bilayer membrane to which EIM is added corresponds to the inside or the outside of the axon. The conductance-voltage curves of alamethicin and monazomycin, on the other hand, differ from the axon conductance-voltage curve, regardless of the sign of membrane potential.

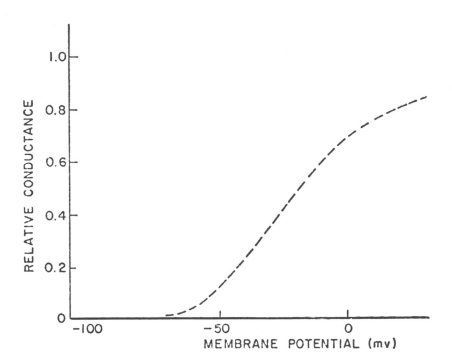

Fig. 21-7 *Steady state conductance-voltage curve for a squid axon, calculated from Hodgkin and Huxley (1952).*

The conductance-voltage curves considered above all apply to steady-state conditions. Thus, the comparison is between the lipid bilayer systems and the system in the axon that is primarily concerned with potassium permeability. Actually, it is the sodium system of the axon that is more intimately related to generation of action potentials. The peak sodium conductance-voltage curve for an axon, however, is quite similar to the steady-state conductance-voltage curve. For simplicity, only steady-state conductance-voltage curves have been presented above.

Another comparison that should be made concerns the method of translocation. As indicated above, it not yet clear whether alamethicin acts as a carrier or a channel, but EIM clearly acts as a channel. What about the axon? Hille (1970) examined a number of estimates for the conductance of a single sodium channel obtained in different manners, and concluded that they are all consistent with a value of 1×10^{-10} mho in frog Ringer's. This is the same order of magnitude found for the conductance of an EIM channel, but it is inconsistent with a carrier model. Of the various estimates considered by Hille, the most compelling is that of Luttgau (1958), who observed quantal behavior in subthreshold responses of nodes of Ranvier. In short, there is good, but not incontrovertible, evidence that ions pass through the axon membrane by means of channels whose conductance is approximately equal to that of an EIM channel.

Note added in proof. The existence of transition bumps in the EIM system makes possible the determination of a number of single-channel properties besides those mentioned above. For example, the relative selectivity of the alkali cations is proportional to the bump heights measured with salts of these cations and relatively impermeable anions. These bump height experiments have recently been performed in our laboratory for lithium, sodium, potassium and cesium, and the results indicate that the permeabilities of the alkali cations are approximately proportional to their free-solution mobilities (Lecar, Latorre, and Ehrenstein, 1971).

Another use of transition bumps is in the determination of the kinetics for opening and closing of EIM channels. It is possible to tabulate the waiting time for a single channel to open or close for a large number of events at a fixed membrane potential. The distribution of these times at a given membrane potential and the variation of the distribution as a function of membrane potential can then be used to test kinetic models. We have been performing experiments of this type in our laboratory and our initial observations indicate that both opening and closing can be described as first order rate processes (Ehrenstein and Lecar, 1961).

REFERENCES

AKASAKI, K., K. KARASAWA, M. WATANABE, H. YONEHARA and H. UMEZAWA. 1963. Monazomycin, a New Antibiotic Produced by a Streptomyces. *J. Antibiotics (Tokyo).* A16:127.

BANGHAM, A. D. 1968. Membrane Models with Phospholipids. *Progr. Biophys. Mol. Biol.* 18:29.

BEAN, R. C., H. CHAN and J. T. EICHNER. 1968. Single Channel Conductance in an Excitable Lipid Bilayer Membrane. *Biophys. J.* 8:A-26.

BEAN, R. C., W. C. SHEPHERD, H. CHAN and J. T. EICHNER. 1969. Discrete Conductance Fluctuations in Lipid Bilayer Protein Membranes. *J. Gen. Physiol.* 53:741.

CASS, A., and A. FINKELSTEIN. 1967. Water Permeability of Thin Lipid Membranes. *J. Gen. Physiol.* 50:1765.

CHANDLER, W. K., and H. MEVES. 1965. Voltage Clamp Experiments on Internally Perfused Giant Axons. *J. Physiol. (London).* 180:788.

CHAPMAN, D., R. J. CHERRY, E. G. FINER, H. HAUSER, M. C. PHILLIPS and G. G. SHIPLEY. 1969. Physical Studies of Phospholipid/Alamethicin Iı.teractions. *Nature.* 224:692.

COLE, K. S. 1968. *Membranes, Ions and Impulses.* University of California Press, Berkeley.

EHRENSTEIN, G., and H. LECAR. 1971. Kinetics of the Opening and Closing of Individual EIM Channels. *Biophys. Soc. Absts.* 11:316a.

EHRENSTEIN, G., H. LECAR and R. NOSSAL. 1970. The Nature of the Negative Resistance in Bimolecular Lipid Membranes Containing Excitability-Inducing Material. *J. Gen. Physiol.* 55:119.

FITZHUGH, R. 1969. Mathematical Models of Excitation and Propagation in Nerve, pp. 1–85. *In* H. P. Schwann [ed.] *Biological Engineering.* McGraw Hill, New York.

HENN, F. A., G. L. DECKER, J. W. GREENAWALT and T. E. THOMPSON. 1967. Properties of Lipid Bilayer Membranes Separating Two Aqueous Phases: Electron Microscope Studies. *J. Mol. Biol.* 24:51.

HILLE, B. 1970. Ionic Channels in Nerve Membranes. *Progr. Biophys. Mol. Biol.* 21:1.

HODGKIN, A. L., and A. F. HUXLEY. 1952. A Quantitative Description of Membrane Current and its Application to Conduction and Excitation in Nerve. *J. Physiol. (London).* 117:500.

HUANG, C., and T. E. THOMPSON. 1965. Properties of Lipid Bilayer Membranes Separating Two Aqueous Phases: Determination of Membrane Thickness. *J. Mol. Biol.* 13:183. Correction, 1966. *J. Mol. Biol.* 16:576.

LECAR, H., R. LATORRE and G. EHRENSTEIN. 1971. Ion Transport Properties of EIM Channels in Lipid Bilayers. *Biophys. Soc. Absts.* 11:317a.

LUTTGAU, H. D. 1958. Sprunghafte Schwankungen unterschwelliger potentiale an markhaltigen Nervenfasern. *Z. Naturforsch.* 13b:692.

MEYER, C. E., and F. REUSSER. 1967. A Polypeptide Antibacterial Agent Isolated from Trichoderma Viride. *Experientia.* 23:85.

MUELLER, P., and D. O. RUDIN. 1963. Induced Excitability in Reconstituted Cell Membrane Structure. *J. Theor. Biol.* 4:268.

―――. 1967. Action Potential Phenomena in Experimental Bimolecular Lipid Membranes. *Nature.* 213:603.

―――. 1968a. Action Potentials Induced in Bimolecular Lipid Membranes. *Nature.* 217:713.

―――. 1968b. Resting and Action Potentials in Experimental Bimolecular Lipid Membranes. *J. Theor. Biol.* 18:222.

―――. 1969. Translocators in Bimolecular Lipid Membranes: Their Role in Dissipative and Conservative Bioenergy Transductions. *Current Topics Bioenergetics.* 3:157.

MUELLER, P., D. O. RUDIN, H. T. TIEN and W. C. WESCOTT. 1964. Formation and Properties of Bimolecular Lipid Membranes. *Recent Progr. Surface Sci.* 1:379.

PRESSMAN, D. C. 1968. Ionophorous Antibiotics as Models for Biological Transport. *Federation Proc.* 27:1283.

ROBERTSON, J. D. 1960. The Molecular Structure and Contact Relationships of Cell Membranes. *Progr. Biophys. Mol. Biol.* 10:343.

ROTHFIELD, L., and A. FINKELSTEIN. 1968. Membrane Biochemistry. *Ann. Rev. Biochem.* 37:463.

STEIN, W. D. 1968. Turnover Numbers of Membrane Carriers and the Action of the Polypeptide Antibiotics. *Nature*. 218:570.

TIEN, H. T., and A. L. DIANA. 1968. Bimolecular Lipid Membranes: A Review and a Summary of Some Recent Studies. *Chem. Phys. Lipids*. 2:55.

VILLEGAS, R., and G. M. VILLEGAS. 1960. Characterization of the Membranes in the Giant Nerve Fiber of the Squid. *J. Gen. Physiol*. Suppl. 43:73.

THE VARIETIES OF EXCITABLE MEMBRANES 22

H. GRUNDFEST

Despite the advances in ultrastructure research and molecular biology, we still know very little about the structure and composition of the cell membrane and nothing at all about the factors that lead to its functional complexities. Such deficiencies are especially obvious in the case of membranes of excitable electrogenic cells. For such membranes, through electrophysiological techniques, a great deal of information is known regarding many varieties of functional manifestations that must be presumed to arise from morphological differences in macromolecular structures. At the present time, we can only speculate on the membrane components that are variously permselective to different ions and how these components are affected by different stimuli so as to alter membrane permeability. Furthermore, we are also deficient in a rigorous electrochemical theory of thin, heterogeneous, charged structures that react with the multifarious properties of excitable membranes. Nevertheless, it is remarkable that a wide range of electrogenic phenomena can be classified, ordered and explained by suitable applications of the ionic theory for spike electrogenesis. The ionic theory was first developed by Bernstein (1902, 1912) and was placed on a realistic basis by Hodgkin and his colleagues in their analysis of the spike of the squid giant axon (Hodgkin and Katz, 1949; Hodgkin and Huxley, 1952a, b, c, d; Hodgkin, Huxley and Katz, 1952).

477

HETEROGENEITY OF THE CELL MEMBRANE

The excitable membrane is probably formed basically of a lipid bilayer matrix with proteinaceous additions in which are dispersed, rather sparsely (Grundfest, 1963), variously permselective aqueous ionic channels. Some of the channels are normally closed in the resting membrane and may be opened (activated) by stimuli. Others are normally open and may be closed (inactivated) in response to stimulation. These changes, which are measured in terms of a conductance increase (activation) or decrease (inactivation), may develop with different kinetics. They may be long lasting or transient. They may be more or less independent of one another and may occur in different time sequences. Numerous combinations and permutations of the different varieties of responsiveness are found in the excitable membranes of different species of cells. Most, but not all, of the large variety of electro-physiological manifestations of these cells can be analyzed in terms of the conductance changes that are induced by appropriate stimuli.

The cell membrane may also be regarded as an electrode system. In the resting state it is often predominantly selective for K. The conductance is highest for this ion and the membrane potential is basically that for a K battery, E_K. In some cells, however, the membrane is an electrode system for several cationic and/or anionic species simultaneously. What appear to be specific components of the cell membrane, so classified by their reactions to physical stimuli or to chemical agents, and by their physiological mani-festations of activity, respond to appropriate excitants by a change in con-ductance for certain ionic species. The membrane then changes its electrode properties. Depending upon the difference in electrode properties and relative conductances between the resting and excited components of the membrane, its response to stimuli may or may not be accompanied by a change in membrane potential. The primary electrogenic response of the excited mem-brane is the change in its conductance. The conductance change can be measured by a variety of methods and, in many cases, the ionic species that are involved in the change of conductance can also be identified.

CLASSIFICATION OF ELECTROGENIC COMPONENTS

Components of excitable membranes which respond to changes in the electrical field across the membrane are called *electrically excitable*. *Electri-cally inexcitable* components respond only to such stimuli as chemical agents, or mechanical, photic or thermal forces. The differently excitable components experimentally exhibit different properties which indicate the existence of specific, though as yet unknown, differences in the macromolecular structures of the various membrane components. The presence of the various components in close proximity within a given membrane raises the possibility of their interplays and leads to a number of functionally important consequences.

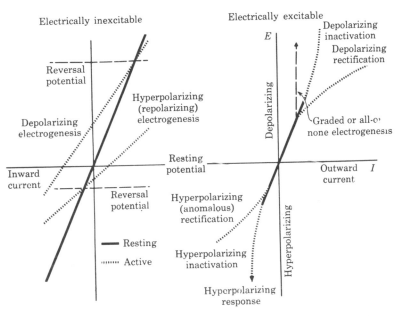

Fig. 22-1 *I-E relations of electrogenic membranes. Electrically inexcitable membranes behave as ohmic resistances, E changing linearly with I. However, the slope of the relation changes during activation of the membrane by an appropriate stimulus. Broken lines represent active membrane of (depolarizing) EPSP's or generator potentials and of (hyperpolarizing) IPSP's respectively. Recorded amplitudes of the responses (given by the difference between the resting I-E line and that during activity) change with the change in membrane potential. Thus a characteristic feature of electrically inexcitable electrogenesis is its change in sign when E exceeds a reversal potential specified by the intersection of the resistance lines for active and passive membrane.*

Electrically excitable membranes exhibit nonlinear behavior characterized by one or by several varieties of conductance change. On stimulation with depolarizing currents many, but not all, cells respond with conductance increase (depolarizing activation) for Na, Ca or Mg. This introduces a previously occult inside-positive emf and causes graded or all-or-none depolarizing electrogenesis (spikes or graded responses). Increased conductance for K or Cl (inappropriately termed "rectification") that is evoked by depolarizing stimuli or for various ions by hyperpolarizing currents is usually not regenerative. However, it can become regenerative when electrochemical conditions permit. Decreased conductance or inactivation may be evoked by depolarizing or hyperpolarizing currents and it can also be regenerative. Various ionic processes which cause nonlinear relations exhibit different degrees of time variance. (Modified from Grundfest, 1961b.)

LINEAR (OHMIC) RESPONSES OF ELECTRICALLY INEXCITABLE COMPONENTS

The different modes of excitability are reflected in different current-voltage (I-E) characteristics (Fig. 22-1). The conductance of the electrically inexcitable elements is not altered by changing the membrane potential and thus the I-E relation is linear and obeys Ohm's Law. However, when some specific agent excites (or activates) the electrically inexcitable components, the conductance increases and there may be an electrogenesis of either sign, depending upon the specific ionic processes. The I-E characteristic is then displaced along the voltage axis, depending upon the sign and magnitude of the electrogenesis and its new slope is a measure of the change in conductance.

At any value of the applied current the separation between resting and active characteristics defines the magnitude of the electrogenesis. The response apparently disappears at the reversal potential, the region in which the two characteristics cross, and reappears as the membrane is polarized further, but it is now reversed in sign. The reversal potential approximates the emf of the electrogenic system and, thus, it is determined by the ionic batteries of the electrogenesis. Generator potentials of sensory neurons and excitatory postsynaptic potentials usually have a reversal potential which is the Thevenin emf of the K and Na batteries, E_K and E_{Na}. Since the latter is strongly inside-positive, the reversal potential for this electrogenesis is well positive to the resting potential. The electrogenesis of inhibitory postsynaptic potentials, on the other hand, is usually due to the emf of a single ionic species, E_K or E_{Cl}. The reversal potential of the IPSP's is therefore closer to the resting potential.*

NONLINEAR EFFECTS IN ELECTRICALLY EXCITABLE CHANNELS

The electrically excitable components have nonlinear I-E characteristics, which reflect the fact that the conductance changes as a function of the membrane potential. The change may be toward activation or inactivation and both of these responses may be effected by depolarizing or by

---

* The remainder of this chapter will be concerned with electrically excitable processes. For detailed expositions of various electrically inexcitable phenomena consult Grundfest, 1957, 1959, 1961a, 1965, 1967a, 1969, 1971. It is worth noting, however, at this time that the linear behavior of electrogenically unreactive as well as of electrically inexcitable electrogenic membrane components, sometimes over a range of about ±200 mv, places some restrictions on the types of models that can be used in proposing membrane structures and particularly, in seeking to account for nonlinear properties of electrically excitable membrane elements.

hyperpolarizing the resting membrane. A decrease in conductance will result in a disproportionate (nonlinear) increase in potential when the current is increased, since a given current develops a higher voltage across a higher resistance. Activation processes may cause a relative decrease in membrane potential since a given current now develops a lower voltage across the lower resistance. Spike electrogenesis or that of graded electrically excitable responses can also result from activation processes that involve certain ionic species. The nature of these ions, however, usually cannot be specified solely by measurements of the I-E characteristics.

EQUIVALENT CIRCUIT OF EXCITABLE MEMBRANES

The behavior of all the electrogenic components that give the different I-E characteristics of Fig. 22-1 can be specified further by extending and generalizing the equivalent circuit which was first employed by Hodgkin and Huxley (1952d) to account for the spike electrogenesis of squid giant axons. This equivalent circuit (Fig. 22-2) is made with reference to the batteries for ions that are commonly present in the medium bathing the excitable cell. However, other varieties of ionic batteries (for trace components) and other electrogenic processes (e.g., electrogenic pumps [Grundfest, 1955]) may be present but are not represented.

Fig. 22-2 also shows voltage clamp presentations of the I-E characteristics. As before (Fig. 22-1), the characteristics of the electrically inexcitable components are linear, with the active elements exhibiting lines of steeper slope appropriately displaced along the voltage axis. The nonlinear characteristics of the electrically excitable components are also dependent upon the emf's of the respective ionic batteries. Increased conductance for K or Cl, whether in response to depolarizing or hyperpolarizing currents, shifts the corresponding characteristic only to the degree that the emf for K or Cl differs from the resting potential. Since the difference is usually small there is little shift along the voltage axis. Inactivation processes for Cl or Na (i.e. without a change in K conductance) would be masked by the usually higher resting conductance for K. Thus, only K inactivation tends to create a nonlinear characteristic signifying a conductance decrease. The emf therefore remains essentially that of the resting membrane. On the other hand, an increased conductance for Na or for a divalent cation k^{++} (Ca or Mg under normal ionic conditions) tends to emphasize the contribution of the corresponding inside-positive emf (E_{Na} or E_{k++}) and the characteristic is then translated toward the intercept on the voltage axis that represents this emf, while the slope of the line represents the high conductance during maximum activation for the specific ion. The intermediate region connecting the low

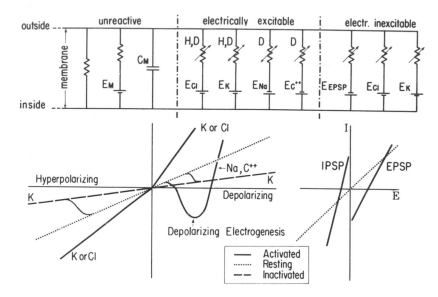

Fig. 22-2 *Generalized equivalent circuit and the voltage-current characteristics of a heterogeneous excitable membrane with a variety of ionic batteries. Above: Equivalent circuit diagram. Emf's contributed by electrogenic ion pumps are omitted. Below: Current-voltage (I-E) characteristics of electrically excitable components (left) and of electrically inexcitable (right), in voltage clamp presentations. Origins are set at resting potential (E_M).*

Membrane capacity (C_M) and the invariant conductive component represent major, unreactive portions of the membrane. The conductive component is subdivided into an ion permselective element with E_M as the average emf and a nonselective element symbolized by a resistance without emf. Reactive components are represented by variable resistances in series with different ionic batteries depending on different perm-selectivities. E_K and E_{Cl} in general are close to E_M, but E_{Na} and E_k^{++} (Ca or Mg in various cells) are shown as inside positive. Permselective electrically excitable channels respond to depolarizing (D) and/or hyperpolarizing (H) stimuli with activation ($\nearrow$) or inactivation ($\swarrow$). Electrically inexcitable depolarizing electrogenesis of receptive and synaptic membrane is indicated by an inside-positive battery (E_{EPSP}). Inhibitory synaptic electrogenesis involves increased conductance for either Cl (E_{Cl}) or K (E_K).

Unreactive electrically inexcitable electrogenic components have linear (ohmic) I-E characteristics, but activation of reactive electrically inexcitable components by specific stimuli increases the slope (indicating higher conductance in voltage clamp presentation). Depolarizing electrogenesis translates the characteristic to the right. The diagram shows inhibitory electrogenesis (IPSP) as hyperpolarizing and the characteristic translated to the left. As the membrane is polarized by applied currents, the resting and active characteristics approach a crossing beyond which the sign of the recorded electrically inexcitable response is reversed relative to the steady membrane potential. The reversal potential approximates the equilibrium (Nernst) potential of ionic batteries that cause the electrogenesis.

The I-E characteristics of electrically excitable components exhibit nonlinearities which result from transition of the resting membrane conductance to higher or lower values. Only the conductance increase caused by Na or k^{++} activation shifts the characteristic significantly along the voltage axis. Three nonlinear regions with negative slope characteristics are shown. They mark transitions from E_M to E_{Na} or E_k^{++} by activation processes and from the resting conductance to lower conductance by depolarizing and hyperpolarizing K inactivation respectively. (From Grundfest, 1967a).

conductance resting characteristic and the high conductance state of activation for the new emf generates an n-shaped curve. In current-clamp measurements this characteristic forms a "forbidden" zone, but when the membrane potential is controlled by voltage clamping this region can be readily measured.

NEGATIVE SLOPE CHARACTERISTICS AND SPIKE ELECTROGENESIS

The slope conductance $(d\mathrm{I}/d\mathrm{E})$ of the intermediate n-shaped region has a negative portion. The analytical expression for this condition is derived from the conductance form of Ohm's Law, $\mathrm{I} = \mathrm{g}\,(\mathrm{E} - \mathrm{E}_o)$, where g is the chord conductance, E the membrane potential and E_o the emf of the ionic battery involved in the conductance increase. The slope conductance then is

$$\frac{d\mathrm{I}}{d\mathrm{E}} = \frac{d\mathrm{g}}{d\mathrm{E}}\,(\mathrm{E} - \mathrm{E}_o) + \mathrm{g}$$

It is negative when g increases with E, provided E_o is sufficiently positive (Cole, 1968).

The n-shaped characteristic also describes electrical devices that can be triggered to yield regenerative responses, analogous to spike electrogenesis. However, the mechanisms by which the regenerative responses are produced are vastly different. This difference can be appreciated clearly when it is realized that it is only the existence of an appropriately inside-positive ionic battery that permits spike electrogenesis. The precise ionic species itself is relatively unimportant.*

The diagram of Fig. 22-3A illustrates the independence of the negative slope characteristic from any specific ionic mechanism. Line 1 shows the ohmic (linear) low conductance of the resting state with the resting potential as the origin. If the membrane develops depolarizing activation for K or Cl (as in Fig. 22-2) the conductance increases and generates the characteristic of line 3. A corresponding increase for Na or k^{++} (as in Fig. 22-2) generates the n-shaped characteristic of line 2. Suppose, however, that $\mathrm{E_K}$ or $\mathrm{E_{Cl}}$ can be shifted to become more positive than the resting potential. Then, depolarizing activation for K or Cl could, in principle, generate line 3′, which also incorporates the n-shaped characteristic. Experimental conditions are known that do provide such a change from a repolarizing (degenerative) electrogenesis to one that provides a regenerative K or Cl spike.

---

* Spike electrogenesis is the *sine qua non* for propagation of an impulse along the highly glossy cable of the axon. However, the change in potential during the spike is itself no guarantee that the response can propagate. Even if the conductance increase is small a large spike can be generated in a "space clamped" axon but the current may be too small to insure excitation of adjacent inactive regions. Thus, invasion of the spike will be blocked.

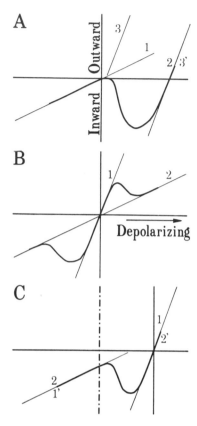

Fig. 22-3 *Comparison of various negative slope characteristics involved in different types of electrogenesis. A.   The shift from the resting state (1) to higher conductance for a more positive ionic battery (2) results in a negative slope characteristic that causes the all-or-none response of normal spike electrogenesis. K or Cl activation (3) causes little change in membrane potential. If E_K or E_{Cl} is made positive to E_M (3') "anomalous" spikes are evoked. B.   Inactivation processes shift the characteristic from (1) to (2), but with little or no change in E_M. In either quadrant, the transition is a negative-slope region giving rise to inactivation responses. C.   Current-voltage characteristic of a cell depolarized by immersion in K-rich medium. The original membrane potential is shown by the broken line. Conductance (1) is higher than in the original state (2). Applying an inward current causes a transition to state (2) and a hyperpolarizing response. If the cell is kept hyperpolarized (1'), K activation is initiated by a depolarizing stimulus and the characteristic shifts temporarily to the high conductance state (2'). The result is a K spike. (Modified from Grundfest, 1966.)*

Cl SPIKES

Skate electroplaques do not normally generate spikes, but they nevertheless have an electrically excitable component of depolarizing activation that is readily demonstrated by the nonlinear I-E characteristic (Fig. 22-4). When the membrane is depolarized by some 15 mv the characteristic exhibits a strongly time-variant deviation from linearity. The deviation increases with

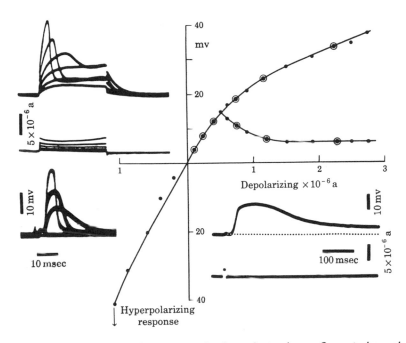

Fig. 22-4 *Electrically excitable processes in skate electroplaque. Current clamp data. Origin of the graph is the resting potential. Large circles represent measurements from the six superimposed records in upper left quadrant. Depolarizations (upper traces), incited by the 3 smallest currents (lower traces), fell on the linear portion of I-E characteristic, but larger currents induced nonlinear time variant changes in potential so that the characteristic developed two branches. In this presentation a downward curvature of the characteristic represents an increased conductance. In the steady-state, during the largest conductance changes, the potential fell to about 7 mv positive relative to E_M (see also Fig. 22-5). Lower left quadrant: Superimposed PSP's in response to increasing neural stimulation. Steps of different amplitude indicate that 3 nerve fibers were activated. Depolarization during the smallest PSP was not sufficient to evoke electrically excitable Cl activation. Larger PSP's did evoke Cl activation and the responses were shortened by the electrogenesis for E_{Cl}. Note their resemblance to the potentials evoked by the 2 largest depolarizing currents (above). Right lower quadrant: An iontophoretically applied jet of acetylcholine (signalled on lower trace) evoked long-lasting depolarization. Peak depolarization was limited to about 20 mv by onset of Cl activation. (Unpublished data from Cohen, Bennett, and Grundfest, 1961.)*

time and the steady-state characteristic becomes nearly parallel to the current axis. This, however, does not indicate that the conductance is increasing continuously toward infinity. Rather, it reflects the change of the membrane to a high conductance for an ionic battery with an emf that is different from, though close to, the resting potential. Voltage clamp measurements (Fig. 22-5) confirm this interpretation. Depolarization by some 15 mv or more induces a slowly increasing rise in outward current and the steady-state characteristic of the depolarized cells then exhibits a conductance some 10 to 15 times higher than that of the resting state. The high conductance line indicates an emf some 7 mv positive to the resting potential.

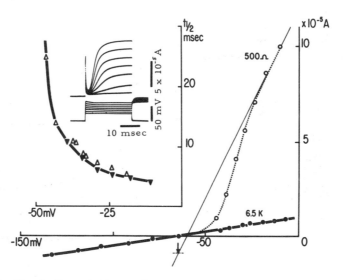

Fig. 22-5 *Electrically excitable activity in skate electroplaque analyzed with voltage clamping. Records (upper left quadrant) show outward currents (top traces) evoked by depolarizing steps of various amplitudes (bottom traces). Increase in current to its steady-state is relatively slow, but half-time ($t_{1/2}$) to reach the latter becomes shorter with increasing depolarization (inset graph). The main graph shows the I-E characteristic in voltage clamp presentation. Heavy line and filled circles; ohmic relation of the resting membrane. Dotted line and open circles: steady-state attained during depolarizing activation. For large depolarization, the slope of the characteristic reaches a limiting value (thin line) that represents a decrease of resistance from 6.5 KΩ to 0.5 KΩ. The intercept of this line on the voltage axis (ca. 7 mv positive to E_M) represents the emf of the Cl battery of this cell. An arrow marks inward current that flowed after the depolarizing pulse was terminated. This is also represented by the undershoot in the voltage clamp records and represents dissipation of positive charge (of E_{Cl}) on the membrane capacity. Note that even for very large depolarizations the initial outward current is merely that of the ohmic "leak" channels. This indicates that the Cl channels are closed in the resting state. Their transformation to the open state (by electrostatically controlled conformational changes ?) is a rather slow as well as a graded process. A range of about 25 mv encompasses the transition from all closed to all open. (Unpublished data from Hille, Bennett, and Grundfest, 1965.)*

The conductance increase of the electroplaques results from depolarizing Cl activation (Cohen, Bennett and Grundfest, 1961; Bennett, 1961; Grundfest, Aljure and Janiszewski, 1962; Hille, Bennett and Grundfest, 1965, and unpublished). Thus, it may be expected that substitution for the Cl in the bathing medium with some impermeant anion will lead to a shift of E_{Cl} toward still more positivity. This, in fact, happens and the cell can then generate a prolonged Cl spike which is abolished when Cl is again introduced (Fig. 22-6). Spike electrogenesis by Cl activation occurs normally in the fresh water algae *Chara* and *Nitella* (Gaffey and Mullins, 1958; Mullins, 1962; Kishimoto, 1965). The intracellular Cl concentration is high relative to the concentration in the fresh water bathing medium and E_{Cl} is appropriately inside-positive.

K SPIKES

A shift of E_K to more positive values usually also involves a parallel shift of the resting potential. Thus, it is generally difficult to fulfill the conditions

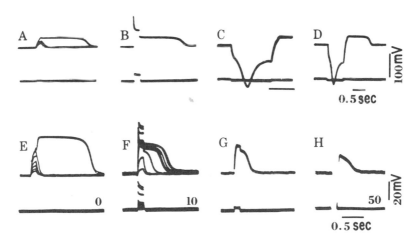

Fig. 22-6 Cl *spikes and hyperpolarizing responses in skate electroplaques bathed in Cl-free saline. A, B, E. Long-lasting all-or-none responses triggered by depolarizing pulses. C, D. Inward currents evoked hyperpolarizing responses and, when the current was terminated, the cells produced "anode break" spikes. The spike triggered by depolarization (E) was diminished and finally abolished when Cl was reintroduced to concentrations of 10 mM (F) and 50 mM (G and H). A very strong stimulus of outward currents was applied in the latter recording. (Unpublished data from Grundfest, Aljure and Janiszewski, 1962.)*

of Fig. 22-3*A* so as to develop an *n*-shaped characteristic for K activation. However, an equivalent change can be produced by inducing a relative "resting" negativity with an applied current as is diagrammed in Fig. 22-3*C*. Application of high K concentration solutions causes a shift of the resting potential (and E_K) and a high conductance state indicated by the conductance line 1. An inward current which brings the membrane potential back to its original resting value (broken line) also returns the conductance to a lower value (line 2). The connecting link between the two conductance lines forms

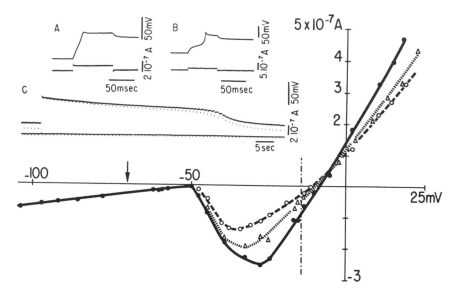

Fig. 22-7 *Regenerative electrogenesis owing to* K *activation in puffer neurons bathed in isosmotic* KCl *solution and repolarized by applied inward currents, A–C: Prolonged* K *spikes evoked by depolarizing pulses. Holding potentials were* −68 mv, A ; −71 mv B ; *and* −60 mv, C. *In the latter recording, small, constant, repetitive, inward pulses were applied. The diminished amplitudes of the hyperpolarizations they induced show that the conductance increased during the* K *spike. Graph: Voltage clamp data from another cell which was depolarized to* −14 mv (*broken vertical line*) *by isosmotic substitution of KCl for NaCl. The holding potential was* −70 mv (*arrow*). *The conductance increase that is due to* K *activation evoked by depolarizing stimuli and that is represented by a large change in slope of the I-E characteristic, was long lasting. Filled circles show peak current shortly after the step changes in voltage. Triangles, measurements 80 msec after onset of depolarization. Open circles, 860 msec after onset. The* K *equilibrium potential, the point at which the 3 curves cross, is about* −5 mv. *The very gradual subsidence of* K *activation is denoted by a decrease in the high slope of the characteristic. The end of the* K *activation and of spike electrogenesis would restore the characteristic to the slope at the holding potential. (Modified from Nakajima and Kusano, 1966 and from Nakajima, 1966.)*

an *n*-shaped characteristic that can be triggered back from 2 to 1 by a depolarizing stimulus. The K-spikes and corresponding voltage clamp measurements for such a condition are shown in Fig. 22-7.

The resting potential of the muscle fibers of the mealworm (*Tenebrio molitor*) is rather insensitive to large changes in external K (Belton and Grundfest, 1962). Thus, the transformation of E_K to relative positivity (Fig. 22-3*A*) can be accomplished in this preparation (Fig. 22-8). The spike in the upper left was evoked with the preparation bathed in solutions with

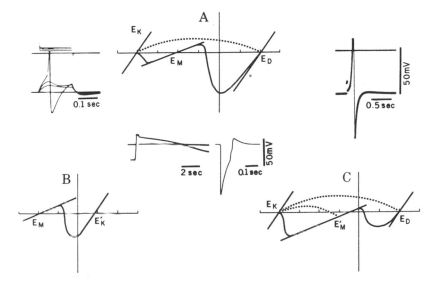

Fig. 22-8 *Intracellularly recorded "anomalous" spikes in muscle fibers of* Tenebrio molitor *and in esophageal cells of* Ascaris lumbricoides, *and explanation of the electrogenesis in diagrammatic voltage clamp presentations. A. Under approximately normal ionic conditions* $E_D > E_M > E_K$, *where* E_D *is the inside-positive ionic battery for depolarizing electrogenesis* (Mg *in* Tenebrio). *A depolarizing stimulus shifts the characteristic from* E_M *toward* E_D, *giving rise to depolarizing electrogenesis of spikes. Depolarization induces K activation and a shift of the characteristic toward* E_K *(dotted line), giving rise to the second phase of the spike. The latter is particularly marked in recording from the* Ascaris *cell (right). Hyperpolarization quenches K activation and the potential slowly returns to* E_M. *B.* Tenebrio *muscle fibers can tolerate the presence of as much as* 340 mM K *without appreciable change in* E_M. *However* $E_K > E_M$ *and the response to a brief stimulus (left center record) is a long-lasting K spike. C.* Ascaris *esophageal cells depolarize when the bathing solution contains an increased concentration of Cl, but* E_K *is relatively unchanged. Thus* $E_M' > E_K$ *and most or all of the electrogenesis is hyperpolarizing, occurring as the characteristic shifts from* E_D *to* E_K *or from* E_M *to* E_K. *The hyperpolarizing spike is brief (as in A) because K activation is quenched by hyperpolarization. (From Grundfest, 1967b,* Tenebrio *data from Belton and Grundfest, 1962,* Ascaris *data from Del Castillo and Morales, 1967.)*

low K concentrations. E_K therefore was considerably negative to the resting potential, E_M. The intracellularly recorded response shifted from the limit E_D, to the limit E_K as described in the voltage clamp diagram. For the response in Fig. 22-8B, K was the only cation and because of its high concentration in the bath, E'_K was positive, whereas E_M remained inside-negative. The response to a depolarizing stimulus was regenerative K activation and a prolonged K spike (center, left).

The foregoing data demonstrated that bringing into play a previously occult emf that is positive to the resting potential is all that is necessary to generate a spike; the ion species forming the battery is immaterial. All that is demanded is that the ionic gradient contribute an inward flow of current, which maintains the membrane depolarized beyond the level required to initiate activation or the conductance increase for this ion species. The negative slope characteristic then causes this change in conductance and the resulting electrogenesis to be regenerative. However, the negative slope is itself a consequence of the increase in conductance.

Once the response is initiated, the stimulating current may be withdrawn. The spike electrogenesis will continue as long as the newly activated generator supplies a current to charge the membrane capacity. Thus, spikes are *auto-genetic* and, in this way, differ from the regenerative inactivation responses. The inactivation processes which are responsible for the latter responses last only while the stimulating current is applied (Grundfest, 1966). It is also evident (Figs. 22-6, 22-7 and 22-8) that spike electrogenesis may be relatively long lasting as compared with the spikes of the squid axon. The pulse shaping processes of Na inactivation and K activation are prominent in the axon. The decay of the high conductance for Cl or K appears to be a slower process and the spikes generated by these ionic batteries are long lasting. Prolonged spikes, however, can also be generated in squid axons (Tasaki and Hagiwara, 1957; Takenaka and Yamagishi, 1969) as well as in many other cells (Grundfest, 1961b, 1966).

Spikes can also be generated by an increased conductance for Na alone (Fig. 22-9). In eel electroplaques the K channels are normally open at rest, but close when the membrane is depolarized by about 40 mv (Nakamura, Nakajima and Grundfest, 1965; Morlock, Benamy and Grundfest, 1968; Ruiz-Manresa, Ruarte, Schwartz and Grundfest, 1970). Thus, during most of the course of spike electrogenesis there are only two conductance components in the system (Fig. 22-10): the leak channels, with an invariant conductance g_L, and the sodium channels with a time variant conductance, g_{Na}. This condition gives rise to impedance changes that are very different from those observed in squid axons (Cole and Curtis, 1939). The initial increase in bridge output (Fig. 22-10A) that results from Na activation subsides as g_{Na} falls during Na inactivation. The bridge output returns to null

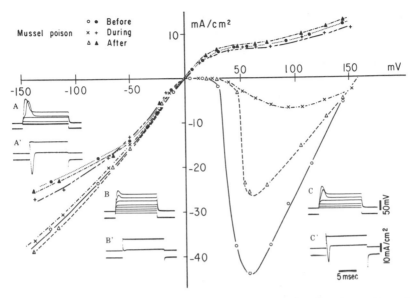

Mussel poison

Fig. 22-9 *Electrogenic components of eel electroplaque. Graph, voltage clamp data. The resting potential is the origin. The large inward Na current was greatly reduced (and would eventually have been abolished) during the application of saxitoxin (mussel poison). Spike electrogenesis was also abolished reversibly (Records A–C). The response of the K channels was not affected by the poison. At rest the steady state conductance was high, but it decreased rapidly on depolarization and more slowly on hyperpolarization, in both cases as a manifestation of K inactivation. A'-C' show the currents during a depolarization of 75 mv lasting 12 msec. The inward current that caused the spike electrogenesis was almost abolished in the presence of saxitoxin (B') so that the cell could generate only small graded responses (B). (Unpublished data from Nakamura and Grundfest, 1965.)*

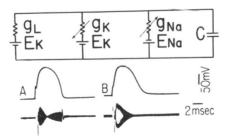

Fig. 22-10 *Equivalent circuit of eel electroplaque and the impedance changes of the cell during spike electrogenesis in the presence of normal depolarizing K inactivation (A) and after pharmacological K inactivation (B). The equivalent circuit is similar to that of the squid axon except that the g_K channels close rather than open during the spike and also that both the g_K channels and the invariant "leak" channels (g_L) have the same emf (E_K). The diphasic change in impedance during the spike (A) becomes monotonic (B) on closing the g_K channels by pharmacological K inactivation with Ba^{++}. Further description in text. (Modified from Morlock et al., 1968, and Ruiz-Manresa, 1970.)*

while the spike is still near its peak. At this point g_{Na} becomes equal to g_K and the total conductance $g_{Na} + g_L$ now equals that of the resting membrane, $g_K + g_L$. As g_{Na} continues to fall toward zero there is another imbalance in the bridge, but this time reflecting a decrease in conductance below that of the resting membrane. The peak represents g_L alone, when both g_{Na} and g_K are zero. The impedance change subsides rapidly when the spike falls below about 40 mv and the g_K channels reopen regeneratively. The correctness of this analysis is confirmed by the measurements of Fig. 22-10B. In this case the cell was exposed to a low concentration of Ba^{++}. This induced pharmacological K inactivation so that the g_K channels were closed off in the resting state as well. In this condition the bridge was nulled for the conductance g_L alone. The bridge output during the spike now became a single elevation reflecting the rise and subsidence of g_{Na}.

CONDUCTANCE INCREASE FOR AN IONIC BATTERY WITH AN INSIDE-NEGATIVE EMF

Since the resting potential is usually close to E_K and E_{Cl}, an electrogenesis that is strongly inside-negative is rare. One case of purely inside-negative electrogenesis is that of frog slow muscle fibers (Fig. 22-11). Like the skate electroplaques (Fig. 22-4 and 22-5) these cells develop only an electrically inexcitable depolarizing electrogenesis (Kuffler and Vaughan-Williams, 1953; Burke and Ginsborg, 1956; Belton and Grundfest, 1961), but the depolarization, whether it is neurally evoked, or is caused by applied outward current, is succeeded by a long-lasting hyperpolarization which is due to depolarizing K activation. The membrane potential shifts from the resting value toward an inside-negative E_K. The hyperpolarizing electrogenesis may be expected to quench the K activation and then the inside-negativity must be dissipated across the higher resistance of the resting membrane. For this reason the decay of the hyperpolarization is much slower than is its rise.

The presence of an inside-negative E_K in conjunction with an inside-positive emf (E_D) results in diphasic intracellularly recorded spikes (Fig. 22-8). This characteristic is particularly prominent in the spike of *Ascaris* esophageal muscle (Del Castillo and Morales, 1967). The transition from positivity to negativity is rapid because the shift is made backward on the characteristic and thus traverses the negative slope region. The return from negativity to the resting potential is slow, however, as described in connection with Fig. 22-11. The depolarizing component may be small or even absent under some conditions in the *Ascaris* cell (Fig. 22-8C) and the response then can be an inside-negative spike. The hyperpolarizing undershoot that is a characteristic of the squid axon spike is also an example of activation for a system that has an emf negative to the resting potential (Hodgkin and Huxley, 1952*d*). In this case, the return from E_K to the resting potential is terminated by a phase of high resistance (Shanes, Grundfest and Freygang, 1953).

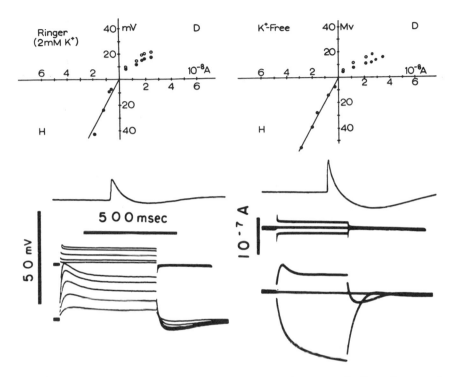

Fig. 22-11 *Example of autogenetic hyperpolarizing electrogenesis. Frog slow muscle fiber in control Ringer's solution (left) and in a K-free saline (right). Top: In both cases the steady state characteristic exhibits a curvature in the depolarizing quadrant which denotes an increase in conductance. Note that this cell does not generate spikes. The conductance change is graded, increasing with greater depolarization and is time variant, increasing with time during the applied current (filled circles). Middle: Neurally evoked depolarizing postsynaptic electrogenesis is succeeded by a prolonged after hyperpolarization. The latter is enhanced when E_K is made more negative by removal of K_o. The hyperpolarization is therefore due to depolarizing K activation. Note that the peak of the hyperpolarization is followed by a slower return to the resting potential. Bottom: Responses to applied currents. The changes in membrane potential during the applied current exhibit development of a plateau as in the middle row of Fig. 12-15. The K activation persisted after the current was terminated and caused hyperpolarization. An inward current (right) caused a large but slowly increasing hyperpolarization. The more rapid change induced by the outward current indicates that depolarizing K activation ("delayed rectification") occurred rapidly. Note also that the hyperpolarizing electrogenesis induced by the outward current subsided more slowly than did the much larger hyperpolarization induced by the inward current. This is indicative of an autogenetic response in which depolarizing K activation produces a hyperpolarizing electrogenesis. (Modified from Grundfest 1961b after unpublished data by Belton and Grundfest, 1961.)*

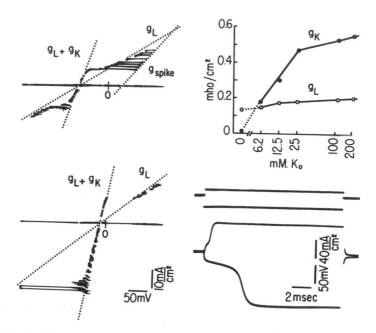

Fig. 22-12 *Depolarizing and hyperpolarizing* K *inactivation responses in the same cell (eel electroplaque). Left: Steady-state characteristics as measured with currents that cause different degrees of change in the membrane potential. Upper graph is for cell in the standard saline, lower, after replacing all Na with K. At rest the conductance is the sum of two components, contributed by the unreactive leak channels,* g_L *and the reactive channels* g_K. *The latter close when the membrane potential is displaced beyond certain values. Elimination of* g_K *causes the membrane potential to shift regeneratively to larger* I · R *values. In the control saline, strong depolarizing currents evoked spikes, the peaks of which are registered as the points on the extreme right. The* g_K *component of the resting cell is increased markedly in high* K, *as denoted by the steeper slope of the characteristic (lower left) and in the graph for another cell (upper right).* g_L *is increased relatively little, however. The negative slope during the transition from the high conductance* ($g_L + g_K$) *to low conductance* (g_L) *in either direction is accentuated, creating a "forbidden zone" of instability. The records (lower right) show the inactivation responses evoked in high* K *by depolarizing and hyperpolarizing currents. Note that the threshold for depolarizing inactivation was lower than that for hyperpolarizing inactivation. This difference is also seen in the characteristic curves (lower left). The records on the right also show that the development of hyperpolarizing* K *inactivation was much slower, despite the higher applied inward current. (Modified from Ruiz-Manresa, 1970.)*

NEGATIVE SLOPE CHARACTERISTICS OF INACTIVATION RESPONSES

Depolarizing and hyperpolarizing K inactivation both cause a decrease in conductance (i.e., dg/dE is negative) and the characteristic then has negative slope in either quadrant (Fig. 22-2, 22-3 and 22-9). The membrane potential shifts regeneratively even though there is no change in the emf of the system

(Fig. 22-12). However, the response is due only to an increase in R and the potential is produced as the product of the applied current and the resistance. Thus, if the current is withdrawn, the inactivation response collapses. This criterion clearly distinguishes the "spike" of depolarizing inactivation responses, or the "negative spikes" of hyperpolarizing inactivation, from the autogenetic spikes that involve a current flow generated by the membrane itself.

Inactivation responses become particularly prominent when the difference between the low resistance of the resting state and the high resistance of the inactivated state is large. This is clearly evident in the records of Fig. 22-12 on an eel electroplaque. When it was bathed in normal saline (upper left) the resting conductance of the cell was $g_L + g_K$ (ca. 0.45 mho/cm^2). Depolarizing or hyperpolarizing inactivation closed off the g_K channels and the steady state conductance became g_L (ca. 0.12 mho/cm^2). The transition from a steady-state potential $I/(g_L + g_K)$ to I/g_L is a regenerative inactivation response. Also seen in the depolarizing quadrant is the regenerative response of spike electrogenesis. The peaks of the spikes fall on a line (g_{spike}) that represents the conductance (ca. 0.22 mho/cm^2) of $g_L + g_{Na}$, where g_{Na} is the conductance of the Na channels at the peaks of the spikes (Ruiz-Manresa et al., 1970). This line crosses the voltage axis at a value which is the amplitude of the overshoot of a neurally or directly evoked spike. The intersection of the line g_{spike} with g_L gives the emf of the Na battery, E_{Na} (+170 mv). Both the g_L and $g_L + g_K$ lines have a common origin, which is the resting potential and is the emf of the K battery, E_K (−80 mv).

When most cells are exposed to increase in K_o, the resting conductance increases and the cells depolarize. In eel electroplaques it has been shown that the conductance increase is not due to the depolarization *per se*, but is a specific increase in the conductance of the g_K channels while g_L increases only to a small degree (Nakamura et al., 1965; Ruiz-Manresa, 1970)*. This is shown (for another cell) in the graph (upper right) of Fig. 22-12. For the lower portion of Fig. 22-12 the electroplaque was exposed to 200 mM KCl. There was no Na present and spike electrogenesis was absent in the cell, which was depolarized to about −15 mv. g_L had increased only to 0.15 mho/cm^2, but g_K had increased to 0.77 mho/cm^2 ($g_L + g_K = 0.92$ mho/cm^2).

---

* It is likely that the increase in conductance of the g_K channels by increasing K_o which is clearly evidenced in eel electroplaques also occurs in other cells, although experimental difficulties limit clear-cut demonstration of this. In eel electroplaques g_K increases in spite of the fact that increasing K_o induces a depolarization, which might be expected to cause depolarizing K inactivation. However, further depolarization, *by an applied current*, does induce K inactivation. Cs or Ba close the g_K channels by pharmacological K inactivation, with the kinetics of competitive action. Furthermore, the increase in g_L is far less than might be expected from the change in K conductance of the high K_o solution. All these findings will need to be incorporated in a comprehensive theory of the structure and function of excitable membranes and thus they provide useful restrictions upon speculations on this subject.

Larger currents were now required to trigger the closure of the g_K channels and the voltages produced across the "leak" resistance of the membrane were therefore larger (lower right). A larger current was needed to cause hyperpolarizing K inactivation and the inactivation developed more slowly than did depolarizing inactivation. The reasons for these differences are not known.

When the inactivation responses involve only a single ionic component they are readily demonstrable (Fig. 22-9 and 22-12.) However, inactivation responses are frequently masked by the occurrence of spike electrogenesis

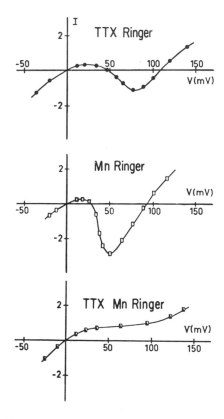

Fig. 22-13 *Three differently reactive ionic channels, one of which is for depolarizing K inactivation in frog atrial muscle. Voltage clamp measurements. An inward current develops even after the Na channels are blocked by TTX (upper graph), indicating the influx of another ion with an inside positive emf. Blockage of Ca influx with Mn still leaves an inward current, but the characteristic for this ion, presumably Na, is somewhat different. When both TTX and Mn are present no inward current develops. The characteristic then exhibits a flattened region in which dg/dE is negative. This indication of depolarizing K inactivation is also found in other cardiac tissue. The K channels reopen with further depolarization. (Modified from Rougier et al., 1969.)*

or of a conductance increase for K or Cl, but appropriate measurements can disclose the occurrence of K inactivation (Reuben and Gainer, 1962; Fig. 22-9, 22-13). In other cells the low conductance may be superceded by a more slowly developing high conductance (Fig. 22-14) as in the hyperpolarizing response of lobster muscle (Reuben, Werman and Grundfest, 1961). The response is a "negative spike" representing a regenerative increase in potential as the resistance increases, followed by a plateau of smaller voltage, during the maintained application of the current. The subsidence of the peak is also a regenerative event, as the potential traverses the characteristic backward and returns through the negative slope region (Fig. 22-3B). The occurrence of secondary time variant conductance changes complicates the analysis of such hyperpolarizing response, particularly if the delayed conductance increase is for an ion species other than K. The complete equivalent

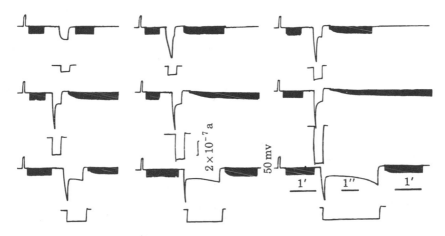

Fig. 22-14 Interplay of inactivation and activation processes during hyperpolarizing responses of lobster muscle fiber and subsequent persistent increased conductance. Each trace (ink-writer records) was registered in a sequence of chart speeds. Initial fast: calibrating pulses of 50 mv. Slow: series of brief hyperpolarizing testing pulses of constant strength. Fast: during which hyperpolarizing current was applied. Slow: during which testing hyperpolarizing pulses were again applied. Second trace in each record monitors the applied current.

After a hyperpolarizing response was evoked (middle record, upper row) the membrane resistance remained lower as is indicated by the decrease in the amplitudes of the hyperpolarizing testing pulses. As the stimulating current was increased, the early pulse-like part of the hyperpolarizing responses became shorter. The membrane resistance following the inactivation response decreased further and remained low for longer times. The conductance increase lasted more than 1 minute with the highest currents used (middle row). Note that only a small degree of after-depolarization is associated with the relatively large conductance changes. During prolonged hyperpolarizing currents (lower row), the membrane resistance began to rise, but after the current was withdrawn the membrane resistance fell below its resting value. (From Reuben, Werman and Grundfest, 1961.)

circuit must then include a change not only in the conductance, but also an emf. The analysis then becomes similar to that required for spike electrogenesis, since it also involves different time variant conductances and different emfs.

EQUIVALENT CIRCUITS FOR TIME VARIANT CONDUCTANCE CHANGES

In order to account for complex electrophysiological manifestations by formalistic equivalent circuits it may be necessary to introduce "anomalous" or "phenomenological" impedance components that are not present physically in the cell membrane (Cole, 1941, 1968; Mauro, 1961; Mauro et al., 1970). Thus, the subthreshold response of a squid giant axon may be analyzed quantitatively with an equivalent circuit that varies in time with the steady-state membrane potential on which the small signals are applied to the axon (Mauro et al., 1970). When the resting membrane is kept at E_K the equivalent circuit has only the conductances of the ionic channels in parallel with the membrane capacity, as in Fig. 22-2. However, when the squid axon is at its usual resting potential, which is positive to E_K, the membrane behaves as if there were an inductance in parallel with the other components.

Such an "inductive" behavior is also exhibited by frog slow muscle fibers that are depolarized by applied currents (Fig. 22-11). As in squid axons, E_K of the muscle fibers is negative to the resting potential and the increase in conductance due to depolarizing K activation develops slowly. Thus, when a depolarizing current is applied that is sufficient to cause K activation there is first an early peak in the membrane potential, which subsides to a steady-state as the conductance increases. Oscillations such as may be seen in squid axons (Mauro et al., 1970) are absent probably because only one ion species (K) is involved in the conductance increase of the muscle fibers. However, when the current is terminated the membrane hyperpolarizes toward E_K and the hyperpolarization subsides slowly as the K activation subsides.

The response to a hyperpolarizing (inward) current of the same intensity is markedly different (Fig. 22-11). The hyperpolarization is large and develops slowly. The time course seems to call for a "two time constant" capacitative equivalent circuit. When the current is terminated, the return to the resting potential also appears to have two time constants but the repolarization is considerably faster than is the decay of the hyperpolarization that is caused by depolarizing K activation.

Somewhat more complex, but still readily susceptible to analysis are the effects in skate electroplaques (Fig. 22-4, 22-5 and 22-6). The I-E characteristic (Fig. 22-4) is linear in the vicinity of the resting potential. Thus, the

3 smallest applied currents evoked simple changes in membrane potential, such as would be represented by an equivalent circuit composed of an ohmic resistor in parallel with a capacity. The next 3 pulses of outward current caused changes that might be described as "inductive." However, the times to the peak depolarization and the subsequent levels of the steady-state varied with current in a manner entirely unlike the changes in Fig. 22-11. Furthermore, when the current was withdrawn the return to the resting potential appeared to be a simple relaxation, with a single time constant. Thus, the "inductance" to be inserted into the formal equivalent circuit would need to have very complex properties, even for a parametric reactance.

The characteristics shown in Fig. 22-4 and 22-5 describe the behavior of the membrane in a simpler way. The resting potential is at or close to E_K. Depolarizing Cl activation is responsible for the conductance increase when the membrane is depolarized by more than 10 mv, but E_{Cl} is very close to the resting potential. Thus, for small signals the cell membrane behaves as a simple R-C network. Cl activation develops slowly (Fig. 22-4 and 22-5), but the rate increases with the applied current or voltage. Thus, as the current of an applied pulse (Fig. 22-4) is increased there is an apparent acceleration to the peak voltage. The peak becomes smaller than would be predicted for a simple "inductive" equivalent circuit, as the conductance rises.

The steady-state membrane potential is a function of the voltages $E_{Cl} + E_K + IR$ where IR is the voltage drop produced by the applied current. The IR term is important only when g_{Cl} is small, prior to full Cl activation. When g_{Cl} reaches its high and constant value (Fig. 22-5), E_{Cl} becomes the dominant term for the lower branch of the characteristic in Fig. 22-4. It is about 7 mv positive to E_K (Fig. 22-5). When the depolarizing current is withdrawn, g_{Cl} subsides gradually, since E_{Cl} is below the threshold for depolarizing Cl activation. Thus, the after-depolarization declines smoothly to the resting potential.

For inward currents (Fig. 22-4) or hyperpolarizing voltages (Fig. 22-5) the I-E characteristic is linear until large currents induce hyperpolarizing inactivation (Fig. 22-6). This response is complex, resembling that of lobster muscle fibers (Fig. 22-14), but the ionic basis of the response in skate electroplaques has not yet been analyzed.

RECAPITULATION

The diagrams of Fig. 22-15 summarize the different varieties of electrically excitable electrogenic processes that have been discussed in the foregoing All involve a conductance change, g(V), as shown in the first column. A decrease in conductance below that of the resting state (upper row) involves

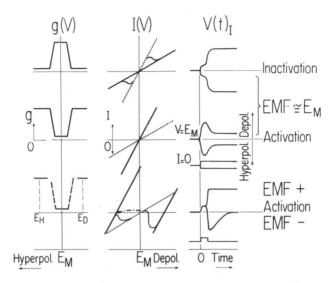

Fig. 22-15 *Diagrammatic representation of the consequences of various electrically excitable conductance changes, g(V), on the current-voltage characteristic, I(V), and on the electrogenic activity, V(t)$_I$.*

Top: Decrease of conductance (inactivation) is usually seen only when the resting conductance is high because in addition to the unreactive "leak" channels an appreciable fraction of the available reactive K channels are open. The inactivation usually does not affect the membrane potential materially. The characteristic (middle column) is merely rotated about the origin (E_M) toward the abscissa. The transition from high to low conductance is a region of negative slope conductance and when a constant current is applied the potential (I/g) shifts regeneratively to a higher value as g decreases. The shift is a depolarizing or hyperpolarizing inactivation response as shown in the last column.

Middle: Increase in conductance (activation) for an ionic species with an emf at or near E_M also does not significantly shift the origin of the characteristic, but rotates the line in the opposite direction. The voltage drop resulting from a constant applied current becomes smaller as g increases. Note that activation, like inactivation, can occur in response to hyperpolarizing currents. The foregoing 4 types of responses are generated only while the current is applied.

Bottom: Increased conductance for species of ionic batteries that have emf's differing from E_M introduces possibilities for negative slope characteristics and responses which are autogenetic. They are triggered by a brief stimulus and follow a time course which is determined by the kinetics of the activation process itself. Further description in text.

a negative slope in the characteristic, I(V) and a regenerative response, V(t)I when a constant current is applied to drive the potential through the "forbidden zone" of instability. Inactivation responses persist only as long as there is an applied current.

An increase in conductance (middle row) for an ion with an emf close to the resting potential does not result in a negative slope region in the characteristic. When a constant current is applied, the membrane potential falls as the conductance increases with time. These activation responses (often called rectifications) also collapse when the current is terminated.

Autogenetic responses result when the conductance increases for an ionic battery that has an emf considerably distant from that of the resting potential, E_M (lower row). The most common response is that which contributes an emf (E_D) considerably positive to E_M. Secondary factors such as Na inactivation and K activation shape the depolarizing electrogenesis of the spike in many cells. In eel electroplaques, however, the shaping is done by Na inactivation alone, since K activation is normally absent and is replaced with K inactivation, which tends to prolong the depolarization. In other cells the interplay of the various electrogenic processes results in graded responses or prolonged spikes, depending upon the relative contributions of the relevant ionic batteries. In various contexts these depolarizing electrogenic events all have specific functions to perform. It is doubtful, however, whether the processes that induce inactivation responses or conductance increases for Cl or K alone have any functional utility. Their existence in excitable membranes is a demonstration of the varieties of electrogenic processes that can occur. In fact, their occurrence provides added insight into the general principles of bioelectric phenomena.

ACKNOWLEDGMENTS

The work in the author's laboratory is supported in part by grants from the Muscular Dystrophy Associations of America, Inc.; by Public Health Service Research Grants (NB03728, NB03270 and Training Grant NB5328) from the National Institute of Neurological Diseases and Stroke; and from the National Science Foundation (GB-6988X).

REFERENCES

BELTON, P., and H. GRUNDFEST. 1961. The Ionic Factors in the Electrogenesis of the Electrically Inexcitable and Electrically Excitable Membrane Components of Frog Slow Muscle Fibers. *Biol. Bull.* 121: 382.

———. 1962. Potassium Activation and K-Spikes in Muscle Fibers of the Mealworm Larva (*Tenebrio molitor*). *Am. J. Physiol.* 203: 588–594.

BENNETT, M. V. L. 1961. Modes of Operation of Electric Organs. *Ann. N. Y. Acad. Sci.* 94: 458–509.

BERNSTEIN, J. 1902. Untersuchungen zur Thermodynamik der bioelektrischen Ströme. *Pflugers Arch. Ges. Physiol.* 92: 521–562.

———. 1912. *Elektrobiologie.* Braunschweig, Fr. Vieweg.

BURKE, W., and B. L. GINSBORG. 1956. The Electrical Properties of the Slow Muscle Fibre Membrane. *J. Physiol. (London)*. 132: 586–598.

COHEN, B., M. V. L. BENNETT and H. GRUNDFEST. 1961. Electrically Excitable Responses in *Raia erinacea* Electroplaques. *Federation Proc.* 20: 339.

COLE, K. S. 1941. Rectification and Inductance in the Squid Giant Axon. *J. Gen. Physiol.* 25:29–51.

COLE, K. S. 1968. *Membranes, Ions and Impulses.* University of California Press, Berkeley.

COLE, K. S., and H. J. CURTIS. 1939. Electric Impedance of the Squid Giant Axon during Activity. *J. Gen. Physiol.* 22: 649–670.

DEL CASTILLO, J., and T. MORALES. 1967. The Electrical and Mechanical Activity of the Esophageal Cell of *Ascaris lumbricoides*. *J. Gen. Physiol.* 50: 603–629.

GAFFEY, C. T., and L. J. MULLINS. 1958. Ion Fluxes during the Action Potential in *Chara*. *J. Physiol. (London).* 144: 505–524.

GRUNDFEST, H. 1955. The Nature of the Electrochemical Potentials of Bioelectric Tissues, pp. 141–166. *In* T. Shedlovsky [ed.] *Electrochemistry in Biology and Medicine.* John Wiley, New York. pp. 141–166.

———. 1957. Electrical Inexcitability of Synapses and Some of its Consequences in the Central Nervous System. *Physiol. Rev.* 37: 337–361.

———. 1959. Synaptic and Ephaptic Transmission, pp. 147–197. *In* J. Field [ed.] *Handbook of Physiology, Section 1, Neurophysiology I.* American Physiological Society, Washington, D.C.

———. 1961a. General Physiology and Pharmacology of Junctional Transmission, pp. 329–389. *In* A. M. Shanes [ed.] *Biophysics of Physiological and Pharmacological Actions.* American Association for the Advancement of Science, Washington, D.C.

———. 1961b. Ionic Mechanisms in Electrogenesis. *Ann. N. Y. Acad. Sci.* 94: 405–457.

———. 1963. Impulse Conducting Properties of Cells, pp. 277–332. *In* D. Mazia and A. Tyler [ed.] *The General Physiology of Cell Specialization.* McGraw-Hill, New York.

———. 1965. Electrophysiology and Pharmacology of Different Components of Bioelectric Transducers. *Cold Spring Harb. Symp. Quant. Biol.* 30:1–14.

———. 1966. Comparative Electrobiology of Excitable Membranes, pp. 1–116. *In* O. E. Lowenstein [ed.] *Advances in Comparative Physiology and Biochemistry.* Vol. 2. Academic Press, New York.

———. 1967a. Synaptic and Ephaptic Transmission, pp. 353–372. *In* G. C. Quarton, T. Melnechuck and F. O. Schmitt [ed.] *The Neurosciences. A Study Program.* Rockefeller University Press, New York.

———. 1967b. The "anomalous" Spikes of *Ascaris* Esophageal Cells. *J. Gen. Physiol.* 50:1955–1959.

———. 1969. Synaptic and Ephaptic Transmission, pp. 463–491. *In* G. H. Bourne [ed.] *The Structure and Function of Nervous Tissue,* Vol. II. Academic Press, New York.

———. 1971. The General Electrophysiology of Input Membrane in Electrogenic Excitable Cells, pp. 135–165. *In* W. R. Loewenstein [ed.] *Handbook of Sensory Physiology, I. Principles of Receptor Physiology.* Springer-Verlag, Berlin.

GRUNDFEST, H., E. ALJURE and L. JANISZEWSKI. 1962. The Ionic Nature of Conductance Increases Induced in Rajid Electroplaques by Depolarizing and Hyperpolarizing Currents. *J. Gen. Physiol.* 45: 598A.

HILLE, B., M. V. L. BENNETT and H. GRUNDFEST. 1965. Voltage Clamp Measurements of the Cl-Conductance Changes in Skate Electroplaques. *Biol. Bull.* 129: 407.

HODGKIN, A. L., and A. F. HUXLEY. 1952a. Currents Carried by Sodium and Potassium Ions through the Membrane of the Giant Axon of *Loligo. J. Physiol.* 116: 449–472.

———. 1952b. The Components of Membrane Conductance in the Giant Axon of *Loligo. J. Physiol. (London).* 116: 473–496.

———. 1952c. The Dual Effect of Membrane Potential on Sodium Conductance in the Giant Axon of *Loligo. J. Physiol. (London).* 116: 497–506.

———. 1952d. A Quantitative Description of Membrane Current and its Application to Conduction and Excitation in Nerve. *J. Physiol. (London).* 117: 500–544.

HODGKIN, A. L., A. F. HUXLEY and B. KATZ. 1952. Measurement of Current-Voltage Relations in the Membrane of the Giant Axon of *Loligo. J. Physiol. (London).* 117: 424–448.

HODGKIN, A. L., and B. KATZ. 1949. The Effect of Sodium Ions on the Electrical Activity of the Giant Axon of the Squid. *J. Physiol. (London).* 108: 37–77.

KISHIMOTO, U. 1965. Voltage Clamp and Internal Perfusion Studies on *Nitella* Internodes. *J. Cell. Comp. Physiol.* 66: 43–54.

KUFFLER, S. W., and E. M. VAUGHAN-WILLIAMS. 1953. Properties of the "Slow" Skeletal Muscle Fibres of the Frog. *J. Physiol. (London).* 121: 318–340.

MAURO, A. 1961. Anomalous Impedance, A Phenomenological Property of Time Variant Resistance. An Analytic Review. *Biophys. J.* 1: 353–372.

MAURO, A., F. CONTI, F. DODGE and R. SCHOR. 1970. Subthreshold Behavior and Phenomenological Impedance of the Squid Giant Axon. *J. Gen. Physiol.* 55: 497–523.

MORLOCK, N. L., D. A. BENAMY and H. GRUNDFEST. 1968. Analysis of Spike Electrogenesis of Eel Electroplaques with Phase Plane and Impedance Measurements. *J. Gen. Physiol.* 52: 22–45.

MULLINS, L. J. 1962. Efflux of Chloride Ions during the Action Potential of *Nitella. Nature.* 196: 986–987.

NAKAJIMA, S. 1966. Analysis of K-inactivation and TEA Action in the Supramedullary Cells of Puffer. *J. Gen. Physiol.* 49: 629–640.

NAKAJIMA, S., and K. KUSANO. 1966. Behavior of Delayed Current under Voltage-Clamp in the Supramedullary neurons of Puffer. *J. Gen. Physiol.* 49: 613–628.

NAKAMURA, Y., and H. GRUNDFEST. 1965. Different Effects of K and Rb on Electrically Excitable Membrane of Eel Electroplaques. *XXIII Int. Physiol. Cong.*, Abst. 167.

NAKAMURA, Y., S. NAKAJIMA and H. GRUNDFEST. 1965. Analysis of Spike Electrogenesis and Depolarizing K-Inactivation in Electroplaques of *Electrophorus electricus*, L. *J. Gen. Physiol.* 49: 321–349.

REUBEN, J. P., and H. GAINER. 1962. Membrane Conductance during Depolarizing Postsynaptic Potentials of Crayfish Muscle Fibre. *Nature.* 193:142–143.

REUBEN, J. P., R. WERMAN and H. GRUNDFEST. 1961. The Ionic Mechanisms of Hyperpolarizing Responses in Lobster Muscle Fibers. *J. Gen. Physiol.* 45: 243–265.

ROUGIER, O., G. VASSORT, D. GARNIER, Y. M. GARGOUIL and E. CORABOEUF. 1969. Existence and Role of a Slow Inward Current during Frog Atrial Action Potential. *Pflüg. Arch.* 308:91–110.

RUIZ-MANRESA, F. 1970. *Electrogenesis of Eel Electroplaques. Conductance Components and Impedance Changes during Activity.* Ph.D. Thesis, Columbia University, New York.

RUIZ-MANRESA, F., A. C. RUARTE, T. L. SCHWARTZ and H. GRUNDFEST. 1970. K-inactivation and Impedance Changes during Spike Electrogenesis in Eel Electroplaques. *J. Gen. Physiol.* 55:33–47.

SHANES, A. M., H. GRUNDFEST and W. H. FREYGANG. 1953. Low Level Impedance Changes Following the Spike in the Squid Giant Axon Before and After Treatment with "Veratrine" Alkaloids. *J. Gen. Physiol.* 37: 39–51.

TAKENAKA, T., and S. YAMAGISHI. 1969. Morphology and Electrophysiological Properties of Squid Giant Axons Perfused Intracellularly with Protease Solution. *J. Gen. Physiol.* 53: 81–96.

TASAKI, I., and S. HAGIWARA. 1957. Demonstration of Two Stable Potential States in the Squid Giant Axon under Tetraethylammonium Chloride. *J. Gen. Physiol.* 40: 859–885.

GLOSSARY OF SYMBOLS

$-a$	subscript for denoting anion
a_i	activity for ionic species, i
A	surface area
A	gain of an amplifier or circuit

c_M	membrane concentration
c_i	concentration of ionic species, i
C	capacitance
C_M	membrane capacity
C	Celsius temperature

d	radial repeat distance
D_o	permittivity of free space
D	diffusion coefficient

E_{AP}	membrane potential at peak of action potential
E_d	Donnan potential
E_i	equilibrium potential for ionic species, i
E_H	membrane holding potential in voltage clamped state
E_M	membrane potential
E_o	membrane potential measured from holding potential to peak of voltage sine wave
E_{rev}	membrane potential at which membrane current component = zero (reversal potential)
E_{RP}	membrane resting potential
E_m	measured potential

f	frequency of a wave
F	Faraday = 96,500 coulombs/mole
f	farad
g_i	conductance for ionic species, i
$\tilde{g}_i$	maximal possible value of g_i
g_L	leakage conductance
g_M	membrane chord conductance
g_o	instantaneous conductance
g_p	membrane conductance at the peak of the initial transient current component
g_{ss}	membrane conductance during the steady-state of the delayed membrane current component
G	generalized conductance
G	Gibbs free energy
h	time and potential dependent parameter used by Hodgkin and Huxley to formulate equations describing membrane currents
h	diffraction order
i	current density
I_C	capacitative current
I_D	delayed current
I_E	early transient current
I_i	ionic current for species, i
I_i	maximum ionic current at clamped voltage
I_L	leakage current
I_o	peak to peak amplitude of current wave
I_{ss}	steady-state current
$I(h)$	X-ray intensity
j_i	molar flux density
j	net flux
$J_{obs}(h)$	set of X-ray intensities on relative scale
$-k$	subscript denoting cation
k	rate constant
k	Boltzmann constant
K	equilibrium constant
K	Debye reciprocal length
K_m	Michaelis constant

$K^{k_1}_{k_2}$	proportionality coefficient describing selectivity of membrane for cation k_2 over cation k_1
K	selectivity coefficient
L_{ii}	coupling coefficient
L_{ij}	cross coupling coefficient
m	time and potential dependent parameter used by Hodgkin and Huxley to formulate equations describing membrane currents
M	average electron density
M	molarity
n	time and potential dependent parameter used by Hodgkin and Huxley to formulate equations describing membrane currents
n_i	number of moles of the ith substance
N	Avogadro's number = 6.02×10^{23} molecules/mole
p	pressure
p_i	membrane permeability of ionic species, i
P	permeability coefficient
Q	electronic charge
Q	constant current threshold
Q_{10}	temperature dependency factor of rate constants
r_x	Donnan ratio
R	resistance
R	index of how closely theoretical and observed diffraction agree
R	universal gas constant = 8.314 joule$/°K \cdot$ mole
R_a	access resistance
R_f	feedback resistance
R_M	membrane resistance
R_s	series resistance
s	surface of membrane
S	entropy

T	temperature in $°K$
t(r)	electron density of triple layered unit
$T(R)$	Fourier transform of t(r)
$T(h)$	Fourier transform of $I(h)$
u	electrical mobility
u	absolute mobility
v	velocity
V	potential
V	volume
V_M	membrane potential
V_o	peak to peak amplitude of voltage wave
V_{out}	voltage output
V_w	partial molal volume of solvent
X_w	mole fraction
x_j	internal molar diffusion force of jth substance
z_i	valence of ionic species, i
$\alpha_{m,n\ or\ h}$	voltage-dependent rate constant for ionic conductance used by Hodgkin and Huxley
$\beta_{m,n\ or\ h}$	voltage-dependent rate constant for ionic conductance used by Hodgkin and Huxley
β	partition coefficient
γ	activity coefficient
δ	membrane thickness
ε	dielectric constant
ε	small potential difference (Chapter 5 only)
ε	proportionality constant
η	coefficient of viscosity
θ	wave propagation velocity
λ	wavelength
λ	space or length constant of axon
λ	specific conductance
λ	accommodation time
μ	ionic strength
$\tilde{\mu}$	electrochemical potential
μ_i	chemical potential
π	3.1416
Π	osmotic pressure
ρ	net charge density in solution

σ	asymmetry of the membrane unit
σ	reflection coefficient
$\bar{\sigma}$	membrane surface charge density
σ_s	total charge density
τ	time constant
ϕ	electrical potential
ϕ'	electrical potential with reference to the average potential in the membrane
ψ	potential energy
ω	$2\pi f$, where f is the frequency of a potential wave

AUTHOR INDEX

AUTHOR INDEX

513

SUBJECT INDEX

SUBJECT INDEX

519